The Multiscale Global Monsoon System

World Scientific Series on Asia-Pacific Weather and Climate

This series will continue the collaboration with the Working Group on Tropical Meteorology Research, a component of the World Weather Research Programme, and other international weather and climate programs in publication projects.

The Editorial Board welcomes book or program/project proposals relevant to weather and climate or related fields, including interdisciplinary topics. Interested scientists please contact the Editor-in-Chief Professor Chih-Pei Chang (cpchang@nps.edu).

Published

Vol. 1: East Asia and Western Pacific Meteorology and Climate: Selected Papers of the Fourth Conference
edited by C.-P. Chang, Guoxiong Wu, Ben Jou and C. Y. Lam

Vol. 2: East Asian Monsoon
edited by C.-P. Chang

Vol. 3: Observation, Theory, and Modeling of Atmospheric Variability: Selected Papers of Nanjing Institute of Meteorology Alumni in Commemoration of Professor Jijia Zhang
edited by Xun Zhu, Xiaofan Li, Ming Cai, Shuntai Zhou, Yuejian Zhu, Fei-Fei Jin, Xiaolei Zou and Minghua Zhang

Vol. 4: Global Perspectives on Tropical Cyclones: From Science to Mitigation
edited by Johnny C. L. Chan and Jeffrey D. Kepert

Vol. 5: The Global Monsoon System: Research and Forecast (2nd Edition)
edited by C.-P. Chang, Yihui Ding, N.-C. Lau, Richard H. Johnson, Bin Wang and Tetsuzo Yasunari

Vol. 6: Climate Change: Multidecadal and Beyond
edited by C.-P. Chang, Michael Ghil, Mojib Latif and John M. Wallace

Vol. 7: Indo-Pacific Climate Variability and Predictability
edited by Swadhin K. Behera and Toshio Yamagata

Vol. 8: Aridity Trend in Northern China
edited by Congbin Fu and Huiting Mao

Vol. 9: The Global Monsoon System: Research and Forecast (3rd Edition)
edited by C.-P. Chang, H.-C. Kuo, N.-C. Lau, R. H. Johnson, B. Wang and M. C. Wheeler

Vol. 10: Bridging Science and Policy Implication for Managing Climate Extremes
edited by Hong-Sang Jung and Bin Wang

Vol. 11: The Multiscale Global Monsoon System
edited by C.-P. Chang, Kyung-Ja Ha, Richard H. Johnson, Daehyun Kim, Gabriel N. C. Lau and Bin Wang

WORLD SCIENTIFIC SERIES ON ASIA-PACIFIC WEATHER AND CLIMATE

Vol. 11

The Multiscale Global Monsoon System

edited by

Chih-Pei Chang
Naval Postgraduate School and National Taiwan University

Kyung-Ja Ha
Pusan National University

Richard H. Johnson
Colorado State University

Daehyun Kim
University of Washington

Gabriel N. C. Lau
The Chinese University of Hong Kong and Princeton University

Bin Wang
University of Hawaii

World Scientific

NEW JERSEY • LONDON • SINGAPORE • BEIJING • SHANGHAI • HONG KONG • TAIPEI • CHENNAI • TOKYO

Published by

World Scientific Publishing Co. Pte. Ltd.
5 Toh Tuck Link, Singapore 596224
USA office: 27 Warren Street, Suite 401-402, Hackensack, NJ 07601
UK office: 57 Shelton Street, Covent Garden, London WC2H 9HE

British Library Cataloguing-in-Publication Data
A catalogue record for this book is available from the British Library.

World Scientific Series on Asia-Pacific Weather and Climate — Vol. 11
THE MULTISCALE GLOBAL MONSOON SYSTEM

ISBN 978-981-121-659-6 (hardcover)
ISBN 978-981-121-660-2 (ebook for institutions)
ISBN 978-981-121-661-9 (ebook for individuals)

For any available supplementary material, please visit
https://www.worldscientific.com/worldscibooks/10.1142/11723#t=suppl

Printed in Singapore

Foreword

For centuries, monsoons have played a vital role in societies' culture, lifestyle and activities. Those living in monsoon-affected regions, which are prone to floods and droughts, have suffered losses of lives, destroyed homes and millions have been displaced. In 2019, the global economy suffered a loss of USD 232 billion due to natural disasters. Most of the damage (USD 229 billion) was due to weather-related disasters. Floods, including seasonal monsoon floods in China and India, accounted for five of the top ten costliest disasters of the year. The World Meteorological Organization, working on improving the well-being of societies of all nations, has always placed utmost importance on its activities relating to monsoons and the application of research results by its Members to better understand and deal with the vagaries of the monsoon.

A flagship activity undertaken by the World Weather Research Programme (WWRP) and the World Climate Research Programme (WCRP), is the quadrennial International Workshop on Monsoons (IWM), providing a unique forum for monsoon researchers and forecasters from all monsoon-affected regions to have an in-depth exchange on the complex problems of monsoons, to discuss recent advances and current issues covering all time scales (mesoscale, synoptic, intraseasonal, climate) that are relevant to the forecast of high-impact weather in the world's monsoon regions. IWM is also a means to transfer new science and technology to National Meteorological and Hydrological Services in these regions.

The present publication, the fourth volume in a series of The Global Monsoon System books, is a collection of peer-reviewed papers presented by leading monsoon experts at the latest IWM-VI (13–17 November 2017, Singapore). The volume is intended to provide a comprehensive review of the current status of knowledge, modeling capability and future directions in research of monsoon systems around the globe.

I expect this publication to serve as a basis to advance the application of monsoon research through the exchange of new ideas and results among research scientists, forecasters, and end-users of monsoon predictions.

On behalf of WMO, I wish to thank all experts who contributed to this publication and devoted their time and knowledge to help in the advancement of global efforts for improved understanding of monsoons. Our gratitude also goes to all the Chair, Professor Chih-Pei Chang, and members of the International and Local Organizing Committees of IWM-VI and to the six editors of this publication for their devoted efforts and excellent collaboration.

Professor Petteri Taalas
Secretary-General, World Meteorological Organization

Preface

This volume represents the fourth in a series of publications stemming from International Workshops on Monsoons (IWMs) carried out under the auspices of the World Meteorological Organization (WMO). These publications, entitled *The Global Monsoon System: Research and Forecast*, present the outcomes of the following workshops: IWM-III (Hangzhou, 2004), IWM-IV (Beijing, 2008), IWM-V (Macao 2013), and IWM-VI (Singapore, 2017). The first in the series was published as WMO/TD No. 1266 while the following three were part of the World Scientific Series on Asia-Pacific Weather and Climate.

Great strides have been made in basic knowledge and prediction of monsoon systems over the nearly two decades of the IWM-based series. These advances are documented and reviewed in these books by experts from the global monsoon community. Highlights of the significant gains in understanding and prediction of the Asian, Maritime Continent, North and South American, and African monsoons during this period are the following:

- the roles of ocean-land-atmosphere coupling in monsoons;
- the internal dynamics of monsoon systems;
- mechanisms involved in winter and summer monsoons;
- monsoon-ENSO interactions;
- the diurnal cycle of monsoons;
- intraseasonal variability of monsoons;
- relationships between monsoons and tropical cyclones;
- monsoons and extreme weather;
- interannual and interdecadal variability of monsoons;
- the role of aerosols in monsoon circulations;
- improved regional weather and hydrometeorological forecasts;
- improved understanding of topographic and vegetation effects, microphysics, radiation, and aerosols in models, along with advances in ensemble and multi-model prediction;
- improved seasonal monsoon prediction; and
- progress in understanding the response of monsoons systems to changing climate.

In addition to these advances in basic research and prediction, the four volumes summarize findings from a series of international monsoon-related field campaigns during the past two decades: the 2004 North American Monsoon Experiment (NAME), the 2006 African Monsoon Multidisciplinary Analysis (AMMA) campaign, the 2008 Southwest Monsoon Experiment/Terrain-induced Monsoon Rainfall Experiment (SoWMEX/TiMREX); the 2011–12 Dynamics of the MJO (DYNAMO) campaign; the 2013–18 South China Monsoon Rainfall Experiment (SCMREX), the 2017–19 Years of the Maritime Continent (YMC) experiment; and the 2018–19 Propagation of Intraseasonal Tropical Oscillations (PISTON) experiment.

The *Global Monsoon System* series has served as both an historical and an up-to-date account of progress in understanding and prediction of global monsoons. The 16-year-long IWM workshops and related book series would not have been possible without the strong support of the WMO and the sustained efforts of several key individuals in academia. Most notably, the extraordinary vision and leadership of Chih-Pei Chang of the Naval Postgraduate School and the National Taiwan University were responsible for the formulation of the plans for these activities and their eventual actuality. His outstanding ability in coordination and organization not only of the scientists in different regions, but also of various governments and international organizations, was a key factor in the success. Prior to this time, there had been no other dedicated forum for the exchange of research results from the various global monsoon regions by the international scientific community. The monsoon book series spearheaded by C.-P. represents a permanent legacy of these international workshops, and continues the heritage of the book *Monsoon Meteorology* edited by him and T. N. Krishnamurti (1987, Oxford University Press). Additional significant contributions to the IWMs and book series throughout this period have been provided by Bin Wang of the University of Hawaii and Ngar-cheung Gabriel Lau of NOAA Geophysical Fluid Dynamics Laboratory and the Chinese University of Hong Kong. The outstanding support provided by the WMO WWRP Office, particularly Ms. Nanette Lormada, is another important factor in the success of this project over the years.

I was fortunate to have joined the IWM effort from the start and therefore have witnessed the full scope and impact of the workshops and book series on advancing monsoon research and prediction. These accomplishments are a credit to C.-P. and his team. The international scientific community expresses its gratitude for this important and lasting contribution to our science.

Richard H. Johnson
Colorado State University

Contents

Part III Intraseasonal Oscillation and Forecast

Part IV Climate Change

Part V Field Experiments

Part I
Regional Monsoons

Chapter 1

Role of Land-Ocean Contrast in the Indian Summer Monsoon Rainfall

Sulochana Gadgil[*,¶], P. A. Francis[†], K. Rajendran[‡], Ravi S. Nanjundiah[*,§] and Suryachandra A. Rao[§]

[*] *Centre for Atmospheric and Oceanic Sciences, Indian Institute of Science, Bangalore 5600012, India*
[†] *Indian National Centre for Ocean Information Services, Ministry of Earth Sciences, Hyderabad 500090, India*
[‡] *CSIR-Fourth Paradigm Institute, Bangalore 560037, India*
[§] *Indian Institute of Tropical Meteorology, Ministry of Earth Sciences, Pune 411008, India*
[¶] *sulugadgil@gmail.com*

For well over 300 hundred years, the monsoon has been considered to be a gigantic land-sea breeze driven by the land-ocean contrast in surface temperature. This popular perception of the monsoon as a gigantic land-sea breeze is not only mentioned in text books and reviews but also used in some studies of the impact of global warming to deduce the expected monsoon variability from the expected land-ocean contrast. If the primary cause of the monsoon is the differential heating between land and ocean, we expect enhanced land-sea contrast to be associated with enhanced monsoon rainfall. It is important to assess whether observations of monsoon variability are consistent with this expectation. Since models are used for deriving the nature of the impact of a change in land-ocean contrast on the monsoon, it is also important to investigate if such a role of land-ocean contrast is simulated by models. Here, we address this problem with an analysis of the relationship between the Indian summer monsoon rainfall (ISMR) and the surface temperature of the land, the surrounding ocean, and the land-ocean contrast in observations and simulations using one high-resolution atmospheric model and one coupled model. We find that in all three cases, the ISMR is negatively correlated with the land surface temperature as well as land-ocean contrast. Thus, for observations and models, the relationship of the variation of land-ocean contrast with that of the monsoon rainfall is opposite to what is expected from the land-sea breeze hypothesis. This clearly suggests that the time has come to abandon the perception of the monsoon system as a gigantic land-sea breeze.

1. Introduction

Halley (1686) first suggested that the primary cause of the annual cycle of monsoon circulation was the differential heating between land and ocean caused by the seasonal march of the sun. The monsoon was, therefore, viewed as a gigantic land-sea breeze. Even today, differential heating between land and ocean is considered to be the primary driver of the monsoon in many papers and most textbooks (e.g., Lau and Li 1984; Webster 1987; Webster *et al.* 1998; James 1994). It has been shown that with enhancement of concentrations of greenhouse gases, the surface temperature rises everywhere, but with more warming over the land than the ocean.

The Multiscale Global Monsoon System
Edited by C. P. Chang et al.

According to Lau and Kim (2017), "a steady increasing warm-ocean–warmer-land (WOWL) trend has been in effect since the 1950s, substantially increasing moisture transport from adjacent oceans, and enhancing rainfall over the Asian monsoon regions." Assuming that land-ocean contrast is one of the main drivers of the monsoon rainfall, in the 5th Assessment Report of the Intergovernmental Panel on Climate Change (IPCC 2013), the likely enhancement of monsoon rainfall was attributed to increased land-sea contrast and more abundant precipitable water in a warmer climate. In a recent study, Jin and Wang (2017) showed that the summer monsoon rainfall over north-central India has increased since 2002 and suggested that *this apparent revival of summer monsoon precipitation is closely associated with a favourable land-ocean temperature gradient, driven by a strong warming signature over the Indian subcontinent and slower rates of warming over the Indian Ocean.* All these deductions of the nature of the impact of deviation in the land-ocean temperature contrast on the monsoon rainfall are based on the assumption that enhanced/decreased land-sea contrast is associated with enhanced/decreased monsoon rainfall. Clearly, it is important to unravel the role of land-ocean contrast in the observed monsoon rainfall as well as the monsoon rainfall simulated by models from which we deduce the impact of this factor. Here, we address this problem for the summer monsoon (June–September) rainfall over the Indian region, i.e., ISMR. We focus only on this factor and do not attempt a review on the monsoon and its variability in view of the constraints on the length of the paper and also since comprehensive reviews are available such as the recent one by Wang *et al.* (2017).

2. Essential Features of the Leading Mode of AM Variability

Gadgil (2018) has shown that the role of land-ocean contrast expected from the land-sea breeze model for the monsoon is not supported by the observations of the variability of the monsoon. In fact, two centuries after Halley, Blanford (1886) had opposed the idea of land-ocean contrast being considered the prime cause of the monsoon. He suggested that heating of the atmosphere above the heated land can lead to only feeble convergence in spring and cannot explain the heavy rainfall associated with the summer monsoon. Blanford (1886) attributed the monsoon to the annual visit of a system which we can now identify as the Intertropical Convergence Zone (ITCZ, Charney 1969) or the equatorial trough (Riehl 1954, 1979). Gadgil (2003) has shown that observations support this alternative theory according to which the basic system responsible for the monsoon is the ITCZ or equatorial trough, and the Indian summer monsoon is a manifestation of the seasonal migration of the ITCZ onto the heated subcontinent in the summer. In fact, this is now accepted by several scientists (e.g., Schneider *et al.* 2014).

In this paper, we have analyzed the relationship of the interannual variation of ISMR with that of the surface temperature over the Indian subcontinent, over the surrounding ocean, equatorial Indian Ocean, and the difference between the land and ocean temperatures, in observations as well as simulations with a high-resolution coupled ocean-atmosphere general circulation model viz. CFS2-T382 (Ramu *et al.* 2015, henceforth referred to as CFS) and a much higher resolution (T_L959) MRI atmospheric general circulation model (Mizuta *et al.* 2006, henceforth referred to as MRI), for the period 1982–2008. For the CFS model, the ensemble average of retrospective predictions with April initial conditions and for the MRI model the run with observed SST are considered. The mean May and June–September air temperatures from a combination of surface observations at India Meteorological Department (IMD, from Srivastava *et al.* 2008) and the HadISST (Rayner *et al.* 2003), and the temperature at 2 m over land and ocean from NCEP reanalysis (Kalnay *et al.* 1996) as well as from the CFS and MRI runs are shown in Fig. 1. The mean patterns of June–September rainfall from CMAP (Xie and Arkin 1997) and from the runs of the CFS and MRI models are also shown in Fig. 1. For May, observations as well as simulations show a high contrast between the high temperatures over the Indian subcontinent and the temperatures over the north and equatorial Indian Ocean. Even during the summer monsoon season, the observed temperature of a large fraction of the Indian landmass north of about 20°N is higher than that over the oceans.

In order to assess the role of land-ocean contrast in the observed variation of ISMR as well as in the variation of the ISMR from the models, we consider the air temperature at 2 m averaged over the region east of 60°E (shown in Fig. 1a and b), designated as ALS, and the air temperature at 2 m averaged over the equatorial Indian Ocean (60°–100°E, 5°S–5°N) designated by AOS for the period 1981–2008. The variation of ISMR with ALS, AOS, and ALS–AOS from observations and the CFS and MRI models is depicted in Fig. 2. The linear Pearson correlation coefficients computed from anomalies obtained by subtracting the corresponding long-term mean from the actual time-series, without removing any trends, are also shown. It is seen that the correlation between ISMR and the air temperature over the land surface, ALS, is negative in observations as well as in the models. However, the scatter is larger and the magnitude of correlation coefficient is smaller for the observed relationship of ISMR with ALS than the relationship of the simulated ISMR with simulated ALS for both the models (Fig. 2).

The spatial variation of the correlation of ISMR with the surface air temperature over the Indian region and the surrounding ocean in observations and the CFS and MRI models is shown in Fig. 3. It is seen that the ISMR is negatively correlated with the air temperature over the Indian region in observations as well as models, suggesting that the results on the relationship to surface temperature over a specific region of land in Fig. 2 are robust. This is consistent with Simpson's (1921) observation that *the average temperature is much greater in years of bad rains than in years of good rains* and with the results of Kothawale and Rupakumar (2002) who showed that the land surface temperature anomaly is positive for droughts and negative for excess rainfall monsoon seasons. Walker *et al.* (2015) also found that the correlation of the SASM index (which is an index for the South Asian Summer Monsoon, SASM, defined as the vertically integrated moisture flux convergence over a representative region, i.e., 60–100°E, 10–30°N) with the temperature at 850 hPa over the land is negative.

The land-sea breeze hypothesis focuses primarily on the impact of the high-land surface temperature on the atmospheric circulation and rainfall. However, the system is an interactive one with the atmosphere also having an impact on the temperature of the land. As pointed out by Trenberth and Shea (2005) *strong negative correlations (between surface temperature and precipitation) occur over continental land in summer in both hemispheres, so that summers tend to be either hot and dry or cool and wet, but not some other combination.* Thus, the relationship between monsoon rainfall and surface temperature over the land in observations and models suggests that, in both, the impact of the atmosphere on land surface temperature is dominant in the land–atmosphere interaction. *It, therefore, appears that land temperature responds to the variation of the monsoon rainfall, rather than determining it.*

The relationship of the AOS with ISMR is rather weak in observations as well as models and hence the relation between ISMR and the land-ocean contrast, ALS–AOS, is similar to that between ISMR and ALS. The correlation of ISMR with the land-ocean contrast, ALS–AOS, is significantly negative for observations

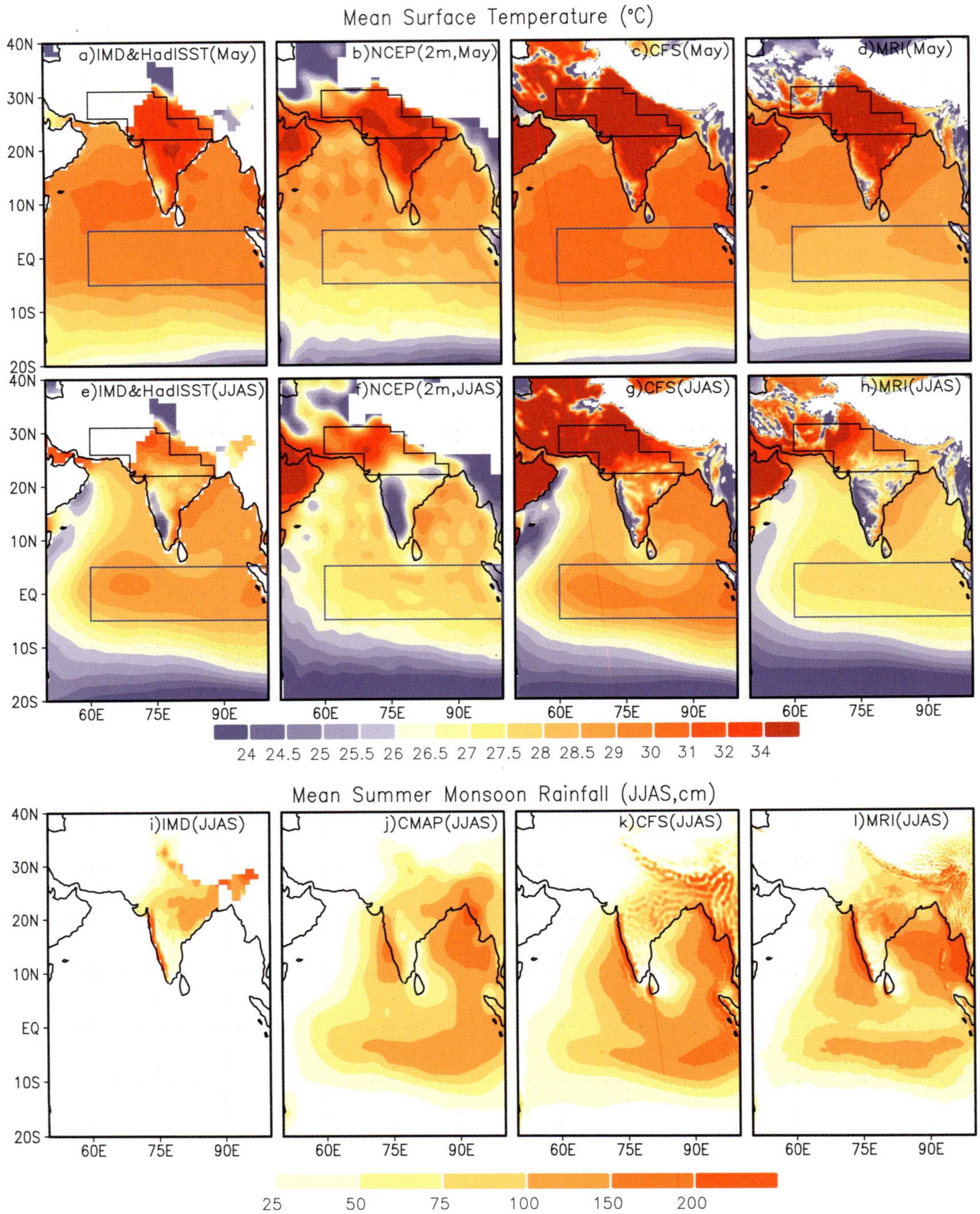

Fig. 1. (a) For the period 1982–2008, the mean surface temperature for May and (b) the mean surface temperature for June–September derived from observations are shown in first column, NCEP reanalysis in second column, CFSv2 in third column, and MRI in last column, and (c) mean rainfall for June-September observations from IMD, from CMAP, and from models CFS and MRI.

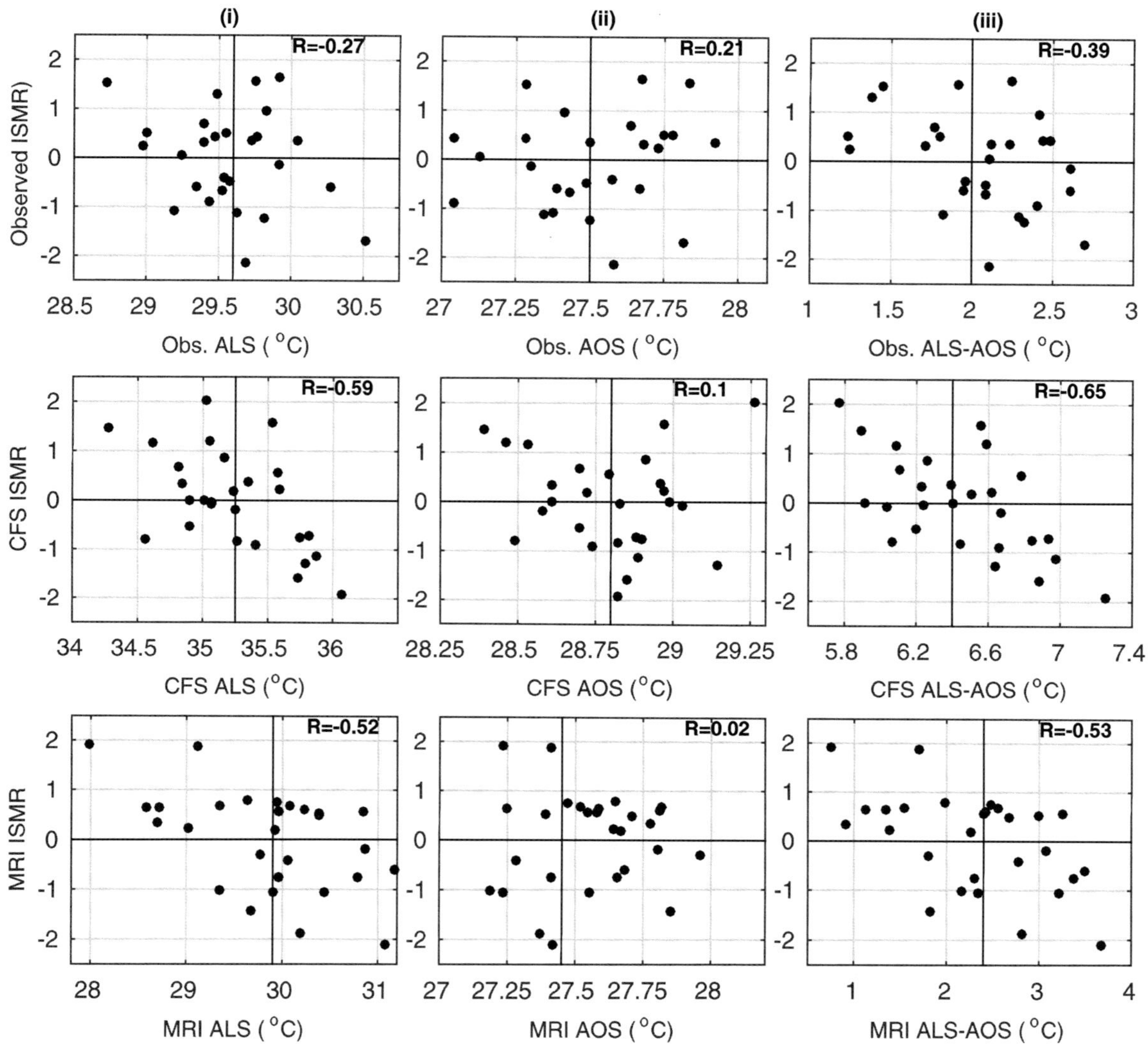

Fig. 2. Variation of ISMR with (i) the temperature above land surface, (ii) temperature above the sea surface, and (iii) the difference between (i) and (ii) for June–September, derived from NCEP 2 m (top row), CFS (middle row), and MRI (bottom row).

(significant at 0.05 level) as well as the models (significant at 0.01 level). The results are similar when the ocean from 5°S to 20°N is used instead of the ocean from 5°S to 5°N for deriving AOS. Since the magnitude of the correlation over land is much higher than that over most of the ocean, the correlation between ISMR and the land-ocean contrast is negative in all the cases.

Thus, for observations and models, the relationship of the land-ocean contrast with the monsoon rainfall is opposite to what is expected from the land-sea breeze hypothesis. Clearly the time has come to abandon the land-sea breeze hypothesis for the monsoon system and accept the alternative hypothesis that the basic system responsible for the monsoon is the ITCZ or equatorial trough, and the Indian summer

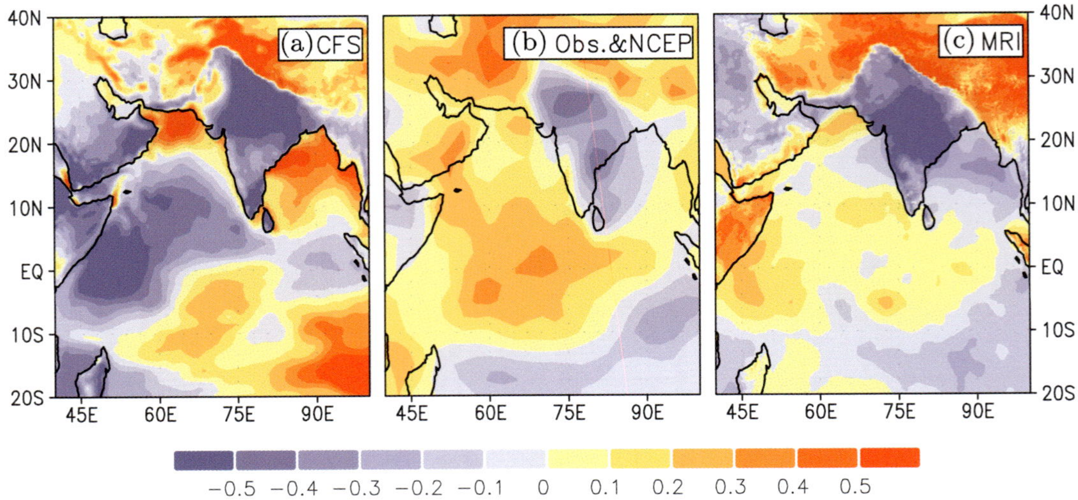

Fig. 3. Correlation between ISMR and surface temperature averaged for June–September for the period 1981–2008, derived from (a) CFS, (b) NCEP reanalysis, and (c) MRI.

monsoon is a manifestation of the seasonal migration of the ITCZ onto the heated subcontinent in the summer (Gadgil 2003).

Since the ITCZ is associated with intense deep convection and hence upper tropospheric heating, a continental ITCZ is associated with a meridional gradient of tropospheric temperature, which is directly related to the intensity of the ITCZ and hence the strength of the monsoon. The importance of this factor has been brought out by several studies using observations (e.g., Xavier *et al.* 2007) and general circulation models (e.g., Sabeerali *et al.* 2015). The interannual variation of ISMR is determined to a large extent by its links with major oscillations over the equatorial Pacific and Indian oceans (Gadgil *et al.* 2007). The link with ENSO (Sikka 1980; Rasmusson and Carpenter 1983 and many subsequent studies) is a natural consequence of the planetary scale of the basic system, which often extends from the Indian longitudes eastward over the tropical Pacific (Gadgil 2003). Interaction between the ITCZ over the subcontinent and that over the warm equatorial Indian Ocean leads to a link of the monsoon with the Equatorial Indian Ocean Oscillation (EQUINOO, Gadgil *et al.* 2004). The link with ENSO explains about 30% of the ISMR variance and the link with EQUINOO explains about 20%, while the composite index from these two modes explains well over 50% of the ISMR variance (Sajani *et al.* 2015).

3. Role of Pre-monsoon Heating of the Land in the Summer Monsoon Rainfall

It is well known that the heating of the subcontinent in April–May, before the summer monsoon season, leads to the development of low pressure there and hence the meridional pressure gradient associated with monsoon circulation. We consider here whether this heating of the land surface before the summer monsoon (which leads to feeble convergence in Spring, as Blanford (1886) had pointed out) plays a role in determining the intensity of the ITCZ over the subcontinent in the forthcoming monsoon season and hence the monsoon rainfall.

It is expected that land surface conditions in May will play some role in the rainfall during the forthcoming monsoon season. For example, Ghosh *et al.* (2016) with an analysis of

AGCM run for 50 years with climatological SST showed that the June-July rainfall over the Indian region is significantly correlated with the moist static energy as well as with the moisture in the top 10 cm of the soil in May. We focus here only on the surface temperature. For the period 1958–2012, the correlation of surface temperature in May with the all-India average rainfall in June and with the ISMR is depicted in Fig. 4. It is seen that the correlation with June rainfall is positive and significant, being larger than 0.3 over a substantial part of the country, and higher over a part of the Indian monsoon zone and particularly high over the northwestern part. The pattern of correlation with ISMR is rather similar to that for June, but the magnitude of the correlation coefficient is somewhat smaller. It therefore appears that the heating over that region plays an important role in setting up the meridional pressure gradient and the low-level circulation of the summer monsoon.

Thus, the land surface temperature in May does play some role in determining the summer monsoon rainfall in the forthcoming season;

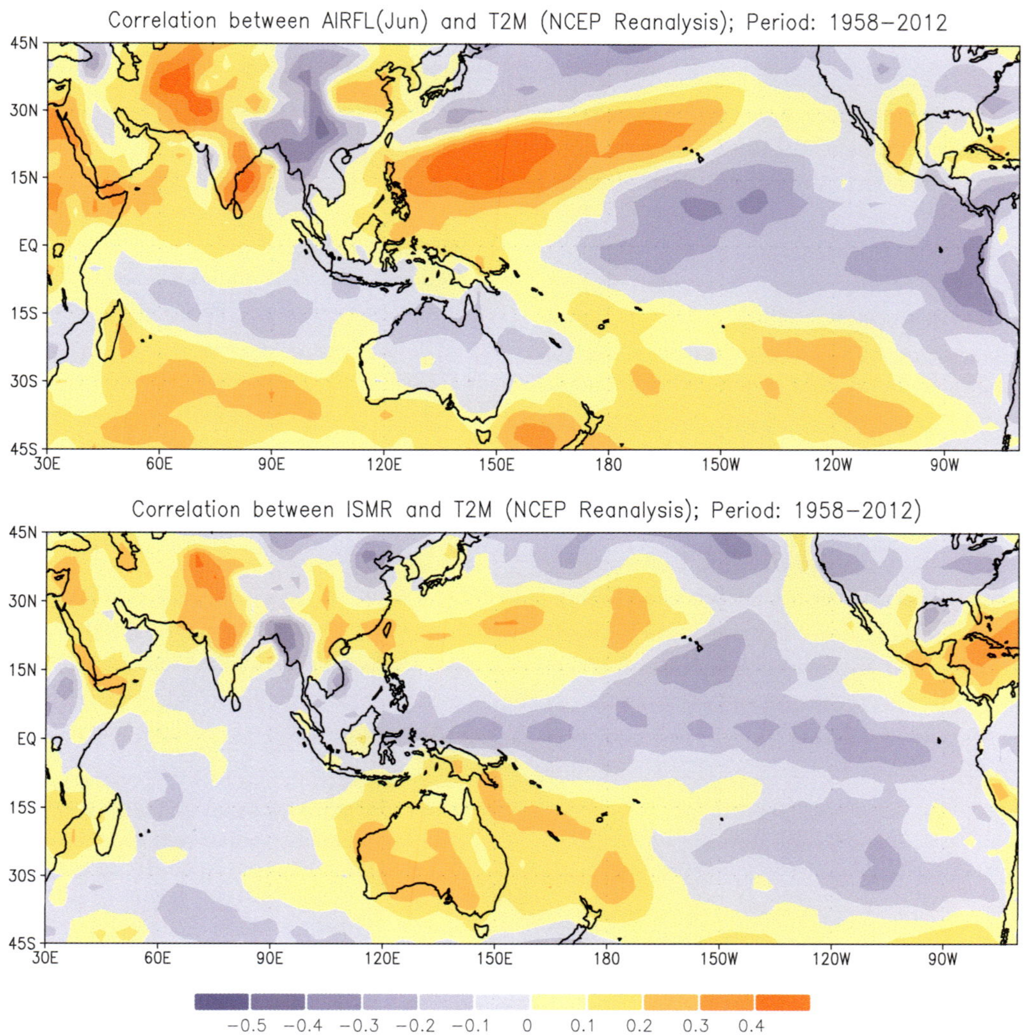

Fig. 4. Correlation between the NCEP temperature at 2 m in May and the all-India rainfall in June (above) and ISMR for the period 1958–2012.

but only about 10% of the variance of ISMR is explained by this factor. However, once the monsoon sets in and the ITCZ gets established over the subcontinent, the land cools and land-ocean temperature contrast plays no role in sustaining the monsoon nor in determining its interannual variation. The role of the pre-monsoon heating in the interannual variation of the monsoon rainfall is seen to be of far less importance than that of the ENSO and EQUINOO modes.

4. Conclusions

It is clear that neither the land surface temperature nor the land-ocean temperature contrast during the summer monsoon season plays a role in determining the Indian summer monsoon rainfall in observations or models. In fact, rather than influencing the monsoon, the land surface temperature responds to the monsoon rainfall, being cooler when the rainfall is high and warmer when it is lacking. Thus, deducing anything relating to the monsoon rainfall on the basis of what is expected regarding land surface temperature or land-ocean contrast of the summer monsoon season, as done in several studies, is not justified. The Indian summer monsoon rainfall does depend to some extent on the land surface temperature in May. However, the correlation is poorer than that with ENSO and EQUINOO, which are known to play important roles in the interannual variation of the monsoon, and the variance of ISMR explained by the May land surface temperature is only about 10%.

Acknowledgments

It is a pleasure to thank Prof. J. Srinivasan, Indian Institute of Science, for critical contributions to our understanding of many aspects of the problem addressed. We are grateful to Prof. Bin Wang and the reviewers for comments which helped us improve the paper significantly. CFS high-resolution model data were generated under the National Monsoon Mission, Ministry of Earth Sciences, Govt. of India. KR and RSN are thankful for financial support from the Department of Environment and Climate Change, Govt. of Kerala (R-1-166) and collaborative research project with MRI (R-8-118) and PAF acknowledges funding support through O-MASCOT project of INCOIS, Ministry of Earth Sciences, Govt. of India.

References

Blanford, H. F., 1886: Rainfall of India. Mem. India Meteorological Department No. 2, pp. 217–448.

Charney, J. G., 1969: The intertropical convergence zone and the Hadley circulation of the Atmosphere. *Proc. WMO/IUGG Intern. Symp. on Numer. Weather Predict. Jpn. Meteorol.* Agency, III, pp. 73–79.

Gadgil, S., 2003: The Indian Monsoon and its variability, *Annu. Rev. Earth. Planet. Sci.*, **31**, 429–467.

Gadgil, S., 2018: The monsoon system: Land–sea breeze or the ITCZ? *J. Earth Sys. Sci., Ind. Acad. Sci.*, **127**(1), 1–29, https://doi.org/10.1007/s12040-017-0916-x.

Gadgil, S., M. Rajeevan, and P. A. Francis, 2007: Monsoon variability: Links to major oscillations over the equatorial Pacific and Indian oceans. *Curr. Sci.*, **93**(2), 182–194.

Gadgil, S., P. N. Vinayachandran, P. A. Francis, and S. Gadgil, 2004: Extremes of the Indian summer monsoon rainfall, ENSO and equatorial Indian Ocean oscillation. *Geophys. Res. Lett.*, **31**(12), L12,213.

Ghosh, R., A. Chakraborty, and R. S. Nanjundiah, 2016: Relative role of pre-monsoon conditions and intra seasonal oscillations in determining early-vs-late Indian monsoon intensity in a GCM. *Theo. Appl. Climatol.*, **131**(1), 319–333, https://doi.org/10.1007/s00704-016-1970-z.

Halley, E., 1686: A historical account of the trade winds and monsoons observable in the seas between and near the tropics with an attempt to assign a physical cause of the said winds. *Phil. Trans. Roy. Soc. London*, **16**, 153–168.

IPCC Climate Change, 2013: The Physical Science Basis, the contribution of Working Group I to the Fifth Assessment Report of the Intergovernmental Panel on Climate Change. (Eds.) Stocker, T. F. *et al.*, Cambridge University Press, 1535p.

James, I. N, 1994: *Introduction to Circulating Atmospheres.* Cambridge Atmospheric and Space Series (9), Cambridge University Press, 444p.

Jin, Q. and C. Wang, 2017: A revival of Indian summer monsoon rainfall since 2002. *Nat. Clim. Change*, **7**, 587–594, https://doi.org/10.1038/nclimate3348.

Kalnay, E. *et al.*, 1996: The NCEP/NCAR 40-year reanalysis project. *Bull. Am. Meteorol. Soc.*, **2**, 437–471.

Kothawale, D. R. and K. Rupakumar, 2002: Tropospheric temperature variation over India and links with the Indian summer monsoon: 1971–2000. *Mausam*, **53**, 289–308.

Lau, W.-K. and M. K. Kim, 2017: Competing influences of greenhouse warming and aerosols on asian summer monsoon circulation and rainfall. *Asia-Pac. J. Atmos. Sci.*, **53**(2), 181–194.

Lau, W.-K. and M. T. Li, 1984: The monsoon of East Asia and its global associations — A survey. *Bull. Am. Meteor. Soc.*, **65**, 114–125.

Mizuta, R., K. Oouchi, H. Yoshimura, A. Noda, K. Katayama, S. Yukimoto, M. Hosaka, S. Kusunoki, H. Kawai, and M. Nakagawa, 2006: 20-km- mesh global climate simulations using JMA-GSM model: Mean climate states. *J. Meteorol. Soc. Jpn.*, **84**(1), 165–185, https://doi.org/10.2151/jmsj.84.165.

Ramu, D. A., C. T. Sabeerali, Rajib Chattopadhyay, D. N. Rao, G. George, A. R. Dhakate, K. Salunke, A. Srivastava, and S. A. Rao, 2015: Indian summer monsoon rainfall simulation and prediction skill in the CFSv2 coupled model: Impact of atmospheric horizontal resolution. *J. Geophys. Res. Atmos.*, **121**, 2205–2221, 10.1002/2015JD024629.

Rasmusson, E. M. and T. H. Carpenter, 1983: The relationship between the eastern Pacific sea surface temperature and rainfall over India and Sri Lanka. *Mon. Weather Rev.*, **111**, 354–384.

Rayner, N. A., D. E. Parker, E. B. Horton, C. K. Folland, L. V. Alexander, D. P. Rowell, E. C. Kent, and A. Kaplan, 2003: Global analyses of sea surface temperature, sea ice, and night marine air temperature since the late nineteenth century. *J. Geophys. Res.*, **108**(D14), 4407, 10.1029/2002JD002670.

Reihl, H., 1954: *Tropical Meteorology*, McGraw Hill. NewYork, 392p.

Reihl, H., 1979: *Climate and Weather in the Tropics*, Academic Press, San Diego, 611p.

Sabeerali, C. T., Rao, S. A., Dhakate, A. R., Salunke, K. and Goswami, B. N. (2015). Why ensemble mean projection of south Asian monsoon rainfall by CMIP5 models is not reliable? *Clim. Dynam.*, **45**(1–2), 161–174.

Sajani, S., S. Gadgil, P. A. Francis, and M. Rajeevan, 2015: Prediction of Indian rainfall during the summer monsoon season on the basis of links with equatorial Pacific and Indian Ocean climate indices. *Environ. Res. Lett.* **10**(9) 094004, https://doi.org/10.1088/1748-9326/10/9/094004.

Schneider, T., T. Bischoff, and G. H. Haug, 2014: Migrations and dynamics of the intertropical convergence zone. *Nature*, **513**, 45–53, https://doi.org/10.1038/nature13636.

Sikka, D. R., 1980: Some aspects of the large-scale fluctuations of summer monsoon rainfall over India in relation to fluctuations in the planetary and regional scale circulation parameters. *Proc. Indian Acad. Sci. (Earth & Planet. Sci.)*, **89**, 179–195.

Simpson, G., 1921: The south-west monsoon. *Q. J. R. Meteorol. Soc.,* **199**(17), 150–173.

Srivastava, A. K., M. Rajeevan, S. R. Krishisagar, 2008: Development of a high resolution daily gridded temperature data set (1969–2005) for the Indian region. National Climate Centre, India Meteorological Department, Pune, Research Report No: 8/2008.

Trenberth, K. E. and D. J. Shea, 2005: Relationships between precipitation and surface temperature. *Geophys. Res. Let.*, **32**, L14703, doi:10.1029/2005GL022760.

Walker, J. M., S. Bordoni, and T. Schneider, 2015: Interannual variability in the large-scale dynamics of the south Asian summer monsoon. *J. Climate*, **28**, 37–31.

Wang Pin Xian, Wang, Bin Wang, Hai Cheng, John Fasullo, Zheng Tang Guo, Thorsten Kiefer, and Zheng Yu Liu, 2017: The global monsoon across time scales: Mechanisms and outstanding issues. *Earth-Sci. Rev.*, **2017**, 173.

Webster, P. J., 1987: The elementary monsoon. In *Monsoons*, Eds., Fein and Stephens, J. Wiley Co., pp. 3–32.

Webster, P. J., V. O. Magaña, T. N. Palmer, J. Shukla, R. A. Tomas, M. Yanai, and T. Yasunari, 1998: Monsoons: Processes, predictability and the prospects for prediction. *J. Geophys. Res.*, **103**, 14,451–14,510.

Xavier, P. K., C. Marzin, and B. N. Goswami, 2007: An objective definition of the Indian summer monsoon season and a new perspective on the ENSO–monsoon relationship. *Quart. J. Roy. Meteorol. Soc.*, **133**(624), 749–764.

Xie, P. and P. A. Arkin, 1997: Global precipitation: A 17-year monthly analysis based on gauge observations, satellite estimates, and numerical model outputs. *Bull. Am. Meteorol. Soc.*, **78**, 2539–2558.

Chapter 2

Interannual Variations of the Asian Monsoon System: Mechanism and Predictability

Bin Wang
Department of Atmospheric Sciences and International Pacific Research Center, University of Hawaii, Honolulu, USA
wangbin@hawaii.edu

Interannual variability (IAV) of the Asian monsoon (AM) involves complex internal dynamics within the coupled atmosphere–ocean–land–ice system and is influenced by various external forcings. El Niño–Southern Oscillation (ENSO) has been recognized as a primary driver. However, this is by no means the whole story. The local monsoon–warm ocean interaction is critically important in sustaining seasonally evolving anticyclonic anomalies over the western North Pacific (WNP) ocean and southern Indian Ocean (IO), which extend El Niño's impacts to the off-equatorial AM regions. In addition, the monsoon annual cycle controls the nature of the monsoon-ocean interaction and strongly modifies the AM response to remote El Niño forcing. Therefore, ENSO, monsoon-ocean interaction, and modulation of the annual cycle are the three major factors governing the leading mode of AM variability. The SST anomalies over the Indo-western Pacific warm pool are the result of the monsoon-ocean interaction, and it is generally inadequate to consider them as a forcing to the AM system.

Considerable progress has been made in understanding the physical factors driving the Asian summer monsoon (ASM) variability, but it remains uncertain to what extent a combination of these factors can predict ASM rainfall variability. Analysis of the ASM precipitation (1979–2016) reveals four physically meaningful modes, including the central Pacific-ENSO forced mode, the Indo-Pacific monsoon-ocean coupled mode, the Indian Ocean dipole (IOD) mode, and the global warming mode, which together explain about 45% of the total variance. It has been demonstrated that the four modes are, to a large extent, predictable, providing an estimate of the lower bounds for the AM rainfall predictability.

1. Introduction

The Asian monsoon (AM) fluctuates from year to year with a large deviation from its climatological mean annual cycle. These year-to-year fluctuations are often referred to as the interannual variability (IAV). The IAV of the AM and associated extreme weather and climate events has a profound impact on water resources, food security, economy, society, and environment. Therefore, the interannual variation of the AM has been a most extensively studied subject and has a long history of research that dates back to the late 19th century (Blandford 1884; Walker 1923; Walker and Bliss 1932). Great efforts have been made to understand the origin of the variability and sources of predictability in order to improve its seasonal prediction (Charney and Shukla 1981; Webster *et al.* 1998, etc.).

Webster (2006) reviewed the progress in monsoon prediction and concluded, "Unfortunately, our abilities to predict (monsoon) variability have not changed substantially over the last few decades." This assertion made a decade ago remains equally valid today. Taking All-Indian Rainfall prediction as an example, the operational statistical forecast and dynamic

The Multiscale Global Monsoon System
Edited by C. P. Chang et al.

models' hindcasts have shown few skills since 1989 (Wang *et al.* 2015a). Seasonal prediction of the land monsoon rainfall (LMR) remains one of the most difficult tasks in climate sciences. The predictability of the land monsoon rainfall remains to be better estimated.

Study of the AM IAV has been a challenging subject in climate research, partially because the IAV of the AM exhibits substantial spatial variability due to its immense spatial scale, unique land–ocean configuration, and existence of the Tibetan Plateau (e.g., Lau and Wu 2001). As such, the AM system may be considered to comprise four major regional components including India (South Asia), East Asia, tropical WNP, and Indonesia (Wang *et al.* 2001). Numerous works have dug into the regional variability. Early studies of the AM were primarily carried out to understand and predict the Indian monsoon. During the past two-to-three decades tremendous efforts have been devoted to understanding monsoon variability outside South Asia. Many studies have documented the indigenous features of the IAV in various sub-regions and locations of the AM (e.g., Shukla and Mooley 1987; Tao and Chen 1987; Meehl 1994), but this chapter will be restricted to the variability and predictability of the*entire AM system.*

The reason to focus on the entire AM system is that the IAV of the AM exhibits strong season dependence due to the modulation of the most powerful monsoon annual cycle. The AM variability should not be perceived as local changes in isolation over specific locations; rather, the AM anomalies are intertwined and migrate from one place to another following the seasonal cycle. While each region or location may have its indigenous features, these regional variations are interconnected by large-scale variation of the AM circulation system. Local changes cannot be fully comprehended without understanding the variability and predictability of the AM system *as an entirety.*

The long-term (interdecadal or longer) climate variability of the coupled Earth's climate system and external natural and anthropogenic forcings all strongly modulate the interannual variations of the AM system. Projection of its future behavior has also attracted great attention in our climate community. This chapter, however, will focus on only the major mechanisms of the IAV of the AM.

This chapter aims to address the following issues: What are the essential features of the leading mode of the IAV of the AM system? What are the major physical processes determining the IAV of the AM system? How predictable is the IAV of the AM precipitation? Addressing these questions may provide useful insight into the major physical processes that are responsible for the IAV and predictability of the AM system.

2. Essential Features of the Leading Mode of AM Variability

Novel analysis approaches are required to reveal the spatially and seasonally evolving variability of the entire AM system. Since the AM anomalies are sturdily regulated by the seasonal march of the solar radiation forcing and the resultant climatological annual cycles, the Season-reliant Empirical Orthogonal Function (S-EOF) analysis is an effective tool to detect seasonally evolving modes of climate variability (Wang and An 2005).

Figure 1 shows the leading S-EOF mode of the low-level (850 hPa) circulation anomalies and the associated precipitation anomalies over land and SST anomalies over ocean. Although the circulation anomalies are derived within the AM domain without information on ENSO, the resultant principal component (PC) is highly correlated with the Niño 3.4 index in the boreal winter (DJF), suggesting that the seasonally evolving AM anomalies are closely allied with the development, and mature and decay phases of an El Niño event. The patterns with opposite polarities tend to be associated with a La Niña event. There are two salient features. First,

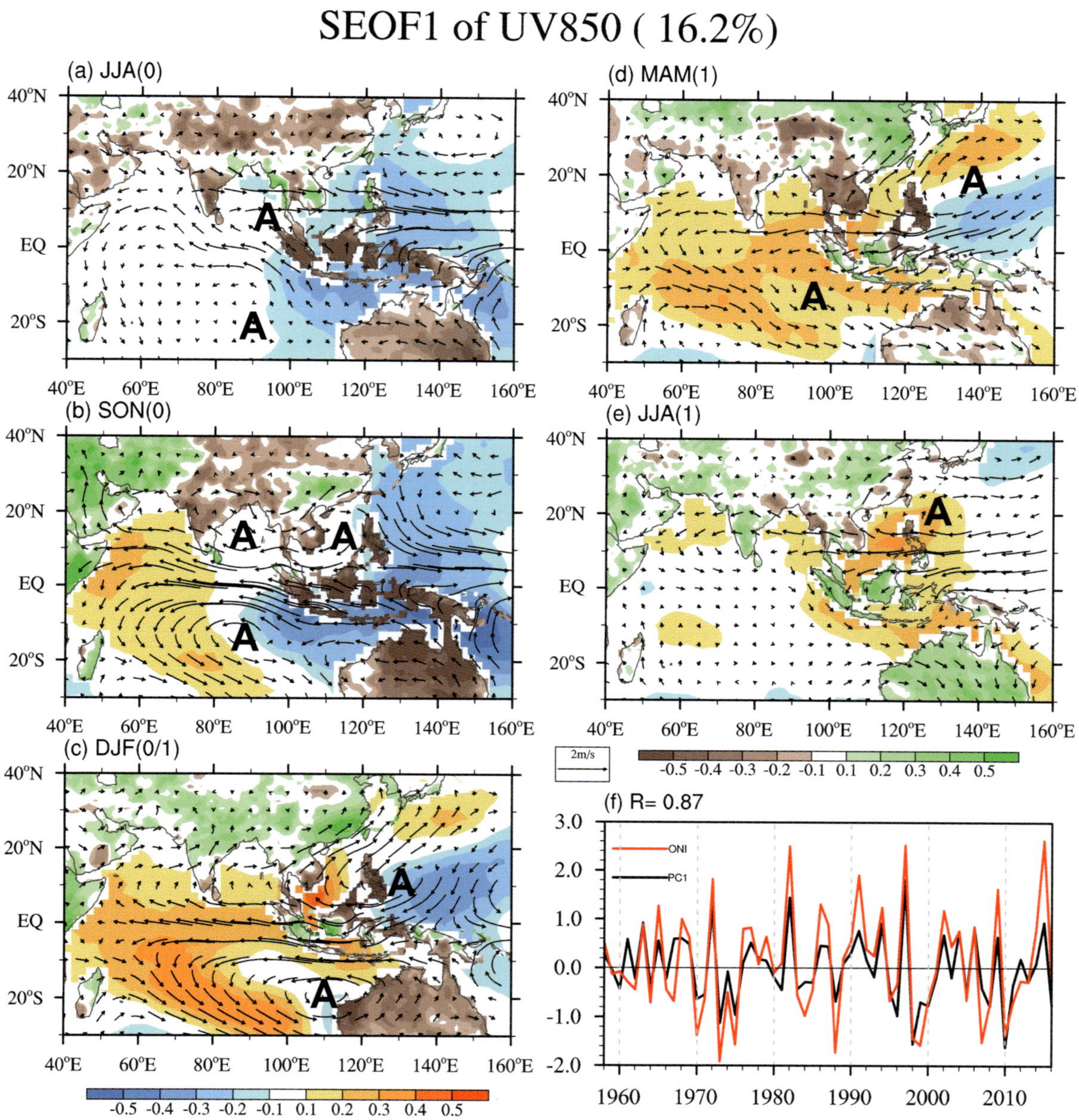

Fig. 1. Spatial patterns of the leading mode of the S-EOF of the 850 hPa winds from five consecutive seasons JJA(0) to JJA(1) for the period of 1958–2017. The color shadings represent the regressed SST anomalies over ocean and the correlation coefficient with rainfall anomalies over land with respect to the principal component (PC1). ERA UV850 data from January 1958 to December 2017 are used. The GPCC precipitation data and a merged SST dataset from ERSST and HadlSST are used.

the AM circulation anomalies are dominated by two off-equatorial low-level anomalous anticyclones, of which one is located in the Northern Hemisphere (NH) and the other in the Southern Hemisphere (SH). The NH anomalous anticyclone initially occurs over South Asia in JJA(0), moves to Southeast Asia in SON(0), and the Philippines in DJF(0/1), and persists in the western North Pacific (WNP) in MAM(1) and even to the ensuing summer (Fig. 1). The SH anomalous anticyclone appears mainly in the southern Indian Ocean (IO) from the development to mature phases (Fig. 1). Second, the circulation anomalies tend to shift eastward from the Indian ocean to western Pacific Ocean, so that the circulation anomalies tend to reverse their signs from JJA(0) to JJA(1) to the east of 80°E, indicating a strong biennial tendency of the AM variability from the development to decay phases of an El Niño event (Chang and Li 2000).

It is note worthy that the Asian land monsoon precipitation anomalies are closely associated with these two anomalous anticyclones. During the development phase of an El Niño from June to October, severe droughts often occur over Indonesia, India, and central northern China; during the mature phase of an El Niño from November to April, reduced rainfall takes place over the Philippines and IndoChina while enhanced precipitation appears over the East Asian subtropical front extending from southern China to western Japan; during the decaying summer, increased rainfall tends to occur over the middle reaches of the Yangtze River and Yellow River Valleys (Fig. 1). It is therefore critical to unravel the origins of the two anomalous anticyclones in the leading mode of AM variability.

3. Fundamental Mechanisms Determining the Dominant Mode of AM Variability

3.1. *ENSO forcing*

Retrospectively, a number of mechanisms have been proposed to explain how ENSO affects the AM (Shukla and Wallace 1983; Palmer *et al.* 1992; Nigam 1994; Ju and Slingo 1995; Webster *et al.* 1998; Lau and Bua 1998; Yang and Lau 1998; Kawamura 1998; Lau and Nath 2000).

The schematic diagrams shown in Fig. 2a illustrate the major processes by which the developing El Niño and La Niña affect Asian monsoon rainfall. When an El Niño develops (Fig. 2a), the changing zonal SST gradients generate anomalous convection in the central Pacific, which stimulates eastward-propagating Kelvin and westward-propagating equatorial Rossby waves. These waves act as efficient agents to adjust atmospheric circulation along the equator, resulting in a zonal shift of the Walker cell. Consequently, the rainfall over the Maritime Continent (MC) is robustly reduced during JJASO(0). The reduced-precipitation heating over the MC further excites descending Rossby waves to its west, forming an elongated anticyclonic ridge extending from the MC to India, and thereby instigating deficient Indian summer rainfall. The deficient Indian precipitation heating further generates an anomalous low in the upper troposphere over central Asia through atmospheric Rossby wave response. The central Asian low further excites a barotropic Rossby wave train along the waveguide provided by the westerly jet stream. This wave train travels all the way to the North Pacific and North Atlantic, showing a pattern called Circum-Global Teleconnection (CGT) (Ding and Wang 2005). One of the anomalous lows on this CGT wave train weakens the northern part of the western Pacific subtropical high (WPSH), reducing moisture transports to northern China and decreasing rainfall there. The part of the CGT from central Asia to the western Pacific was coined the "Silk Road Teleconnection" (Enomoto *et al.* 2003). For a more detailed discussion of the mechanisms of the CGT, the readers are referred to Ding *et al.* (2011) and Wu (2017).

Figure 2b illustrates that during a developing La Niña, the cooling in the equatorial central Pacific shifts the Walker cell, reducing convection around 160°E and strengthening convection over the MC. The suppressed convection around 160°E can directly strengthen the WPSH by

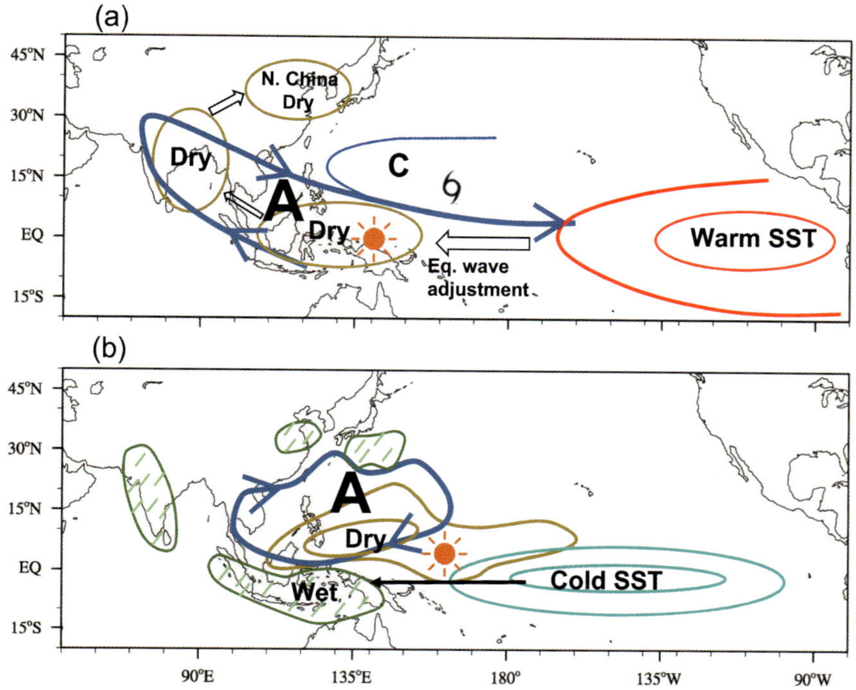

Fig. 2. A schematic diagram showing how ENSO affects Asian monsoon by the developing El Niño (a) and La Niña (b).

emanation of descending Rossby waves. Meanwhile, the reinforced MC convection can also enhance the WPSH via inducing equatorial easterlies over the western Pacific as a Kelvin wave response, which generates off-equatorial anticyclonic shear vorticity over the Philippine Sea. For a developing central Pacific El Niño, the anomalous patterns of circulation and precipitation tend to be the opposite.

3.2. *Monsoon-ocean interaction*

The notion of treating monsoon–warm ocean as a coupled system has been elaborated in a number of previous studies (e.g., Meehl 1994; Webster *et al.* 1998, 1999; Lau and Nath 2000; Wang *et al.* 2003). The monsoon-ocean interaction is often triggered by remote ENSO forcing or other factors external to the AM system; conversely, the monsoon anomalies can significantly rectify the effects of the original forcing. Several different types of monsoon-ocean interaction mechanisms have been proposed, which are reviewed in this subsection.

The equatorial Bjerknes (1969) *positive* feedback mechanism, which was employed to explain ENSO growth in the eastern Pacific, was applied to the equatorial IO to elucidate the development of the Indian Ocean Dipole (IOD) mode (Saji *et al.* 1999) or the Indian Ocean Zonal Mode (IOZM) (Webster *et al.* 1999) from boreal summer to fall (Fig. 1). Assuming that the sinking motion initially occurs over the MC, easterly anomalies can be generated along the west coast of Sumatra and the eastern equatorial IO. The easterly would induce coastal and equatorial upwelling, which produces a cold SST anomaly over the equatorial eastern IO. The cold SST further induces westward pressure gradient force and easterly anomalies, enhancing eastern IO cooling through the wind–upwelling–SST feedback. This positive feedback contributes to development of the IOD from summer to autumn. Some of the IODs are related to ENSO, but not all of them (Saji *et al.* 1999).

The second mechanism is a *negative* monsoon-ocean feedback through either monsoon-induced surface heat flux exchanges (Lau and Nath 2000) or monsoon-induced Ekman transport of ocean heat (Webster *et al.* 2002; Loschnigg *et al.* 2003). By diagnosis of the surface heat fluxes from a coupled model, Lau and Nath (2000) found that El Niño forcing tends to weaken the Indian Summer Monsoon (ISM), and the weakened monsoon winds and reduced cloudiness tend to decrease the surface latent heat fluxes and increase downward shortwave radiative fluxes, thereby warming the IO, which, in turn, induces a local atmospheric response that tends to reinforce the ISM. As such, the local atmosphere–ocean interaction is inclined to offset the original remote impacts of ENSO or to provide a negative feedback. Monsoon-induced anomalous ocean heat transport and heat exchanges can also provide a damping mechanism for the variations of the ISM. This mechanism was illustrated by Webster *et al.* (2002) and Loschnigg *et al.* (2003). When ISM winds are enhanced, the wind-induced Ekman flow transports heat from the NH to the SH, thus reducing the meridional thermal contrast between the NH and SH and thereby weakening the ISM. This negative feedback mechanism can operate on both the annual and interannual time scales. This partially explains why the interannual variability of the ISM is relatively small. The negative feedback along with the ocean memory provides a mechanism for the biennial oscillation of the AM system (Meehl 1994, 1997; Meehl and Arblaster 2001). The atmosphere–ocean interaction plays an active role in the Tropical Biennial Oscillation (TBO) theory (Chang and Li 2000).

The third mechanism is an off-equatorial anomalous anticyclone–SST dipole feedback mechanism (Wang *et al.* 2000), which can provide either a *positive or a negative feedback* to initially forced monsoon anomalies depending on the climatological background flows. Different from the equatorial Bjerknes feedback, this mechanism emphasizes the interaction between the off-equatorial atmospheric Rossby waves and underlying SST dipole anomalies through the thermodynamic feedback, i.e., the wind–evaporation/entrainment–SST feedback and the cloud–radiation–SST feedback.

The anomalous anticyclone–SST dipole feedback mechanism is critically important in understanding the evolution of the two dominant anomalous anticyclones seen in Fig. 1. The *anomalous* WNP anticyclone conveys the impacts of El Niño on East Asian climate during the mature and decay phases of an El Niño (from winter to the ensuing summer). The WNP anticyclone forms in the fall about one season prior to the peak El Niño and its development is attributed to the combined effects of the remote El Niño forcing, tropical–extratropical interaction, and monsoon-ocean interaction (Wang and Zhang 2002). From the El Niño developing fall to the decaying summer, the WNP anticyclone is coupled with a pronounced anomalous SST dipole: a cooling to its east in the WNP and a warming to its west over the northern IO and East Asia (EA) marginal seas (Fig. 1). Note that the anomalous anticyclone not only controls the WNP and affects EA but also extends into the northern Indian Ocean (NIO); as a result, easterly anomalies prevail over the NIO from SON(0) to JJA(1).

The anomalous WNP anticyclone–SST dipole interaction provides a positive thermodynamic feedback, which amplifies the WNP anticyclone from fall to spring, and maintains it to the El Niño decaying summer during strong El Niño events (Wang *et al.* 2000; Lau 2003; Lau *et al.* 2005; Lee *et al.* 2013). The positive feedback mechanism that sustains the WNP anticyclonic anomaly is illustrated in Fig. 3.

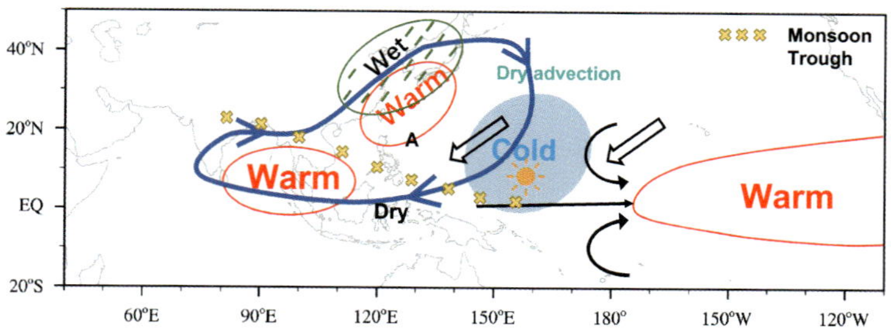

Fig. 3. A schematic diagram showing anomalous WNP anticyclone and Indo-Pacific SST dipole SSTA feedback according to Wang *et al.* (2013).

Over the WNP, to the east of the anticyclone the anomalous northeasterly is superposed on the mean northeasterly winds, increasing wind speeds and the associated surface evaporation and entrainment, thus cooling the ocean. On the contrary, the anomalous northeasterly transports dry air to the southeast of the anomalous WNP anticyclone. Both the SST cooling and dry advection reduce convective instability and suppress convection to the east and southeast of the anticyclone. The reduced convective heating there can then generate descending Rossby waves, which reinforce the WNP anticyclone in their westward journey (Wang *et al.* 2000, 2013). This positive thermodynamic feedback between the anomalous WNP anticyclone and the underlying warm pool dipole SST anomaly has been demonstrated by numerical experiments with coupled atmosphere–ocean models (Lau and Nath 2000; Lau *et al.* 2005; Lau and Wang 2006; Chowdary *et al.* 2010; Xiang *et al.* 2013).

Over the northern IO, the ridge of the WNP anticyclone extends westward to the Bay of Bengal, such that the associated anomalous subsidence and easterlies penetrate into the northern IO (Fig. 1), increasing downward solar radiation flux and reducing evaporation cooling, thus warming the northern IO (Wang *et al.* 2003, Du *et al.* 2009). On the contrary, the IO warming stimulates a warm atmospheric Kelvin wave response to generate anticyclonic shear vorticity and further to enhance the WNP anticyclone (Xie *et al.* 2009; Wu *et al.* 2009). More recently, the atmosphere–ocean feedback mechanism maintaining the WNP anticyclone has generally been considered as a coupled ocean-atmosphere mode across the Indo-Pacific warm pool (Wang *et al.* 2013, 2017; Xie *et al.* 2016; Li *et al.* 2017).

This anomalous WNP anticyclone–SST dipole feedback mechanism is thought to be strong during spring and early summer and gradually weaken toward late summer as the cooling to the east of the anticyclone wanes. However, the dry advection and atmospheric stabilization to the southeast of the anticyclone remain in action in late summer. In addition, the atmospheric Rossby response is more sensitive to SST anomalies during the late summer and the response can be significant even to a moderate SST anomaly of 0.2–0.3°C because the background SST is over 29°C and the mean precipitation is high Xiang *et al.* 2013).

3.3. *Important roles of the annual cycle*

The annual cycle of the AM can strongly modify the AM response to the remote El Niño forcing and control the nature of the monsoon-ocean feedback, thereby playing a "hidden" yet critical role in formulating AM anomalies.

A persuasive example of the role of monsoon annual cycle is the evolution of the IOD, which is an equatorial manifestation of the southern IO anticyclone. The IOD and southern IO anticyclone, which affect climate conditions over the IO, eastern Africa, and India, emerge during summer, reach peak intensity in fall, and swiftly decay after November. The drastic amplification of the southern IO anticyclone in the El Niño developing summer and the ensuing decay during mature phases of El Niño cannot be explained solely by the remote El Niño forcing.

Such a life cycle has its roots in the local monsoon-ocean interaction as illustrated in Fig. 4 and it is discussed hereafter following the arguments in Wang *et al.* (2003). From boreal summer to fall, the positive monsoon-ocean feedback can amplify the monsoon and SST dipole anomalies over the IO. On the one hand, the negative SST anomalies to the east of the southern IO anticyclone can generate descending atmospheric Rossby waves that enhance the low-level anticyclone in their decaying journey to the west. On the other hand, under the summer mean flow conditions the anomalous southern IO anticyclone can boost the southeasterly winds off the coast of Sumatra, which can further intensify coastal and equatorial upwelling (Saji *et al.* 1999; Murtugudde *et al.* 2000) (Fig. 4a). However, from boreal fall to winter, the monsoon flows reverse direction and westerly prevails to the north of the southern IO convergence zone (Fig. 4b); as a result, the aforementioned IO anticyclone–SST dipole interaction turns from the positive feedback to a negative feedback, causing rapid decay of the southern IO anticyclone and IOD.

Another example of the role of the mean monsoon flows is seen during the summer of El Niño development. According to the Matsuno (1966) and Gill (1980) theory, suppressed convection over the MC would excite an equatorially symmetric Rossby wave response (double anticyclones). The observed pattern in Fig. 1a, however, shows an obvious equatorial asymmetry with the NH anticyclone being stronger and extending from the MC to India (Fig. 1a). This equatorially asymmetric response arises from the effects of the vertical shear of monsoon mean flows on the equatorial Rossby wave response. Wang and Xie (1996) and Xie and Wang (1996) extended Matsuno's equatorial wave theory and demonstrated the effects of three-dimensional background circulations on equatorial waves. They explained why easterly vertical shears can drastically amplify moist Rossby waves while westerly shears tend to reduce them. Under the summer monsoon easterly vertical shears, the Rossby waves are more trapped in the lower troposphere, and they

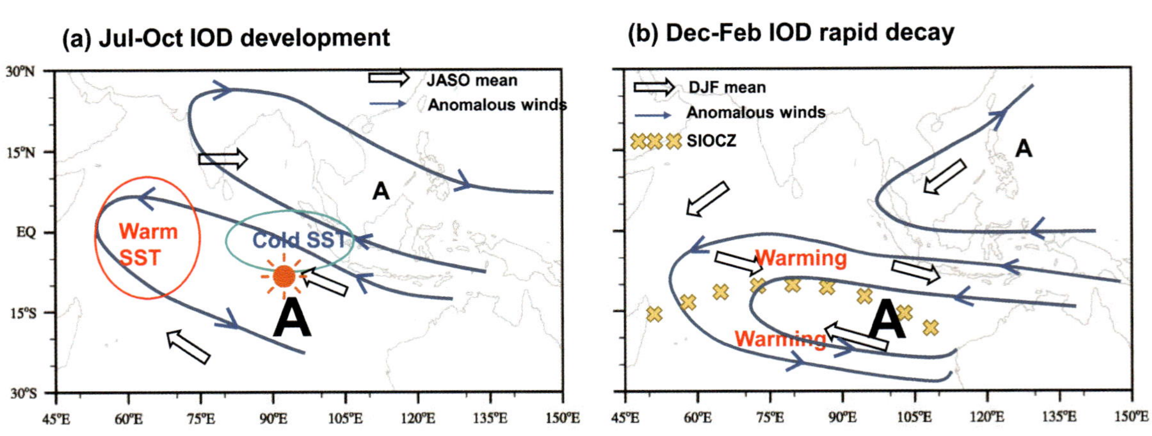

Fig. 4. A schematic diagram showing how mean flow affects IOD evolution (a) during July–October and (b) during December–February according to Wang *et al.* (2003).

interact more vigorously with boundary layer Ekman flow-induced moisture convergence. As such, the moist equatorial Rossby wave response in the presence of easterly vertical shear is substantially stronger than that under westerly shears. The large easterly vertical shear located in South Asia can enhance Rossby wave response, leading to a stronger anticyclonic ridge to the north of the equator. Numerical model experiments have confirmed the aforementioned theoretical results (Wang *et al.* 2003).

In sum, the monsoon mean flows can not only effectively modify the teleconnection caused by remote ENSO forcing but also determine the nature of the local atmosphere–ocean interaction, thus strongly modifying the evolution of AM anomalies.

4. Predictability of Asian Summer Monsoon Precipitation Anomalies

Considerable progress has been made in figuring out the physical factors that drive ASM variability, but it remains uncertain what the relative contributions of these factors are and to what extent the combination of these factors can predict ASM rainfall variability. A better method of estimating the predictability limit for monsoon rainfall remains challenging and controuersial.

To determine the predictable part of the total variance, Kumar *et al.* (2007) used a mean square error (MSE) method based on multi-dynamical models' simulations. The optimal estimate of the noise variance is obtained from the smallest MSE among many models' results at each geographical location. The signal variance is estimated by subtracting the smallest MSE (which is the estimated internal variance) from the observed total variance. This method requires perfect models and large ensembles. Wang *et al.* (2015b) applied a predictable mode analysis (PMA) method for estimating potential seasonal predictability (Lee *et al.* 2011). The PMA method first identifies the most important, physically meaningful modes of variability; it then uses retrospective predictions with dynamical models or physical–empirical models (PEMs) to identify the so-called predictable modes. The potential predictability is then estimated by the fractional variance of the sum of the predictable modes.

Analysis of the ASM precipitation (1979–2016) reveals four physically meaningful modes (Fig. 5): the central Pacific (CP)-ENSO forced mode, the Indo-Pacific monsoon-ocean coupled mode, the IOD mode, and the global warming mode, which together explain about 44% of the total variance. It is shown that sources of predictability for those four modes primarily come from ENSO (Mode 1), monsoon-ocean interactions over the Indo-Pacific warm pool (Modes 2 and 3) and the long-term global warming effects (Mode 4). It has been demonstrated that the four modes are, to a large extent, predictable with the PEMs and dynamical models' MME. The dynamical MME prediction skills largely originate from the four modes. Thus, these four modes are identified as predictable modes and the other higher modes are "noises". The dynamical and PEM prediction skills are complimentary. The results suggest that about half of the total variance of the ASM rainfall is potentially predictable. Keep in mind, however, that this estimate is based on limited data of 38 years. Longer datasets are obviously needed for reassessing the potential predictability. The PMA method, however, is shown to provide not only an effective approach for estimating the seasonal potential predictability but also an instrument for better understanding sources of predictability.

5. Discussion

The IAV of the AM is generally subject to secular changes. Webster (2006) pointed out, "Combination of modeling problems and empirical non-stationarity has plagued monsoon prediction on interannual time scales. Empirical

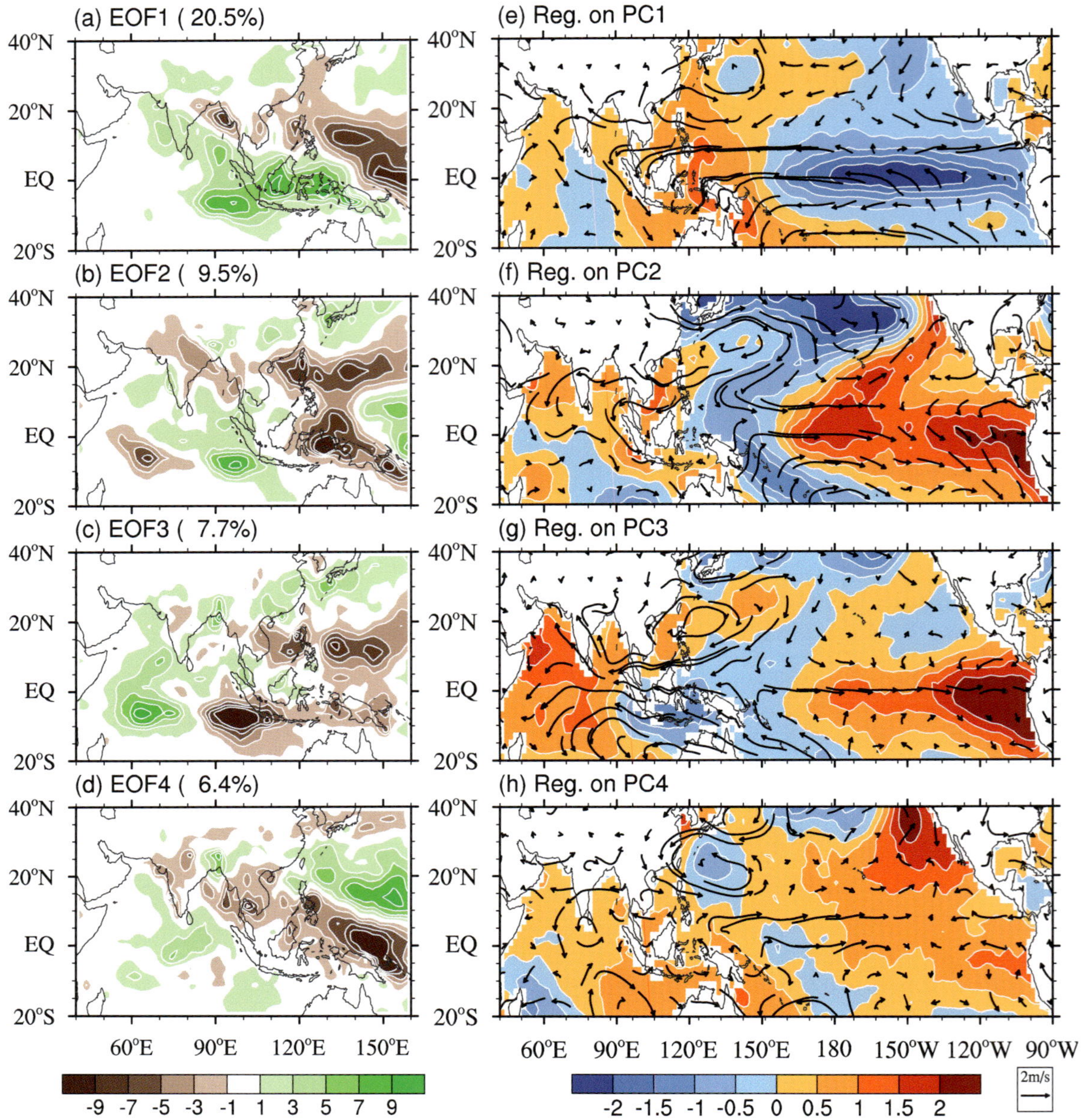

Fig. 5. The four leading EOF modes of Asian monsoon precipitation during JJA. The left panels show the spatial patterns, and the right panels show the regressed wind anomalies (vectors, m/s) and SSTA (shading, °C) with regard to the corresponding four principal components.

forecasts have to contend with the specter of statistical non-stationarity. . . ." The Pacific Decadal Oscillation, North Atlantic Oscillation, and Atlantic Multidecadal Oscillation have been shown to be able to modulate the AM interannual variability (e.g., Lee *et al.* 2013).

Conventionally, local SST anomalies in the monsoon oceans such as the NH western Pacific and the IO have been considered as drivers for AM variability (Nitta 1987; Shen and Lau 1995; Kawamura *et al.* 2001). Following this impression, many numerical experiments were conducted using stand-alone atmospheric general circulation models (AGCMs). However, it was surprisingly found that 10 AGCMs forced by the observed SST anomalies failed simulation of monsoon rainfall anomalies even given the strongest 1997–1998 El Niño forcing (Wang *et al.* 2004). The models' failure is ascribed to the modeling strategy, i.e., treating monsoon as a slave to prescribed SST in the monsoon ocean regions. This finding was further confirmed by the observed negative correlation between precipitation and SST anomalies over the summer monsoon oceans. This negative correlation is more significant when precipitation anomalies precede SST anomalies by one month, which suggests that over the precipitating summer monsoon regions, the SST is generally a passive response to the atmosphere rather than a forcing (Wang *et al.* 2005). Therefore, it is largely inadequate to attribute AM anomalies to the impacts of local warm pool SST anomalies. The "Indian Ocean capacitor" mechanism (Xie *et al.* 2009) has led researchers to consider the northern IO SST warming as a delayed "forcing" to the atmosphere. In fact, the northern IO warming is also generally forced by anomalous atmospheric circulations, and consequently the role of the IO warming should be understood from the standpoint of atmosphere–ocean interaction. Coupled climate model experiments are thus requested to test this type of hypothesis. Note also that the tropical atmospheric circulation is largely driven by precipitation latent heat. Positive SST anomalies over the IO are not necessarily a sign of atmospheric heating. If the IO warming is not accompanied by positive precipitation anomalies, the IO warming itself cannot force the atmosphere. In that case, use of AGCM would yield misleading interpretation.

Therefore, the traditional wisdom regarded SST anomalies in the Indian and western Pacific Oceans as a cause of the AM variability should be questioned. The SST anomalies in these warm ocean regions are, to a large extent, a result of the impacts of anomalous monsoons. Thus, it is the monsoon–warm ocean interaction that can significantly modify the impacts of the remote El Niño forcing and should be considered as one of the physical factors determining the variability of the AM system.

This chapter focuses on the IAV of the ASM system as a whole, but understanding regional differences and finding contiguous factors in regional monsoon predictions are equally important. Responses to ENSO forcing in different regions are closely dependent on the ENSO intensity and the location and evolution of the SST anomalies associated with different types of ENSO events (e.g., Yu *et al.* 2017). Those are active areas of research. Other mechanisms that can affect AM variability, especially on the regional scales, include, but are not limited to, land surface processes and the atmosphere–land interaction, Eurasian and Tibetan Plateau snow anomalies, Tibetan Plateau warming, and spring Arctic sea ice anomalies.

With the aim of stimulating future studies on this subject, a number of questions and controversial issues are listed here to spur further exploration. (i) Why do current climate models have notorious deficiencies and encounter enormous difficulties in correct simulations of the Western Pacific Subtropical High and the Meiyu/Baiu front zones? (ii) Is the TBO a robust and intrinsic feature of the AM variability? What physical processes are essential for the biennial tendency of the AM system? (iii) To what extent do the mid–high-latitude circulation anomalies prior to the summer monsoon affect the ASM? How are they generated and maintained? (iv) How are land surface anomalies generated and maintained? How do they feed back to the seasonal variability of precipitation? (v) How do radiative impacts of clouds and aerosols affect monsoon evolution and intensity? (vi) How are the boreal summer teleconnection (the PJ and circumglobal teleconnection) modes associated with the ASM excited and maintained? Are those modes intrinsic to the low-frequency variability of the boreal summer mean states? (vii) What are the potential predictabilities of the regional components of the ASM during the years when ENSO is in a near-normal state? (viii) Are there fundamental differences between seasonal predictions of mean climate and the total number of extreme climate events? What are the physical bases for seasonal prediction of extreme events? (ix) How will IAV of the AM likely change in a warming world under anthropogenic forcing? These issues call for new ideas and breakthroughs.

References

Bjerknes, J., 1969: Atmospheric teleconnections from the equatorial Pacific. *Month. Weath. Rev.*, **97**, 163–172.

Blanford, H. F., 1884: On the connexion of the Himalaya snowfall with dry winds and seasons of drought in India. *Proc. Roy. Soc. Lond.*, **37**, 3–22.

Chang, C. and T. Li, 2000: A theory for the tropical tropospheric biennial oscillation. *J. Atmosph. Sci.*, **57**, 2209–2224.

Chang, C.-P., Y. Zhang, and T. Li, 2000: Interannual and interdecadal variations of the East Asian summer M-monsoon and tropical Pacific SSTs. Part I: Roles of the subtropical ridge. *J. Climate*, **13**, 4310–4325.

Charney, J. G. and J. Shukla, 1981: *Predictability of Monsoons. Monsoon Dynamics*, J. Lighthill, and R.P. Pearce, Eds., Cambridge University Press, pp. 99–108.

Chowdary, J., S. P. Xie, J. Y. Lee, Y. Kosaka, and B. Wang, 2010: Predictability of summer northwest Pacific climate in 11 coupled model hindcasts: Local and remote forcing. *J. Geophys. Res. — Atmos.*, **115**, D22121.

Ding, Q. and B. Wang, 2005: Circumglobal teleconnection in the Northern Hemisphere summer. *J. Clim.*, **18**, 3483–3505.

Ding, Q., B. Wang, J. M. Wallace, and G. Branstator, 2011: Tropical-extratropical teleconnections in boreal summer: Observed interannual variability. *J. Clim.*, **24**, 1878–1896.

Du, Y., S.-P. Xie, G. Huang, and K. Hu, 2009: Role of air–sea interaction in the long persistence of El Niño–induced north Indian Ocean warming. *J. Climate*, **22**, 2023–2038.

Enomoto, T., B. J. Hoskins, and Y. Matsuda, 2003: The formation mechanism of the Bonin high in August. *Quart. J. Roy. Meteorol. Soc.*, **129**, 157–178.

Gill, A. E., 1980: Some simple solutions for heat-induced tropical circulation. *Quart. J. Roy. Meteorol. Soc.*, **106**, 447–462.

Ju, J. and J. Slingo, 1995: The Asian summer monsoon and ENSO. *Quart. J. Roy. Meteorol. Soc.*, **121**, 1133–1168.

Kawamura, R., 1998: A possible mechanism of the Asian summer monsoon-ENSO coupling. *J. Meteorol. Soc. Jpn. Ser. II*, **76**, 1009–1027.

Kawamura, R., T. Matsuura, and S. Iizuka, 2001: Role of equatorially asymmetric sea surface temperature anomalies in the Indian Ocean in the Asian summer monsoon and El Niño-Southern Oscillation coupling. *J. Geophys Res — Atmos.*, **106**, 4681–4693.

Kumar, A., B. Jha, Q. Zhang, and L. Bounoua, 2007: A new methodology for estimating the unpredictable component of seasonal atmospheric variability. *J. Clim.*, **20**, 3888–3901.

Lau, K. and H. Wu, 2001: Principal modes of rainfall–SST variability of the Asian summer monsoon: A reassessment of the monsoon–ENSO relationship. *J. Clim.*, **14**, 2880–2895.

Lau, K. and H. Weng, 2001: Coherent modes of global SST and summer rainfall over China: An assessment of the regional impacts of the 1997–98 El Niño. *J. Clim.*, **14**, 1294–1308.

Lau, K., K. Kim, and S. Yang, 2000: Dynamical and boundary forcing characteristics of regional components of the Asian summer monsoon. *J. Clim.*, **13**, 2461–2482.

Lau, K., J. Lee, K. Kim, and I. Kang, 2004: The North Pacific as a regulator of summertime climate over Eurasia and North America. *J. Clim.*, **17**, 819–833.

Lau, K.-M. and W. Bua, 1998: Mechanisms of monsoon-Southern Oscillation coupling: Insights from GCM experiments. *Clim. Dynam.*, **14**, 759–779.

Lau, N.-C. and M. J. Nath, 2000: Impact of ENSO on the variability of the Asian–Australian monsoons as simulated in GCM experiments. *J. Clim.*, **13**, 4287–4309.

Lau, N.-C., 2003: Atmosphere–ocean variations in the Indo-Pacific sector during ENSO episodes. *J. Clim.*, **16**, 3–20.

Lau, N.-C. and B. Wang, 2006: Interactions between the Asian monsoon and the El Niño/ Southern Oscillation. *The Asian Monsoon*, Springer, pp. 479–512.

Lau, N.-C., A. Leetmaa, M. J. Nath, and H.-L. Wang, 2005: Influences of ENSO-induced Indo–western Pacific SST anomalies on extratropical atmospheric variability during the boreal summer. *J. Clim.*, **18**, 2922–2942.

Lee, E.-J., K.-J. Ha, and J.-G. Jhun, 2014: Interdecadal changes in interannual variability of the global monsoon precipitation and interrelationships among its subcomponents. *Clim. Dynam.*, **42**, 2585–2601.

Lee, J.-Y. and Coauthors, 2011: How predictable is the northern hemisphere summer upper-tropospheric circulation? *Clim. Dynam.*, **37**, 1189–1203.

Lee, S.-S., Y.-W. Seo, K.-J. Ha, and J.-G. Jhun, 2013: Impact of the western North Pacific subtropical high on the East Asian monsoon precipitation and the Indian Ocean precipitation in the boreal summertime. *Asia-Pacific J. Atmosph. Sci.*, **49**, 171–182.

Li, T., B. Wang, B. Wu, T. Zhou, C.-P. Chang, and R. Zhang, 2017: Theories on formation of an anomalous anticyclone in western North Pacific during El Niño: A review. *J. Meteorol. Res.*, **31**, 987–1006.

Loschnigg, J., G. A. Meehl, P. J. Webster, J. M. Arblaster, and G. P. Compo, 2003: The Asian monsoon, the tropospheric biennial oscillation, and the Indian Ocean zonal mode in the NCAR CSM. *J. Clim.*, **16**, 1617–1642.

Matsuno, T., 1966: Quasi-geostrophic motions in the equatorial area. *J. Meteorol. Soc. Jpn. Ser. II*, **44**, 25–43.

Meehl, G. A., 1994: Coupled land-ocean-atmosphere processes and South Asian monsoon variability. *Science*, **266**, 263–267.

Meehl, G. A., 1997: The south Asian monsoon and the tropospheric biennial oscillation. *J. Clim.*, **10**, 1921–1943.

Meehl, G. A. and J. M. Arblaster, 2001: The tropospheric biennial oscillation and Indian monsoon rainfall. *Geophys. Res. Lett.*, **28**, 1731–1734.

Murtugudde, R., J. P. McCreary, and A. J. Busalacchi, 2000: Oceanic processes associated with anomalous events in the Indian Ocean with relevance to 1997–1998. *J. Geophys. Res.: Oceans*, **105**, 3295–3306.

Nigam, S., 1994: On the dynamical basis for the Asian summer monsoon rainfall-El Niño relationship. *J. Clim.*, **7**, 1750–1771.

Nitta, T., 1987: Convective activities in the tropical western Pacific and their impact on the Northern Hemisphere summer circulation. *J. Meteorol. Soc. Jpn. Ser. II*, **65**, 373–390.

Palmer, T., Č. Branković, P. Viterbo, and M. Miller, 1992: Modeling interannual variations of summer monsoons. *J. Clim.*, **5**, 399–417.

Saji, N., B. Goswami, P. Vinayachandran, and T. Yamagata, 1999: A dipole mode in the tropical Indian Ocean. *Nature*, **401**, 360–363.

Shen, S. and K.-M. Lau, 1995: Biennial oscillation associated with the East Asian summer monsoon and tropical sea surface temperatures. *J. Meteorol. Soc. Jpn. Ser. II*, **73**, 105–124.

Shukla, J. and J. Wallace, 1983: Numerical simulation of the atmospheric response to equatorial Pacific sea surface temperature anomalies. *J. Atmosph. Sci.*, **40**, 1613–1630.

Shukla, J. and D. Mooley, 1987: Empirical prediction of the summer monsoon rainfall over India. *Month. Weath. Rev.*, **115**, 695–704.

Tao, S. and L. Chen, 1987: A review of recent research on East Asian summer monsoon in China. *Monsoon Meteorology*, Eds. C.-P. Chang, and T. N. Krishnamurti, Oxford University, pp. 60–92.

Walker, G. T., 1923: Correlation in seasonal variations of weather, VIII: A preliminary study of world weather. *Mem. Indian Meteor. Dept.*, **24**, 75–131.

Walker, G. T. and E. W. Bliss, 1932: World Weather. *Mem. Roy. Meteorol. Soc.*, **4**, 53–84.

Wang, B. and X. Xie, 1996: Low-frequency equatorial waves in vertically sheared zonal flow. Part I: Stable waves. *J. Atmosph. Sci.*, **53**, 449–467.

Wang, B. and Q. Zhang, 2002: Pacific–east Asian teleconnection. Part II: How the Philippine Sea anomalous anticyclone is established during El Nino development. *J. Clim.*, **15**, 3252–3265.

Wang, B. and S. I. An, 2005: A method for detecting season-dependent modes of climate variability: SEOF analysis. *Geophys. Res. Lett.*, **32**, L15710.

Wang, B., R. Wu, and X. Fu, 2000: Pacific–East Asian teleconnection: How does ENSO affect East Asian climate? *J. Clim.*, **13**, 1517–1536.

Wang, B., R. Wu, and K.-M. Lau, 2001: Interannual variability of the Asian summer monsoon:

Contrasts between the Indian and the western North Pacific–East Asian monsoons. *J. Clim.*, **14**, 4073–4090.

Wang, B., R. Wu, and T. Li, 2003: Atmosphere–warm ocean interaction and its impacts on Asian Australian monsoon variation. *J. Clim.*, **16**, 1195–1211.

Wang, B., I.-S. Kang, and J.-Y. Lee, 2004: Ensemble simulations of Asian–Australian monsoon variability by 11 AGCMs. *J. Clim.*, **17**, 803–818.

Wang, B., B. Xiang, and J.-Y. Lee, 2013: Subtropical high predictability establishes a promising way for monsoon and tropical storm predictions. *Proc. Nat. Acad. Sci.*, **110**, 2718–2722.

Wang, B., Q. Ding, X. Fu, I. S. Kang, K. Jin, J. Shukla, and F. Doblas-Reyes, 2005: Fundamental challenge in simulation and prediction of summer monsoon rainfall. *Geophys. Res. Lett.*, **32**, L15711.

Wang, B., B. Xiang, J. Li, P. J. Webster, M. N. Rajeevan, J. Liu, and K.-J. Ha, 2015a: Rethinking Indian monsoon rainfall prediction in the context of recent global warming. *Nat Commun.*, **6**, 7154.

Wang, B., J.-Y. Lee, and B. Xiang, 2015b: Asian summer monsoon rainfall predictability: A predictable mode analysis. *Clim. Dynam.*, **44**, 61–74.

Wang, B., J. Li, and Q. He, 2017: Variable and robust East Asian monsoon rainfall response to El Niño over the past 60 years (1957–2016). *Adv. Atmosph. Sci.*, **34**, 1235–1248.

Webster, P. and Coauthors, 2002: The JASMINE pilot study. *Bull. Am. Meteorol. Soc.*, **83**, 1603–1630.

Webster, P. J., 2006: The coupled monsoon system. *The Asian Monsoon*, Springer, pp. 3–66.

Webster, P. J., A. M. Moore, J. P. Loschnigg, and R. R. Leben, 1999: Coupled ocean–atmosphere dynamics in the Indian Ocean during 1997–98. *Nature*, **401**, 356–360.

Webster, P. J., V. O. Magana, T. Palmer, J. Shukla, R. Tomas, M. Yanai, and T. Yasunari, 1998: Monsoons: Processes, predictability, and the prospects for prediction. *J. Geophys. Res.: Oceans*, **103**, 14451–14510.

Wu, B., T. Zhou, and T. Li, 2009: Seasonally evolving dominant interannual variability modes of East Asian climate. *J. Clim.*, **22**, 2992–3005.

Wu, R., 2017: Relationship between Indian and East Asian summer rainfall variations. *Adv Atmos. Sci.*, **34**, 4–15.

Xiang, B., B. Wang, W. Yu, and S. Xu, 2013: How can anomalous western North Pacific subtropical high intensify in late summer? *Geophys. Res. Lett.*, **40**, 2349–2354.

Xie, S.-P., Y. Kosaka, Y. Du, K. Hu, J. S. Chowdary, and G. Huang, 2016: Indo-western Pacific Ocean capacitor and coherent climate anomalies in post-ENSO summer: A review. *Adv. Atmosph. Sci.*, **33**, 411–432.

Xie, S.-P., K. Hu, J. Hafner, H. Tokinaga, Y. Du, G. Huang, and T. Sampe, 2009: Indian Ocean capacitor effect on Indo–western Pacific climate during the summer following El Niño. *J. Clim.*, **22**, 730–747.

Xie, X. and B. Wang, 1996: Low-frequency equatorial waves in vertically sheared zonal flow. Part II: Unstable waves. *J. Atmosph. Sci.*, **53**, 3589–3605.

Yang, S. and K. Lau, 1998: Influences of sea surface temperature and ground wetness on Asian summer monsoon. *J. Clim.*, **11**, 3230–3246.

Yu, J. and S. W. Fang, 2018: The distinct contributions of the seasonal footprinting and Charge discharged mechnisms to ENSO compexity. *Geophys. Res. Letter*, Accepted.

Chapter 3

The Nature of the Thermal Forcing of the Asian Summer Monsoon

Guoxiong Wu*, Yimin Liu and Bian He

State Key Laboratory of Numerical Modeling for Atmospheric Sciences and Geophysical Fluid Dynamics, Institute of Atmospheric Physics, Chinese Academy of Sciences, Beijing 100029, China

*gxwu@lasg.iap.ac.cn

Monsoon is characterized by the reversal of prevailing surface wind between winter and summer accompanied by the occurrence of heavy precipitation. Traditionally, monsoon is considered as a consequence of the response of atmospheric circulation to the seasonal change of land-sea thermal contrast induced by the annual cycle of solar radiation. Here, it is demonstrated that land-sea thermal contrast is only one part of the monsoon story. The idealized experiments presented in this study reveal that the seasonal change of land-sea thermal contrast alone is not a sufficient condition for the existence of the monsoon. The seasonal change of thermal forcing of the Tibetan–Iranian Plateau plays a significant role in generating the continental Asian summer monsoon. The nature of the Asian monsoon is indeed the response of atmospheric circulation to the seasonal changes of land-sea thermal contrast and thermal status of large-scale topography induced by the annual cycle of solar radiation.

In this chapter, results based on numerical experiments are used to reveal the nature of the Asian summer monsoon (ASM). It starts with a set of aqua-planet experiments to understand the roles of large-scale mountains in the formation of the ASM in contrast to those of land-sea distribution. Then, a series of experiments based on an atmospheric general circulation model is employed to demonstrate the significance of the thermal forcing of the Tibetan–Iranian Plateau (TIP) in forcing the vertical transport of water vapor, large-scale air ascent, and horizontal transport of water vapor from the ocean to the inland area in the ASM area. Discussions on different hypotheses are also provided.

1. Aqua-Planet Experiment

Wu *et al.* (2012) developed a series of idealized numerical experiments based on an atmospheric general circulation model SAMIL-R42L9 (Wu *et al.* 2003) to understand how the land-sea distribution and large-scale mountain ranges, the Tibetan–Iranian Plateau (TIP), affect the formation of the ASM. To conduct sensitivity experiments, the entire surface of the model Earth is first covered with water to form an aqua-planet experiment (Exp AQU). The sea surface temperature (SST) used is the zonally averaged climatological SST provided by the Second Atmospheric Model Intercomparison Program (AMIP-II, Fiorino 2000), which has seasonal variation. The zonal means of the model climatology of wind, temperature, humidity, and surface pressure from Exp AQU are taken as initial values for other idealized experiments. Four types of land distributions with different geometries are embedded in the aqua-planet for different experiments (Table 1). In Exp MID, a continent located over 0–120°E and 30–90°N is investigated for its influence on circulations at middle and high latitudes. In Exp SUB, the southern boundary of the continent in Exp MID is extended 10° southward into the subtropics, mimicking the main part of the Eurasian continent.

The Multiscale Global Monsoon System
Edited by C. P. Chang et al.

Table 1. Experiment design of the idealized aqua-planet experiments.

Experiment	Distribution of land
MID	Higher-latitude continent (0°–120°E, 30°–90°N)
SUB	Subtropical continent (0°–120°E, 20°–90°N)
TRO	Tropical continent (SUB land and 0°–50°E, 35°S–20°N; 75°–85°E, 5°–20°N; and 95°–105°E, 9°S–20°N)
TIP	TRO and Tibetan–Iranian Plateau (elliptic topography TP with maximam altitude of 5000 m, centered at (87.5°E, 32.5°N), and elliptic topography IR with maximam altitude of 3000 m, centered at (53.4°E, 32.5°N)).

In Exp TRO, three square-shaped tropical lands over (0–50°E, 35°S–20°N), (75°–85°E, 5°–20°N), and (95°–105°E, 9°S–20°N) are introduced to represent the tropical African, Indian, and Indochinese Peninsula subcontinents, respectively, which are then integrated into the main Eurasian continent as in Exp SUB to form the "Afro-Eurasian continent". Exp TIP uses the same continent as that in Exp TRO, but adds an idealized TIP to investigate its influence on the monsoon. All of these experiments were integrated for 10 years, and the means of the last 8 years were used for the analyses.

In Exp AQU, the July mean precipitation largely occurs near the equator along the "Inter-tropical convergent zone (ITCZ)", as obtained in Liang *et al.* (2006). For a continent located at extratropical latitudes, the experiments (Exps MID and SUB) produce insignificant monsoon rainfall. A very weak monsoon rainband along the southern continental boundary coexists with the ITCZ along the equator, and tropical water vapor substantially converges to the ITCZ as in Exp AQU, which effectively prohibits the interaction of circulation systems between the two hemispheres. The inclusion of a tropical land (Fig. 1c) produces cross-equatorial flow, which destroys the local ITCZ, pushes it to about 10°–15°N across the Africa–Indian Ocean sector, making it a tropical convergent zone (TCZ, Fig. 1c), and forms the tropical summer monsoon there termed as the southern branch of the South Asian summer monsoon (SASM). It also enhances the monsoon over the southern and southeastern boundary of the extratropical continent as shown in Fig. 1b.

However, the land-sea distribution alone does not produce significant precipitation over the continent in all the above experiments. The uplifted orographic thermal forcing in summer generates a cyclonic circulation in the lower troposphere, with a southerly to its east and a northerly to its west. Consequently, TIP forcing strengthens the circulation coupling between the subtropics and tropics, and between the lower and upper troposphere. Abundant water vapor is transported from the tropical ocean to the subtropical and extratropical continent, where it condenses during the course of northward travel (Fig. 1d). The difference between the experiments with and without the TIP as shown in Fig. 1e reveals that the TIP forcing in summer is to produce the northern branch of the SASM and the Eastern ASM (EASM), and reduce the precipitation of the southern branch of the SASM west of 80°E. This implies that only when the large-scale mountain, the TIP, is presented in the experiment can the northern branch of the SASM and the EASM be generated (Fig. 1d).

2. Sensible Heat-Driven Air-Pump (Shap) of Large-Scale Mountains

The TIP is located in the northern subtropics across most of the Eurasian continent, characterized by prominent atmospheric elevated heating in summer and cooling in winter. The resultant air descent in winter and ascent in summer above the TIP occur remarkably over the vast subtropical areas in the Eurasian continent year after year, working as a huge air pump standing in situ, and regulating the seasonal change of the Asian circulation.

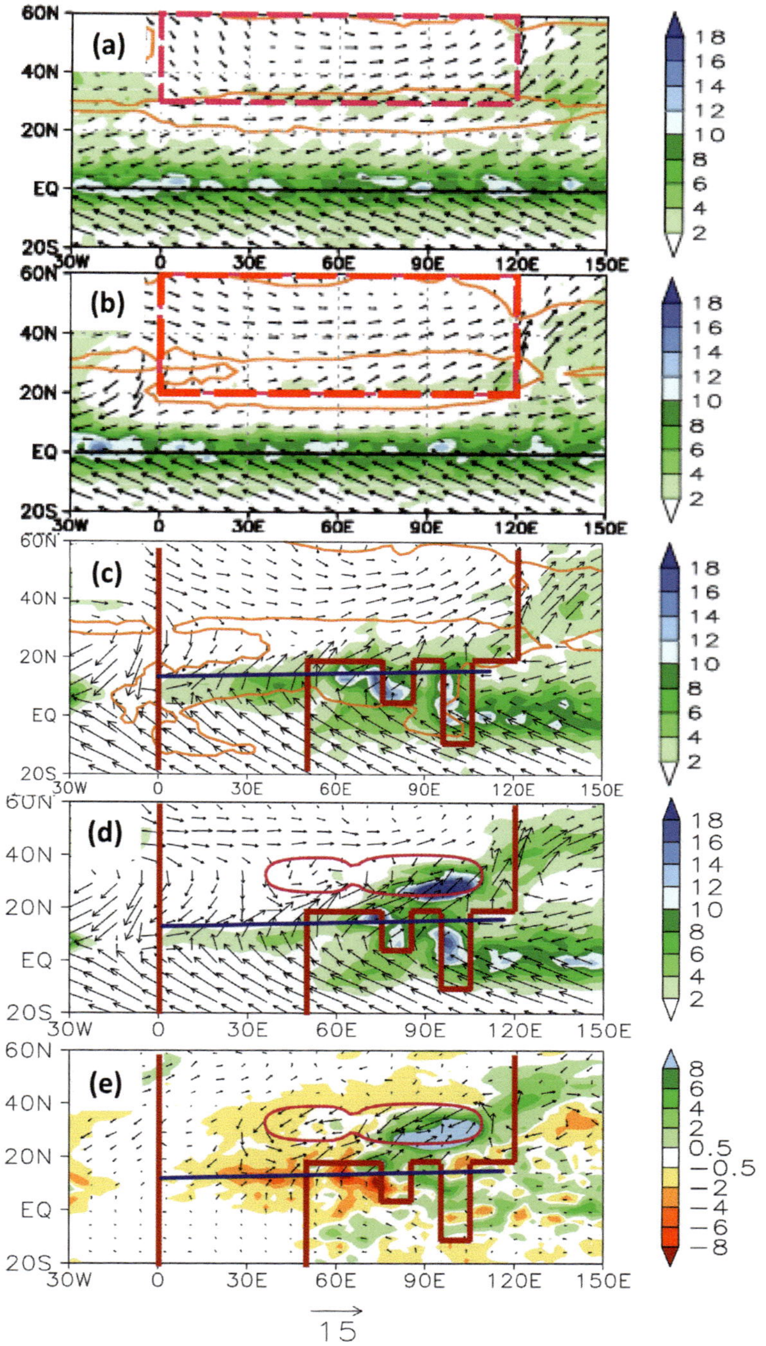

Fig. 1. July mean wind vectors at $\sigma = 0.991$ (arrows; in $\mathrm{m \cdot s^{-1}}$) and precipitation (shading; in $\mathrm{mm \cdot day^{-1}}$) in Exp of (a) MID, (b) SUB, (c) TRO, (d) TIP, and (e) the difference (TIP–TRO). The orange curve in (a-c) indicates the area where the surface wind reversal between Jan and Jul is larger than 120 degrees; the heavy curve denotes the orographic contour at 700 m, the heavy red dashed line denotes the continent boundary, and the heavy blue dashed line indicates the location of the southern branch of the SASM.

Wu *et al.* (2007) demonstrated that the surface sensible heating (SH) on the sloping lateral surfaces of the TP is the major driver of the air pump. According to the thermodynamic equation, in the presence of surface SH on the sloping surface of a mountain, an air parcel moving along an isentropic surface that impinges on the topography is heated as it encounters the SH along the sloping surface of the mountain, and is able to penetrate the isentropic surface and ascend. In the area surrounding the topography, air from lower levels is lifted to the plateau region, producing strong ascent and heavy rainfall over the TP. In the case of no SH on the slope surface of the TP, air parcels travelling in the lower layer and impinging on the TP must remain on the same isentropic surface. The air parcel moves around the TP on an approximately horizontal isentropic surface without ascending and has no significant impact on rainfall. It appears that air pumping is driven by the surface SH on the TP slopes, and is therefore referred to as TP-SHAP (Wu *et al.* 1997). Results from the numerical experiments of Wu *et al.* (2012) demonstrate that such thermal pumping also exists in the Iranian Plateau forcing, and the pumping effect of TIPs forms a TIP-SHAP. Since more than 85% of the atmospheric water vapor is concentrated in a surface layer below 3 km, such vertical pumping of water vapor is important for the development of clouds and precipitation over the slopes of the plateau, which forms a significant atmospheric heating source in the subtropics.

The pumping effect of the surface sensible heating of the TIP in the control experiment (CON) causes the surface moist air flow to climb up the TIP on its southeast (Fig. 2a) where the streamlines of the surface current are almost perpendicular to the plateau contours. On the contrary in the experiment TIP_NS, in which although the surface sensible heating still

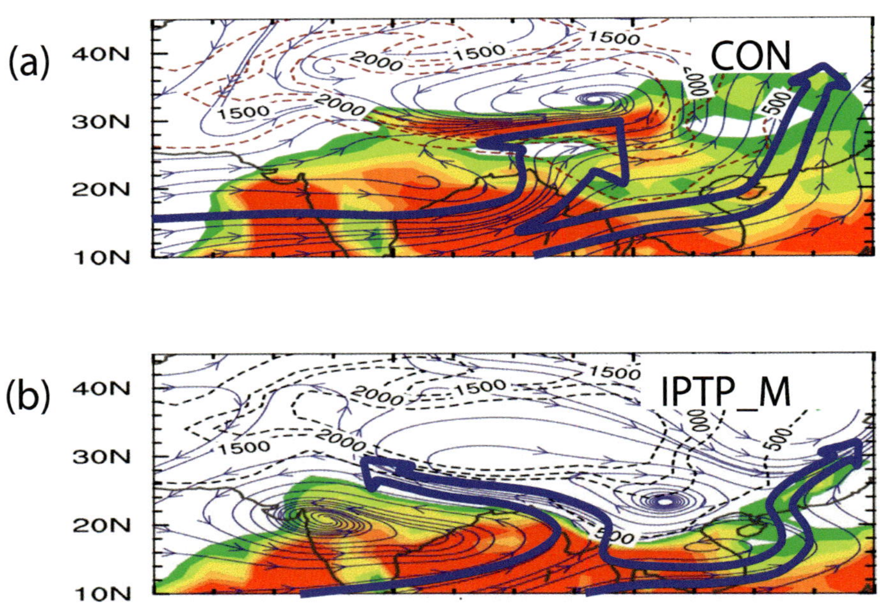

Fig. 2. JJA mean precipitation rate (color shading, unit mm d^{-1}) and streamlines at the $\sigma = 0.89$ level for (a) the CON experiment and (b) TIP_NS experiment. Dashed contours indicate elevations above sea level, with red and black colors, respectively, indicating with and without surface sensible heating of the mountains on the atmosphere. Dark blue open arrows denote the main atmospheric flows impinging on the TP, either climbing up the plateau (a) or moving around the plateau, parallel to orographic contours (b) (reproduced from Wu *et al.*, 2012).

exists over the TIP's surface, it is not allowed to heat the *in situ* atmosphere (Fig. 2b), the surface air currents cannot climb up the TIP, and just go around the TIP horizontally with their streamlines parallel to the TIP's contours. All the above experimental results indicate that the pumping of the TIP SHAP is significant for the formation of the northern branch of the SASM and the EASM.

3. Large-Scale Air Ascent

In the tropics, the planetary vorticity is small and the Rossby deformation radius is large. The atmospheric response to an axisymmetric forcing usually takes a regime of angular momentum conservation (AMC, Plumb and Hou 1992). This AMC regime can be divided into two types (Wu *et al.* 2016): one is ITCZ-induced Hadley-type meridional circulation (H-AMC) with its rising/sinking arm located in the equatorial/subtropical region (Fig. 3a); the other is the monsoonal-type meridional circulation (M-AMC) with its rising arm located in the subtropical region and sinking arm in the southern hemisphere (Fig. 3b). Since the atmospheric vertical velocity w is proportional to the vertical differential potential vorticity (PV = q) advection $\{w \propto f\frac{\partial}{\partial z}[-\vec{\mathbf{V}}\cdot\bar{\mathbf{\nabla}}q]\}$ (Hoskins *et al.* 2003), and since PV advection decreases with height along the H-AMC meridional circulation and increases with height along the M-AMC meridional circulation, the H-AMC is thus accompanied by descending air and a vertical westerly shear; on the contrary, the M-AMC is accompanied by ascending air and a vertical easterly shear. Therefore, the large-scale background of ascending motion in the ASM area must be related to the monsoonal-type meridional circulation, M-AMC.

To reveal the cause by which the M-AMC-type meridional circulation is generated, the WRF model version 3.4.1 (Skamarock *et al.* 2008) with 45-km horizontal resolution and 35 vertical layers was used to execute a pair of numerical experiments, each incorporated six summers (2003–2008) with the initial conditions set at 0000 UTC May 1 (Wu *et al.* 2015). The integration time was 4 months, i.e., every experiment ended at 1800 UTC August 31, and output from the final 3 months (June–August, JJA) was analyzed. The normal experiment with the WRF model was named the control run (CTL). The other experiment is the sensitivity experi-

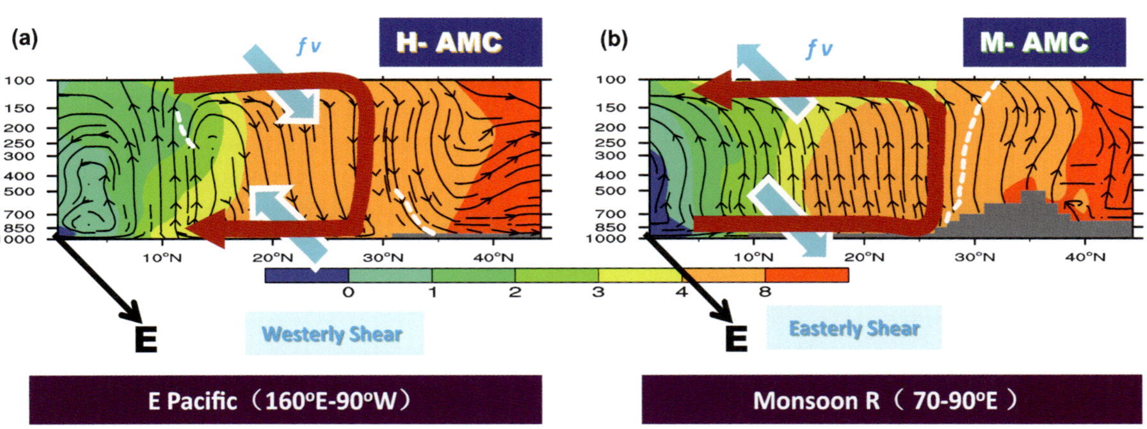

Fig. 3. Cross sections of climate July mean (1979–1998) meridional circulation (lines with arrows), absolute vorticity (color shading; units: $10^{-5}\,\mathrm{s}^{-1}$), and the zonal wind (u) line where u = 0 (white dashed line) for the (a) East Pacific (160°E–90°W) regional mean and (b) Asian monsoon area (70°–90°E) regional mean. Gray shading denotes topography, heavy red curve denotes the main meridional circulation, and the blue arrow indicates the zonal wind component *fv*.

ment (TP_NS), which has the same setting as the CTL run, except that the surface SH over the TP area where terrain is higher than 2 km over the main TP (23°–40°N, 70°–105°E) was not allowed to heat the atmosphere. The results obtained from the difference between CTL and TP_NS can be considered as the influence of the surface sensible heating over the main part of the TP. The TP surface SH generates remarkably different anticyclone circulation and a warm center over the TP at 300 hPa (Fig. 4b), and a much stronger anticyclone circulation but a cold temperature center at 100 hPa (Fig. 4a). Thus, the static stability at the center of the different anticyclone is minimum (Fig. 4c). It was proved (Wu *et al.* 2016) that there exists a critical pressure level p_C, where the horizontal temperature gradient vanishes ($\nabla_{p_c} T(x, y) \equiv 0$) and the circulation possesses the strongest anticyclone vorticity. Therefore, at this critical level, both the absolute vorticity and potential vorticity in the anticyclone area are minimum compared to its upper and lower levels. These imply that the TIP heating changes the atmospheric thermal structure aloft, forming minimum absolute vorticity and potential vorticity in the subtropics near the tropopause. Subject to the constraint of AMC, a monsoonal type of meridional circulation is produced in the ASM area with its ascending arm located in the subtropics (Fig. 4d). Since the PV advection is increased with height within this monsoonal type of meridional circulation, large-scale air ascent thus prevails over the vast Asian monsoon area,

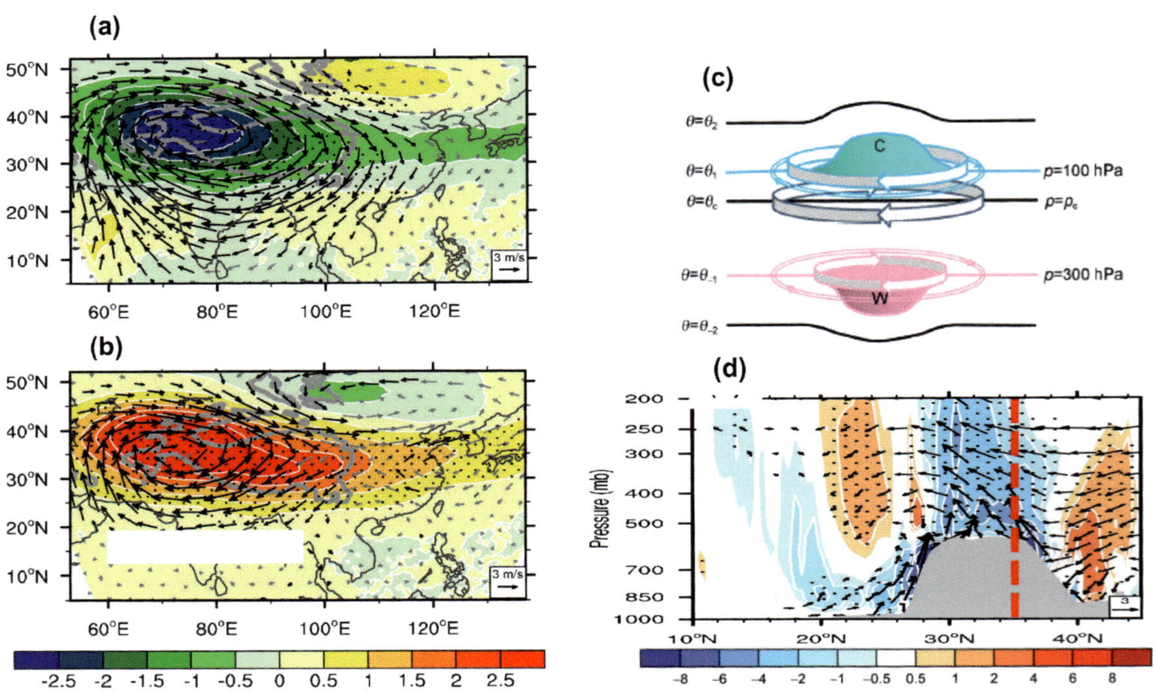

Fig. 4. Summer mean differences in air temperature (shading, unit: K) and wind field (vectors, unit: m s^{-1}) between CTL and TP_NS at (a) 100-hPa and (b) 300-hPa; (c) Schematic diagram indicating the formation of the area of minimum PV forcing near the tropopause due to thermal forcing over the main TP, Vector indicates anticyclone circulation; "C" and blue color denote cold temperature, and "W" and pink color denote warm temperature; and (d) pressure–latitude cross sections of the vertical circulation (vectors, v and $-50^*\omega$) and vertical velocity (shading, unit: 0.02 Pa s^{-1}) averaged from 85°E to 95°E. Dotted regions denote statistical significance of the difference above the 95% level. Only vertical circulation difference with statistical significance above the 95% level is plotted (reproduced from Wu *et al.*, 2016).

providing a favorable background for monsoon development.

4. Water Vapor Transport from Ocean to Inland

Continental monsoon development requires tremendous water vapor supply which cannot be fulfilled by local evaporation from the land surface. Instead, it depends significantly on the *in situ* convergence of water vapor flux. Thus, water vapor transport from the source area is critical for the development of continental monsoon. The key problem here is how the circulation, which is responsible for the water vapor transport, is formed.

Due to the PV conservation constraint, a 10°C warm anomaly on the surface can initiate a cyclonic circulation near the surface (Hoskins *et al.* 1985). Based on the AGCM SAMIL, He *et al.* (2015) conducted a pair of numerical experiments in which the control run was the normal climate integration, whereas the sensitivity run was the same except the SH over the TIP was not allowed to heat the atmosphere. The differences between the two experiments demonstrate that the surface SH over the TIP can generate a strong surface warm cyclonic circulation surrounding the TIP of more than 8°C (Fig. 5a). In another parallel set of experiments in which the specific humidity was set to zero, similar results were obtained (Fig. 5b), implying the essential role of the strong TIP surface SH in initiating the warm surface cyclonic circulation. It is due to this TIP-SHAP-induced strong cyclonic circulation that abundant water vapor is transported from the ocean surface to the inland area to feed the Asian continental monsoon rainfall.

Some recent studies (e.g. Boos and Kuang 2010) hypothesized that the Himalayas act as a thermal insulator that prohibit cold and dry northerly air intrusion from middle latitudes so that the lower-layer atmosphere with high-level "moist static energy" over India can couple

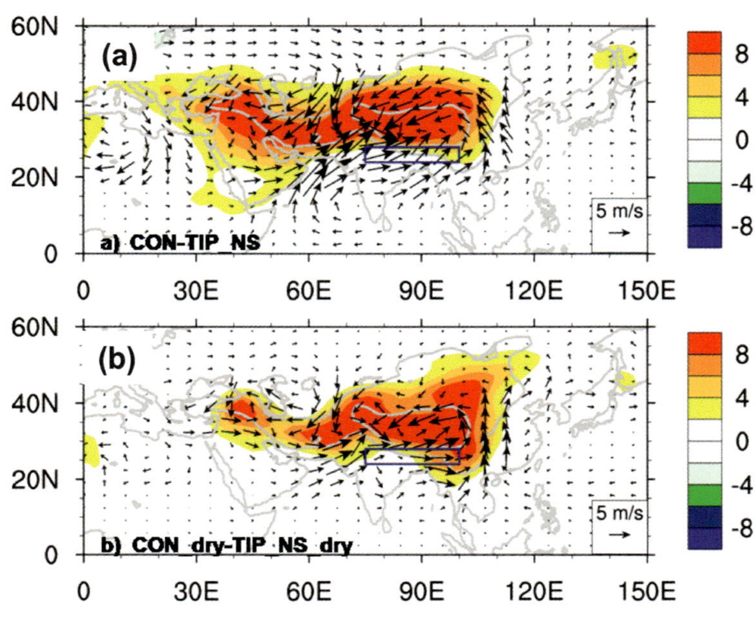

Fig. 5. JJA-mean differences of near-surface (sigma = 0.991) potential temperature θ (shading, K) and circulation (vectors, m s^{-1}) between (a) CON and TIP_NS, (b) as for (a) but for a dry atmosphere. The square indicates the SASM region of (24–28°N, 75–100°E) (reproduced from He *et al.* 2015).

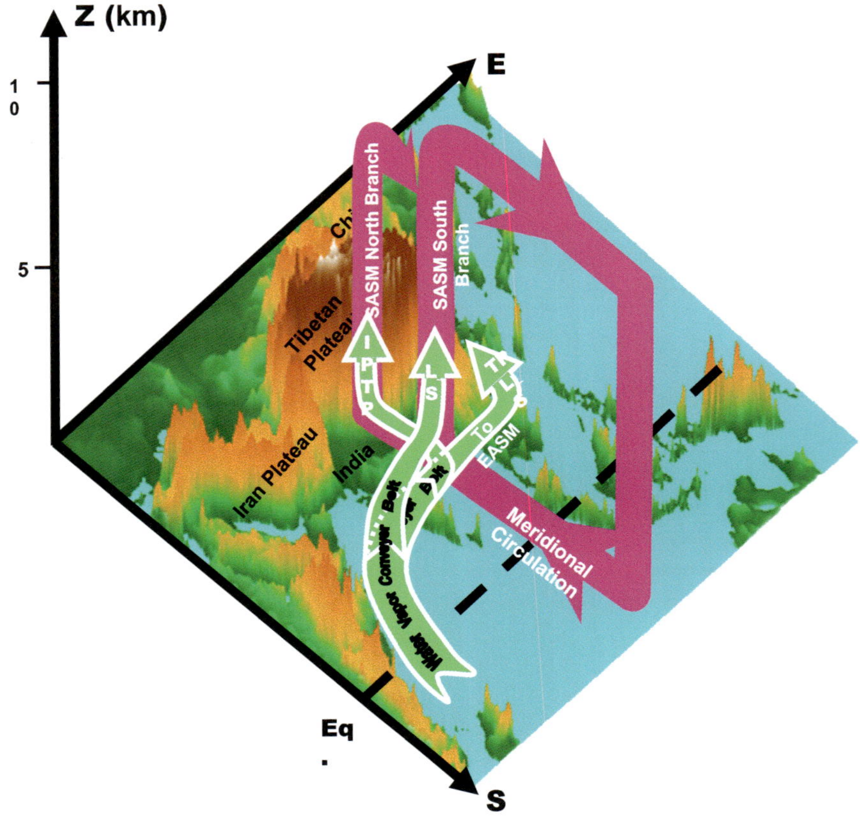

Fig. 6. Schematic diagram showing the gross structure of the Asian summer monsoon. For the southern branch, water vapor along the conveyer belt is lifted up due mainly to land-sea thermal contrast in the tropics; for the northern branch the water vapor is drawn away from the conveyer belt northward toward the foothills and slopes of the TP, and is uplifted to produce heavy precipitation that is controlled mainly by TIP-SHAP; the rest of the water vapor is transported northeastward to sustain the East Asian summer monsoon, which is controlled by the land-sea thermal contrast as well as thermal forcing of the TIP (Wu *et al.* 2012).

with the warm center in the upper troposphere through monsoonal convection, and the "physical" insulation influence of the Himalayas is equal to or more important than the thermal influence of the TP for the SASM maintenance. However, in summer there does not exist such a northerly air intrusion either in observation or in modeling and the "thermal insulator" is not required; whereas in winter the Himalayas cannot protect India from cold and dry northerly intrusion (He *et al.* 2015). This is because in winter the northerly flow can go around the TIP, causing northwesterly or northeasterly flows and intrusion into India. In summer compared to India, the solar radiation at the top of the atmosphere is stronger and the cloud cover is diminished to the north of the TP. There, the surface shortwave radiation is much stronger, resulting in the development of a hot and low-pressure surface to the north of the TP. More importantly, in summer under the continental-scale thermal forcing, a continental-scale surface cyclonic circulation dominates the Eurasian continent (Wu *et al.* 2009). Thus, the prevailing wind in the lower troposphere is southerly across the TP. Therefore, the development of the SASM does not require a thermal insulator to its north. As to the high "moist static energy" over the Indian Subcontinent, it is basically due to the high moisture content there (He *et al.* 2015). As discussed above, it is due to the TIP thermal forcing that the water vapor can be transported from the ocean area to the inland area to create high "moist static energy" near the surface over India. Nevertheless, it does not mean that the TIP's mechanical forcing can be ignored. As shown in Fig. 2a, when the moist southerly flow converges toward the TIP, in addition to the flow climbing up the mountain, there exists a branch of air flow deflected by the mountain, contributing to the formation of the monsoon trough over the Indian Subcontinent and affecting the detailed distribution and variation of the ASM.

5. Conclusion

Results based on numerical experiments demonstrate that the Asian summer monsoon is thermally controlled (Fig. 6). While land-sea thermal contrast significantly determines the southern branch of the South Asian summer monsoon in the tropics, the large-scale mountain TIP's thermal pumping is fundamental to the formation of the northern branch of the South Asian summer monsoon and the East Asian summer monsoon. It is therefore concluded that the nature of the Asian summer monsoon is a consequence of the response of atmospheric circulation to the seasonal changes in land-sea thermal contrast and in the thermal status of the large-scale plateau TIP which is induced by the annual cycle of solar radiation.

References

Boos, W R. and Z. M. Kuang. 2010: Dominant control of the South Asian monsoon by orographic insulation versus plateau heating, *Nature*, **463**, doi:10.1038/nature0870.

Fiorino M. 2000: AMIP II sea surface temperature and sea ice concentration observations. PCMDI Report, Lawrence Livermore National Laboratory.

He, B., G. X. Wu, Y. M. Liu, and Q. Bao, 2015: Astronomical and hydrological perspective of mountain impacts on the Asian summer monsoon. *Sci. Rep.*, **5**, 17586; doi: 10.1038/srep17586.

Hoskins, B. J., M. E. Mclntyre, and A. W. Robertson, 1985: On the use and significance of isentropic potential vorticity maps. *Quart. J. Roy. Meteor. Soc.*, **111**, 877–946.

Hoskins, B. J., M. Pedder, and D. W. Jones, 2003: The omega equation and potential vorticity. *Q. J. R. Met. Soc.*, **129**, 3277–3303.

Liang, X. Y., Y. Liu, and G. X. Wu, 2006: Roles of tropical and subtropical land-sea distribution and the Qinghai-Xizang Plateau in the formation of the Asian summer monsoon. *Chin J. Geophys.-Ch.*, **49**(4), 983–992 (in Chinese).

Plumb, R. A. and A. Y. Hou, 1992: The response of a zonally symmetric atmosphere to subtropical thermal forcing: Threshold behavior. *J. Atmos. Sci.*, **49**, 1790–1799.

Skamarock, W. C., J. B. Klemp, J. Dudhia, D. O. Gill, D. M. Barker, M. G. Duda, X. Y. Huang, W. Wang, and J. G Powers, 2008: A Description of the Advanced Research WRF Version 3. NCAR Technical Note NCAR/TN- 475+STR.

Wu, G. X., W. Li, and H. Guo *et al.*, 1997: Sensible heat driven air-pump over the Tibetan Plateau and its impacts on the Asian summer monsoon. In: Ye, DZ (ed.). Collections on the Memory of Zhao Jiuzhang. Beijing: Science Press, 116–26. (in Chinese).

Wu, G. X., Y. M. Liu, and Q. Zhang *et al.*, 2007: The influence of mechanical and thermal forcing by the Tibetan Plateau on Asian climate. *J. Hydrometeorol*, **8**: 770–89.

Wu, G. X., Liu, Y. M., Zhu, X. Y., Li, W. P., Ren, R. C., Duan, A. M., and X. Y. Liang, 2009: Multi-scale forcing and the formation of subtropical desert and monsoon. *Ann. Geophys.*, **27**, 3631–3644.

Wu, G. X., Y. Liu, B. Dong, X. Liang, A. Duan, Q. Bao, and J. Yu, 2012: Revisiting Asian monsoon formation and change associated with Tibetan plateau forcing: I. Formation. *Climate Dyn.*, **39**(5), 1169–1181.

Wu, G. X., A. Duan, Y. Liu, J. Mao, R. Ren, Q. Bao, BianHw, B. Liu, and W. Hu, 2015: Tibetan Plateau climate dynamics: Recent progress and outlook. *Nat. Sc. Rev.*, **2**, 100–116.

Wu, G., H. Bian, L. Yimin, B. Qing, R. Rongcai, and L. Boqi, 2016: Recent progresses on dynamics

of the Tibetan Plateau and Asian summer monsoon. *Chin. J. Atmos. Sci.*, **40**(1), 1–11.
Wu, G. X., H. F. Zhuo, Z. Q. Wang, and Y. M. Liu, 2016: Two types of summertime heating over the Asian large-scale orography and excitation of potential-vorticity forcing I. Over Tibetan Plateau. *Sci. Chin. Earth Sci.*, **59**(10), 1996–2008.
Wu, T. W. *et al.*, 2003: The performance of atmospheric component model R42L9 of GOALS/LASG. *Adv. Atmos. Sci.*, **20**, 726–742.

Chapter 4

Multifaceted Intraseasonal Modes in the East Asian-Western North Pacific Summer Monsoon Climate

Kyung-Ja Ha* and Hyoeun Oh

Center for Climate Physics, Institute for Basic Science, Busan, Korea
Department of Atmospheric Sciences, College of Natural Sciences, Pusan National University, Busan, Korea
**kjha@pusan.ac.kr*

The East Asian–western North Pacific summer monsoon (EA-WNPSM) has multifaceted features involving complex dynamic interactions between local and large-scale circulations and nonlinear intraseasonal variability. These complexities cause difficulties in summer monsoon prediction, presenting a challenge in monsoon regions. In this chapter, we review dynamic and thermodynamic characteristics of distinguished intraseasonal modes in the EA-WNPSM, their irregularities in occurrences and patterns conducive to climate extremes, and asymmetric responses to El Niño–Southern Oscillation (ENSO). First, we briefly review the identification of the irregular four major intraseasonal modes. The four major modes are classified as the pre-Meiyu–Baiu, Changma–Meiyu, WNPSM, and monsoon gyre modes. The WNPSM and monsoon gyre modes occur preferentially in late summer and have the potential to trigger more frequent extreme rainfall in accordance with typhoon-related precipitation over the western North Pacific and the Korean Peninsula. Moreover, the asymmetric responses to ENSO provide sources of predictability that mainly originate from an interactive coupling over the Indian Ocean and western Pacific. Major predictors in terms of the modes have been obtained from persistent signals and trends of the ocean and land surface anomalous conditions reflected on a physical basis. Ultimately, these studies provide primary sources for prediction in the context of the intraseasonal variability of the EA-WNPSM, and the results have important implications for reducing extreme climate disasters in the EA-WNPSM region.

1. Introduction

Although more than half of the world's population is affected by monsoon rainfall, the causes remain heavily debated (Qiu *et al.* 2013). Moreover, extreme natural disasters in Asia demonstrate its high vulnerability to the changing climate. Therefore, the hydrological cycle in the region is of critical scientific interest in terms of the socio-economic impacts, especially on the intraseasonal time scale. However, the intraseasonal variability (ISV) of the East Asian–western North Pacific summer monsoon (EA-WNPSM) is characterized by complex and irregular torrential rainfall, and it is not always phase locked. Because of its irregular structure and propagation, the intraseasonal fluctuation of summertime monsoon characteristics, along with the dynamic nonlinear interactions between local and large-scale phenomena is poorly understood (Yoo *et al.* 2010). Thus, statistical analysis using a linear approach to identify characteristics of the ISV is inadequate. Several studies have used a cluster analysis to assess nonlinear ISV characteristics, which has revealed significant distinctions among the classified intraseasonal modes (Yoo *et al.* 2010; Chu *et al.* 2012; Oh and Ha, 2015). It is generally assumed that regional monsoons are created by nonlinear coupled systems between circulation and rainfall on the intraseasonal time scale. In recent studies, this assumption has

The Multiscale Global Monsoon System
Edited by C. P. Chang et al.

been explored by applying methods based on self-organizing maps (SOMs; Kohonen 1990) to six daily dynamic monsoon indices representing the western North Pacific (WNP) high, upper-level vorticity, vertical wind shear, and low-level southeasterlies (Wang *et al.* 1998, 2001; Lau *et al.* 2000; Ha *et al.* 2005). Chu *et al.* (2012) and Oh and Ha (2015) identified the prevailing features of the irregular intraseasonal system over the EA-WNPSM region using the SOM analysis to reveal four major modes, pre-Meiyu–Baiu (SOM1), Changma–Meiyu (SOM2), WNP summer monsoon (WNPSM, SOM3), and monsoon gyre (SOM4), and they examined those distinct thermodynamic structures and further explored extreme climate events and an occurrence of tropical cyclones (TCs) associated with the four intraseasonal modes (Chu *et al.* 2012; Oh and Ha 2015).

The EA-WNPSM involves complex natural atmosphere–ocean–land–ice interactions, and it is the most representative region to account for the interactions of the earth's climate system in predictions (Wang *et al.* 2013). Thus, the accurate prediction of monsoon onset and duration and its response to anthropogenic forcing are vital to the world's economy and sustainable development. The intraseasonal modes in the EA-WNPSM are crucial for predicting and understanding this system, which exhibits asymmetric responses to El Niño–Southern Oscillation (ENSO) (Yun *et al.* 2010; Oh and Ha 2016). The ENSO could affect the EA-WNPSM through various climate systems such as snow cover over the Tibetan Plateau, North Atlantic Oscillation (NAO), Indian Ocean (IO), and WNP subtropical high (WNPSH) (Wu *et al.* 2012a, 2012b; Yim *et al.* 2010), which can be used for precursors. Moreover, the changes in extreme rainfall over the EA-WNP region have distinctive characteristics depending on various spatial and temporal scales (Oh *et al.* 2018). Thus, studying TCs over the WNP is important because it is the most active region for TC formation (Chen *et al.* 2004; Chang *et al.* 2012; Choi *et al.* 2015). Here, we include the following questions: When did these changes start occuring frequently? Where did these changes take place? It is a crucial part of preventing extreme climate disasters caused by rainfall. Now we review identified intraseasonal modes and their underlying characteristics in terms of dynamic and thermodynamics, then address their predictability.

2. Multifaceted Intraseasonal Phases

Many previous studies have shown the impact of the ISV on the timing and duration of monsoon onset and retreat over the EA-WNPSM region (Yun *et al.* 2008; Lee *et al.* 2013). Ha and Ha (2006) noted that the ISV can effectively explain up to 40% of the total variance over the EA-WNPSM region. Following Chu *et al.* (2012), the four major intraseasonal modes using a nonlinear pattern recognition method, SOM, together account for more than 60% of the total frequency in the EA-WNPSM (Fig. 1). The first SOM pattern represents the zonally elongated rainband from southeastern China to southern Japan (Fig. 1a). The spatial pattern and phase-locked annual cycle obtained from the number of clustered days correspond well to the early summer, which is named the pre-Meiyu–Baiu mode. This mode is commonly known as the seasonal mean structure of the East Asian summer monsoon. The anomalous circulation field shows dipole structures of the mid-latitude cyclone over southern Japan and the subtropical anticyclone over the South China Sea. In the second pattern, the Changma–Meiyu mode, the zonally elongated rainband is shifted northward along the western flank of the WNPSH, which links the tropics to the mid-latitudes (Fig. 1b), and it has a phase-locked characteristic from mid-June to late July. The third pattern, the WNPSM mode, displays not only a large amount of convective

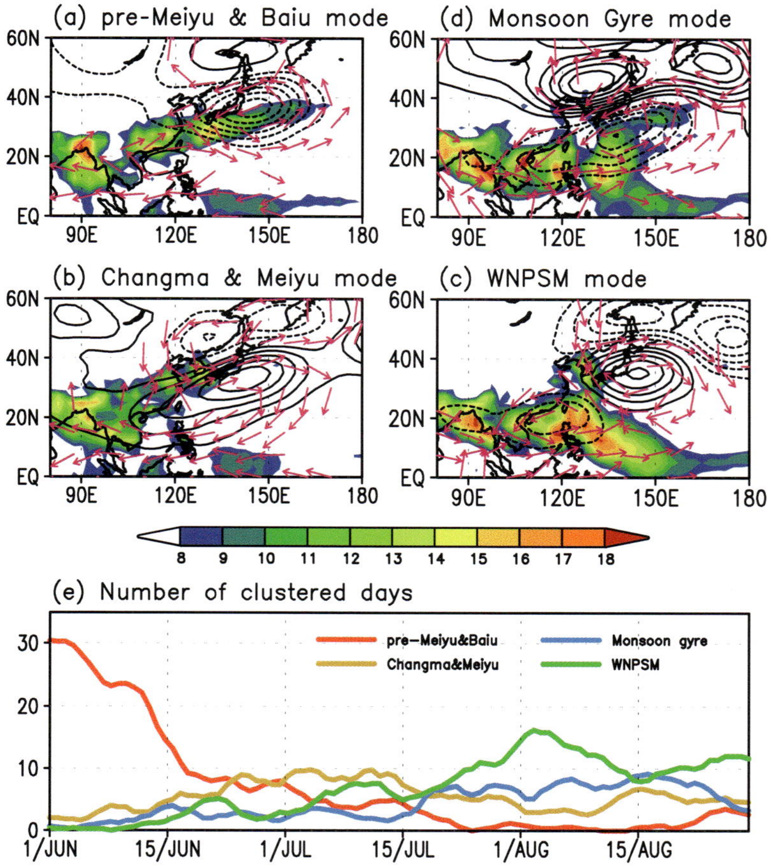

Fig. 1. Self-organizing map patterns of precipitation (shading), 500-hPa geopotential height anomaly (contour), and 850-hPa wind anomaly (vector) obtained by compositing daily data corresponding to the days clustered at the (a) pre-Meiyu–Baiu, (b) Changma–Meiyu, (c) WNPSM, and (d) monsoon gyre modes. The resultant frequency time series are shown as the number of days for June–July–August in (e). According to time sequence in (e), (a)–(d) are arranged in a counterclockwise direction.

rainfall over the WNP but also localized precipitation over the Korean Peninsula (Fig. 1c). The anomalous circulation field is the opposite of the pre-Meiyu–Baiu mode, with both a mid-latitude anticyclone over southern Japan and a subtropical cyclone over the South China Sea. The last pattern has a single pole of anomalous cyclonic circulation with precipitation located in the center of the cyclone (Fig. 1d), and it is known as the monsoon gyre mode, which tends to occur in late summer. The four major SOM patterns and their annual cycles show the multifaceted monsoon ISV, which is simply considered a northward-propagating monsoonal wind driven by the land-sea thermal contrast. On the interannual time scale, the pre-Meiyu–Baiu (Changma–Meiyu) and WNPSM (monsoon gyre) modes have a negative correlation, with the distinct intraseasonal phases.

To clarify the physical mechanisms of the multifaceted intraseasonal modes, Fig. 2 shows the large-scale thermodynamics in the context of the temperature and moisture budget equation. In the pre-Meiyu–Baiu mode, a strong meridional temperature gradient forms a strong polar jet that induces secondary circulation, with rising warm air and sinking cold air (Sampe and Xie 2010). In other words, both the cold and

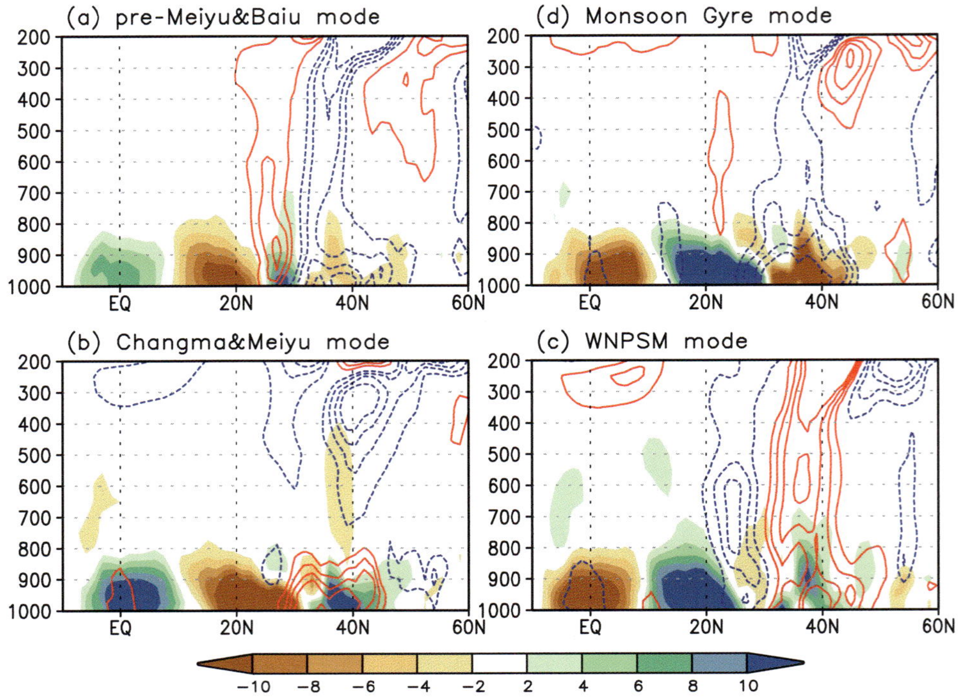

Fig. 2. Latitude–height cross sections of moisture flux convergence (shading unit: day^{-1}) and thermal advection anomalies (contour unit: K day^{-1}) for the (a) pre-Meiyu–Baiu, (b) Changma–Meiyu, (c) WNPSM, and (d) monsoon gyre modes averaged longitudinally (120°–140°E). Green shading represents moisture flux convergence, while yellow shading indicates divergence. Red solid and blue dashed lines represent warm and cold advections, respectively. The shading and contour intervals are 2.0 and 0.1, respectively (adapted from Oh and Ha, 2015).

warm advections, which are located in the north and south of the region, respectively, contribute to the rainfall pattern by modulating a baroclinic instability. In the Changma–Meiyu mode, the low-level moisture-laden warm flow along the WNPSH and upper-level dry cold air advection create convectively unstable conditions and tend to have a low-centered echo (Kar and Ha, 2003). Also, the Changma–Meiyu mode has a much stronger connection with the tropical–subtropical system through the WNPSH than the pre-Meiyu–Baiu mode.

The rainfall in the WNPSM and monsoon gyre modes is characterized by a strong subtropical moisture flux convergence, since it relies on the tropics. The WNPSM mode has a pressure system by the extratropical Rossby wave response to tropical convection, such as the Pacific–Japan pattern. Also, the localized heavy precipitation over the Korean Peninsula is caused by a humid southeasterly flow along the anomalous anticyclone circulation over the southern part of Japan. The thermodynamic structures of the monsoon gyre mode show the opposite sign as compare to the Changma–Meiyu mode. The low-level dry and cold air advection over the northern flank of the cyclone stabilizes the atmosphere, leading to suppressed convection. In addition, the northeasterly terminates the southwesterly moisture supply, and it forms relatively dry conditions over northeast Asia.

3. Climate Extremes

The distinctive instabilities of the four major modes, induced by the dynamic and thermodynamic structures, can contribute to triggering extreme precipitation events. It was confirmed that more extreme precipitation (more than 50 mm/day) occurred in late summer than in early summer (figure not shown). Also, in terms of typhoon activity, it can be inferred that precipitation in the WNPSM mode shows northward-shifting features in accordance with the typhoon track. The monsoon gyre mode is responsible for a large proportion of extreme precipitation events with anomalous cyclone circulation over the WNP, which can create favorable conditions for typhoon formation (Chen *et al.* 2004). Based on the typhoon strength, typhoons can be categorized as intense typhoons, typhoons, and tropical storms. The WNPSM mode is strongly correlated with typhoon activity at the 99% confidence level with a correlation coefficient of 0.46. For tropical storms, the monsoon gyre mode has a positive significant relationship at the 95% confidence level since this mode has anomalous cyclonic circulation over the WNP, thus favoring typhoon genesis. However, in early summer, only the Changma–Meiyu mode has a significant negative correlation with typhoon activity and occurrence. The total number of typhoons exhibits a significant opposite relationship with the Changma–Meiyu and monsoon gyre mode at the 99% confidence level with a correlation coefficient of −0.51 and 0.58, respectively. The genesis potential index (GPI) reported by Murakami and Wang (2010) has been used to show the influences of the significant contributors on TC formation (Fig. 3). The Changma–Meiyu mode is negatively related to the GPI over the WNP region, whereas the WNPSM and monsoon gyre modes are positively related to it. Moreover, the GPI over the South China Sea is positively correlated with the Changma–Meiyu

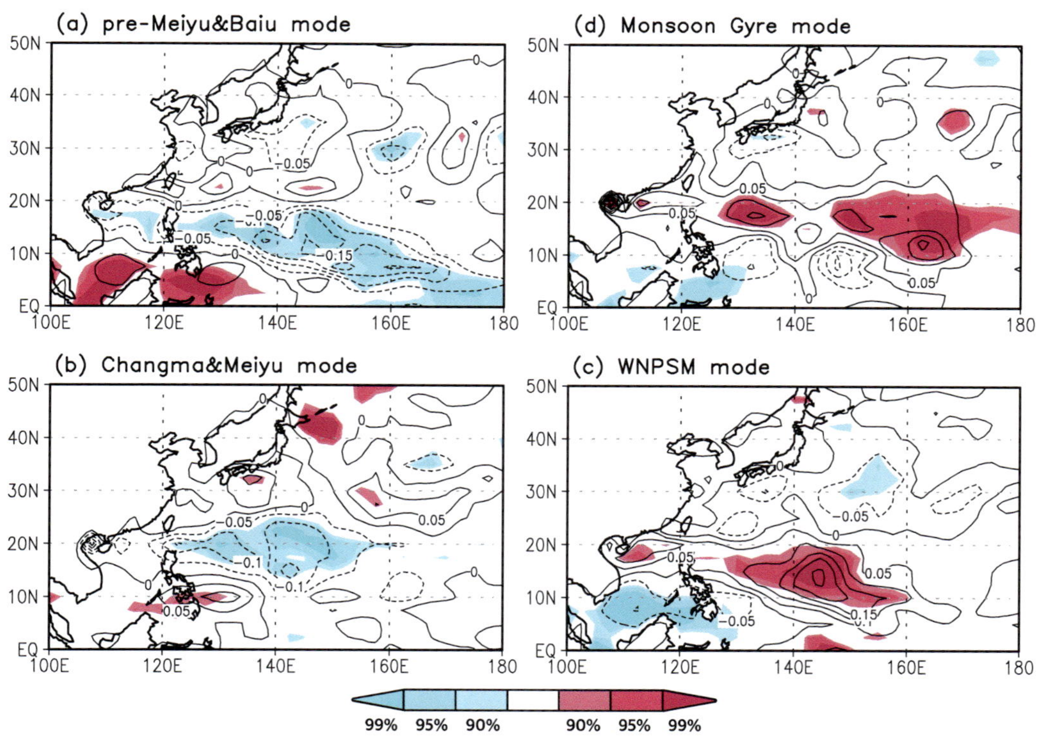

Fig. 3. Regressed fields of the genesis potential index (GPI) against the mean annual number of days clustered at the (a) pre-Meiyu–Baiu, (b) Changma–Meiyu, (c) western North Pacific summer monsoon (WNPSM), and (d) monsoon gyre modes. The shadings indicate significant GPI above the 90%, 95%, and 99% confidence levels.

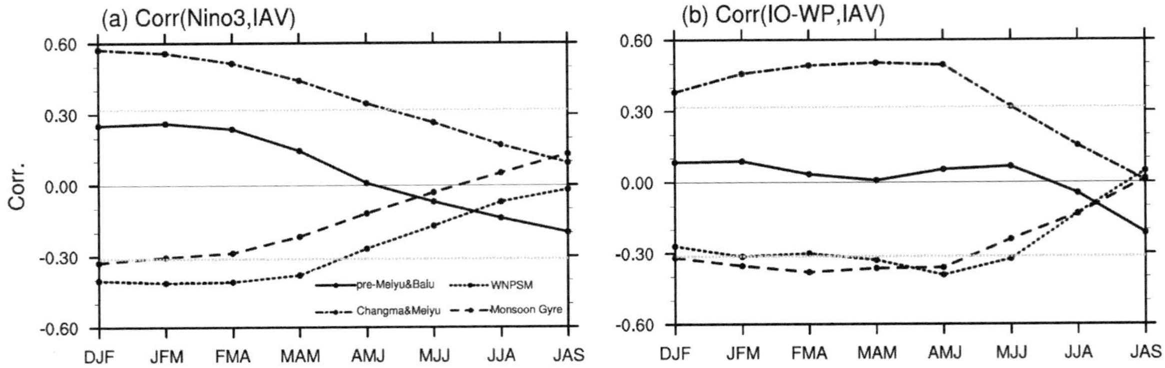

Fig. 4. Relationship between EA-WNPSM and tropical SST showing the lead–lag correlation coefficients of interannual variability of self-organizing map patterns with regards to the (a) Niño3 index (adapted from Chu *et al.* 2012) and (b) SST difference between the IO and WP. The IO SST is averaged over 40°–100°E, 20°S–20°N. The WP SST is averaged over 125°–150°E, EQ–10°N.

mode, while the opposite trends occur for the two late summer modes. The TCs in the WNPSM (monsoon gyre) mode are somewhat associated with the occurrence of the pre-Meiyu–Baiu (Changma–Meiyu) mode, since the relationship between them is opposite on the interannual time scale. These significant relationships enable TC prediction (Wang *et al.* 2013), and could provide a statistical basis for predicting TCs by understanding the occurrence of the early summer mode.

Given the highly complex and inhomogeneous characteristics, it is challenging to interpret the regional monsoon from a global perspective and to determine sources of nonlinear ISV predictability. One possible approach is to investigate the associations among tropical systems such as ENSO, IO warming, and western Pacific (WP) air-sea coupling.

4. Monsoon–ENSO Relationship

Because the regional monsoons are modulated by diverse external forcings and internal dynamics, the monsoon–ENSO teleconnection is significantly correlated at certain times (Wu and Wang 2002; Ha *et al.* 2017). To quantify the relationship between ENSO and EA-WNPSM, Chu *et al.* (2012) investigated the lead–lag correlation coefficients between the Niño3 index (150°–90°W, 5°S–5°N) and the four major modes (Fig. 4). The Changma–Meiyu mode is strongly related to the preceding winter El Niño, and the correlation coefficient peak in December–January–February gradually decays until the following spring. The ENSO signal during boreal winter is not statistically significant with the pre-Meiyu–Baiu mode, while it is negatively correlated with the WNPSM and monsoon gyre modes, which are marginally significant at the 95% confidence level. When it comes to the frequency of each intraseasonal mode, the pre-Meiyu–Baiu and Changma–Meiyu (WNPSM and monsoon gyre) modes are more (less) frequent through the WNPSH by the preceding El Niño (Table 2 in Oh and Ha, 2015). In other words, the two modes in early summer are favorable to the occurrence of El Niño even though the pre-Meiyu–Baiu mode is not significantly related. As for the pre-Meiyu-Baiu mode, there has been a significant interdecadal change in the responses of La Niña (Wang *et al.* 2011), but its influence on the East Asian monsoon could not be considered as symmetric with El Niño (Xue and Zhao 2017). Thus, the complexities related to La Niña should be investigated in the near future.

The significant relationship between the Changma–Meiyu mode and tropical SST is not only confined to the eastern Pacific (EP) but also positively (negatively) related to equatorial the IO (WP). Interestingly, there is a strong negative relationship between the monsoon gyre mode and IO. To date, remote (i.e. IO) SST warming (Yun *et al.* 2008; Xie *et al.* 2009; Ding *et al.* 2010), and local air-sea interactions over the WNP (Wang *et al.* 2000) have been regarded as important contributing factors to the monsoon intraseasonal oscillation–ENSO relationship. Considering that the Changma–Meiyu and monsoon gyre modes are closely linked to the tropical system, including the IO and WP, the lead–lag relationship of the four SOM patterns and the IO minus WP (IO-WP) SSTs was newly performed (Fig. 4b). The IO SST is averaged over 40°–100°E, 20°S–20°N and that for WP SST is 125°–150°E, EQ–10°N. The IO-WP SST tends to prolong the effects of tropical SST. With the exception of the first mode, the relationships among the other modes and IO-WP SSTs are consistently maintained from the preceding winter to the following spring, and decay abruptly after April–May–June. This phenomenon can be interpreted as the IO capacitor effect, which prolongs the El Niño effect on the WP region via atmospheric Kelvin waves generated by IO warming, similar to the recharging and discharging of a battery (Xie *et al.* 2009). Thus, distinctive associations of the intraseasonal patterns in the EA-WNPSM are present.

Several scholars have proposed the local air-sea interactions over the WP and IO or SST over the North Atlantic triggered by EP warming as a key driving factor of the WNPSH(Wang *et al.* 2000; Li *et al.* 2007; Wu *et al.* 2012a; Xiang *et al.* 2013; Lee *et al.* 2014, Yun *et al.* 2015). Although theories for the monsoon–ENSO teleconnection have been proposed (Li *et al.* 2007; Wang *et al.* 2008), the physical mechanisms underlying ENSO teleconnection are still debated.

5. Prediction by Physical–Empirical (PE) Model

A "predictable mode analysis (PMA)" method is proposed to estimate the potential predictability statistically and physically through the leading empirical orthogonal function modes, which together explain a large portion of the total variability (Wu *et al.* 2009; Wang *et al.* 2015). The predictors of the leading Asian monsoon modes are classified into the central Pacific (CP) ENSO developing, IO-WP coupled by being sustained by a positive thermodynamic feedback, IO dipole, NAO, and warming trend mode, and these are used to adjust a (PE) model using the linear regression method. Several studies have investigated an association between regional monsoons and climate variability on a bimonthly time scale of the ISV (Table 1). The PE model was established to predict summer monsoon over East Asia with ENSO and spring NAO (Wu *et al.* 2009). Yim *et al.* (2014) showed that May–June East Asian rainfall variability is primarily attributed to positive feedback between the anomalous WNPSH and the underlying Indo-Pacific warm ocean forcing associated with the Pacific Decadal Oscillation-like SST anomaly, development of CP SST anomalies, and Tibetan Plateau and Eurasian land temperature anomalies. Moreover, Xing *et al.* (2016) showed that peak-summer (July and August) rainfall is affected by an equatorial WP–northern East Asia teleconnection, WNPSH dipole SST feedback, CP-ENSO, and Eurasian wave train.

Recently, Oh and Ha (2016) proposed the sources of predictability for the four major modes using distinctive spatial structures and temporal variabilities of the preceding ENSO and established the PE model by developing a physical understanding of the sources. Major predictors in terms of the four dominant modes have been obtained from persistent signals and trends of the ocean and land surface anomalous

Table 1. Description of each predictor for the East Asian summer monsoon rainfall using the PE model.

Reference	Defining variables	Regions	Target region and period
Yim *et al.* (2014)	February-minus-December Indo-Pacific warm pool dipolar SST	[5°S–10°N, 70°E–100°E] + [15°N–25°N, 120°E–160°E] – [0°–15°N, 150°E–160°W]	May–June south China [21.25°N–28.75°N, 110°E–120°E], 1979–2012
	February-minus-December North Atlantic tripolar SST	[0°–15°N, 60°W–20°W] + [35°N–50°N, 70°W–30°W] – [20°N–30°N, 80°W–50°W]	
	February-minus-December Siberia T2M	[55°N–70°N, 80°E–140°E]	
Yim *et al.* (2015)	MA-minus-JF Western north pacific SST	[15°N–25°N, 130°E–180°E]	May–June Taiwan (0-month lead) [21°N–26°N, 119°E–123°E], 1979–2012
	MA-minus-JF Tripolar North Atlantic SST	[0°–20°N, 60°W–15°W] + [45°N–55°N, 55°W–40°W]–[25°N–40°N, 65°W–40°W]	
	MA-minus-JF East Asia T2M	[30°N–60°N, 110°E–140°E]	
Oh and Ha (2016)	JFMA IO-WP SST	[20°S–5°N, 50°E–100°E]–[0°–10°N, 125°E–150°E]	SOM2 in EA-WNPSM, 1979–2014
	April-minus-March Kuroshio SST	[25°N–35°N, 130°E–160°E]	
Xing *et al.* (2016)	MAMJ IO-WNP SST	[5°N–20°N, 50°E–100°E]–[5°N–20°N, 125°E–180°E]	July–August southern East Asia EOF1 [5°N–26.5°N, 100°E–140°E], 1979–2013
	MJ-minus-MA EP SLP	[20°S–20°N, 150°W–70°W]	
Yim *et al.* (2016)	MAMJ equatorial WP SST	[10°S–15°N, 120°E–160°E]	July–August northern East Asia EOF1 [26.5°N–50°N, 100°E–140°E], 1979–2013
	MAMJ North Asian dipole SLP	[50°N–65°N, 155°E–170°W]–[50°N–70°N, 70°E–110°E]	
	MJ-minus-MA western IO SST	[20°S–0°, 40°E–60°E]–[0°–20°N, 50°E–70°E]	
	MJ-minus MA south Asian SLP	[10°N–30°N, 80°E–130°E]	
Xing *et al.* (2017)	JFMA equatorial zonal SST gradient	[5°S–10°N, 120°W–80°W] + [10°S–20°N, 80°E–140°E]–[5°N–15°N, 160°E–160°W]	May–June East Asia EOF1 [20°N–45°N, 100°E–130°E], 1979–2015
	JFMA North Atlantic meridional tripolor SST	[0°–15°N, 60°W–15°W] + [40°N–55°N, 60°W–30°W]–[25°N–35°N, 80°W–45°W]	
	MA-minus-JF thermal tendency contrast	T2m [40°N–60°N, 50°E–120°E]–SST[15°N–30°N, 120°E–180°E]	

conditions reflected on a physical basis. In the first (pre-Meiyu–Baiu) mode, a persistent land warming over northern Asia plays an important role in leading an anomalous cyclonic circulation over southern Japan, and a warming over the southern IO leads to an anomalous anticyclone over the WNP. Also, the SST cooling tendency lasting until summer over the WNP is one of the causes that modulates a strong wind shear which is relevant for the baroclinic instability. The Changma–Meiyu mode is strongly associated with the WNPSH caused by the difference between the IO and WP. In addition, two other modes can be explained by the tropical SST (Oh and Ha, 2016). The selected predictors based on stepwise regression supported the statistical significance and relative independence of this model. In order to confirm the prediction skill of the four major modes using the PE model, the cross-validated method to remove 3 years from each step was assessed. The cross-validated correlation skills of the four modes with their precursors are significant with significant skills ranging from 0.54 to 0.64.

6. Conclusions

This review provides evidence that the multi-faceted intraseasonal monsoon variability over the EA-WNPSM region can be distinguished by performing cluster analysis of daily dynamic monsoon indices. Oh and Ha (2015) have found that the pre-Meiyu–Baiu rain belt and WNPSM are dominated by thermal advection induced by the upper-level jet at the mid-latitudes. However, the Changma–Meiyu (monsoon gyre) mode is dominated by low-level moisture flux convergence (divergence) and warm (cold) advection, which links the tropics to the mid-latitudes and simultaneously shows opposing moisture and thermal features at the upper level. These vertical differences in thermodynamic characteristics create each instability.

Moreover, Oh and Ha (2015) newly examined the extreme climate and occurrence of TCs as extreme precipitation events and considered the role of the modified GPI, which is useful for examining the factors that are significant contributors to TC formation. For example, there is a striking difference between the early summer modes (pre-Meiyu–Baiu and Changma–Meiyu) and the late summer modes (WNPSM and monsoon gyre). Also, the SOM method enables the estimation of distinct responses to ENSO, providing a source of predictability with intraseasonal variability in the EA-WNPSM. The Changma–Meiyu and monsoon gyre modes are closely linked to the tropics through the subtropical pressure system, whereas the pre-Meiyu–Baiu and WNPSM modes are characterized by compartmental mid-latitude systems. The sources of predictability for regional monsoon rainfall on the intraseasonal time scale over the EA-WNPSM can be determined by coupling to the tropical–subtropical system and convective coupling between circulation and convection in terms of the four major modes. Understanding the physical process for the subseasonal time scale is important for improving and assisting in predictions. Asymmetric responses to the ENSO for the four major modes have provided sources of predictability. In addition, tropical forcing helps explain the fact that the EP warming affects the ISV in the EA-WNPSM through the WNPSH produced by air-sea coupling over the IO and WP.

Acknowledgments

We thank Professor Bin Wang from the University of Hawaii for his constructive comments on an early version of the manuscript. This work was supported by IBS-R028-D1 and Global Research Laboratory grant (NRF-2011-0021927) of the National Research Foundation funded by the Korean Government. The data used for the analysis of large-scale circulation characteristics, including geopotential height and the horizontal components of wind at selected pressure levels, were collected from 1979 to 2016 by the National Centers for Environmental Prediction–National Center for Atmospheric Research reanalysis (Kalnay *et al.* 1996). The data used for the analysis of summer monsoon rainfall were collected from 1997 to 2014 by the daily Global Precipitation Climatology Project (Adler *et al.* 2003).

References

Adler, R. F., G. J. Huffman, A. Chang, R. Ferraro, P. Xie, J. Janowiak, B. Rudolf, U. Schneider, S. Curtis, D. Bolvin, A. Gruber, J. Susskind, and P. Arkin, 2003: The version 2 global precipitation climatology project (GPCP) monthly precipitation analysis (1979–present). *J. Hydrometeor.*, **4**, 1147–1167.

Chang, C.-P., Y. Lei, C.-H. Sui, X. Lin, and F. Ren, 2012: Tropical cyclone and extreme rainfall trends in East Asian summer monsoon since mid-20th century. *Geophys. Res. Lett.*, **39**, L18702.

Chen, T.-C., S.-Y. Wang, M.-C. Yen, and W. A. Gallus Jr., 2004: Role of the monsoon gyre in the interannual variation of tropical cyclone formation over the Western North Pacific. *Wea. Forecasting*, **19**, 776–785.

Choi, Y., K.-J. Ha, C.-H. Ho, and C. E. Chung, 2015: Interdecadal change in Typhoon genesis condition over the western North Pacific. *Clim. Dyn.*, **45**(11), 3243–3255.

Chu, J.-E., S. N. Hameed, and K.-J. Ha, 2012: Nonlinear, intraseasonal phases of the East Asian summer monsoon: Extraction and analysis using self-organizing maps. *J. Climate*, **25**, 6975–6988.

Ding, R., K.-J. Ha, and J. Li, 2010: Interdecadal shift in the relationship between the East Asian summer monsoon and the tropical Indian Ocean. *Climate Dyn.*, **34**, 1059–1071.

Ha, K.-J., S.-K. Park, and K.-Y. Kim, 2005: On interannual characteristics of climate prediction center merged analysis precipitation over the Korean Peninsula during the summer monsoon season. *Int. J. Climatol.*, **25**, 99–116.

Ha, K.-J. and E. Ha, 2006: Climatic change and interannual fluctuations in the long-term record of monthly precipitation for Seoul. *Int. J. Climatol.*, **26**(5), 607–618.

Ha, K.-J., Y.-W. Seo, J.-Y. Lee, R. H. Kripalani, and K.-S. Yun, 2017: Linkages between the South and East Asian summer monsoons: a review and revisit. *Clim. Dyn.*, doi:10.1007/s00382-017-3773-z.

Kalnay, E., Coauthors, 1996: The NCEP/NCAR 40-year reanalysis project. *Bull. Amer. Meteor. Soc.*, **77**, 437–471.

Kar, S. K. and K.-J. Ha, 2003: Characteristic differences of rainfall and cloud-to-ground lightning activity over south Korea during the summer monsoon season. *Mon. Wea. Rev.*, **131**, 2312–2323.

Kohonen T., 1990: The self-organizing map. *Proc. IEEE*, **78**, 1464–1480.

Lau, K.-M., K.-M. Kim, and S. Yang, 2000: Dynamical and boundary forcing characteristics of regional components of the Asian summer monsoon. *J. Climate*, **13**, 2461–2482.

Lee, E.-J., K.-J. Ha, and J.-G. Jhun, 2014: Interdecadal changes in interannual variability of the global monsoon precipitation and interrelationships among its subcomponents. *Clim. Dyn.*, **42**(9), 2585–2601.

Lee, J.-Y., B. Wang, M. C. Wheeler, X. Fu, D. E. Waliser, and I.-S. Kang, 2013: Real-time multivariate indices for the boreal summer intraseasonal oscillation over the Asian summer monsoon region. *Clim. Dyn.*, **40**, 493–509.

Li, Y., R. Lu, and B. Dong, 2007: The ENSO–Asian monsoon interaction in a coupled ocean–atmosphere GCM. *J. Climate*, **20**, 5164–5177.

Murakami H. and B. Wang, 2010: Future change of North Atlantic tropical cyclone tracks: projection by a 20-km-mesh global atmospheric model. *J. Climate*, **23**, 2699–2721.

Oh, H. and K.-J. Ha, 2015: Thermodynamic characteristics and responses to ENSO of dominant intraseasonal modes in the East Asian summer monsoon. *Clim. Dyn.*, **44**, 1751–1766.

Oh, H. and K.-J. Ha, 2016: Prediction of dominant intraseasonal modes in the East Asian-western North Pacific summer monsoon. *Clim. Dyn.*, **47**, 2025–2037. doi:10.1007/s00382-015-2948-8.

Oh, H., K.-J. Ha, and A. Timmermann, 2018: Disentangling impacts of dynamic and thermodynamic components on late summer rainfall anomalies in East Asia. *J. Geophys, Res.*, **123**, 8623–8633.

Qiu, J., 2013: Monsoon Melee. *Science*, **340**(6139), 1400–1401.

Sampe, T. and S. Xie, 2010: Large-scale dynamics of the Meiyu–Baiu rainband: Environmental forcing by the westerly jet. *J. Climate*, **23**, 113–134, doi:10.1175/2009JCLI3128.1.

Wang, B., R. Wu, and X. Fu, 2000: Pacific–East Asian teleconnection: How does ENSO affect East Asian climate? *J. Climate*, **13**, 1517–1536.

Wang, B., R. Wu, and K.-M. Lau, 2001: Interannual variability of the Asian summer monsoon: Contrasts between the Indian and the western North Pacific–East Asian monsoons. *J. Climate*, **14**, 4073–4090.

Wang, B., Z. Wu, J. Li, J. Liu, C.-P. Chang, Y. Ding, and G. Wu, 2008: How to measure the strength of the East Asian summer monsoon. *J. Climate*, **21**, 4449–4463.

Wang, B., B. Xiang, and J.-Y. Lee, 2013: Subtropical High predictability establishes a promising way for monsoon and tropical storm predictions. *PNAS*, **110**(8), 2718–2722.

Wang, B., B. Xiang, and J.-Y. Lee, 2015: Asian summer monsoon rainfall predictability: A predictable mode analysis. *Clim. Dyn.*, **125**(2), 281–290.

Wang, Q., Y. Ding, and Y. Jiang, 1998: Relationship between Asian monsoon activities and the precipitation over China mainland. *Q. J. App. Meteorol.*, **9**, 84–89.

Wu, R. and B. Wang, 2002: A contrast of the East Asian summer monsoon-ENSO relationship between 1962–77 and 1978–93. *J. Climate*, **15**, 3266–3279.

Wu, Z., B. Wang, J. Li, and F.-F. Jin, 2009: An empirical seasonal prediction model of the East Asian summer monsoon using ENSO and NAO. *J. Geophys. Res.*, **114**, D18120.

Wu, Z., J. Li, Z. Jiang, and T. Ma, 2012a: Modulation of the Tibetan plateau snow cover on the ENSO teleconnections: from the East Asian summer monsoon perspective. *J. Climate*, **25**, 2481–2489.

Wu, Z., J. Li, Z. Jiang, J. He, and X. Zhu, 2012b: Possible effects of the North Atlantic oscillation on the strengthening relationship between the East Asian summer monsoon and ENSO. *Int. J. Climatol.*, **32**, 794−800.

Xiang, B., B. Wang, W. Yu, and S. Xu, 2013: How can anomalous western North Pacific subtropical high intensify in late summer? *Geophys. Res. Lett.*, **40**, 2349–2354.

Xie, S.-P., K. Hu, J. Hafner, H. Tokinaga, Y. Du, G. Huang, and T. Sampe, 2009: Indian ocean capacitor effect on Indo–Western Pacific climate during the summer following El Niño. *J. Climate*, **22**, 730–747.

Xing, W., B. Wang, and S.-Y. Yim, 2016: Peak-summer East Asian rainfall predictability and prediction part I: Southeast Asia. *Clim. Dyn.*, **47**(1–2), 1–13.

Xing, W., B. Wang, S.-Y. Yim, and K.-J. Ha, 2017: Predictable patterns of the May-June rainfall anomaly over East Asia. *J. Geophys. Rea. Atmos.*, **122**, 2203–2217.

Xue, F., and J.-J. Zhao, 2017: Intraseasonal variation of the East Asian summer monsoon in La Niña years. *Atmospheric and Oceanic Science Letters*, **10:2**, 156–161.

Yim, S.-Y., J.-G. Jhun, R. Lu, and B. Wang, 2010: Two distinct patterns of spring Eurasian snow cover anomaly and their impacts on the East Asian summer monsoon, *J. Geophys, Res.*, **115**, D22113.

Yim, S.-Y., B. Wang, and W. Xing, 2014: Prediction of early summer rainfall over South China by a physical-empirical model, *Clim. Dyn.*, **43**, 1883–1891.

Yim, S.-Y., B. Wang, W. Xing, and M.-M Lu, 2015: Prediction of Meiyu rainfall in Taiwan by multi-lead physical-empirical models. *Clim. Dyn.*, **44**, 3033–3042.

Yim, S.-Y., B. Wang, and W. Xing, 2016: Peak-summer East Asian rainfall predictability and prediction part II: extratropical East Asia. *Clim. Dyn.*, **47**, 15–30.

Yoo, J.-H., A. W. Robertson, and I.-S. Kang, 2010: Analysis of intraseasonal and interannual variability of the Asian summer monsoon using a hidden markov model. *J. Climate*, **23**, 5498–5516.

Yun, K.-S., K.-H. Seo, and K.-J. Ha, 2008: Relationship between ENSO and northward propagating intraseasonal oscillation in the East Asian summer monsoon system. *J. Geophys. Res.*, **113**, D14120, doi:10.1029/2008JD009901.

Yun, K.-S., K.-H. Seo, and K.-J. Ha, 2010: Interdecadal change in the relationship between ENSO and the intraseasonal oscillation in East Asia. *J. Climate*, **23**, 3599–3612.

Yun, K.-S., K.-J. Ha, S.-W. Yeh, B. Wang, and B. Xiang, 2015: Critical role of boreal summer North Pacific subtropical highs in ENSO transition. *Clim. Dyn.*, **44**(7–8), 1979–1992.

Chapter 5

South and North American Monsoons: Characteristics, Life Cycle, Variability, Modeling, and Prediction

Alice M. Grimm[*,‡‡], Francina Dominguez[†], Iracema F. A. Cavalcanti[‡], Tereza Cavazos[§], Manoel A. Gan[‡], Pedro L. Silva Dias[¶], Rong Fu[§], Christopher Castro[||], Huancui Hu[**] and Marcelo Barreiro[††]

[*] *Federal University of Paraná, Brazil*
[†] *University of Illinois, USA*
[‡] *National Institute for Space Research, Brazil*
[§] *Centro de Investigación Científica y de Educación Superior de Ensenada, México*
[¶] *University of São Paulo, Brazil*
[||] *University of Texas, USA*
[**] *University of California, Los Angeles, USA*
[††] *Universidad de la República, Uruguay*
[‡‡] *grimm@fisica.ufpr.br*

The American monsoons are important components of the global monsoon system. Since the annual precipitation over most of South America is mainly concentrated in the summer monsoon season, the economy, agriculture, water/energy resources and, consequently, the livelihoods of the great majority of population are heavily dependent on the South American monsoon (SAM). On the other hand, the North American monsoon (NAM) is the predominant influence on the boreal summer climate of the Southwestern United States and Northwestern Mexico, providing between 40% and 80% of the total precipitation in this region. This chapter summarizes the weather and climatic aspects of the American monsoons, using observational and modeling studies, with focus on their life cycles, mutual influence, variability on a wide range of temporal scales, extreme events, modeling, and prediction. Ongoing and future projections of climatic changes are also addressed. SAM and NAM are the result of land/atmosphere/ocean coupling and characterized by multi-scale interactions that are not completely known or understood. Many challenges still remain to improve understanding and prediction.

1. Introduction

The American monsoons are important components of the global monsoon system. Over most of South America (SA), including part of the subtropical regions, the annual precipitation is mainly concentrated during the summer monsoon (Fig. 1). The economy, agriculture, water/energy resources, and consequently the livelihoods of the great majority of the SA population are heavily dependent on the South American monsoon (SAM). On the contrary, the North American monsoon (NAM) is the predominant influence on the boreal summer climate of Southwestern (SW) United States (US) and Northwestern (NW) Mexico, providing between 40% and 80% of the total precipitation that falls within this region (Douglas *et al.* 1993; Adams and Comrie 1997) (Fig. 2a). In this chapter the weather and climatic aspects of the monsoon are summarized, with focus on variability at a range of temporal scales, modeling, and prediction.

While NAM and SAM are part of the global monsoon system, it still remains largely unknown whether they influence each other.

The Multiscale Global Monsoon System
Edited by C. P. Chang et al.

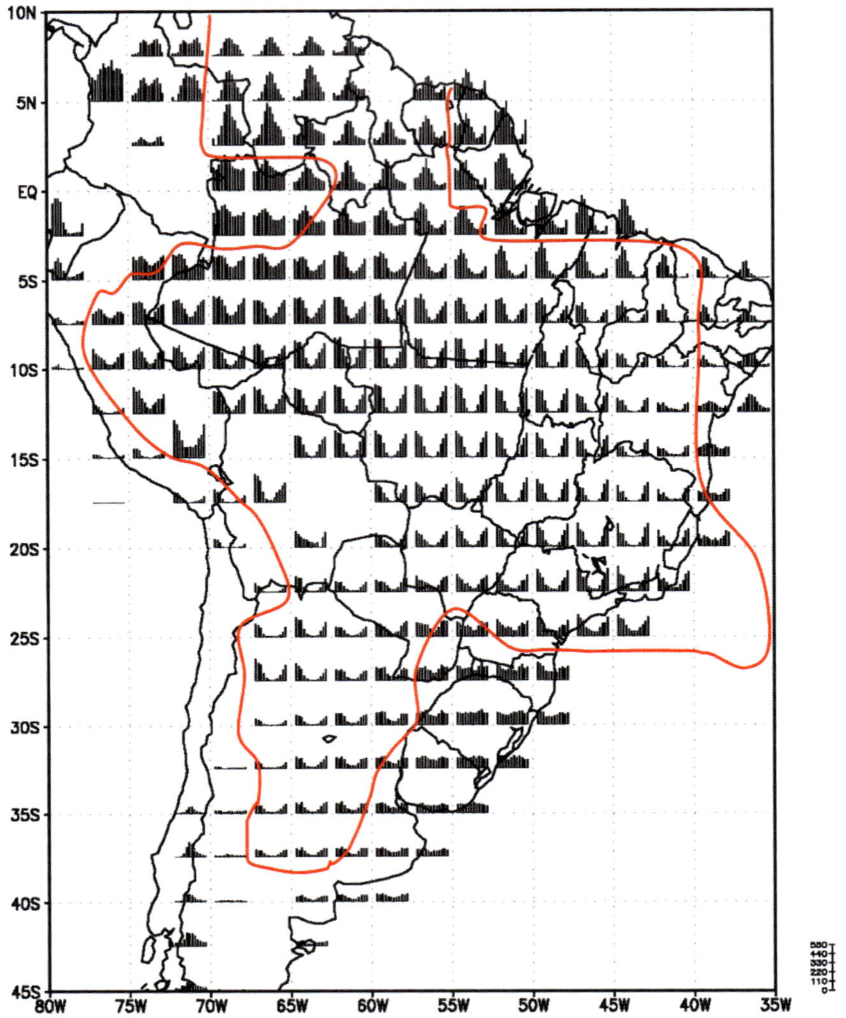

Fig. 1. Annual cycles of precipitation over South America. The monsoon domain is delimited in red, according to the regions defined by Wang *et al.* (2012) (adapted from Grimm 2011).

Global model simulations suggest that the removal of the Amazon rain forest can influence the climate of North America (NA) (Avissar and Werth 2005). Fu *et al.* (2016) found a lagged statistical link suggesting the influence of wind anomalies over parts of the SAM on anomalies in NAM and vice versa. They hypothesize that the onset of SAM (NAM) can weaken moisture transport to NAM (SAM), thus facilitating its retreat. Besides, equatorial penetration of extreme cold front incursions from extratropical SA can induce an anomalous shallow cross-equatorial circulation and increase low-tropospheric geopotential height over tropical and subtropical North Atlantic (Bowerman *et al.* 2017), strengthening the North Atlantic subtropical high (NASH), which can weaken NAM circulation (Arias *et al.* 2012).

2. Main Features and Life Cycle

A monsoon system is generally characterized by the total wind reversal in the lower troposphere, but this is not observed in SAM due to the

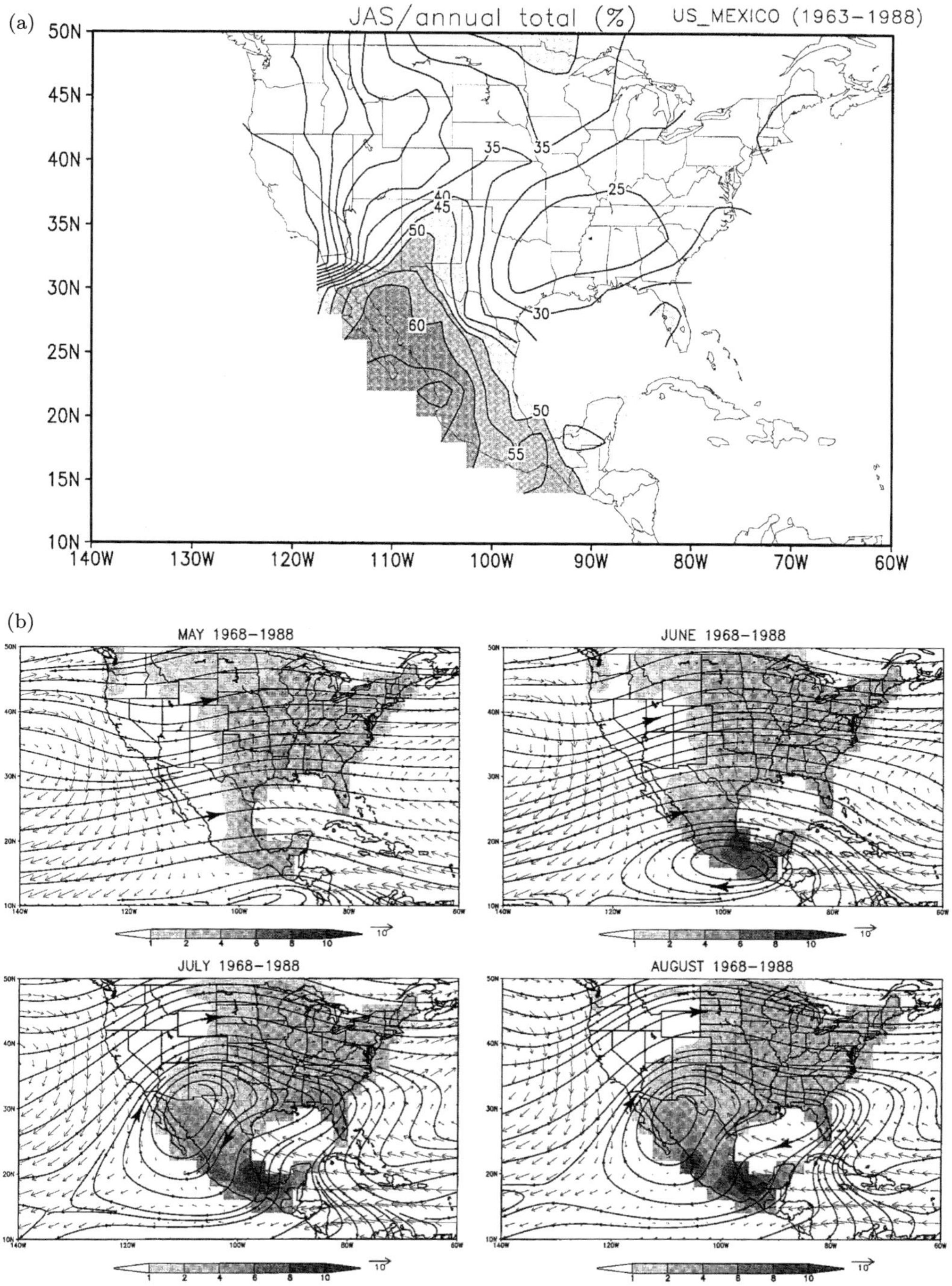

Fig. 2. (a) Contribution of the precipitation during July–September to the annual total, expressed in percent, from the US–MEXICO merge precipitation analysis. The contour interval is 5% and values greater than 40% are shaded. (b) Mean (1968–88) monthly 925-hPa vector wind (ms^{-1}), 200-hPa streamlines and US–MEXICO precipitation (shading) for May, June, July, and August. Circulation data are from NCEP-NCAR Reanalysis. Precipitation amounts are in mm day^{-1} and values greater than 1 mm day $^{-1}$ are shaded. The characteristic vector length is 10 ms^{-1}. (*Source*: Higgins *et al.* 1999).

narrowness of the continent in extratropical latitudes. While the zonal wind changes from easterly (dry season) to westerly (rainy season), the meridional wind only changes its intensity (Gan *et al.* 2004). The wind reversal is observed when the annual mean is removed from the seasonal mean (Zhou and Lau, 1998). Before the monsoon onset in the austral spring over Central Brazil, the atmosphere starts destabilizing through surface sensible heat flux and moisture transport from the Amazon. Intensified ascending motion transports moisture upward and induces a higher-level heat source through latent heat release. Low-level northeasterly flow from the Atlantic Ocean enters tropical SA, turns southward near the Andes producing northerly/northwesterly winds (sometimes a low-level jet) east of the Andes, which turn around a low-pressure system (Chaco Low) in the central part of the continent (Fig. 3a). The low-level flow transports heat and moisture from the Amazon region to Central Brazil, contributing to the monsoon onset over this region in October/November, and the formation of the South Atlantic Convergence Zone (SACZ). Soil moistening increases the latent heat flow, convective cells develop, and the rainy season starts over that region. Sensible heat flux is reduced and latent heat flux is increased (Fig. 3b). At upper levels, a "monsoon high" (Bolivian High) develops southwest of the maximum precipitation due to latent heat release. This latent heat source also enhances the low-pressure system to the north of the thermally induced Chaco Low (Silva Dias *et al.* 1983; Figueroa *et al.* 1995). During the demise in early autumn the low-level pressure over the continent increases, the low-level zonal winds return to easterlies, and the moisture flux from the Amazon region to Southeast Brazil is reduced (Gan *et al.* 2004; Raia and Cavalcanti 2008).

Preceding the NAM, in late boreal spring, the land-sea thermal contrast and heating of the elevated Sierra Madre Occidental (SMO) Mountains in NW Mexico (Turrent and Cavazos 2009) result in a middle/upper-tropospheric "monsoon high" that is analogous to the Bolivian high over SA (Fig. 2b). During the NAM, the dry westerly mid-level winds are replaced by moist southeasterly winds that help initi-

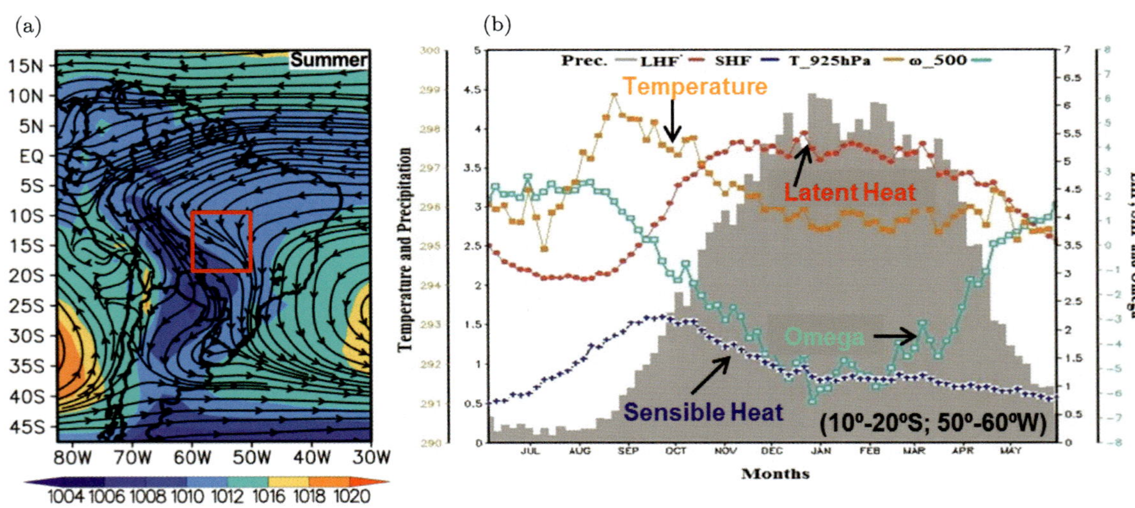

Fig. 3. (a) 850-hPa streamlines and sea level pressure during austral summer (DJF), with the red box indicating the region analyzed in the right panel (adapted from Grimm 2018). (b) Annual evolution of precipitation (in gray bars), temperature, sensible heat, latent heat, and omega, averaged over the region marked on the left.

ate monsoon thunderstorms (Adams and Comrie 1997). Monsoon onset occurs in early June in southern Mexico and rainfall advances northward along the western slopes of the SMO, arriving to the US in early July (Douglas *et al.* 1993) (Fig. 2b). The main source of moisture for the core monsoon is the eastern Tropical Pacific, and is channeled at low levels through the Gulf of California into the NAM region (Turrent and Cavazos 2009), while upper-level moisture comes preferentially from the Gulf of Mexico or the Caribbean (Schmitz and Mullen 1996; Dominguez *et al.* 2016). Terrestrial moisture sources are also important, especially in late summer (Hu and Dominguez 2015). The monsoon ends in early autumn and rainfall gradually retreats toward southern Mexico.

3. Variability

3.1. *Diurnal, mesoscale, and synoptic variability*

During the SAM season, the daily precipitation variability results mainly from equatorial trades and easterly disturbances, as well as equatorward incursions of mid-latitude synoptic systems. The diurnal precipitation cycle is controlled by the development of mesoscale systems frequently triggered by differential heating due to surface heterogeneities in a thermodynamically unstable environment (Ramos da Silva *et al.* 2010). Deep convection undergoes mesoscale organization in certain regions. Mesoscale convective systems (MCSs) are strongly influenced by both transient synoptic systems and mesoscale effects such as jets (e.g., the low-level jet east of the Andes, Marengo *et al.* 2002), and other circulations forced by topography and surface–atmosphere interactions modulated by the diurnal cycle. Models have been unable to reproduce the observed diurnal cycle of precipitation in tropical SA (Machado *et al.* 2014). Betts and Silva Dias (2010) point to the processes for coupling surface energy transfer, shallow convection, and deep precipitating systems. Negative precipitation bias over the Amazon forest is ascribed to failure in producing rainfall in shallow and warm clouds. The steep Andes orography contributes to poor simulation of precipitation in the Amazon Basin (Figueroa *et al.* 2016).

During the NAM season, thunderstorms initiate by air rising over mountain ranges in a conditionally unstable environment. The transition from shallow to deep convection occurs in the mid-afternoon (Nesbitt *et al.* 2008) (Fig. 4).

On occasion, monsoon thunderstorms may grow and organize into MCSs that propagate off the mountain ranges of NW Mexico and Arizona (Lang *et al.* 2007). Within the SW US, Maddox (1995) showed that a stronger and northward-displaced monsoon high enhances atmospheric instability and advection of upper-level moisture, favoring MCSs. "Monsoon bursts", intense thunderstorm activity, are associated with synoptic-scale phenomena: upper-level inverted troughs (IVs), gulf surges, tropical easterly waves (TEW), and tropical cyclones (TCs). IVs propagate westward on the southern periphery of the monsoon high, facilitating MCS-type convection (Finch *et al.* 2010). TEWs moving across Mexico and Central America favor the development of organized convection (Seastrand *et al.* 2015). Gulf of California moisture surges are stronger when related with TCs (Higgins and Shi 2005). TCs from the eastern tropical Pacific contribute to late summer–early autumn rainfall (Wood and Ritchie 2013).

3.2. *Intraseasonal variability*

There are diverse temporal bands of intraseasonal variability, but the predictability is higher in the 30–70-day band. The focus is on the Madden–Julian Oscillation (MJO), for which there is prediction skill out to 3 weeks in some

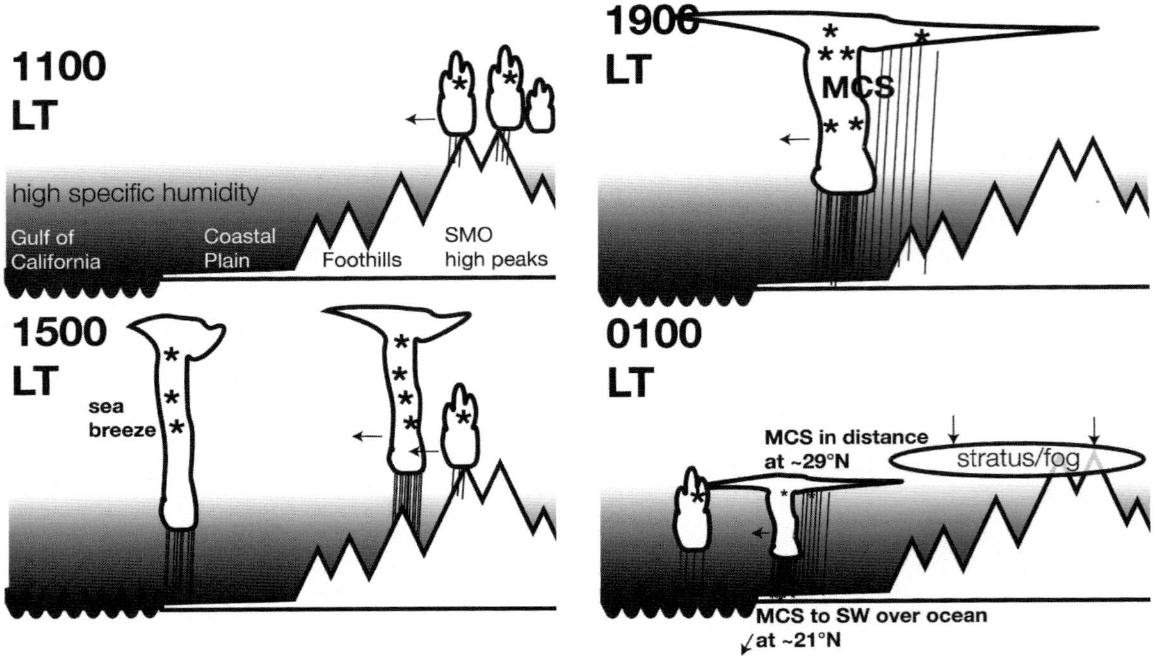

Fig. 4. Schematic of observed diurnal mechanisms along the Sierra Madre Occidental at 25°N. Cloud type indicates relative height attained by clouds, shading indicates specific (not relative) humidity contrasts, asterisks indicate mixed-phase microphysical processes, and the density of the vertical streaks indicates the locations and relative intensities of precipitation (*Source*: Nesbitt *et al.* 2008).

models (Wang *et al.* 2014). Therefore, information on impacts associated with each MJO phase is useful to enrich subseasonal monsoonal prediction.

The SAM precipitation intraseasonal variability is maximum in northeast SA and central-east SA (CESA), but is also significant in subtropical southeast SA (SESA). The MJO affects these regions (Alvarez *et al.* 2015; Grimm 2019), which include SACZ (Carvalho *et al.* 2004) and equatorial Brazil (Souza and Ambrizzi 2006). The contribution of MJO alone in increasing average daily monsoon precipitation can reach 30% of the climatological amount near the core monsoon region, during Phase 1 of MJO. The frequency of extreme events can double during this phase over densely populated areas in the SACZ region, and increase more than 1.6 times in South Brazil, part of SESA (Fig. 5, Grimm 2019). Hirata and Grimm (2015) showed that the MJO enhances wave accumulation in the SACZ region, favoring occurrence of extreme precipitation events. The evolution of the most intense MJO-related SAM precipitation anomalies depends on the interplay of tropics–tropics and tropics–extratropics teleconnections (Grimm 2019).

Spectral analysis of daily summer rainfall in the NAM reveals a significant peak in the 12–18-day band, and another near 40 days. Thus, monsoon bursts and breaks seem to be modulated by low-frequency variability (Cavazos *et al.* 2002). The MJO affects monsoonal precipitation by amplifying easterly waves in the East Pacific, and making it more likely for these waves to trigger gulf surges (Lorenz and Hartmann 2006; Barlow and Salstein 2006). MJO also provides a favorable environment for TCs, which can generate gulf surges (Lorenz and Hartmann 2006).

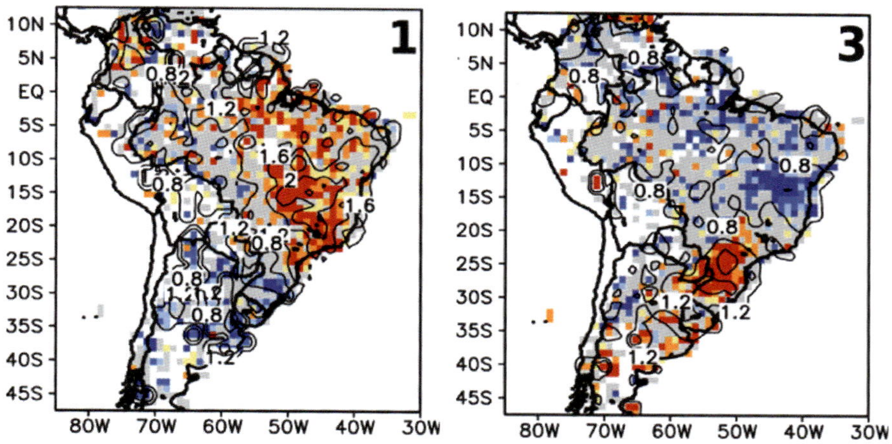

Fig. 5. Ratio between the probability of extreme precipitation events (above 90th percentile) in MJO phases 1 and 3, and the climatological probability. Contour interval is 0.4 and contour 1.0 is omitted. Colors indicate levels of confidence above 90% for the difference between the probability for each MJO phase and the climatological probability (adapted from Grimm 2019).

3.3. *Interannual variability*

Most of the interannual variability of monsoonal precipitation in the SAM and NAM stems from El Niño Southern Oscillation (ENSO), in combination with regional forcing. During the SAM the impacts change from early to peak monsoon season over CESA (including SACZ in its southern half) and part of SESA, suggesting that remote influences prevail in spring, while in peak summer regional land–atmosphere processes dominate, linked to previous soil moisture conditions and ocean-atmosphere interactions (Grimm 2003). For instance, during El Niño CESA (SESA) displays below (above) normal rainfall in November, but above (partially below) normal rainfall in the next January. ENSO-related land–atmosphere interactions are also stressed in Barreiro and Diaz (2011). Although stronger in ENSO, the tendency toward inverse anomalies from early to peak monsoon over CESA also exists in other years (Grimm *et al.* 2007; Grimm and Zilli 2009). Different types of ENSO, with highest sea surface temperature (SST) anomalies in Central or East Pacific, can produce diverse impacts, but the most remarkable differences occur in autumn (Tedeschi *et al.* 2015, 2016). Despite its influence, ENSO explains less than 40% of the interannual variance of the monsoon onset over southern Amazon; subtropical jets and thermodynamic instability in the dry season explain 26%, adding predictability (Yin *et al.* 2014).

During the NAM, large-scale atmospheric teleconnections modulate the positioning and strength of the monsoon high (Higgins and Shi 2000). The ENSO-associated teleconnection emanating from the western Pacific occurs predominately in early summer and modulates monsoon onset (Castro *et al.* 2007) (Fig. 6). El Niño (La Niña) favors southward (northward) shift of the monsoon high during late June–early July and a late (early) onset of the monsoon in the SW US and NW Mexico. Interannual variability of the NAM precipitation has also been linked to land-sea temperature contrasts (Turrent and Cavazos 2009) (Fig. 7) and antecedent soil moisture. While Zhu *et al.* (2007) concluded that early (late) NAM onset in this region is linked to the antecedent dry (wet) winter season, Zhu *et al.* (2009) found a positive feedback with positive (negative) soil moisture anomalies leading to more (less) rainfall.

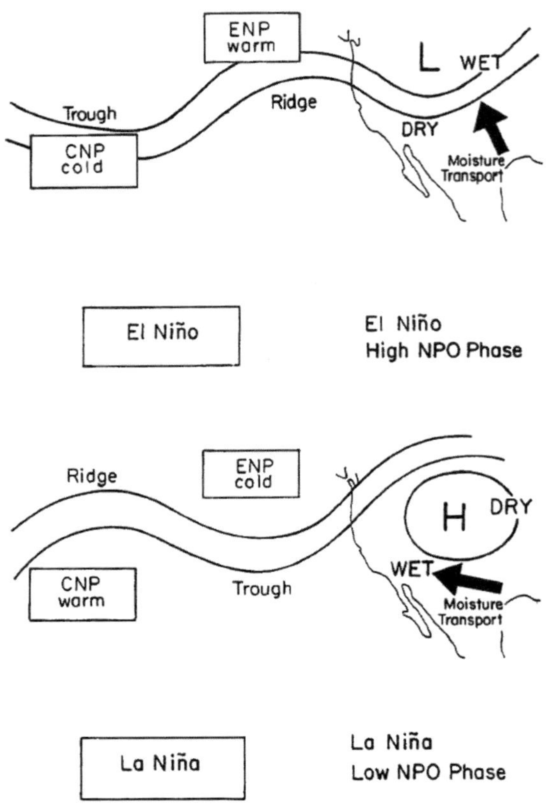

Fig. 6. Idealized relationship of monsoon ridge position and mid-level moisture transport to Pacific SSTs at NAM onset (*Source*: Castro *et al.* 2001).

3.4. *Decadal/interdecadal variability*

SAM precipitation shows strong interdecadal variability, in connection with atmospheric or SST variations (Grimm and Saboia 2015; Grimm *et al.* 2016). As in interannual time scales, the first modes are dipole-like, displaying opposite anomalies in CESA and SESA (Fig. 8a), and tend to reverse polarity from spring to summer. However, the summer second mode and its related spring fourth mode (not shown), which affect the core monsoon region in central Brazil and NW Argentina, show anomalies of same sign during both seasons, describing the "climatic shift" of the mid-1970s. The main interdecadal modes are related with more than one climatic index, stressing combined influence, e.g., the first mode (Fig. 8a for summer, but similar for spring) is negatively (positively) correlated with the Interdecadal Pacific Oscillation (IPO) and Atlantic Multidecadal Oscillation (AMO) in spring (summer).

During the NAM season, there is a teleconnection driven by North Pacific SST variability, associated with Pacific decadal variability (Castro *et al.* 2001), and also influence from the AMO. It affects NAM precipitation through modulation of the NASH. The warm (cold) phase of AMO is related to below (above) average summer rainfall in Arizona and NW Mexico (Arias *et al.* 2012) and in NW Mexico (Mendez and Magaña 2010). Drought over northern Mexico may be also related to changes in transient activity in the Caribbean Sea. According to Mendez and Magaña (2010) during the negative phase of the Pacific decadal oscillation (PDO), the Caribbean low-level jet (CLLJ) weakens and TEW activity increases, leading to more tropical convection over Mesoamerica and less moisture flux into northern Mexico, including the NAM region.

3.5. *Modulation of extreme events by climate variability*

There is a remarkable influence of climate variability in all time scales on the frequency of extreme events during SAM. In extensive regions of SA the influence on the extreme events is even stronger than on the average daily, monthly, or seasonal precipitation (Grimm and Tedeschi 2009; Grimm *et al.* 2016; Grimm 2018; 2019). Figure 5 shows the influence of MJO, while Fig. 8a demonstrates the significant difference in the frequency of extreme events between opposite phases of the first interdecadal variability mode of summer rainfall (not shown, but very similar to Fig. 8a). Figure 8b shows that the influence of this mode is greater on the frequency of extreme events than on moderate and weak rainfall.

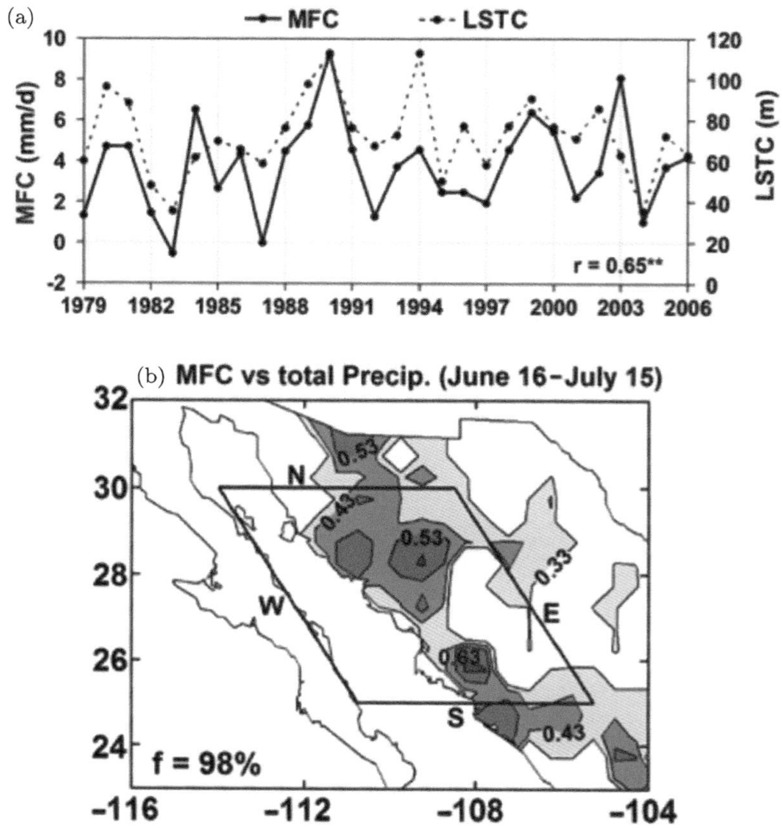

Fig. 7. (a) Time series of moisture flux convergence (MFC) and land-sea thermal contrast (LSTC) indices; the correlation coefficient (r) is statistically significant at the 99% level. (b) Correlation between the MFC index and accumulated precipitation (June 16–July 15) during the period 1979–2004; contour interval is 0.1; only significant contours ($p < 0.05$) are plotted; f is the level of field significance (*Source*: Turrent and Cavazos 2009).

In NAM, the atmospheric moisture and instability in the SW US have increased since 1990 (Luong *et al.* 2017). This more favorable thermodynamic environment has led to stronger though less frequent MCS-type convection since the environment is dynamically less favorable for convection (Lahmers *et al.* 2016). A significant increasing trend of extreme precipitation in the core of the monsoon region, linked to TCs, has also been reported (Cavazos *et al.* 2008; Arriaga-Ramirez and Cavazos 2010). Other significant annual trends of extreme daily precipitation in the NAM are partially linked to natural variations resulting from the combined effects of positive ENSO and positive PDO phases during winter and are, in most cases, explained by the positive PDO (Arriaga-Ramirez and Cavazos 2010).

3.6. *Paleoclimate and climate change*

Paleoclimatic reconstructions of SAM in different time scales, and their relationship with climatic modes, have been enabled by speleothems and tree rings, together with lacustrine, marine sedimentary records, and corals from key places of SA (Cruz *et al.* 2005; Pessenda *et al.* 2010; Apaéstegui *et al.* 2014; Chiessi *et al.* 2009) and nearby oceans under the influence of the SAM and associated mechanisms (ITCZ, SACZ). PMIP3 model reconstructions (Silva Dias *et al.*

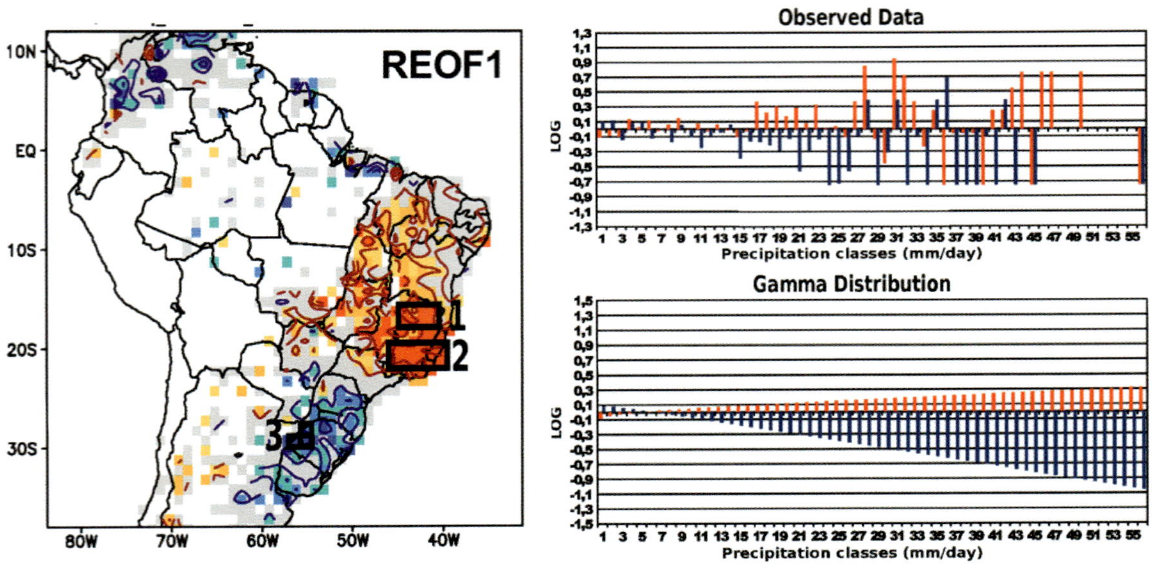

(a) (b)

Fig. 8. (a) Differences between frequency of extreme rainfall events in positive and negative phases of the summer REOF1 of interdecadal precipitation variability. Colors represent level of confidence better than 90%. (b) For region 1 (marked on the map), logarithm of the ratio fpos:fneu in red bars, and fneg:fneu in blue bars, for each daily precipitation interval (fpos = frequency in positive phase; fneg = frequency in negative phase; fneu = frequency in neutral phase). The upper panel shows the values for observed daily precipitation, while the lower panel shows the values for Gamma distribution fitted to the observed daily data in each phase (adapted from Grimm *et al.* 2016).

2009) and TRACE21K have indicated a strong interaction between the SAM and the Atlantic ITCZ positioning (Marshall *et al.* 2014), controlled by changes in the thermohaline circulation heat flux and in the Agulhas leakage (Castelhanos *et al.* 2014).

A 470-year reconstruction of NAM precipitation in SW NA from tree-ring latewood width chronologies shows that past drought events were more severe and persistent than any in the instrumental record (Griffin *et al.* 2013). These "Megadroughts" tend to be driven by dual-season drought, when both monsoon and cool season precipitation tend to be below average over several decades (Carrillo *et al.* 2016). Metcalfe *et al.* (2015) used information from both continental and marine records and reviewed extensive evidence on the Holocene history of the NAM. They indicate that the middle part of the Holocene between ~8,000 and 4,000 cal yr BP was characterized by increased La Niña-like conditions which would favor a more northerly ITCZ and increased monsoonal rainfall in NW Mexico and Arizona (e.g., Barron *et al.* 2012). Monsoon moisture gulf surges probably diminished after ~3,000 cal yr BP due to the increased expression of El Niño events which would have caused the Pacific ITCZ to position further to the south (Perez-Cruz 2013). There is a correspondence of periods of solar minima with episodes of drought especially in the central and southern portions of the NAM (e.g., Metcalfe *et al.* 2010).

Regarding possible indications of climate change, the length of the dry season has increased over southern Amazon, mainly due to the delay of its end (Fu *et al.* 2013), implying a long-term delay of SAM onset over this region, which does not follow interdecadal oceanic variability. It seems to be due to poleward shift of the southern hemispheric subtropical jets and increased atmospheric stability during the dry season (Fu *et al.* 2013). In addition, land use and the westward displacement of the NASH may contribute to this delaying (Arias *et al.* 2015). Observations show that the late onset of SAM over South Amazon is significantly correlated with an earlier retreat of the NAM, and this is attributed to the contraction of the American–Atlantic ITCZ (Arias *et al.* 2015). Although the cause remains unknown, this contraction is projected by climate models forced by increasing atmospheric CO_2 (Lau and Kim 2015). Arias *et al.* (2012) also found that the drying over NW Mexico during the positive AMO phase is more persistent and stronger in more recent periods (1991–2005), associated with the westward shift of the NASH, possibly due to anthropogenic warming (Li *et al.* 2011).

However, trends in SAM life cycle depend on the method and analyzed areas. The life cycle obtained from the LISAM index, which takes into account other atmospheric variables besides precipitation, considering the SAM system, indicates an earlier onset (Jones and Carvalho 2013), opposite to the onset delay in southern Amazon. On the contrary, a delayed demise was found in southern Amazon and also in the SAM system. In climate change projections by CMIP5 GCMs for the end of the 21st century, when the whole SA monsoon region is analyzed using a precipitation threshold criterion, there is little change in the life cycle, compared to the twentieth century (Kitoh *et al.* 2013). However, if the LISAM index is used, CMIP5 models project earlier onsets and late demises (Jones and Carvalho 2013).

The projections also suggest delays in NAM onset and retreat. During the autumn, the models show a northward migration of the North Pacific ITCZ and moisture transport into the NAM region, thus delaying the monsoon retreat (Torres-Alavez *et al.* 2014). Projected changes in total monsoon precipitation, however, are insignificant (Cook and Seager 2013; Torres-Alavez *et al.* 2014). The relationship between the demise of the NAM and onset of the SAM under climate change conditions needs to be further investigated. Meyer and Jin (2016) suggest, however, that the coarse simulations might be underestimating future precipitation, and thus regional climate modeling may help improve these shortcomings and other relevant physical processes.

4. Modeling Studies

The SAM and NAM general features have been well simulated by some global (GCMs) and regional climate models (RCMs). For SA, the main differences between summer and winter precipitation and circulation are represented in CMIP3 and CMIP5 models (Figs. 9a and b, Carvalho and Cavalcanti 2016). Jones and Carvalho (2013) showed that the CMIP5 models represent more adequately low-level fields associated with SAM and that some CMIP5 models have significantly improved their SAM representation compared to their CMIP3 versions, but there are persistent deficiencies. Overall, models have systematic errors in precipitation intensity in the SAM main areas. CMIP5 models reproduce the observed negative (positive) trend in the SAM onset (demise) dates when the LISAM index is used (Jones and Carvalho 2013). Atmospheric models forced by observed SST reproduce the onset and demise close to the observations, as ECHAM 4.5 AGCM (Liebmann *et al.* 2007) and CPTEC/INPE AGCM (Cavalcanti and Raia 2017). This latter model simulates observed characteristics of the SAM, such as the moisture flux changes during onset and demise (Figs. 9c–f). Improvements in parameterization schemes (mainly convection) and understanding of the interaction between clouds, aerosols, gases, and radiation; land surface and boundary layer fluxes; and regional air-sea interaction can contribute to better

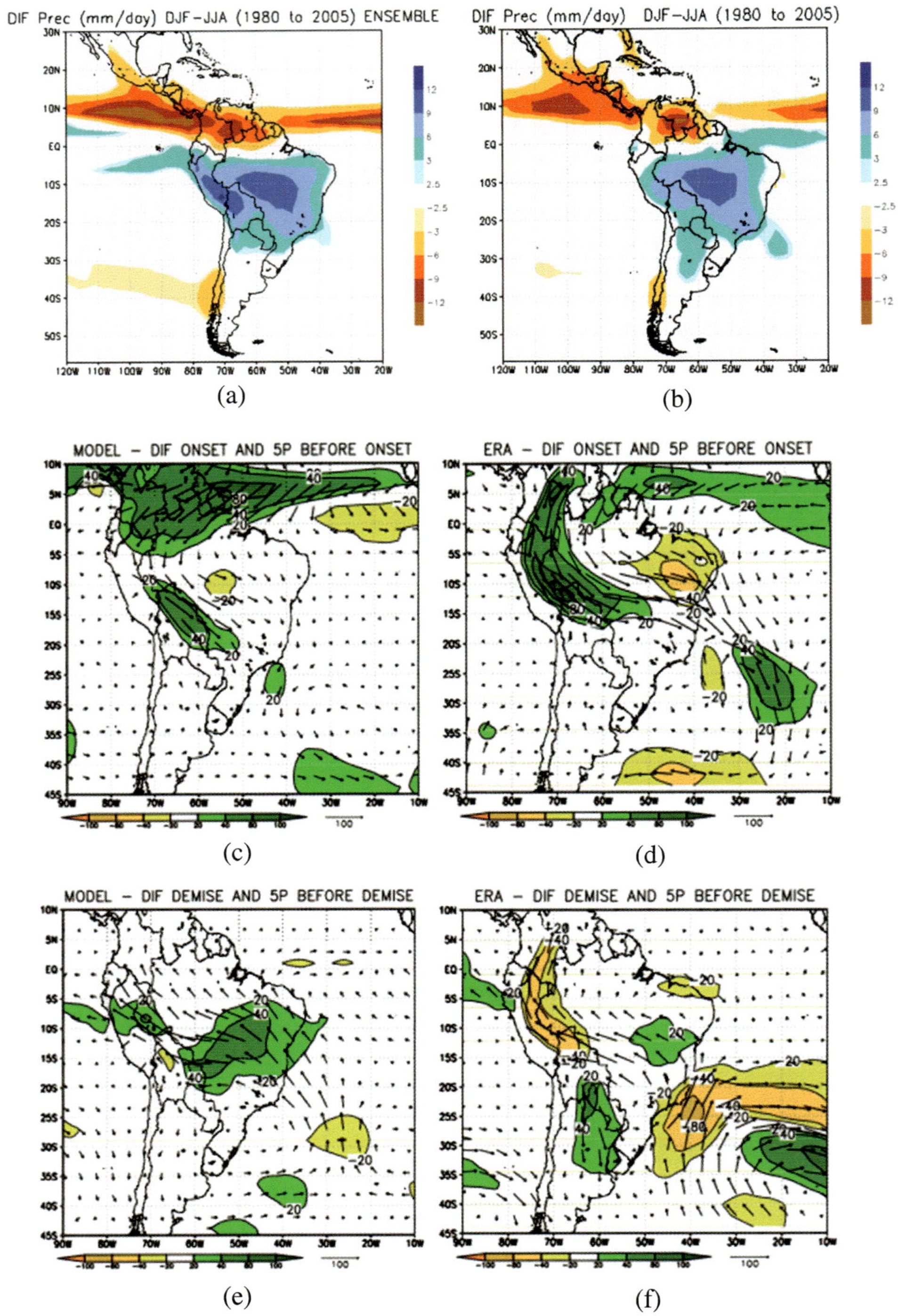

Fig. 9. Difference between precipitation in austral summer (DJF) and winter (JJA) in (a) ensemble of 5 CMIP5 models: HadGEM2-ES, CSIRO, IPSL, MRI, and MIROC; (b) GPCP (*Source*: Carvalho and Cavalcanti 2016). Moisture flux and its divergence during: (c), (d) Onset and (e), (f) Demise of SAM. (c), (e) AGCM and (d), (f) ERA reanalysis (*Source*: Cavalcanti and Raia 2017).

reproducing monsoon features. Coupled regional models have similar deficiencies as global coupled models, such as the double Atlantic ITCZ problem (Sitz *et al.* 2017).

When compared to CMIP3, the CMIP5 models show a significant improvement in the phasing of the seasonal cycle of the NAM, but no improvement in the magnitude of the mean annual cycle of monthly precipitation (Geil *et al.* 2013). Most GCMs have problems representing the NAM climatology and variability. In particular, autumn and winter precipitation in the core NAM region is overestimated, and the NAM retreat in the autumn is not reproduced by some models because of issues in low-level geopotential height gradients (Geil *et al.* 2013). An alternative is to downscale the GCM data using regional climate models, as done in the North American Regional Climate Change Assessment Program (NARCCAP) and the Coordinated Regional Dynamical Downscaling (CORDEX) global effort. NARCCAP, NA-CORDEX, and Central America (CAM-) CORDEX simulations improve the spatial representation of precipitation, and even sometimes represent the Gulf of California low-level jet (Bukovsky *et al.* 2013), extreme events, and TCs (Fuentes-Franco *et al.* 2014).

5. Prediction and Predictability

The prediction of the SAM and NAM onset, demise, active/break periods, and the amount of precipitation requires the correct simulation of variability on all time scales. Remote influences like those from the MJO or ENSO, as well as local land–atmosphere and regional air-sea coupling, dominate the predictability of the monsoon on different time scales and need to be correctly simulated for a successful monsoon prediction. To improve the prediction it is necessary to better understand its modulation by different processes and how they combine to produce observed anomalies. This can be done using hierarchy of models, tailored numerical experiments with stand-alone or coupled regional and GCMs, observational field campaigns in data-sparse regions, and analysis of hindcasts from projects such as the Subseasonal to Seasonal (S2S) project.

Recent studies have shown the feasibility of using current models for actual predictions. Coelho *et al.* (2017) showed a statistical prediction system based on the Niño3 index with skill in predicting the rainy season onset in São Paulo, Brazil. Bombardi *et al.* (2017) used hindcasts of the S2S project to analyze the rainy season onset in SA and found skillful forecast with 1-month lead time. The Climate Forecast System model of NCEP (CSF/NCEP) shows useful skill to forecast SAM features, as onset, demise, and duration (Jones *et al.* 2012). Hirata and Grimm (2017) demonstrated that predictions of heavy rainfall in the SACZ region are possible up to two weeks by calibrating the precipitation hindcasts from CFS-v2. Regarding the regional SST influence on the SACZ, atmospheric model simulations in Barreiro *et al.* (2002) have shown that on interannual time scales, the SACZ responds to a subtropical SST dipole by shifting toward the warmer SSTs. Chaves and Nobre (2004) showed that on seasonal time scales there is a negative feedback between the SACZ and the SST.

NAM S2S forecasting is an area of active research. At present, the operational global models used as part of the North American Multimodel Ensemble (NMME) demonstrate little to no skill in forecasting monsoon precipitation at S2S timescales. Castro *et al.* (2012) tested the potential to improve operational seasonal forecasts of the NAM by dynamical downscaling of the CFS ensemble reforecast data for the period 1982–2000. They found marginal improvements in seasonal forecast and a better representation of the dominant precipitation modes associated with warm season atmospheric teleconnections.

6. Challenges

SAM and NAM are the result of land/atmosphere/ocean coupling and characterized by multiscale interactions that are not completely known or understood. It is still unclear, for example, if the strong droughts in the southern Amazon and shorter seasons of SAM and NAM in recent decades are part of interannual and interdecadal variability or anthropogenic forced changes. The large spatial and temporal variability and the scarcity of observations in extensive regions dominated by the American monsoons hinder, in part, the advances in our understanding of these systems. Models need to represent correctly the impact of large-scale modes of variability and their interactions, as well as local and regional processes. For instance, simulations suggest that coupled climate–dynamic vegetation models can better simulate the magnitude of the recent droughts. However, regional models still represent poorly the diurnal cycle of convection and convective organization. The steep Andean topography may result in incorrect simulation of the SAM Low-Level Jet and the associated moist convective and rainfall processes. Something similar may occur with the complex topography of the NAM.

Models need improvements to reduce errors in prediction of the onset and demise dates, and the active/break periods of the monsoons. Observations, through field campaigns and continuous monitoring, can help improve the validation of models and their general understanding since some of the physical processes involved in the monsoons are not well understood, such as the role of aerosols in cloud formation and precipitation in the Amazon basin. The analysis of global and regional models' hindcasts and the numerical experiments regarding the monsoon features and variability also contribute to this improvement and the assessment of the predictability in different time scales. Continued efforts on S2S prediction and improved understanding of the dynamical changes of the monsoon circulation under warmer conditions are critical. Statistical techniques can also be applied to relate the large-scale conditions and regional features to the life cycle prediction.

Although the American monsoons are still seen as two separate systems, it is becoming increasingly clear that they are one dynamical system. So we should, as a community, combine efforts to understand this dominant feature of the climate of the Americas and the processes by which they influence each other.

Acknowledgments

This work was carried out with the support of the National Council for Scientific and Technological Development (CNPq-Brazil); the Inter-American Institute for Global Change Research (IAI) grant CRN3035, which is supported by the US National Science Foundation (Grant GEO-1128040); and the Department of Physical Oceanography, CICESE.

References

Adams, D. and A. Comrie, 1997: The North American monsoon. *Bull. Amer. Meteor. Soc.*, **78**, 2197–2213.

Alvarez, M. S., C. S. Vera, G. N. Kiladis, and B. Liebmann, 2015: Influence of the Madden Julian Oscillation on precipitation and surface air temperature in South America. *Clim. Dyn.*, **46**, 245–262.

Apaéstegui, J., and Coauthors, 2014: In: Molina, S. G., and J. J. Vacher (eds.), *Cambios Climáticos del Holoceno*. El Perú frente al cambio climático. IRD.

Arias, P. A., R. Fu, and K. C. Mo, 2012: Decadal variation of rainfall seasonality in the North American Monsoon region and its potential causes. *J. Climate*, **25**, 4258–4274.

Arias, P. A., R. Fu, C. Vera, and M. Rojas, 2015: A correlated shortening of the North and South American monsoon seasons in the past few decades. *Clim. Dyn.*, **45**, 3183–3203.

Arriaga-Ramirez, S. and T. Cavazos, 2010: Regional trends of daily precipitation indices in northwest Mexico and southwest United States. *J. Geophys. Res. — Atmos.*, **115**, D14111.

Avissar, R. and D. Werth, 2005: Global hydroclimatological teleconnections resulting from tropical deforestation. *J. Hydrometeorol.*, **6**(2), 134–145.

Barlow, M., and D. Salstein, 2006: Summertime influence of the Madden–Julian oscillation on daily rainfall over Mexico and Central America. *Geophys. Res. Lett.*, **33**, L21708.

Barreiro, M., P. Chang, and R. Saravanan, 2002: Variability of the South Atlantic Convergence Zone as simulated by an atmospheric general circulation model. *J. Climate*, **15**, 745–763.

Barreiro, M. and N. Diaz, 2011: Land-atmosphere coupling in El Nino influence over South America atmos. *Sci. Lett.*, **12**, 351–355.

Barron, J. A., S. E. Metcalfe, and J. A. Addison, 2012: Response of the North America monsoon to regional changes in ocean surface temperature. *Paleoceanography*, **27**(3), PA3206.

Betts, A. K. and M. A. F. Silva Dias, 2010: Progress in understanding land-surface-atmosphere coupling from LBA research. *J. Adv. Model. Earth Sys.*, **2**, 1–20.

Bombardi, R. J., K. V. Pegion, J. L. Kinter, B. A. Cash, and J. M. Adams, 2017: Sub-seasonal predictability of the onset and demise of the rainy season over monsoonal regions. *Front. Earth Sci.*, **5**, 14.

Bowerman, A. R., R. Fu, L. Yin, D. N. Fernando, P. A. Arias, and R. E. Dickinson, 2017: An influence of southern hemispheric cold surge on the North Atlantic through a shallow atmospheric circulation. *J. Geophys. Res.-Atmos.*, **122**, 10133–10556.

Bukovsky, M. S., D. Gochis, and L. O. Mearns, 2013: Towards assessing NARCCAP regional climate model credibility for the North American monsoon: current climate simulations. *J. Climate*, **26**, 8802–8826.

Carrillo, C. M., C. L. Castro, C. A. Woodhouse, and D. Griffin, 2016: Low-frequency variability of precipitation in the North American monsoon region as diagnosed through earlywood and latewood tree-ring chronologies in the southwestern US. *Int. J. Climatol.*, **36**, 2254–2272.

Carvalho, L. M. V., C. Jones, and B. Liebmann, 2004: The South Atlantic convergence zone: Intensity, form, persistence, and relationships with intraseasonal to interannual activity and extreme rainfall. *J. Climate*, **17**, 88–108.

Carvalho, L. M. V. and I. F. A. Cavalcanti, 2016: The South American monsoon system. In: Carvalho, L. M. V., and C. Jones (eds.) *The Monsoons and Climate Change: Observations and Modeling.* Springer, pp. 121–148.

Castellanos, P., E. J. D. Campos, J. Piera, O. T. Sato, and M. A. F. Silva Dias, 2017: Impacts of Agulhas Leakage on the Tropical Atlantic western boundary systems. *J. Climate*, **30**, 6645–6649.

Castro, C. L., T. B. McKee, and R. A. Pielke Sr, 2001: The relationship of the North American monsoon to Tropical and North Pacific sea surface temperatures as revealed by observational analyses. *J. Climate*, **14**, 4449–4473.

Castro, C. L., R. A. S. Pielke, J. O. Adegoke, S. D. Schubert, and P. J. Pegion, 2007: Investigation of the summer climate of the contiguous United States and Mexico using the regional atmospheric modeling system (RAMS). Part II: Model climate variability. *J. Climate*, **20**, 3866–3887.

Castro, C. L., H. I. Chang, F. Dominguez, C. Carrillo, J.-K. Schemm, and H.-M. Henry Juang, 2012: Can a regional climate model improve the ability to forecast the North American monsoon? *J. Climate*, **25**, 8212–8237.

Cavalcanti, I. F. A. and A. Raia, 2017: Lifecycle of South American monsoon system simulated by CPTEC/INPE AGCM. *Int. J. Climatol.*, **37,** 878–896.

Cavazos *et al.*, 2002: Intraseasonal variability associated with wet monsoons in Southeast Arizona. *J. Climate*, **15**, 2477–2490.

Cavazos, T., C. Turrent, and D. P. Lettenmaier, 2008: Extreme precipitation trends associated with tropical cyclones in the core of the North American monsoon. *Geophys. Res. Lett.*, **35**, D05109–5.

Chaves, R. R. and P. Nobre, 2004: Interactions between sea surface temperature over the South Atlantic Ocean and the South Atlantic Convergence Zone. *Geophys. Res. Lett.*, **31**, L03204.

Chiessi, C. M., S. Mulitza, J. Pätzold, and G.Wefer, 2009: Possible impact of the Atlantic Multidecadal Oscillation in the South American summer monsoon. *Geophys. Res. Lett.*, **36**, L21707.

Coelho, C. A. S., M. A. F. Firpo, A. H. N. Maia, and C. MacLachlan, 2017: Exploring the feasibility of empirical, dynamical and combined probabilistic rainy season onset forecasts for São Paulo, Brazil. *Int. J. Climatol.*, **37**, 398–411.

Cook, B. I. and R. Seager, 2013: The response of the North American Monsoon to increased greenhouse gas forcing. *J. Geophys. Res. — Atmos.*, **118**, 1690–1699.

Cruz Jr., F. W., S. J. Burns, I. Karmann, W. D. Sharp, M. Vuille, A. O. Cardoso, J. A. Ferrari, P. L. Silva Dias, O. Viana Jr., 2005: Insolation-driven changes in atmospheric circulation over the past 116,000 years in subtropical Brazil. *Nature*, **434**, 63–66.

Dominguez, F., G. Miguez-Macho, and H. Hu, 2016: WRF with water vapor tracers: A study of moisture sources for the North American monsoon. *J. Hydrometeor.*, **17**, 1915–1927.

Douglas, M. W., R. A. Maddox, K. Howard, and S. Reyes, 1993: The Mexican monsoon. *J. Climate*, **6**, 1665–1677.

Figueroa, S. N., P. Satyamurty, and P. L. Silva Dias, 1995: Simulations of the summer circulation over the South American region with an eta coordinate model. *J. Atmos. Sci.*, **52**, 1573–1584.

Figueroa, S. N. and Coauthors, 2016: The Brazilian Global Atmospheric Model (BAM). Performance for tropical rainfall forecasting and sensitivity to convective scheme and horizontal resolution. *Wea. Forecast.*, **31**, 1547–1572.

Finch, Z. O., R. H. Johnson, and Z. O. Finch, 2010: Observational analysis of an upper-level inverted trough during the 2004 North American Monsoon experiment. *Mon. Wea. Rev.*, **138**, 3540–3555.

Fu, R., L. Yin, W. H. Li, P. A. Arias, R. E. Dickinson, L. Huang, K. Fernandes, Sudip Chakrarborty, B. Liebmann, R. Fisher, and R. B. Myneni, 2013: Increased dry season length over southern Amazonia in recent decades. *Proc. Natl. Acad. Sci.*, **110**, 18110–18115.

Fu, R., P. A. Arias, and H. Wang, 2016: The connection between the North and South American monsoons. In: Carvalho, L. M. V., and C. Jones (eds.), *The Monsoons and Climate Change: Observations and Modeling.* Springer, pp. 187–206.

Fuentes-Franco, R., E. Coppola, F. Giorgi, F. Graef, and E. G. Pavia, 2014: Assessment of RegCM4 simulated interannual variability and daily-scale statistics of temperature and precipitation over Mexico. *Clim. Dyn.*, **42**, 629–647.

Gan, M. A., V. E. Kousky, C. F. Ropelewski, 2004: The South America monsoon circulation and its relationship to rainfall over West-Central Brazil. *J. Climate*, **17**, 47–66.

Geil, K. L., Y. L. Serra, and X. Zeng, 2013: Assessment of CMIP5 model simulations of the North American monsoon system. *J. Climate*, **26**, 8787–8801.

Grantz, K., B. Rajagopalan, M. Clark, and E. Zagona, 2007: Seasonal shifts in the North American Monsoon. *J. Climate*, **20**, 1923–1935.

Griffin, D. and Coauthors, 2013: North American monsoon precipitation reconstructed from tree-ring latewood. *Geophys. Res. Lett.*, **40**, 954–958.

Grimm, A. M., 2003: The El Niño impact on the summer monsoon in Brazil: regional processes versus remote influences. *J. Climate*, **16**, 263–280.

Grimm, A. M., J. Pal, and F. Giorgi, 2007: Connection between spring conditions and peak summer monsoon rainfall in South America: Role of soil moisture, surface temperature, and topography in eastern Brazil. *J. Climate*, **20**, 5929–5945.

Grimm, A. M. and R. G. Tedeschi, 2009: ENSO and extreme rainfall events in South America. *J. Climate*, **22**, 1589–1609.

Grimm, A. M. and M. T. Zilli, 2009: Interannual variability and seasonal evolution of summer monsoon rainfall in South America. *J. Climate*, **22**, 2257–2275.

Grimm, A. M. and J. P. J. Saboia, 2015: Interdecadal variability of the South American precipitation in the monsoon season. *J. Climate*, **28**, 755–775.

Grimm, A. M., N. C. Laureanti, R. B. Rodakoviski, and C. B. Gama, 2016: Interdecadal variability and extreme precipitation events in South America during the monsoon season. *Climate Research*, **68**, 277–294.

Grimm, A. M., 2018: South American monsoon and its extremes. In Vuruputur, V., J. Sukhatme, R. Murtugudde, R. Roca (eds.), *Tropical Extremes: Natural Variability and Trends*, Elsevier, Amsterdam, 51–93. ISBN 978-0-12-809248-4, https://doi.org/10.1016/B978-0-12-809248-4.00003-0.

Grimm, A. M., 2019: Madden-Julian Oscillation impacts on South America summer monsoon season: precipitation anomalies, extreme events, and teleconnections. *Clim. Dyn.*, doi: https://doi.org/10.1007/s00382-019-04622-6

Higgins, R. W. and W. Shi, 2000: Dominant factors responsible for interannual variability of the summer monsoon in the southwestern United States. *J. Climate*, **13**, 759–776.

Higgins, R. W. and W. Shi, 2005: Relationships between Gulf of California moisture surges and tropical cyclones in the eastern Pacific basin. *J. Climate*, **18**, 4601–4620.

Hirata, F. E. and A. M. Grimm, 2015: The role of synoptic and intraseasonal anomalies in the life cycle of summer rainfall extremes over South America. *Clim. Dyn.*, **46**, 3041–3055.

Hirata, F. E. and A. M. Grimm, 2017: Extended-range prediction of South Atlantic convergence zone rainfall with calibrated CFSv2 reforecast. *Clim. Dyn.*, doi: https://doi.org/10.1007/s00382-017-3836-1.

Hu, H. and F. Dominguez, 2015: Evaluation of oceanic and terrestrial sources of moisture for the North American monsoon using numerical models and precipitation stable isotopes. *J. Hydrometeor.*, **16**, 19–35.

Jones, C., L. M. V. Carvalho, and B. Liebmann, 2012: Forecast skill of the South American monsoon system. *J. Climate*, **25**, 1883–1889.

Jones, C. and L. M. V. Carvalho, 2013: Climate change in the South American monsoon system: present climate and CMIP5 projections. *J. Climate*, **26**, 6660–6678.

Kitoh, A., H. Endo, K. K. Kumar, I. F. A. Cavalcanti, P. Goswami, and T. J. Zhou, 2013: Monsoons in a changing world: A regional perspective in a global context. *J Geophys. Res. Atmos.*, **118**, 3053–3065.

Lahmers, T. M., C. L. Castro, D. K. Adams, Y. L. Serra, J. J. Brost, and T. Luong, 2016: Long-term changes in the climatology of transient inverted roughs over the North American monsoon region and their effects on precipitation. *J. Climate*, **29**, 6037–6064.

Lang, T. J., D. A. Ahijevych, S. W. Nesbitt, R. E. Carbone, S. A. Rutledge, and R. Cifelli, 2007: Radar-observed characteristics of precipitating systems during NAME 2004. *J. Climate*, **20**, 1713–1733.

Lau, W. K. M., and K.-M. Kim, 2015: Robust Hadley circulation changes and increasing global dryness due to CO2 warming from CMIP5 model projections. *Proc. Natl. Acad. Sci.*, **112**(12), 3630–3635.

Li, W., L. Li, R. Fu, Y. Deng, and H. Wang, 2011: Changes to the North Atlantic subtropical high and its role in the intensification of summer rainfall variability in the southeastern United States. *J. Climate*, **24**, 1499–1506.

Liebmann, B., S. J. Camargo, A. Seth, J. A. Marengo, L. M. Carvalho, D. Allured, R. Fu, and C. S. Vera, 2007: Onset and end of the rainy season in South America in observations and the ECHAM 4.5 atmospheric general circulation model. *J. Climate*, **20**, 2037–2050.

Lorenz, D. J. and D. L. Hartmann, 2006: The effect of the MJO on the North American monsoon. *J. Climate*, **19**, 333–343.

Luong, T. M., C. L. Castro, H.-I. Chang, T. Lahmers, D. K. Adams, and C. A. Ochoa-Moya, 2017: The more extreme nature of North American monsoon precipitation in the southwestern United States as revealed by a historical climatology of simulated severe weather events. *J. Appl. Meteorol. Climatol.*, **56**, 2509–2529.

Machado, L. A. T. and Coauthors, 2014: The CHUVA Project — how does convection vary across Brazil? *Bull. Amer. Meteor. Soc.*, **95**, 1365–1380.

Maddox, R. A., D. M. McCollum, and K. W. Howard, 1995: Large-scale patterns associated with severe summertime thunderstorms over central Arizona. *Wea. Forecast.*, **10**, 763–778.

Marengo, J. A., M. W. Douglas, and P. L. Silva Dias, 2002: The South American low-level jet east of the Andes during the LBA-TRMM and LBA — WET AMC campaign. *J. Geophys. Res.*, **107**, 47.1–47.11.

Marshall, J., A. Donohoe, D. Ferreira, and D. McGee, 2014: The ocean's role in setting the mean position of the inter-tropical convergence zone. *Clim. Dyn.*, **42**, 1967–1979.

Méndez, M. and V. Magaña, 2010: Regional aspects of prolonged meteorological droughts over Mexico and Central America. *J. Climate*, **23**, 1175–1188.

Metcalfe, S. E. and Coauthors, 2010: Climate variability over the last two millennia in the North American Monsoon region recorded in laminated sediments from the Laguna de Juanacatlan, Mexico. *The Holocene*, **20**, 1195–1206.

Metcalfe S. E., J. A. Barron, and S. J. Davies, 2015: The Holocene history of the North American Monsoon: 'known knowns' and 'known unknowns' in understanding its spatial and temporal complexity. *Quat. Sci. Rev.*, **120**, 1–27.

Meyer, J. D. D. and J. Jin, 2016: The response of future projections of the North American monsoon when combining dynamical downscaling and bias correction of CCSM4 output. *Clim. Dyn.*, **49**, 433–447.

Nesbitt, S. W., D. J. Gochis, and T. J. Lang, 2008: The diurnal cycle of clouds and precipitation along the Sierra Madre Occidental observed during NAME-2004: Implications for warm season precipitation estimation in complex terrain. *J. Hydrometeor.*, **9**, 728–743.

Perez-Cruz, L., 2013: Hydrological changes and paleoproductivity in the Gulf of California during middle and late Holocene and their relationship with ITCZ and North American Monsoon variability. *Quat. Res.*, **79**, 138–151.

Pessenda, L. C. R., S. E. M. Gouveia, A. S. Ribeiro, P. E. Oliveira, and R. Aravena, 2010: Late Pleistocene and Holocene vegetation changes in northeastern Brazil determined from carbon isotopes and charcoal records in soils. *Palaeogeogr. Palaeoclimatol. Palaeoecol.*, **297**, 597–608.

Raia, A. and I. F. A. Cavalcanti, 2008: The life cycle of the South American monsoon system. *J. Climate*, **21**, 6227–6246.

Ramos da Silva, R., A. W. Gandu, L. D. A. Sá, and M. A. F. Silva Dias, 2011: Cloud streets and land water interactions in the Amazon. *Biogeochemistry* (Dordrecht), 1–11.

Schmitz, J. and S. Mullen, 1996: Water vapor transport associated with the summertime North American monsoon as depicted by ECMWF analyses. *J. Climate*, **9**, 1621–1634.

Seastrand, S., Y. Serra, C. Castro, and E. Ritchie, 2015: The dominant synoptic-scale modes of North American monsoon precipitation. *Int. J. Climatol.*, **35**, 2019–2032.

Silva Dias, P. L., W. H. Schubert, and M. DeMaria, 1983: Large-scale response of the tropical atmosphere to transient convection. *J. Atmos. Sci.*, **40**, 2689–2707.

Silva Dias, P. L., B. Turcq, M. A. F. Silva Dias, P. Braconnot, and T. Jorgetti, 2009: Mid-Holocene climate of tropical South America: a model-data approach. In: Vimeux, F., F. Sylvestre, and M. Khodri (eds.), *Past Climate Variability from the Last Glacial Maximum to the Holocene in South America and Surrounding Regions.* Springer-Verlag, pp. 259–281.

Sitz, L. E. and Coauthors, 2017: Description and evaluation of the Earth System Regional Climate Model (RegCM-ES). *J. Adv. Model. Earth Syst.*, doi:10.1002/2017MS000933.

Souza, E. B. and T. Ambrizzi, 2006: Modulation of the intraseasonal rainfall over tropical Brazil by the Madden–Julian oscillation. *Int. J. Climatol.*, **26**, 1759–1776.

Tedeschi, R. G., A. M. Grimm, and I. F. A. Cavalcanti, 2015: Influence of central and east ENSO on extreme events of precipitation in South America during austral spring and summer. *Int. J. Climatol.*, **35**, 2045–2064.

Tedeschi, R. G., A. M. Grimm, and E I. F. A. Cavalcanti, 2016: Influence of Central and East ENSO on precipitation and its extreme events in South America during austral autumn and winter. *Int. J. Climatol.*, **36**, 4797–4814.

Tirabassi, G., C. Masoller, and M. Barreiro, 2014: A study of the air–sea interaction in the South Atlantic Convergence Zone through Granger Causality. *Int. J. Climatol.*, **35**, 3440–3453.

Torres-Alavez, A., T. Cavazos, and C. Turrent, 2014: Land–sea thermal contrast and intensity of the North American monsoon under climate change conditions. *J. Climate*, **27**, 4566–4580.

Turrent, C. and T. Cavazos, 2009: Role of the land–sea thermal contrast in the interannual modulation of the North American monsoon. *Geophys. Res. Lett.*, **36**, L02808.

Wang, B., J. Liu, H. J. Kim, P. Webster, and S. Y. Yim, 2012: Recent change of the global monsoon precipitation (1979–2008). *Clim. Dyn.*, **39**, 1123–1135.

Wang, W., S. J. Weaver, M.-P. Hung, A. Kumar, and X. Fu, 2014: MJO prediction in the NCEP Climate Forecast System version 2. *Clim. Dyn.*, **42**, 2509–2520.

Wood, K. M. and E. A. Ritchie, 2013: An updated climatology of tropical cyclone impacts on the southwestern United States. *Mon. Wea. Rev.*, **141**, 4322–4336.

Yin, L., R. Fu, Y. F. Zhang, P. A. Arias, D. N. Fernandoa, W. H. Li, K. Fernandes, and A. R. Bowerman, 2014: What controls the interannual variation of the wet season onsets over the Amazon? *J. Geophys. Res. — Atmos.*, **119**, 2314–2328.

Zhou, J. and K.-M. Lau, 1998: Does a monsoon climate exist over South America? *J. Climate*, **11**, 1020–1040.

Zhu, C., T. Cavazos, and D. P. Lettenmaier, 2007: Role of antecedent land surface conditions in warm season precipitation over northwestern Mexico. *J. Climate*, **20**, 1774–1791.

Zhu, C., L. R. Leung, D. Gochis, Y. Qian, and D. P. Lettenmaier, 2009: Evaluating the influence of antecedent soil moisture on variability of the North American monsoon precipitation in the coupled MM5/VIC modeling system. *J. Adv. Model. Earth Syst.*, **2**, 13–22.

Chapter 6

Variability of the Rain Belt in Western Central Africa

André Lenouo*,||, Wilfried Pokam†, Appolinaire Derbetini Vondou‡, Serge Janicot§ and François Kamga Mkankam¶

*Department of Physics, Faculty of Science, University of Douala, Cameroon
†Department of Physics, ENS, University of Yaoundé I, Cameroon
‡Department of Physics, Faculty of Science, University of Yaoundé I, Cameroon
§LOCEAN/IPSL IRD 4, Place Jussieu, 75252 Paris cedex 05, France
¶Université des Montagnes, Bangangté, Cameroon
||lenouo@yahoo.fr

This chapter provides a state-of-the-art assessment of recent research addressing the western Central African rain belt and its variability at different timescales. Multiple sources of observations, satellite products, and atmospheric reanalyses have been used to document and interpret this variability and to evaluate regional climate models (RCMs). The diurnal cycle of convection displays a morning minimum in the mesoscale convective system (MCS) number and a late afternoon to early evening maximum, peaking around 1500–1800 LT. Generally, RCM simulations show an unrealistic representation of the diurnal cycle of rainfall. At the synoptic scale, while African easterly waves have been identified as fundamental synoptic weather systems in the West and Central African summer monsoons, other atmospheric systems as convectively coupled Kelvin waves have been shown to account for a substantial amount of organization of rainfall and MCS activity over Central Africa. At the intraseasonal timescale, the Madden–Julian oscillation (MJO) also impacts convective activity. Interannual rainfall variability appears to be significantly controlled through SST anomalies of the different ocean basins in the tropics through atmospheric teleconnections. The RCM successfully reproduces the annual cycle of the tropical rain belt over Central Africa with the double peak of rainfall due to its double passage during the year. Finally, climate change scenarios can differ depending on the climate models used, indicating a need for large-enough ensembles of independent climate projections to allow for reliable climate change assessment. Last but not least, this region, which is one out of the three key convective regions on the planet, is highly under-documented and international support for more *in situ* observations and for a targeted field campaign is now essential.

1. Introduction

Central Africa is the least studied region in Africa and a region that has been much less investigated than any other large-scale convective region in the tropics (Washington *et al.* 2013). However, convective activity in this area is one of the most intense in the world. Moreover, due to the location of the tropical rain belt over the equator during a significant part of the year, the interactions with atmospheric convectively coupled equatorial waves are potentially strong (Laing *et al.* 2011) as a source for large teleconnections within the tropics from intraseasonal to interannual timescales. Climate change will also have an impact in Central Africa on both natural and human systems and the various ways in which it will modify convective and atmospheric variability need to be addressed thoroughly. So, despite the region's failure to qualify as a "monsoonal" region (see the discussion on the definition of monsoons in Wang *et al.* 2017), it is the site of intense convection, like other global monsoon regions. In this chapter, we will also avoid

The Multiscale Global Monsoon System
Edited by C. P. Chang et al.

the use of the term intertropical convergence zone (ITCZ) and prefer the term tropical rain belt as the structure of the motion field during the two rainy seasons over Central Africa shows little similarity to the classic ITCZ paradigm (see Nicholson 2018 for a more detailed discussion).

The drainage basin of the Congo River is located in this region, characterized by the humid forest, which in the source of many vegetables and much vegetative and animal biodiversity (Fig. 1). Water resources, agriculture, and health in Central Africa are also dependent on the seasonal cycle of precipitation, both in terms of timing (for instance the onset of wet seasons) and intensity. The topography of this region also influences the annual rainfall cycle.

However, this region remains largely underdocumented and more *in situ* observations are greatly needed. This chapter presents an assessment of some research conducted on this region over the last 15 years in terms of the annual cycle and the large-scale circulation, including the diurnal cycle, synoptic to intraseasonal timescales, interannual timescales, and climate simulation and projection.

2. The Annual Cycle and Large-Scale Circulation

Tchakoutchio *et al.* (2012) showed that the annual mean rainfall over Central Africa consists of a zone of intense convective activity centered on the equator, which extends southward and covers almost all the Congo forest, making this area one of the three key convective regions on the planet in which the transition seasons dominate global tropical rainfall (Washington *et al.* 2013). The peak of the rain belt shifts from the north to the south of the equator during November–December and returns to the north during March–April. Around the equator, the mean annual rainfall is divided into four periods: December–February (DJF), March–May (MAM), June–August (JJA), and September–November (SON). During the SON (long rains) season, precipitation events tend to be heavier and longer in duration, while the MAM (short rains) season (regions 2, 3, 4, and 5 in Fig. 2) is characterized by less intense precipitation events of shorter duration (Tchakoutchio *et al.* (2012). The highest precipitation over Central Africa is recorded over the East African highlands during the boreal winter (November–April) where the

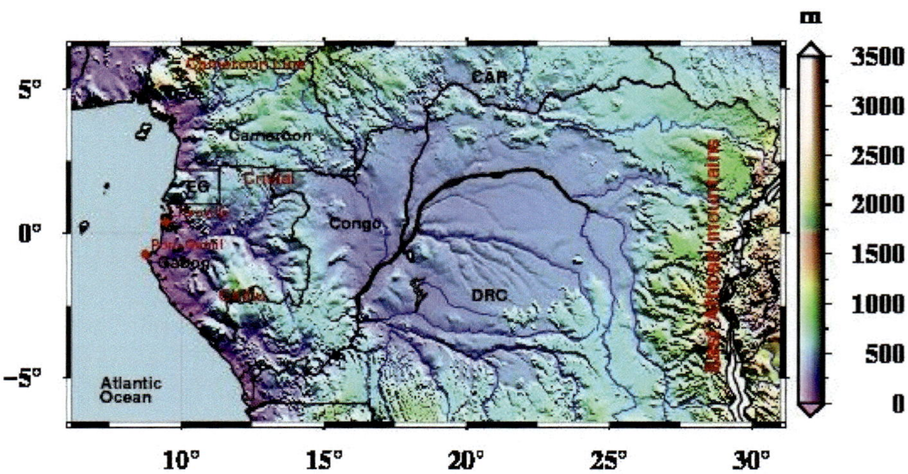

Fig. 1. The map of Central Africa shows the topography, the Congo Basin characterized by the Congo River, and the different countries marked by solid lines.

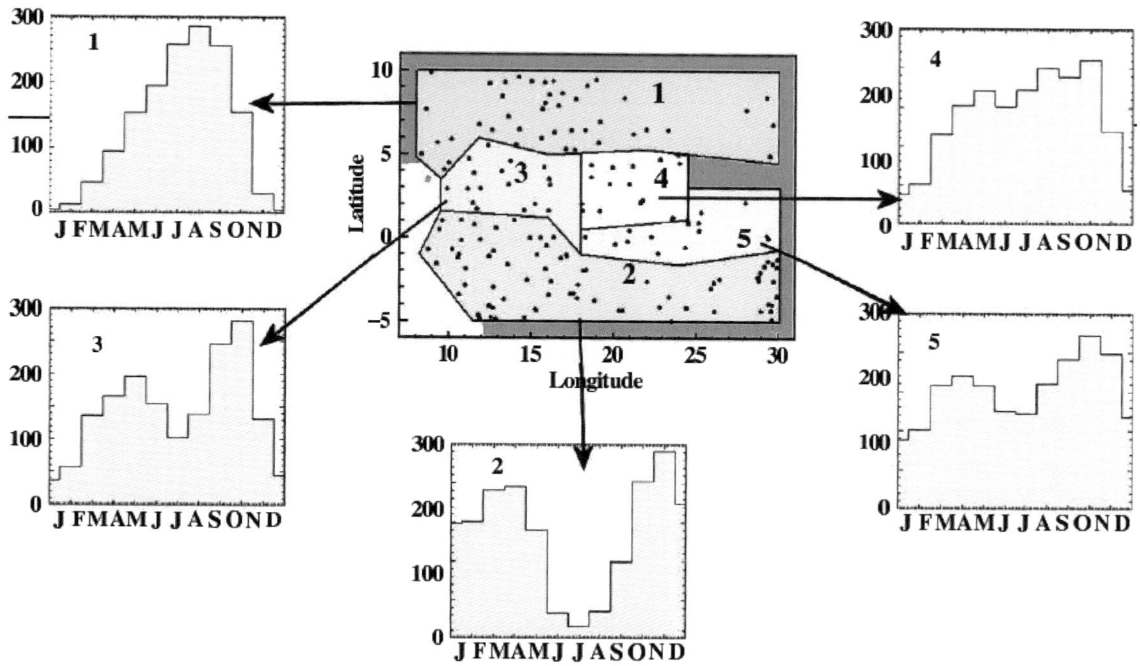

Fig. 2. Rainfall regions and stations. The mean annual cycle of each region is represented in a histogram (rainfall in mm) from Balas *et al.* (2007).

precipitation rate exceeds 10 mm/day (Fig. 2). In the north of the Congo Basin (region 4 in Fig. 2), the rainfall is unimodal where the dry season is less than 3 months (DJF). In contrary, the southern region (2) has a bimodal regime with a long dry season from May to September.

From the view of atmospheric circulation in the lower troposphere, moist air from the Atlantic Ocean, known as low-level equatorial westerlies, is associated with the southeasterly trades on the northeastern flank of the Saint Helena (South Atlantic) high. The low-level westerlies (LLW) are defined throughout the year (Nicholson and Grist 2003). Over west Central Africa, the LLWs are well developed from July to September on the northern side of the equator and in October–February on the southern side. In the vicinity of the equator, LLW are defined from August to January, and for the rest of the year, westerlies disappear and easterlies dominate throughout the atmospheric column. Pokam *et al.* (2014) showed that LLW is controlled by contrasted processes, north and south of the equator, involving the rotational flow driven by the Saharan heat low in the north and by zonal ocean–land heating contrast in the south, inducing a divergent Walker-type circulation (Fig. 3). The Walker circulation usually includes upward branches over the western Pacific warm pool (95°–150°E), eastern Indian Ocean, and equatorial Africa (e.g., the Congo Basin) and a downward branch over the western boundary of the Indian Ocean around 45°–60°E or associated with the formation of the Atlantic cold tongue (Hua *et al.* 2016; Cook and Vizy 2016). Hua *et al.* (2016) also found that an increase in subsidence across the western edge of Indian Ocean and a decrease in convection over the Congo Basin led to the reduction in moisture transport and rainfall across Central Equatorial Africa (Pokam *et al.* 2011). Dezfuli *et al.* (2015) highlighted a similar regional zonal atmospheric circulation linked to daily precipitation

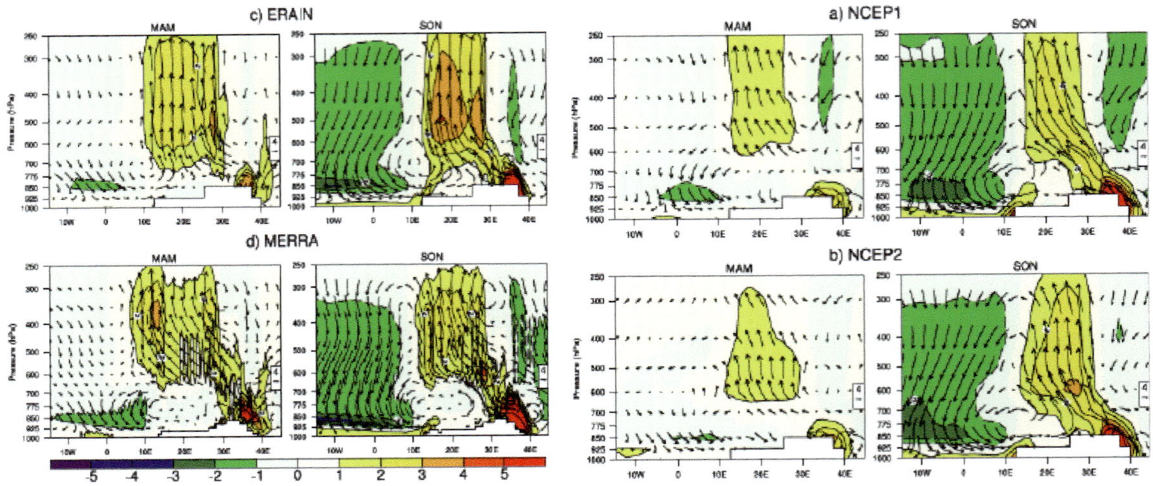

Fig. 3. Diabatic heating (K; day^{-1}) and divergent zonal circulation in longitude pressure section averaged over 0°–10° S (Pokam *et al.* 2014).

events that affected eastern and western equatorial Africa.

3. Diurnal Cycle

Jackson *et al.* (2009) found that there is a morning minimum in mesoscale systems (MCS) number and a late afternoon to early evening maximum, and the maximum in MCS frequency occurs around 1500–1800 LT over Central Africa. These pronounced diurnal variations have been found over the highlands of Cameroon and along the Biafra bay (Fig. 4). The diurnal cycle of rainfall and convective activity is one fundamental mode of variability in the global tropics. Over Central Africa the amplitude and phase of the diurnal cycle of convective systems and the intensity of their updrafts are generally linked with the energy budget through the radiation budget (Vondou *et al.* 2010; Tanessong *et al.* 2014; Embolo *et al.* 2015; Igri *et al.* 2015). Properties of diurnal evolution of convection and precipitation are closely associated with the land-sea distribution and topography, and depend on the local dynamics and thermodynamics of each region. Several studies have highlighted the importance of the diurnal cycle over Central Africa (e.g., Abdou *et al.* 2010; Jackson *et al.* 2009; Laing *et al.* 2011; Gang *et al.* 2016). Comprehensive studies by Vondou *et al.* (2010), Laing *et al.* (2011), and Pokam *et al.* (2014) provided a very detailed view of the diurnal cycle over Central Africa (e.g., Fig. 4). Infrared radiation and satellite rainfall data have been used to examine the diurnal variations in convective activity and rainfall over the Central African region (Laing *et al.* 2011).

4. From Synoptic to Intraseasonal Timescales

Some work has already been completed on rainfall variability through studies of dominant synoptic timescales (Efon *et al.* 2016) and MCSs (Laing *et al.* 2011; Jackson *et al.* 2009). Convectively coupled Kelvin waves (CCKWs) have been shown to account for a substantial amount of organization of tropical rainfall. Due to the importance of Kelvin waves to tropical weather, their spatial structure and propagation have been extensively revealed from observations and the wave theory (Matsuno 1966; Takayabu 1994; Wheeler and Kiladis 1999). The observed CCKWs propagate at a slower phase speed than

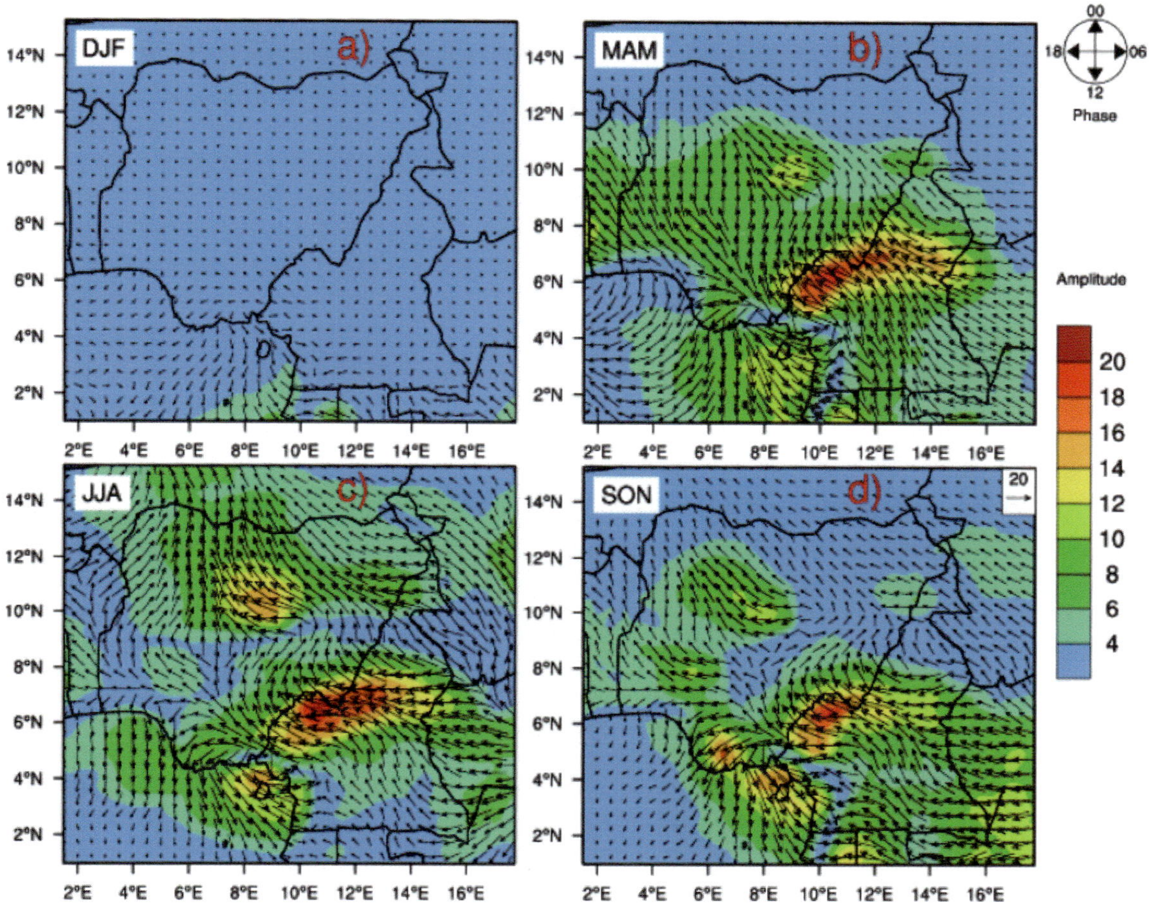

Fig. 4. Seasonal distribution of amplitude and phase of the diurnal cycle of convection over west Central Africa (Vondou *et al.* 2010).

dry Kelvin waves as derived from linear theory (Matsuno 1966). Straub and Kiladis (2002) found that CCKWs typically have a zonal wavelength between 3,000 and 6,000 km and a phase speed between 15 and 25 m/s.

Indeed, the equatorial location of the tropical rain belt may efficiently reinforce Kelvin waves that are theoretically associated with convective heating perturbations centered at the equator. Some studies have been carried out over Central Africa to study synoptic to intraseasonal variability, which is *a priori* strong when the tropical rain belt is close to or over the equator, favoring interactions between equatorial atmospheric dynamics and convection. Figure 5 from Kamsu-Tamo *et al.* (2014) shows westward-propagating MCS-periodicities where precipitation is averaged between 5°W and 5°E. While African easterly waves have been identified as fundamental synoptic weather systems in the West and Central African summer monsoon (see Lenouo *et al.* 2008), other atmospheric systems exist on the synoptic timescale in the tropics.

Nguyen and Duvel (2008) documented large oscillations of convective activity in periods of 3–6 days over equatorial Africa. In March and April, when the tropical rain belt migrates

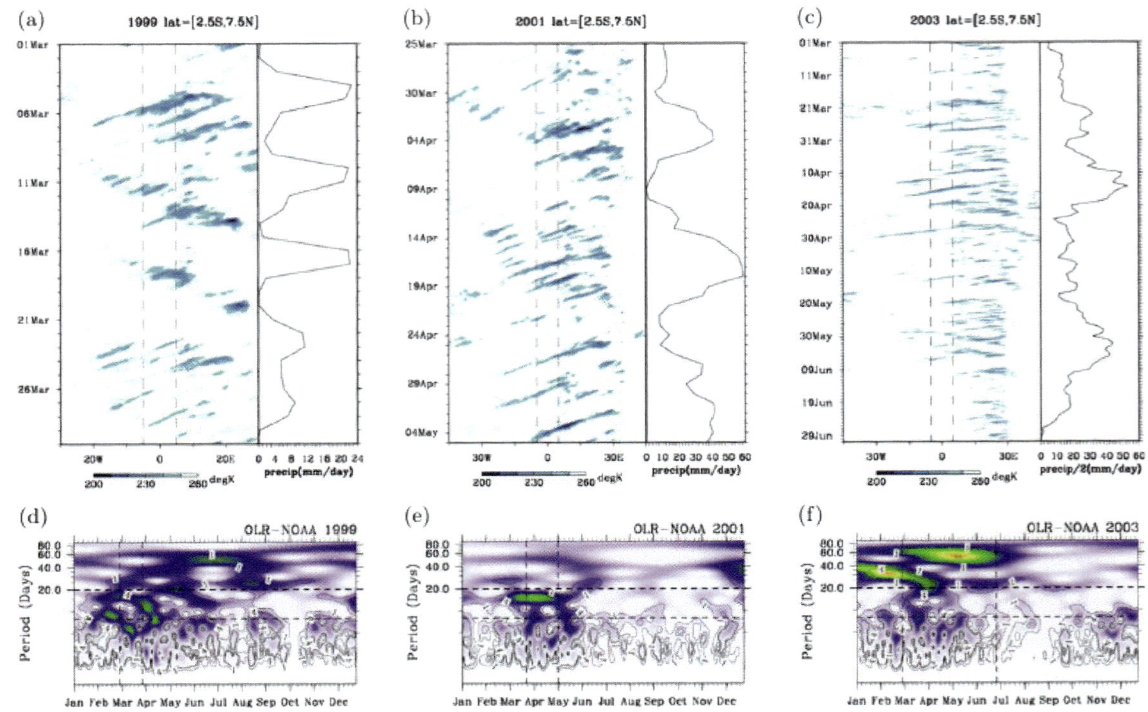

Fig. 5. Examples of characteristic (a) synoptic, (b) medium-range, and (c) intraseasonal time sequences of convection and precipitation over the Atlantic–Africa domain. The time–longitude cross sections show the CLAUS Tb values (°K) along 45°W–45°E and averaged over 2.5°S–7.5°N; it highlights westward-propagating convective systems. Curves represent the TRMM rainfall values (mm/day) averaged over 2.5°S–7.5°N/5°W–5°E. The 5°W–5°E domain is displayed on the cross sections. Corresponding wavelet diagrams of an NOAA OLR index computed as an average over 2.5°S–7.5°N/5°W–5°E are shown below (d)–(f). Three time sequences are presented: (a) the sequence from 1 to 30 March 1999, where the synoptic timescale signal is well defined and daily rainfall evolution is smoothed by a moving sum over 2 days; (b) the sequence from 25 March to 5 May 2001, where the medium-range timescale signal is well defined and daily rainfall evolution is smoothed by a moving sum over 5 days; and (c) the sequence from 1 March to 1 July 2003, where the intraseasonal timescale signal is well defined and daily rainfall evolution is smoothed by a moving sum over 10 days (Kamsu-Tamo *et al.* 2014).

northward and crosses the equator, this periodic behavior is highly pronounced with a marked peak at 5–6 days. Kamsu-Tamo *et al.* (2014) detected and identified the main modes of variability in terms of convection and precipitation over West and Central Africa during spring and fall (Fig. 5). A reference regional index based on the first EOF component of 2–90-day filtered convection and precipitation data was defined, and deseasonalized atmospheric fields were regressed onto this index to get a first view of the dominant atmospheric patterns and dynamics. Within the 2–90-day band, three main bands, 2–8, 8–22, and 20–90 day, have been highlighted. Within these three bands, it has been shown that eastward-propagating signals from CCKWs and from Madden–Julian oscillation (MJO) are highly dominant. So, while individual MCSs are moving westward, their activities are highly modulated by subregional- and regional-scale envelopes moving to the east. Zebaze *et al.* (2017) examined the moisture transport variability and its interaction with rainfall and CCKW events over Equatorial Africa using 1979–2010 ERA-interim reanalysis data, precipitation from Global Precipitation

Climatology Project (GPCP), and outgoing longwave radiation (OLR), and showed that the CCKW event's influence in the Congo Basin varies within each rainy season as the tropical rain belt moves through the region.

Tchakoutio *et al.* (2012, 2014) explored the MJO signal over Central Africa in more detail. They showed that three leading modes explain about 45% of the total intraseasonal variability corresponding to a peak around 45–50 days, with high loadings over Northern Congo, Southern Ethiopia, and South-western Tanzania. They also revealed less interannual modulation of intraseasonal oscillations for the Congo mode, while the Ethiopian and Tanzanian modes exhibited strong interannual variations. Strong MJO activity is observed during La Niña years or during ENSO-neutral years, while weak or absent MJO activity is typically associated with strong El Niño episodes. The mean MJO strength and fluctuations were highest during 1981–1990, corresponding to the highest fluctuations of the ENSO signal.

5. Interannual Timescales

Using a combination of rainfall observations from meteorological stations from the Climate Research Unit (CRU; 1950–2009) and satellite observations of the Tropical Rainfall Measuring Mission (TRMM; 1998–2010), Asefi-Najafabady and Saatchi (2013) showed that Central Africa experienced strong negative water deficit (WD) anomalies over the last decade, particularly in 2005, 2006, and 2007. The Global Precipitation Climatology Project (GPCP) monthly precipitation also exhibited a significant negative trend during 1979–2004 over southern tropical Africa from the Congo River Basin to the east coast. This trend appears as a more than 20% drop beginning in 1992 in the 6-year and 9-year averages of the areal mean GPCP satellite–gauge precipitation, the magnitude of which is largely determined by the gauge analyses (Yin and Gruber 2010). Zhou *et al.* (2014) also found a negative trend in rainfall after the 1960s (Fig. 6), which was markedly deficient from the mid-1980s onward. This total amount rainfall deficit is associated with a reduction of some regional statistics such as occurrence and intensity of heavy precipitation events. This negative trend has also been described by Diem *et al.* (2014) over the eastern edge of central equatorial Africa. The relationship of rainfall variability to sea surface temperature (SST) is highly seasonal and regionally specific (Dezfuli and

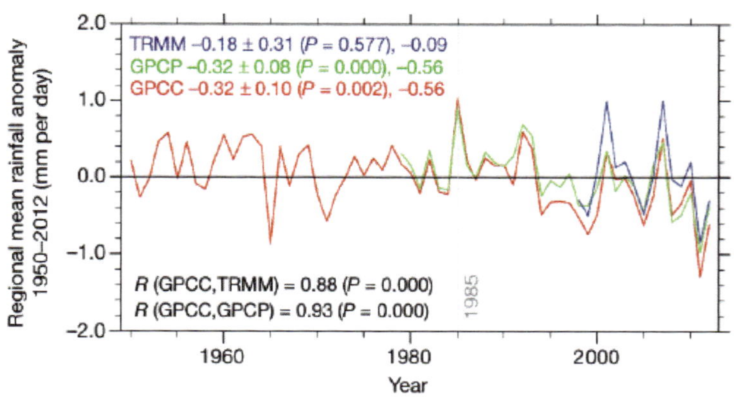

Fig. 6. Interannual variability of April–May–June regional mean anomalies from GPCC, GPCP, and TRMM datasets. The linear trend (one standard deviation) and its significance level P for the period 1985–2012, the decadal difference (2003–2012 minus 1985–1994 averages) for GPCC and GPCP, and the 3-year difference (2009–2012 minus 1998–2001 averages) for TRMM are shown. The correlation coefficient R between GPCC and TRMM and between GPCC and GPCP is also shown (Zhou *et al.* 2014).

Nicholson 2013). This long-term drying trend results primarily from SST variations over the Indo-Pacific associated with enhanced and westward extended tropical Walker circulation (Hua *et al.* 2016).

The Pacific Ocean also contributes to rainfall variability over Central Africa through the ENSO phenomenon with noticeable impact early in the year over western Central Africa both during boreal spring and boreal autumn (Dezfuli and Nicholson 2013). ENSO events are also associated with a prolonged and strengthened summer (July–August) dry season. The Atlantic Ocean through SSTs along the Gulf of Guinea and Benguela current contributes to interannual variability by directly influencing the location and/or intensity of maximum rainfall, in particular over western Central Africa, whereas the Indian Ocean's influence is prominent over eastern Central Africa (Dezfuli and Nicholson 2013). SSTs also modulate the rainfall variability via their effects on moisture transport, which can be examined by the changes in moisture flux (Pokam *et al.* 2011; Hua *et al.* 2016). Decreased convection over the Congo Basin leads to reduced moisture transport and thus reduced rainfall (Fig. 6). These results reinforce the notion that tropical SSTs have great impact on rainfall over equatorial Africa and highlight the need to further distinguish the contribution of SST changes, due to natural variability and anthropogenic forcing, to the drought (Fig. 7, Hua *et al.* 2016).

6. Climate Simulation and Projection over Central Africa

Recently, a lot of effort has been put into performing and evaluating climate simulations over Africa, and especially regional climate simulations over Central Africa, due to the CORDEX-Africa Climate Systems Analysis Group.[1] Only the main results from a limited number of papers are presented here. Hernández-Díaz *et al.* (2013) have used the new fifth-generation Regional Climate Model (CRCM5) driven by ERA reanalyses for the period 1984–2008 over the African continent following the CORDEX experimental

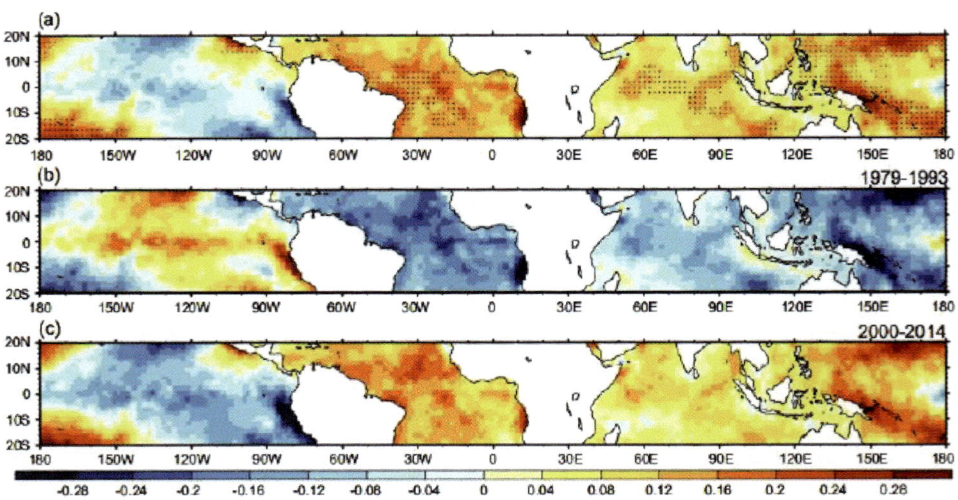

Fig. 7. (a) Linear trend of AMJ SST (°C decade^{-1}) for 1979–2014. (b) AMJ SST anomalies (relative to 1979–2014) for 1979–1993 and (c) for 2000–2014 (Hua *et al.* 2016).

[1]See www.csag.uct.ac.za/cordex-africa/.

protocol. They found that the model succeeds in reproducing the main features of the geographical distribution and seasonal cycle of temperature and precipitation, the diurnal cycle of precipitation, and the West African Monsoon (WAM). In the African regions near the equator, they also found that the model successfully reproduces the double peak of rainfall due to the double passage of the tropical rain belt, although it better simulates the magnitude and timing of the second peak of precipitation. Rainfall peaks are present over the highland regions of Gabon. The CRCM5 simulates rainfall peaks of 8–12 mm/day in the Guinea coastal region.

The Met Office Hadley Centre's PRECIS regional climate modeling system has been used to generate a five-member ensemble of climate projections for Africa over the 50-km resolution Coordinated Regional climate Downscaling Experiment-Africa domain by Buontempo *et al.* (2014). The ensemble comprises the downscaling of a subset of the Hadley Centre's perturbed physics global climate model (GCM) ensemble chosen to exclude ensemble members unable to represent the African climate realistically and then to capture the spread in outcomes from the projections of the remaining models. They found that the regional climate model (RCM) ensemble captures the annual cycle of temperatures well both for Africa as a whole and the subregions. Present-day simulations of the RCM ensemble are similar to each other compared to those of the driving GCM ensemble, which indicates that their climatologies are influenced significantly by the RCM formulation and less so by their driving GCMs. Consistent with this, these authors found that the spread and magnitudes of the large-scale responses of the RCMs are often different from the driving GCMs and arguably more credible given the improved performance of the RCM.

Washington *et al.* (2013) suggested an approach, via a targeted field campaign, for generating useful climate information with which to confront rainfall products and climate models. They found that the Congo Basin is one of the three key convective regions on the planet which, during the transition seasons, dominates global tropical rainfall. There is little agreement as to the distribution and quantity of rainfall across the basin with datasets differing by an order of magnitude in some seasons. Further analysis shows that over some regions the results of the climate change assessment clearly depend on the size of the analyzed ensemble (Haensler *et al.* 2013). We can note that Nicholson and Klotter (2018) described three new rainfall datasets developed for equatorial Africa. The development relies on acquisition of recent gauge data from the relevant countries and statistical methods to fill in gaps in coverage. Two of the three datasets are gridded with spatial resolutions of 2.5° and 5.0°, and the third is regionally aggregated and based purely on gauge data. Their work is based on a total of 1826 gauge records in the analysis sector, of which only several hundred have been operational in recent years.

Central Africa is also one of the areas most vulnerable to climate change due to the increase in extreme weather events. Variability and change in precipitation are mainly invoked as sources of uncertainties. This susceptibility increases while taking into account some ecosystems, water supply, foods production, increase of population, and socio-political security (Busty *et al.* 2014). For instance, using 35 years (1982–2016) of infrared brightness temperature, Raghavendra *et al.* (2018) noted changes in the structure of thunderstorms characterized by areal extent and intensity increase. Cretat *et al.* (2014) showed that RCM simulations of intense rainfall events performed at least as well as the more accurate GCMs over Africa. They accurately captured the spatial and temporal characteristics of intense events while they tended to overestimate their number and underestimate their intensity. Climate extremes over Central Africa will become more frequent, which suggests increased flood risk. Nowadays, many countries are damaged from flooding almost every season. Analyzing 20 GCMs, Fotso *et al.* (2018) found that a tendency toward more extreme rainfall events can be expected throughout the 21st century. Multi-model ensemble projections (Pokam *et al.* 2018) show statistically robust decreasing total rainfall while consecutive dry days are projected to increase significantly, with consistent changes of a maximum of 5 days of rainfall together with variations in the rainfall amounts during the wettest days. Investigating the potential effects of 1.5°, 2° of global warming on the regional climates within Africa, Nikulin *et al.* (2018), Pokam *et al.* (2018), and Diedhiou *et al.* (2018) found that there is a consistent increase in daily precipitation intensity of wet days over Central Africa with high consensus among models.

Acknowledgments

The authors thank WMO for providing the funding to the first author for his participation in the Sixth WMO International Workshop on Monsoons, 13–17 November 2017, Singapore.

References

Asefi-Najafabady, S. and S. Saatchi, 2013: Response of African humid tropical forests to recent rainfall anomalies. *Phil. Trans. R Soc. B*, **368**, 20120306.

Abdou, K., D. J. Parker, B. Brooks, N. Kalthoff, and T. Lebel, 2010: The diurnal cycle of lower boundary-layer wind in the West African monsoon. *Q. J. Roy. Meteor. Soc.*, **136**, 66–76, doi:10.1002/qj.536.

Balas, N., S. E. Nicholson, and D. Klotter, 2007: The relationship of rainfall variability in west central Africa to sea-surface temperature fluctuations. *Int. J. Climatol.*, **27**, 1335–1349.

Buontempo, C., C. Mathison, R. Jones, K. Williams, C. Wang, and C. McSweeney, 2014: An ensemble climate projection for Africa. *Clim. Dyn.*, doi:10.1007/s00382-014-2286-2.

Busby, J. W., K. H. Cook, E. K. Vizy, T. G. Smith, and M. Bekalo, 2014: Identifying hot spots of security vulnerability associated with climate change in Africa. *Climatic Change*, **124**, 717–731.

Cook, K. H. and E. K. Vizy, 2016: The Congo Basin walker circulation: Dynamics and connections to precipitation, *Clim. Dynam.*, **47**, 1–21.

Cretat, J., E. K. Vizy, K. H. Cook, 2014: How well daily intense rainfall events captured by current climate models over Africa? *Clim. Dynam.*, **42**, 2691–2711, doi:10.1007/s00382-013-1796-7.

Dezfuli, A. K. and S. E. Nicholson, 2013: The relationship of rainfall variability in western equatorial Africa to the tropical oceans and atmospheric circulation. Part I: The boreal spring. *J. Clim.*, **26**, 66–84.

Dezfuli, A. K., B. F. Zaitchik, and A. Gnanadesikan, 2015: Regional atmopsheric circulation and rainfall variability in south equatorial Africa. *J. Clim.*, **28**, 809–818, doi:10.1175/JCLI-D-14-00333.1.

Diem, J. E., S. J. Ryan, J. Hartter, and M. W. Palace, 2014: Satellite-based rainfall data reveal a recent drying trend in central equatorial Africa. *Clim. Change*, **126**, 263–272, doi:10.1007/s10584-014-1217-x.

Diedhiou, A., A. Bichet, R. Wartenburger, S. I. Seneviratne, D. P. Rowell, M. B. Sylla, I. Diallo, S. Todzo, N. E. Toure, M. Camara, B. Ngounou Ngatchah, N. A. Kane, L. Tall, and F. Affholder, 2018: Changes in climate extremes over West and Central Africa at 1.5°C and 26°C global warming. *Environ. Res. Lett.*, **13**, https://doi.org/10.1088/1748-9326/aac3e5.

Efon, E., A. Lenouo, D. Monkam, and D. Manatsa, 2016: Cloud properties during active and break spells of the West African Summer Monsoon from CloudSat-CALIPSO measurements. *J. Atmos. Solar-Terrest. Phys.*, **145**, 1–11, doi:10.1016/j.jastp.2016.04.001.

Embolo Embolo, G. B., A. Lenouo, A. T. Nzeukou, D. A. Vondou, and F. M. Kamga, 2017: On convection and static stability during the AMMA SOP3 campaign. *Theor. Appl. Climatol.*, **127**, 197–211, doi:10.1007/s00704-015-1627-3.

Fotso-Nguemo, T. C., R. Chamani, Z. D. Yepdo, D. Sonkoué, C. N. Matsaguim, D. A. Vondou, and R. S. Tanessong, 2018: Projected trends of extreme rainfall events from CMIP5 models over Central. *Africa Atmos. Sci. Lett.*, **19**, e803.

Zhang, G., K. H. Cook, and E. K. Vizy, 2016: The diurnal cycle of warm season rainfall over West Africa. Part I: Observational analysis. *J. Clim.*, **29**(23), 8423–8437.

Haensler, A., F. Saeed, and D. Jacob, 2013: Assessing the robustness of projected precipitation changes over central Africa on the basis of a multitude of global and regional climate projections. *Clim. Change*, **121**, 349, doi:10.1007/s10584-013-0863-8.

Hua, W., L. Zhou, H. Chen, S. E. Nicholson, A. Raghavendra, and Y. Jiang, 2016: Possible causes of the Central Equatorial African long-term drought. *Environ. Res. Lett.*, **11**, 124002, doi:10.1088/1748-9326/11/12/124002.

Igri, P. M., R. S. Tanessong, and D. A. Vondou *et al.*, 2015: Added-value of 3DVAR data assimilation in the simulation of heavy rainfall events over West and Central Africa. *Pure Appl. Geophys.*, **172**, 2751, doi:10.1007/s00024-015-1052-7.

Jackson, B., S. E. Nicholson, and D. Klotter, 2009: Mesoscale convective systems over western equatorial Africa and their relationship to large-scale circulation. *Mon. Weather Rev.*, **137**, 1272–1294, doi:10.1175/2008MWR2525.1.

James, R., R. Washington, and D. P. Rowell, 2013: Implications of global warming for the climate of African rainforests. *Philosophi. Trans. Roy. Soc. B: Biolo. Sci.*, **368**(1625), 20120298, doi:10.1098/rstb.2012.0298.

Kamsu Tamo, P. H., S. Janicot, D. Monkam, and A. Lenouo, 2014: Convection activity over the Guinean coast and Central Africa during northern spring from synoptic to intra-seasonal timescales. *Clim. Dynam.*, doi:10.1007/s00382-014-2111-y.

Laing, A. G., R. E. Carbone, and V. Levizzani, 2011: Cycles and propagation of deep convection over equatorial Africa. *Mon. Weather Rev.*, **139**, 2832–2853, doi:10.1175/2011MWR3500.1.

Laprise, R., L. Hernández-Díaz, K. Tete, L. Sushama, L. Šeparović, A. Martynov, K. Winger, and M. Valin, 2013: Climate projections over CORDEX Africa domain using the fifth-generation Canadian Regional Climate Model (CRCM5). *Clim. Dynam.*, **41**, 3219–3246.

Lenouo, A. and F. M. Kamga, 2008: Sensitivity of African Easterly Waves to boundary layer condition. *Ann. Geophys.*, **26**, 1355–1363, doi:10.5194/angeo-26-1355-2008 — ISSN: 0992-7689.

Matsuno, T. 1966: Quasi-geostrosphic motion in the equatorial area. *J. Meteorol. Soc. Jpn.*, **44**, 25–42.

Nguyen, H. and J.-P. Duvel, 2008: Synoptic wave perturbations and convective systems over equatorial Africa. *J. Clim.*, **21**, 6372–6388, doi:10.1175/2008JCLI2409.1.

Nicholson, S. E. and J. P. Grist, 2003: The seasonal evolution of the atmospheric circulation over West Africa and equatorial Africa. *J. Clim.*, **16**, 1013–1030.

Nicholson, S. E. and A. K. Dezfuli, 2013: The relationship of rainfall variability in western equatorial Africa to the tropical oceans and atmospheric circulation. Part II: The boreal autumn. *J. Clim.*, **26**, 45–65.

Nicholson, S. E., 2018: The ITCZ and the seasonal cycle over equatorial Africa. *Bull. Amer. Meteor. Soc.*, 337–348, doi:10.1175/BAMS-D-16-0287.1.

Nicholson, S. E. and D. Klotter, 2018: New rainfall datasets for the Congo Basin and surrounding regions. *J. Hydrometeorol.*, **19**(10), https://doi.org/10.1175/JHM-D-18-0015.1.

Nikulin, G., C. Lennard, A. Dosio, E. Kjellström, Y. Chen, A. Hänsler, M. Kupiainen, R. Laprise, L. Mariotti, M. C. Fox, E. van Meijgaard, H.-J. Panitz, J. F. Scinocca, and S. Somot, 2018: The effects of 1.5 and 2 degrees of global warming on Africa in the CORDEX ensemble. *Environ. Res. Lett.*, https://doi.org/10.1088/1748-9326/aab1b1.

Pokam, M. W., T. L. A. Djiotang, and K. F. Mkankam, 2011: Atmospheric watervapor transport and recycling in Equatorial Central Africathrough NCEP/NCAR reanalysis data. *Clim. Dynam.*, **38**, 1715–1729.

Pokam, M. W., C. L. Bain, R. S. Chadwick, R. Graham, D. J. Sonwa, and K. F. Mkankam, 2014: Identification of processes driving low-level westerlies in west equatorial Africa. *J. Clim.*, **27**, 4245–4262.

Pokam, M. W., G. N. Longandjo, W. Moufouma-Okia, J. P. Bell, R. James, D. A. Vondou, A. Haensler, T. C. Fotso-Nguemo, G. M. Guenang, A. L. Djiotang Tchochou, P. H. Kamsu-Tamo, R. R. Takong, G. Nikulin, C. J. Lennard, and A. Dosio, 2018: Consequences of 1.5°C and 2°C global warming levels for temperature and precipitation changes over Central Africa. *Environ. Res. Lett.*, **13**, https://doi.org/10.1088/1748-9326/aab048.

Raghavendra, A., L. Zhou, Y. Jiang, and W. Hua, 2018: Increasing extent and intensity of thunderstorms observed over the Congo Basin from 1982 to 2016. *Atmos. Res.*, **213**, 7–26.

Straub, K. H. and G. N. Kiladis, 2002: Observations of convectively coupled Kelvin waves in the eastern Pacific ITCZ. *J. Atmos. Sci.*, **59**, 30–53.

Takayabu, Y. N., 1994: Large scale cloud disturbances associated with equatorial waves: Part I spectral feature of the cloud disturbances. *J. Meteorol. Soc. Jpn.*, **72**, 433–448.

Tanessong, R. S., P. M. Igri, D. A. Vondou, and K. F. Mkankam, 2014: Evaluation of probabilistic precipitation forecast determined from WRF forecasted amounts. *Theor. Appl. Climatol.*, **116**, 649, doi:10.1007/s00704-013-0965-2.

Tchakoutio Sandjon, A., A. Nzeukou, and C. Tchawoua, 2012: Intraseasonal atmospheric variability and its interannual modulation in Central Africa. *Meteorol. Atmos. Phys.*, **117**, 167–179, doi:10.1007/s00703-012-0196-6.

Tchakoutio Sandjon, A., A. Nzeukou, S. B. Tchawoua, and T. C. Siddi, 2014: Comparing the patterns of 20–70 days intraseasonal oscillations over Central Africa during the last three decades. *Theor. Appl. Climatol.*, **118**, 319–329, doi:10.1007/s00704-013-1063-1.

Vondou, D. A., A. Nzeukou, A. Lenouo, and F. Mkankam Kamga, 2010: Seasonal variations in the diurnal patterns of convection in Cameroon-Nigeria and their neighboring areas. *Atmos. Sci. Lett.*, **11**, 290–300, doi:10.1002/asl.297.

Wang, P. X., B. Wang, H. Cheng, J. Fasullo, Z. Guo, T. Kiefer, and Z. Y. Liu, 2017: The global monsoon across time scales: Mechanisms and outstanding issues. *Earth-Sci. Rev.*, doi:10.1016/j.earscirev.2017.07.006.

Washington, R., R. James, H. Pearce, W. M. Pokam, and W. Moufouma-Okia, 2013: Congo Basin rainfall climatology: Can we believe the climate models? *Philos. Trans. R. Soc. London, Ser. B*, **368**(1625), 20120296.

Wheeler, M. and G. N. Kiladis, 1999: Convectively coupled equatorial waves: Analysis of clouds and temperature in the wavenumber–frequency domain. *J. Atmos. Sci.*, **56**, 374–399.

Yin and A. Gruber, 2010: Validation of the abrupt change in GPCP precipitation in the Congo River Basin. *Int. J. Climatol.*, **30**, 110–119.

Zebaze, S., A. Lenouo, C. Tchawoua, A. T. Gaye, and F. M. Kamga, 2017: Interaction between moisture transport and Kelvin waves events over Equatorial Africa through ERA-interim. *Atmos. Sci. Let.*, **18**, 300–306.

Zhou *et al.*, 2014: Widespread decline of Congo rainforest greenness in the past decade. *Nature*, **508**, 86–104, doi:10.1038/nature13265.

Chapter 7

Prediction, Predictability, and Multi-Decadal Variability of the South Asian Monsoon

B. N. Goswami*,‡ and Neena Joseph Mani†

*Cotton University, Guwahati, India
†Indian Institute of Science Education and Research, Pune, India
‡bhupengoswami100@gmail.com

Advances made in seasonal prediction and predictability of the south Asian monsoon (SAM) over the past decade is briefly reviewed. To achieve potential skill of prediction of SAM rainfall, a major targeted effort to improve biases of simulation of precipitation over the Asian monsoon region in a coupled model is described. A series of retrospective forecasts by improved versions of the model indicate that skill of SAM rainfall beyond the potential predictability limit (PPL) estimated from signal to noise (S/N such as ANOVA) is achievable, indicating that S/N method is inadequate in estimating the PPL for SAM rainfall. A new model based method of estimating PPL shows that the SAM rainfall is much more predictable than thought previously. Scientific basis for this enhanced predictability of the SAM rainfall is linked to the emergence of the extratropical SSTs associated with the AMO (in addition to the tropical SSTs associated with the ENSO) as a major contributor to the predictability and the recognition that the statistics of the high frequency fluctuations (e.g. variance of the synoptic disturbances) are tied to predictable drivers and hence partly predictable. We shall also review advances made in understanding and modeling of the SAM multi-decadal variability as it not only modulates the skill of prediction and predictability of the south Asian Monsoon but also makes the human induced green house influence on the SAM difficult to be isolated.

1. Introduction

The All India Summer Monsoon rainfall (ISMR), a leading indicator of the South Asian Monsoon (SAM), significantly influences the Gross Domestic Product (GDP) and economy of the country (Gadgil and Gadgil 2006). As a result, for nearly 100 years, attempts have been made to predict it one season in advance. For want of a reliable dynamical framework for seasonal prediction of ISMR, until recently most of these efforts have been based on statistical models using a variety of regression techniques (Walker, 1924; Rajeevan *et al.* 2007; Wang *et al.* 2015). Statistical models, however, build artificial skill (Delsol and Shukla, 2009) and fail to perform satisfactorily in real-time forecasts. In particular, statistical model predictions are biased toward the "mean" and fail to predict the extremes (Gadgil *et al.* 2005). Therefore, an ocean-atmosphere coupled climate model is essential for seasonal prediction of ISMR (Wang *et al.* 2005). While the skill of seasonal prediction of the ISMR by climate models has improved from an older generation of models (Krishna Kumar *et al.* 2005) to a newer generation of models (Rajeevan *et al.* 2012), it still remains "low" (correlation between observation and predictions being $R \sim 0.4$) and significantly below the conventional potential predictability limit (PPL). Unfortunately, the estimate of PPL ($R \sim 0.65$) for ISMR by "perfect" model correlations (Kumar *et al.* 2005; Rajeevan *et al.* 2012) and by analysis of variance (ANOVA, Kang and Shukla 2006) is itself low. Even if the ISMR prediction skill reaches close to the PPL, it will be

The Multiscale Global Monsoon System
Edited by C. P. Chang et al.

barely useful, demonstrating a grim picture of the future of ISMR prediction. In addition to continuing to work on improving the coupled models, we need to ask if there is a scientific basis for achieving model skill beyond the conventional estimate of PPL.

The scenario has changed significantly during the past decade with climate models beginning to make useful seasonal prediction of ISMR one season in advance. While the seasonal mean ISMR is useful for the policymakers, it is only marginally useful to the farmers as the ISMR has significant temporal variability within the season. Prominent among them is the monsoon intraseasonal oscillation (MISO) with characteristic active and break spells. Therefore, an effective monsoon prediction system should have, in addition to the seasonal prediction of ISMR, extended range prediction of the MISOs at more frequent intervals (e.g., every 5 days) throughout the season. Thus, skillful prediction of the active and break spells 2–3 weeks in advance is highly desirable, and fortunately notable advances have been made in this direction during the past couple of decades. However, we shall refrain from reviewing the advances made in MISO predictions and extended range prediction of ISMR as this is being done in another chapter in the book. We shall confine ourselves to reviewing some of the recent advances made in seasonal prediction of ISMR by climate models and understanding the limit of the predictability.

The scientific basis for seasonal prediction of ISM was set by the pioneering study of Charney and Shukla (1981) and Shukla (1981) with the long-range memory of the climate coming from its association with slowly varying predictable climate modes like the El Ninõ and Southern Oscillation (ENSO). Later, it was recognized that the actual limit of predictability arises from relative contributions to the seasonal mean by predictable modes (e.g., ENSO) and by the "climate noise" or "internal" interannual variability arising from contributions to the seasonal mean from all high-frequency sub-seasonal fluctuations (synoptic and super-synoptic). Various estimates of "internal" variability (Goswami 1998; Goswami and Ajaya Mohan; 2001) indicate that the predictability of ISM is limited by the "internal" variability to explaining 40–50% of interannual variability of the ISM. This potential limit on predictability is supported by perfect model experiments using a variety of models (Kumar *et al.* 2005; Rajeevan *et al.* 2012) showing that the potential limit of achievable correlation between prediction and observation of ISM rainfall is ~0.65. Is this limit sacrosanct? With improvement of models could the actual skill of prediction of seasonal mean monsoon go beyond this limit? Also, ENSO has remained the single dominant predictable climate signal that influences the ISM. However, the ENSO–monsoon relationship goes through its ups and downs, at its peak explaining at most 40% of ISM variability and at its worst only about 10% of its variability (Krishnamurthy and Goswami 2000). In order to improve ISM prediction skill, it is very important to identify other predictable signals that influence ISM and quantify their contribution. Advances in such questions in ISM predictability will also be addressed.

The multi-decadal variability of the dominant drivers of ISMR (e.g., the ENSO) has the potential to lead to multi-decadal variability of the seasonal prediction skill as well its predictability (Krishnamurthy and Goswami 2000). There are enough indications that the ISMR has a significant multi-decadal variability (Kripalani and Kulkarni 1997). Quantification (e.g., period and amplitude) of this low-frequency variability of the ISMR is critical not only for seasonal prediction and predictability but also for disentangling the forced variability due to global warming from the natural variability of ISMR. Toward this goal, reconstruction of ISMR during the past millennium indicates the existence of such a natural multi-decadal mode (MDM) of variability (50–70-year period) of substantial amplitude (Goswami *et al.* 2015; Sinha *et al.* 2015). For skillful seasonal prediction of ISMR, a coupled model must be able to capture the phases of the MDM correctly. Understanding the physical mechanism producing the MDM of ISMR is, therefore, a prerequisite. Interactions between three processes are proposed to be at the heart of the quasi-oscillatory mode of this variability. In Section 4, we present a mechanistic model to explain the quasi-periodic nature of the ISMR MDM, consisting of *local* unstable ocean-atmosphere interactions involving the Indian ocean and the Indian land mass, *remote* ocean-atmosphere interactions involving SAM and the Pacific Ocean, and teleconnections with other multi-decadal climate modes like the Atlantic multi-decadal oscillation (AMO).

2. Challenges of Simulating Mean Indian Summer Monsoon Climate

The ISM rainfall is a manifestation of the northward seasonal migration of the rainband or the ITCZ in the region and of the seasonal mean results in two zonal rainfall maxima, one over the continent and the other over the equatorial Indian Ocean, in addition to a meridional maximum west of the Western Ghats. Indian monsoon precipitation is convectively coupled with low-level circulation characterized by the low-level westerly jet and the upper-level circulation characterized by the monsoon easterly jet. As the monsoon rainy season is defined by the "onset" and the "withdrawal", they also represent important parameters to be simulated by models. The Indian monsoon system, in turn, is a result of coupled ocean-atmosphere interaction involving the Indian Ocean (Webster *et al.* 1998). Therefore, a coupled ocean-atmosphere model is required to simulate the mean monsoon annual cycle and its variability (Wang *et al.* 2005). Hence, a discussion on the performance of atmosphere-only GCMs in simulating the Indian monsoon (e.g., Gadgil and Sajani 1998; Kang *et al.* 2002) is not relevant here. An examination of simulation of Indian monsoon in the Coupled Model Intercomparison Project, phase 3 (CMIP3) and phase 5 (CMIP5), indicates that the majority of coupled models made very little progress in simulating the critical Indian monsoon features (Sperber *et al.* 2013) from CMIP3 to CMIP5, highlighting the challenging nature of the problem. The major bias in most models (as a result manifesting also in multi-model ensemble mean) is a large dry bias extending from north Bay of Bengal (BoB) across central India together with a large wet bias over equatorial Indian Ocean extending to the maritime continent and western Pacific. The pattern and magnitude of these major biases have not improved from CMIP3 to CMIP5 (Fig. 2 of Sperber *et al.* 2013) even though coupled models in CMIP5 have certainly shown improved simulation of other climate modes like ENSO and PDO. The main reason for this pattern of bias in simulating the mean summer monsoon is related to the fact that the seasonal migration of the ITCZ in most coupled models is limited to about 15°N against about 25°N in observation together with an overestimation of the strength of the oceanic rainband (oceanic ITCZ). A fraction of the dry bias over the Indian continent is also related to the underestimation of the high-frequency component of daily precipitation variance by most coupled models (Goswami and Goswami 2017), which may be amenable for improvement with high-resolution models.

The lack of visible progress in the simulation of the Indian monsoon by coupled GCMs over the past two decades against the backdrop of significant improvement of other global climate modes by the same models is likely to be due to the fact that the development of these models has failed to address and improve some issues generic to the Indian monsoon system. Under the "Monsoon Mission" of the Ministry of Earth Sciences, India, we targeted some Indian monsoon-specific issues and addressed model developments required to improve them using

Climate Forecast System version 2 (CFSv.2) as the base model. This targeted model development has led to pathbreaking improvements in simulating the Indian monsoon in the model. For example, in addition to the ENSO, Eurasian snow cover and depth are known to influence the Indian summer monsoon. The CFSv.2 had a strong positive snow depth bias in large part of Eurasia leading to delayed warming of the land, weak north–south temperature gradient, and a weak monsoon. The cause of this bias was identified to be associated with the land surface model having only two snow layers. The land surface model was modified to have five snow layers and a 1D version of the model was first tested offline and the 3D version was eventually added to the CFSv.2 and a long-climate simulation was conducted. Saha *et al.* (2017) demonstrated that the new land surface model removed the positive snow bias from Eurasia and led to notable improvement of simulation of the monsoon annual cycle.

The other Indian monsoon-related issue we targeted was to improve the simulation of the MISOs in the model because it has now been established that the MISOs represent a building block for the Indian monsoon contributing to the seasonal mean on the one hand while clustering the synoptic activity during its active spells on the other (Goswami *et al.* 2003, 2006; Goswami and Ajayamohan 2001). Therefore, in order to simulate the mean monsoon climate correctly, the coupled climate models must also simulate the space–time characteristics of MISOs with fidelity. It so happens that CMIP5 models fail to simulate these properties of MISOs correctly (Sabeeali *et al.* 2013). Most models underestimate the MISO variance and poorly simulate its northward propagation. It has been realized recently that these biases in simulating the MISOs by coupled models may be related to another major bias of the models, namely, overestimation of convective fraction of precipitation and underestimation of the stratiform fraction (Sabeerali *et al.* 2014). It has been found that in observations, the stratiform fraction modifies the vertical profile of heating and helps organize convection and precipitation at the MISO scale (Kumar *et al.* 2017). Inspired by these findings, an attempt has been made to improve the simulation of MISO in the CFSv.2 model by modifying the existing cloud microphysics scheme (Zhao and Carr 1997) to facilitate more ice- and mixed-phase hydrometeors. It has been demonstrated that such improvement of cloud microphysics leads to improvement of the simulation of the convective to stratiform rain fraction. This, in turn, leads to substantial improvement in simulation of space–time characteristics of MISO, together with a significant improvement in simulation of the mean annual cycle of the Indian monsoon (Hazra *et al.* 2017). The fact that better formulation of cloud microphysics parameterization is important for better simulation of the Indian monsoon is demonstrated in another study (Abhik *et al.* 2017) where they replaced the existing two-class cloud microphysics parameterization in the CFSv.2 model (Zhao and Carr 1997) with a more elaborate six-class WSM6 scheme and demonstrated improved simulation of MISO and the mean Indian monsoon. We consider these to be significant advances in modeling of the mean Indian monsoon.

3. Progress in Predictability and Seasonal Prediction of Indian Monsoon

Potential predictability studies of the Indian summer monsoon have indicated that roughly 40–50% of Indian summer monsoon rainfall variability is predictable (Goswami 1998; Webster *et al.* 1998; Kang and Shukla 2006; Krishna Kumar *et al.* 2005; Rajeevan *et al.* 2012). This is consistent with findings that the potentially achievable highest skill of models in terms of correlation between predictions and observations is limited to about 0.65 (Krishna Kumar *et al.* 2005; Rajeevan *et al.* 2012). However, the skill of climate models in predicting the seasonal mean ISM rainfall has remained significantly below this potentially achievable skill. The reasons behind this are partly the biases of the models in simulating the climatological annual cycle of the ISM rainfall (e.g., the dry bias over land and wet bias over the ocean as mentioned eariler) and partly the "internal" interannual variability simulated by models. As mentioned earlier about 50% of the observed interannual variability of the ISM is due to unpredictable "internal" variability (Goswami 1998). Many models tend to overestimate the "internal" variability, thereby underestimating the predictable component and hence the skill of the model. Therefore, it is imperative to understand the source of "internal" variability (in observations as well as in models) so that one could use that knowledge for improved prediction of the seasonal mean ISM. The predictable component, largely considered to come from the ENSO–monsoon relationship explains only about 25–30% of ISM variability. Where does the remaining 20% of predictable variability come from? There is a need to identify these non-ENSO predictable drivers of ISM in order to make sure that models simulate them correctly, so that we have improved skill in predicting the ISM. In this context, a recent study by Chattopadhyay *et al.* (2015) is notable in which it was shown that the extra-tropical north Pacific and north Atlantic SSTs have a significant influence on ISM rainfall when the ENSO is weak or in transition from El Ninõ to La Niña (or vice versa). In order to improve the skill of prediction of ISMR, coupled models must improve simulation of the multi-decadal ENSO and AMO in the models as the slow variations of the north Pacific and north Atlantic SST are related to these natural modes of climate variability. In another development, we find (Goswami and Goswami 2017) that a fraction of the dry bias in simulating monsoon precipitation over continental India is related to underestimation of high-frequency (HF) variance of daily precipitation simulated by the models. Therefore, in order to improve the dry bias over land, for a fixed set of physical parameterizations, a coupled model with a high-resolution atmosphere simulates reduced dry bias and improved seasonal mean through improved simulation of HF variance and its contribution to the mean. While the dry bias arising from the displacement of the ITCZ could not be improved simply by high resolution, the bias in the amplitude (intensity) is improved through improvement of HF variance.

According to conventional wisdom, we may not be able to achieve prediction skill of models beyond the PPL. The basis for this conclusion is the fundamental assumption that the "internal" variability limiting the predictability is unpredictable. If, however, the whole or even a fraction of the "internal" variability is predictable, it will change the equation fundamentally and we may be able to achieve model skill beyond the conventional PPL. We note that a host of studies (Goswami and Ajaya Mohan 2001; Goswami and Xavier 2005; Goswami *et al.* 2006) during the past 20 years have established that the MISOs contribute to a significant fraction of the "internal" interannual variability of ISM. If the statistics of MISOs are modulated by (or tied to) some slowly varying predictable climate modes, the fraction of "internal" variability contributed by MISOs will be predictable. Indeed, in a recent study, it was found that the statistics of the MISOs are modulated by the ENSO (Dwivedi *et al.* 2015), so that the fraction of the "internal" variability is not unpredictable. This finding opens up the possibility that with improved models and observations, we would be able to push the skill of monsoon prediction systems even beyond the PPL, making them extremely reliable and valuable to all stakeholders.

Based on this understanding, for improved skill of seasonal prediction of ISM, the coupled model must simulate the MISO with high fidelity, and if it has a high atmospheric resolution it would be the icing on the cake. It has been found that the standard version of the CFSv.2

model (with atmospheric resolution T126) simulates much improved space–time characteristics of MISO (Sharmila *et al.* 2013) compared to its previous version, namely, the CFSv.1. Therefore, under the Monsoon Mission, we have set up a seasonal prediction system with CFSv.2 with atmosphere at higher resolution (T382). Based on a large ensemble of reforecasts, the skill of the Indian Operational version of the seasonal prediction system is found to be notably improved compared to that of a prediction system with the standard CFSv.2 (Fig. 1). The skill of another seasonal prediction system with the standard CFSv.2 (T126) with improved physics (as mentioned in Section 2) has been tested with a series of reforecast experiments and found to be significantly better than even the high-resolution version of standard CFSv.2 (results not shown in Fig. 1 as they are under review for publication elsewhere). We believe that these are significant advances in operational seasonal prediction of ISM. It is also established that, while improved coupled ocean and atmosphere observations are needed for better skill of ISMR prediction, they are useless if the coupled models are not good enough to capture the leading drivers of ISMR and the teleconnections between important global modes and the ISMR. This leads to the conclusion that improving the coupled climate models through targeted model development is key and critical for improving the skill of ISMR prediction and provides a framework for further incremental improvement toward achieving the potential limit or beyond.

4. Air-Sea Interaction and the Multi-decadal Variability

The instrumental record of ISMR does indicate the existence of a multi-decadal variability

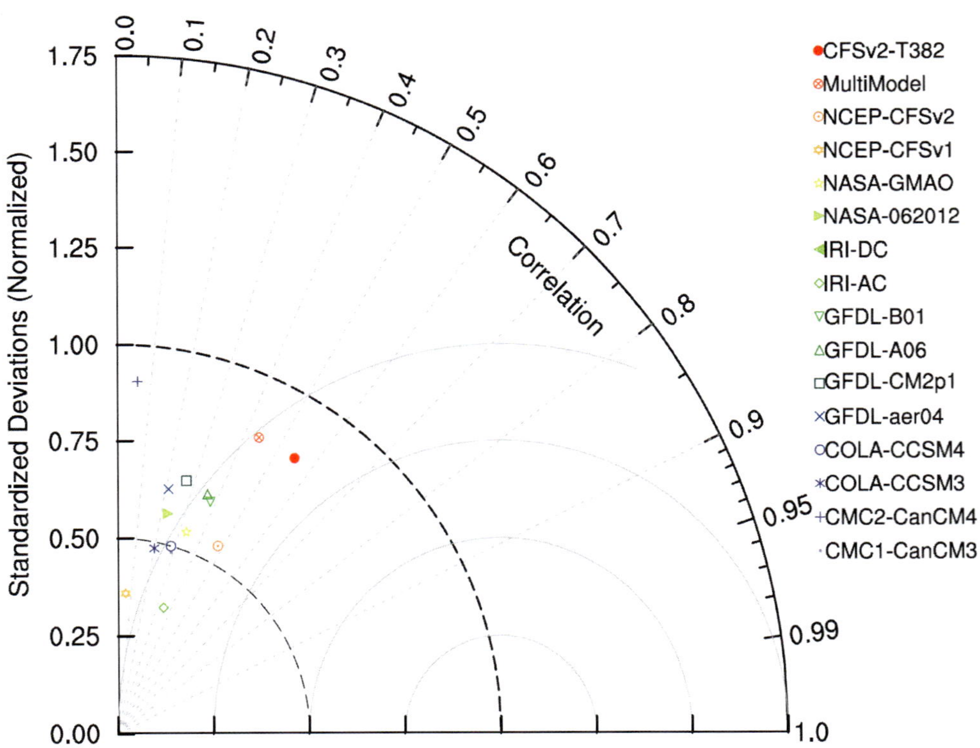

Fig. 1. Taylor diagram showing skill of ISMR prediction from a number of seasonal prediction systems including the Monsoon Mission model CFSv.2-T382.

(e.g., Kripalani and Kulkarni 1997) which was quantified to be a 50–70-year mode of oscillation from several long paleoclimate proxies of ISMR (Goswami *et al.* 2015; Sinha *et al.* 2016). This modulation of the amplitude of ISMR (the signal) at this timescale by the multi-decadal variability of the SAM climate (dynamics and thermodynamics) implies that the predictability of the ISMR also undergoes a multi-decadal variability with the seasonal mean ISMR being more predictable in the epochs at the peak of multi-decadal variability compared to the epochs at the trough of it. Multi-decadal variability of predictability also arises from the fact that the predictable drivers like ENSO (as well as AMO) have multi-decadal variability (e.g., Zhang and Delworth 2006) leading to multi-decadal variations in the ENSO–Indian monsoon relationship (Krishnamurthy and Goswami 2000). The poor skill of some current climate models in predicting the ISMR is partly related to the inability of most climate models to simulate the observed multi-decadal variability of ISMR and other climate modes like ENSO and PDO. with adequate fidelity.

For climate models to be able to simulate the multi-decadal variability of ISMR, the models must strive to incorporate the correct physical processes responsible for the variability. Unfortunately, our understanding of the critical drivers of ISMR multi-decadal variability is still in its infancy. Here, we provide a plausible mechanistic air-sea interaction model of the ISM multi-decadal variability.

Recent examination of what drives the increasing trend of equatorial SST over the Indian Ocean (IO) and how it influences the Indian monsoon has led to the conclusion that indeed large-scale ocean-atmosphere feedback could potentially lead to a multi-decadal variability of the ISM. The decreasing trend of ISMR over the past 5–6 decades has been driven by the increasing trend of equatorial IO SST (Roxy *et al.* 2015) which causes the north–south gradient of tropospheric temperature (TT, see Xavier *et al.* 2007) to decrease. The equatorial SST does this by causing the TT over the southern oceanic box (as in Xavier *et al.* 2007) to increase at a much faster rate than that over the northern box. But what is driving the increasing trend of equatorial IO SST? It has been shown that the weakening surface winds associated with weakening strength of the ISMR change the heat flux and ocean circulation in such a way as to increase the SST further (Swapna *et al.* 2013). Therefore, it is clear that the large-scale positive ocean-atmosphere interaction is responsible largely, if not fully, for the current decreasing trend of the ISM rainfall, indicating that such regional air-sea interactions could contribute to the multi-decadal variability of the ISMR. The observed multi-decadal variability of the ISMR is a result of contributions from the local air-sea interactions together with contributions from teleconnections with multi-decadal ENSO and AMO. The intrinsic timescale associated with the multi-decadal variability arising from air-sea interaction involving IO SST and monsoon is not yet well established. Analysis of the recent decreasing trend of ISMR and increasing trend of IO SST indicates that this timescale would be longer than 60 years. The Pacific decadal oscillation (PDO) and the multi-decadal ENSO are related with a time scale of 50–70 years. The AMO also has a timescale of about 50–70 years. As a result of interactions between the timescales involved with the three different drivers, the multi-decadal ISM mode assumes a broadband 50–70-year periodicity.

Assuming that the current decreasing trend of ISM, arising from a regional positive air-sea interaction, is part of a multi-decadal variations, a negative feedback is required that would arrest this positive feedback and makes the ISM oscillate on this timescale. There is some evidence to suggest that associated with the decreasing ISM, there is tendency of increased convection over the maritime continent (Fig. 2a). The response of the atmosphere to this tendency of increased

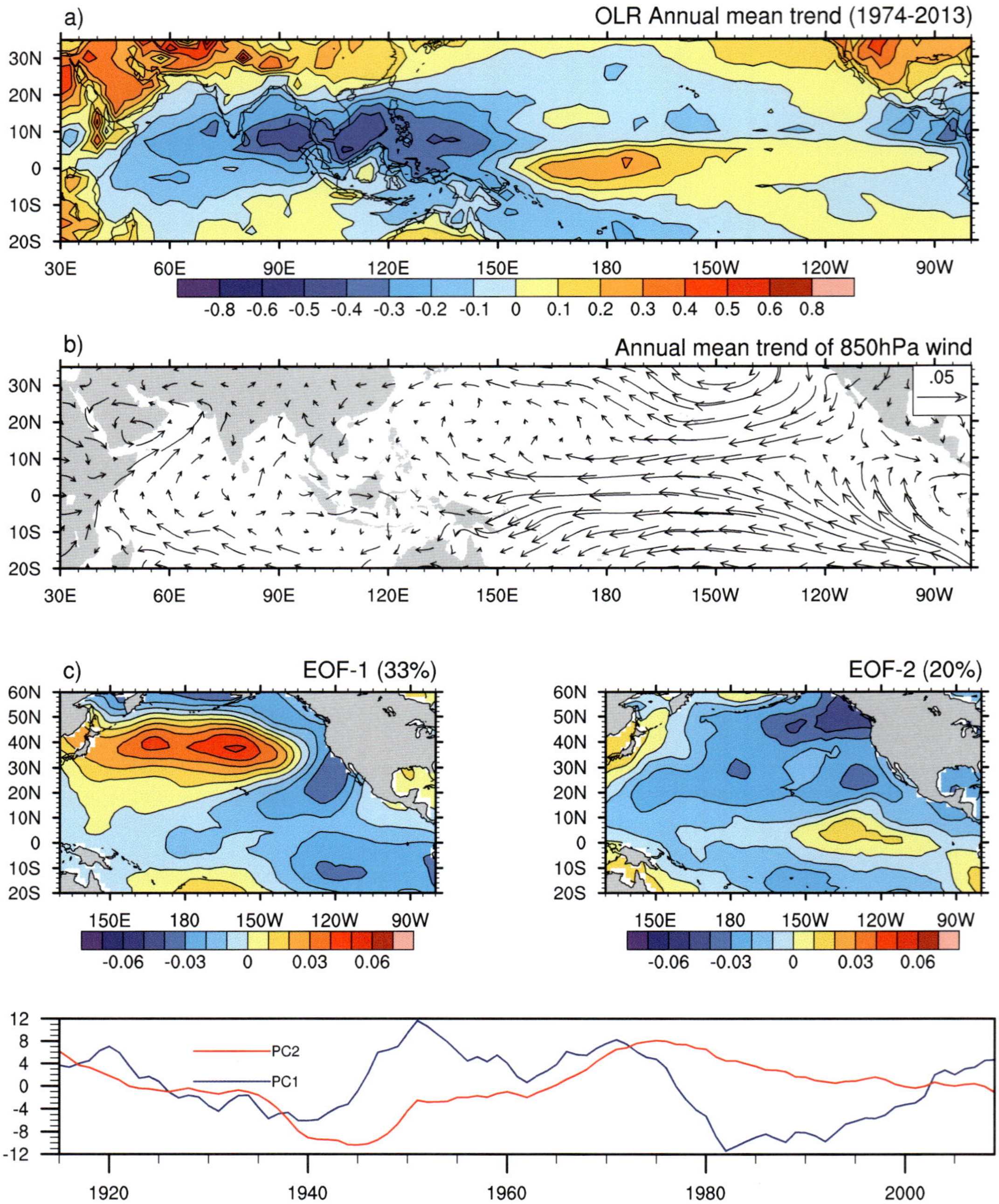

Fig. 2. (a) Trend of Annual Mean OLR during 1974–2013 indicating increasing trend of deep convection over the maritime continent in association with decreasing trend of monsoon. (b) Increasing trend of equatorial Pacific easterlies in association with increased heating over the maritime continent. (c) Low-frequency modes of Pacific SST (1910–2014) after removing the global mean SST trend. EOF1 explaining 33% variance is the multi-decadal ENSO (or PDO) mode, while the EOF2 explaining 20% variance represents the NPGO (North Pacific Gyre oscillation).

convection in this region leads to strengthening easterly trades over central and western Pacific. The response of the equatorial Pacific to such increased background easterly wind (Fig. 2b) would be the deepening of the mean thermocline in the west, shoaling it in the east, and increasing the east–west gradient of the mean thermocline. This oceanic background is conducive for more frequent La Niña conditions on interannual timescales leading to a multi-decadal La Niña phase. Analysis of the low-frequency component of the Pacific SST indeed shows that the multi-decadal ENSO is getting established in a La Niña phase (Fig. 2c). The multi-decadal La Niña is expected to increase positive TT anomaly over the north box (see Xavier *et al.* 2007), increase the north–south TT gradient, and start strengthening the ISMR, eventually taking it to an above-normal state. We propose that this feedback is the key to the oscillating ISM multi-decadal mode. This mechanistic model of ISMR MDM is schematically shown in Fig. 3 involving three drivers, namely, (a) a local air-sea interaction between IO SST and ISMR, (b) a remote air-sea interaction involving ISMR — maritime continent rainfall and tropical Pacific SST, and (c) teleconnections with multi-decadal ENSO and AMO.

In the absence of any closed model to explain the origin of the ISMR multi-decadal variability so far, the model presented in Fig. 3 provides a zero-order model framework for the same. However, further research is required to establish the robustness of the model. Gray areas include whether the increasing trend of easterlies over the central and western Pacific could be entirely forced by the increasing trend of precipitation over the maritime continent or could have been contributed by an increasing trend of SST over the Atlantic over the same period (Li *et al.* 2016; McGregor *et al.* 2014). As strong air-sea interactions are involved, sensitivity experiments with an AGCM may not provide the desired answer. A fully coupled model is required that simulates the observed ISMR multi-decadal variability with fidelity and analysis of long simulations, which may provide confidence on our model framework.

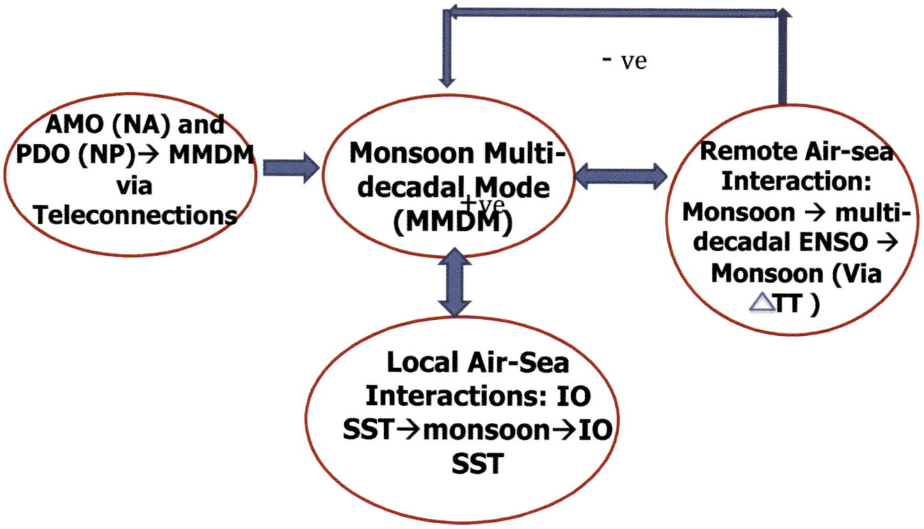

Fig. 3. A schematic representation of a conceptual framework for the Monsoon multi-decadal mode (MMDM) involving three drivers, (a) a local air-sea interaction involving the Indian monsoon and the IO, (b) a remote air-sea interaction involving Indian monsoon/maritime continent precipitation and the tropical Pacific, and (c) teleconnection with other multi-decadal modes like AMO and multi-decadal ENSO.

References

Abhik S., R. P. M. Krishna, M. Mahakur, Malay Ganai, P. Mukhopadhyay, and J. Dudhia, 2017: Revised cloud processes to improve the mean and intraseasonal variability of Indian summer monsoon in climate forecast system: Part 1, *J. Adv. Model. Earth Syst.* **9**, 1002–1029, https://doi.org/10.1002/2016MS000819.

Charney J. G. and J. Shukla, 1981: Predictability of Monsoons, in *Monsoon Dynamics,* J. Lighthill and R. P. Pearce, Eds. (Cambridge Univ. Press, New York, 1981), pp. 99–109.

Chattopadhyay, R., R. Phani, C. T. Sabeerali, A. R. Dhakate, K. Salunke, S. Mahapatra, A. Suryachandra Rao, and B. N. Goswami, 2015: Influence of extra-tropical sea surface temperature on the Indian summer monsoon: An unexplored source of seasonal predictability. *Quart. J. Roy. Met. Soc.,* doi:10.1002/qj.2562.

Delsol, T. and J. Shukla, 2009: Artificial skill due to predictor screening. *J. Clim.*, **22**, 331–345, doi: 10.1175/2008JCLI2414.1.

Dwivedi, S., B. N. Goswami, and F. Kucharski, 2015: Unraveling the missing link of ENSO control over the Indian monsoon rainfall, *Geophys. Res. Lett.*, **42**, doi:10.1002/2015GL065909.

Gadgil, S. and S. Gadgil, 2006: The Indian monsoon, GDP and agriculture. *Econ. Polit. Wkly.*, **XLI**, 4887–4895.

Gadgil, S. and S. Sajani, 1998: *Clim. Dynam.*, **14**, 659, https://doi.org/10.1007/s003820050248.

Gadgil, S., M. Rajeevan, and R. Nanjundiah, 2005: Monsoon prediction — Why yet another failure, *Curr. Sci.* **88**, 1389–1400.

Goswami, B. N. and P. K. Xavier, 2005: Dynamics of 'internal' inter-annual variability of Indian summer monsoon in a GCM, *J. Geophys. Res.*, **110**, D24104, doi:10.1029/2005JD006042.

Goswami, B. B. and B. N. Goswami, 2017: A road map for improving dry-bias in simulating the South Asian monsoon precipitation by climate models. *Clim. Dynam.* **49**, 2025–2034, https://doi.org/10.1007/s00382-016-3439-2.

Goswami, B. N. and R. S. Ajaya Mohan, 2001: Intraseasonal oscillations and inter-annual variability of the Indian summer monsoon, *J. Clim.*, **14**, 1180–1198.

Goswami, B. N., R. S. Ajaya Mohan, P. K. Xavier, and D. Sengupta, 2003: Clustering of low pressure systems during the Indian summer monsoon by intraseasonal oscillations. *Geophys. Res. Lett.* **30**(8), 1431, doi: 10.1029/2002GL016734.

Goswami, B. N., G. Wu, and T. Yasunari, 2006: Annual cycle, Intraseasonal Oscillations and Roadblock to seasonal predictability of the Asian summer monsoon, *J. Clim.*, **19**, 5078–5099.

Goswami, B. N., R. H. Kriplani, H. P. Borgaonkar, and B. Preethi, 2015: Multi-decadal variability of Indian summer monsoon rainfall using proxy data, in *Climate Change: Multi-decadal and Beyond*, Chapter 21, Eds. C. P. Chang, M. Ghil, M. Latif, and M. Wallace, World Scientific, New Jersey, London, Singapore, Beijing, Chennai, pp. 327–346.

Goswami, B. N., 1998: Inter-annual variation of Indian summer monsoon in a GCM: External conditions versus internal feedbacks. *J. Clim.*, **11**, 501–522.

Hazra, A., H. S. Chaudhari, S. K. Saha, S. Pokhrel, and B. N. Goswami, 2017: Progress towards achieving the challenge of Indian summer monsoon climate simulation in a coupled ocean–atmosphere model, *J. Adv. Model. Earth Syst.*, doi:10.1002/2017MS000966.

Kang, I.-S. and J. Shukla, 2006: Dynamic seasonal prediction and predictability of the monsoon. In *The Asian Monsoon*, Chatper 15, (Ed) Bin Wang, Praxis, Springer, Berlin, Heidelberg.

Kang, I. S. *et al.*, 2002: Intercomparison of atmospheric GCM simulated anomalies associated with the 1997/98 El Niño, *J. Clim.*, **15**, 2791–2805.

Kripalani, R. H. and A. Kulkarni, 1997: Climatic impact of El Niño/La Niña on the Indian monsoon: A new perspective. *Weather*, **52**, 39–46, https://doi.org/10.1002/j.1477-8696.1997.tb06267.x.

Krishnamurthy, V. and B. N. Goswami, 2000: Indian monsoon–ENSO relationship on interdecadal timescale, *J. Clim.*, **13**, 579–595.

Krishna Kumar, K. *et al.*, 2005: Advancing dynamical prediction of Indian monsoon rainfall. *Geophys. Res. Lett.* **32**, L08704.

Kumar, S., A., Arora, and R. Chattopadhyay, *et al.*, 2017: *Clim. Dynam.* **48**, 999, https://doi.org/10.1007/s00382-016-3124-5.

Li, X., S. P. Xie, S. T. Gille, and C. Yoo, 2016: Atlantic-induced pan-tropical climate change over the past three decades. *Nat. Clim. Change*, **6**, 275.

McGregor, S., A. Timmermann, M. F. Stuecker, M. H. England, M. Merrifield, F. F. Jin, and Y. Chikamoto, 2014: Recent Walker circulation strengthening and Pacific cooling amplified by Atlantic warming. *Nat. Clim. Change*, **4**, 888.

Rajeevan, M., D. S. Pai, R. Anil Kumar, *et al.* 2007: *Clim. Dynam.* **28**, 813, https://doi.org/10.1007/s00382-006-0197-6.

Rajeevan, M., C. K. Unnikrishnan and B. Preethi, 2012: Evaluation of the ENSEMBLES multi-model seasonal forecasts of Indian summer monsoon variability. *Clim. Dynam.* **38**, 2257–2274.

Roxy Mathew, R. Kapoor, T. Pascal, M. Raghu, A. Karumuri, and B.N. Goswami, 2015: Drying of Indian subcontinent by rapid Indian ocean warming and a weakening land–sea thermal gradient, *Nat. Commun.*, doi:10.1038/ncomms842.

Sabeerali, C. T., A. Ramu Dandi, A. Dhakate, K. Salunke, S. Mahapatra, and S. A. Rao, 2013: Simulation of boreal summer intraseasonal oscillations in the latest CMIP5 coupled GCMs. *J. Geophys. Res. (Atmosph.)* **118**, 4401–4420, https://doi.org/10.1002/jgrd.50403.

Sabeerali, C. T., S. A. Rao, A. R. Dhakate, K. Salunke, and B. N. Goswami, 2014: Why ensemble mean projection of south Asian monsoon rainfall by CMIP5 models is not reliable? *Clim. Dynam.*, doi:10.1007/s00382-014-2269-3.

Saha, S. K., K. Sujith, S. Pokhrel, H. S. Chaudhari, and A. Hazra, 2017. *J. Adv. Model. Earth Sys.* **9**, 271–290, https://doi.org/10.1002/2016MS000845.

Sharmila, S., P. A. Pillai, S. Joseph, M. Roxy, R. P. M. Krishna, R. Chattopadhyay, S. Abhilash, A. K. Sahai, and B. N. Goswami, 2013: Role of ocean-atmosphere interaction on northward propagation of Indian summer monsoon intraseasonal oscillations (MISO). *Clim. Dynam.* **41**(1), 651–669.

Shukla, J., 1981: Dynamical predictability of monthly means, *J. Atmos. Sci.*, https://doi.org/10.1175/1520-0469(1981)038%3C2547:DPOMM%3E2.0.CO;2.

Sinha, A., G. Kathayat, H. Cheng, S. F. M. Breitenbach, M. Berkelhammer, M. Mudelsee, J. Biswas, and R. L. Edwards, 2015: Trends and oscillations in the Indian summer monsoon rainfall over the last two millennia. *Nat. Commun.* **6**, 6309, doi:10.1038/ncomms7309.

Sperber, K. R., H. Annamalai, I.-S. Kang, A. Kitoh, A. Moise, A. Turner, B. Wang, and T. Zhou, 2013: The Asian summer monsoon: an intercomparison of CMIP5 vs. CMIP3 simulations of the late 20th century. *Clim. Dynam.* **41**, 2711–2744. doi:10.1007/s00382-012-1607-6.

Swapna, P., R. Krishnan, and J. M. Wallace, 2013: Indian Ocean and monsoon coupled interactions in a warming environment. *Clim. Dynam.* doi:10.1007/s00382-013-1787-8.

Walker, G. T. 1924: Correlation in seasonal variation of weather, X. Application to seasonal forecasting in India. *Mem. Indian Meteorol. Dept.*, **24**, 333–345.

Wang, B., Q. Ding, X. Fu, I.-S. Kang, K. Jin, J. Shukla, and F. Doblas-Reyes, 2005: Fundamental challenge in simulation and prediction of summer monsoon rainfall, *Geophys. Res. Lett.*, **32**, L15711, doi:10.1029/2005GL022734.

Wang, B. *et al.*, 2015: Rethinking Indian monsoon rainfall prediction in the context of recent global warming. *Nat. Commun.* **6**, 7154, doi:10.1038/ncomms8154.

Webster, P. J., V. O. Magana, T. N. Palmer, J. Shukla, R. T. Tomas, M. Yanai, and T. Yasunari, 1998: Monsoons: Processes, predictability and the prospects of prediction. *J. Geophys. Res.* **103**(C7), 14,451–14,510, http://dx.doi.org/10.1029/97JC02719.

Xavier, P., K. Charline Marzin, and B. N. Goswami, 2007: An objective definition of the Indian summer monsoon season and a new perspective on the ENSO–monsoon relationship. *Q. J. R. Meteorol. Soc.* **133**, 749–764.

Zhang, R. and T. L. Delworth, 2006: Impact of Atlantic multidecadal oscillations on India/Sahel rainfall and Atlantic hurricanes, *Geophys. Res. Lett.*, **33**, L17712, doi:10.1029/2006GL026267.

Zhao Q. and F. H. Carr, 1997: A prognostic cloud scheme for operational NWP models. *Mon. Wea. Rev.*, **125**, 1931–1953, https://doi.org/10.1175/1520-0493(1997) 125<1931:APCSFO>2.0.CO;2.

Chapter 8

A Review of Mechanisms for Formation of an Anomalous Anticyclone in Western North Pacific During El Niño

Tim Li[*,†,§], Bin Wang[*,†] and C.-P. Chang[‡]

[*] *Department of Atmospheric Sciences,*
School of Ocean and Earth Science and Technology,
University of Hawaii at Manoa, Honolulu, HI, USA
[†] *College of Atmospheric Science,*
Nanjing University of Information Science and Technology,
Nanjing, China
[‡] *Naval Postgraduate School, Monterey, CA, USA*
[§] *timli@hawaii.edu*

The western North Pacific anomalous anticyclone (WNPAC) is an important circulation system that conveys the El Niño impact on the East Asian climate. This chapter summarizes research progress made in the past 20 years regarding the physical understanding of the WNPAC formation. Various theories have been proposed. They include the local atmosphere–ocean interaction, the Indian Ocean capacitor, the combination mode, the moist enthalpy advection/Rossby wave modulation, and the central Pacific SST forcing. The major physical processes associated with these theories as well as their limitations are discussed. The authors conclude that the local atmosphere–ocean interaction and the moist enthalpy advection/Rossby wave modulation are the major mechanisms responsible for the formation of the WNPAC during the El Niño mature winter. The Indian Ocean capacitor mechanism does not contribute to the earlier development but helps maintain the WNPAC throughout the El Niño decaying summer. A self-sustained interbasin coupled mode across the Indo-western Pacific warm pool, the central Pacific cold SST anomaly, and the tropical Atlantic capacitor acts as an additional mechanism for the occurrence of the WNPAC in the boreal summer.

1. Introduction

The western North Pacific anomalous anticyclone (WNPAC) is an important low-level atmospheric circulation system that conveys the El Niño effect on the East Asian climate. Two sets of pioneering observational analyses were among the first to discover the El Niño–East Asia linkage through the WNPAC. The first set was based on the interannual variability of summer rainfall over the Yangtze River Valley. For instance, Chang *et al.* (2000) found a simultaneous WNPAC and the signal could be traced back to the preceding spring and winter when pronounced low-level westerly anomalies and warm SST anomalies (SSTAs) were observed over the equatorial Pacific. The second set was based on the El Niño composite (e.g., Zhang *et al.* 1996; Wang *et al.* 2000). A statistically significant feature of the El Niño composite is the occurrence and persistence of an anomalous anticyclone over the Philippine Sea from the El Niño mature winter to the subsequent summer.

Recently, Li *et al.* (2017) carried out an extensive review of past studies on the development and evolution of WNPAC during the El Niño and summarized the relevant mechanisms and effects. This chapter is an attempt to present an outline of Li *et al.*'s (2017) review.

The Multiscale Global Monsoon System
Edited by C. P. Chang et al.

2. Theories for WNPAC Formation

In the past 20 years, various theories have been proposed to explain the development and maintenance of the WNPAC. To help readers better understand how these theories were formed and what were the motivations behind them, we will discuss them one by one following a time sequence.

2.1. *Local air-sea interaction theory*

Wang *et al.* (2000) proposed a local thermodynamic air-sea interaction mechanism to explain the formation of the WNPAC in the northern winter (see schematic diagram Fig. 1). In response to the El Niño heating over the central equatorial Pacific, a Rossby wave response with low-level cyclonic gyre appears in the WNP. Anomalous northeasterlies to the west of the Rossby gyre induce cold SSTA due to enhanced local surface evaporation (because the mean wind in the region is also northeasterly). The cold SSTAs further suppress local convection and induce an anomalous anticyclone to its west as a Rossby wave response to the local negative heating anomaly. Anomalous northeasterlies to the east of the anticyclone further strengthen the cooling due to enhanced wind–evaporation–SST feedback. As a result of this positive air-sea feedback, the WNPAC forms and is maintained from the winter to the subsequent spring.

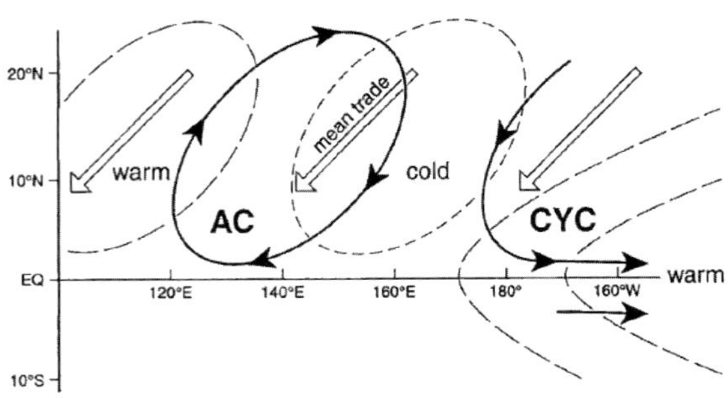

Fig. 1. A schematic diagram illustrating the effect of a positive air-sea feedback between the WNPAC and cold SSTA in WNP. The double arrow denotes the background mean trade wind, and heavy lines with black arrows represent the anomalous wind. The long (short) dashed lines represent a positive (negative) SSTA. From Wang *et al.* (2000). © Copyright 2000 American Meteorological Society (AMS).

2.2. *Indian Ocean capacity theory*

The local positive air-sea feedback proposed by Wang *et al.* (2000) works from the El Niño mature winter to the following spring, but when the summer comes, the mean flow changes its direction and the positive feedback becomes a negative feedback, which leads to the dissipation of the cold SSTA in the WNP during the El Niño decaying summer. Meanwhile, a basin-wide warming appears in the tropical Indian Ocean (IO), from the El Niño mature winter to the succeeding summer. Does this basin-wide IO warming exert a remote impact on WNP circulation? Addressing this question led to the development of the second theory, namely, the IO capacitor effect (Xie *et al.* 2009; Wu *et al.* 2009).

An important issue regarding the IO SSTA effect is whether or not the IO basin warming plays a role in triggering the WNPAC during the El Niño mature winter. An observational analysis shows that during the El Niño peak winter, rainfall and vertical motion anomalies exhibit a clear zonal dipole pattern with anomalous ascending (descending) motion in the western (eastern) IO, even though the SSTA in the tropical IO is approximately uniform (Chen *et al.* 2016). This indicates that the warm SSTA in the western IO plays an active role in driving the atmosphere, while the warm SSTA in the eastern IO plays a passive role and is influenced by the atmosphere.

To resolve the issue above, Chen *et al.* (2016) conducted two sensitivity numerical experiments using ECHAM4. In the first sensitivity experiment, the model was forced by the observed SSTA field over the tropical IO. In this case an anomalous anticyclone in the WNP with pronounced easterly anomalies at the equator was simulated. However, the model simulated a basin-wide anomalous heating field, which does not agree with the observed rainfall pattern. In the second sensitivity experiment, the observed anomalous heating field was specified. Under this forcing scenario, the model simulates an anomalous cyclone in the WNP and pronounced westerly anomalies at the equator. Thus, the numerical model experiments indicate that the basin-wide IO warming does not contribute to the formation of the WNPAC during the El Niño mature winter.

While the rainfall pattern in the IO is characterized by a zonal dipole in the winter, it shows a north–south dipole pattern in the subsequent spring. Thus, the IO basin warming becomes effective in affecting the WNPAC only during the El Niño decaying summer, when a basin-wide large-scale convection and ascending motion anomaly occurs. Such a season-dependent IO forcing effect was first pointed out by Wu *et al.* (2009).

The mechanism through which the IO basin warming affects the WNP circulation in the El Niño decaying summer is summarized below (see Fig. 2). In response to the basin-wide anomalous heating, anomalous easterlies associated with

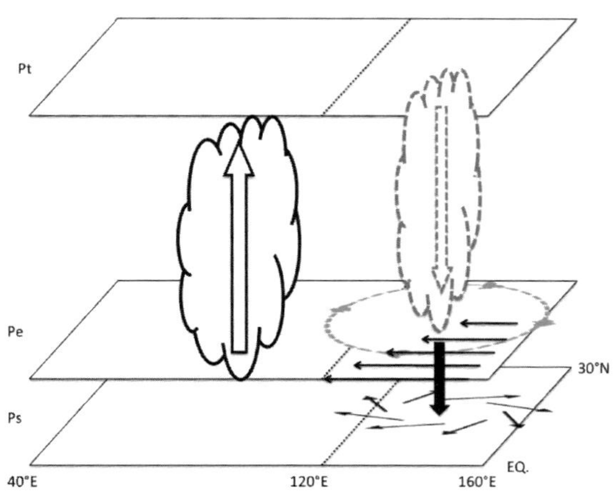

Fig. 2. Schematic diagram illustrating the impact of Indian Ocean basin-wide heating on WNPAC during El Niño decaying summer. From Wu *et al.* (2009). © Copyright 2009 American Meteorological Society (AMS).

Kelvin wave response appear to the east of the heat source. The Kelvin wave easterlies have a maximum at the equator and decrease with latitudes. The so-generated anticyclonic shear leads to boundary layer divergence in the off-equatorial region. The descending motion at the top of the boundary layer due to the Ekman pumping effect decreases local specific humidity and suppresses the mean convection over the WNP monsoon trough region. The negative diabatic heating anomaly further induces the WNPAC.

Another issue is whether or not the local cold SSTAs contribute to the maintenance of the WNPAC during the El Niño decaying summer. Idealized model experiments by Wu *et al.* (2010) showed that both the local cold SSTA in the WNP and the remote SSTA in the tropical Indian Ocean help maintain the WNPAC throughout the El Niño decaying summer. They argued that significant cold SSTAs in the WNP prior to the monsoon onset could have a negative impact on the subsequent monsoon, even though the negative air-sea feedback dissipates the cold SSTA. From the monthly SSTA evolution, one can see clearly that the cold SSTAs in the WNP decrease gradually so that a robust cold SSTA signal is still seen in June and even July.

2.3. *Indo-western Pacific coupled mode theory*

While the IO capacitor effect considered the IO as an external forcing factor, Wang *et al.* (2013) discarded this concept by considering the Indo-Pacific warm pool as an internal coupled system. Their EOF analysis of the summer mean 850 hPa geopotential height (H850) field for the period of 1979–2009 showed that the leading EOF mode accounts for 31% of the total H850 variance. The horizontal pattern of this mode is characterized by an intense southwest–northeast oriented WNPAC, which concurs with suppressed rainfall on its southeastern flank and enhanced rainfall in Korea, Japan, and eastern central China (Fig. 3a).

Wang *et al.* (2013) argued that a positive feedback between the WNPAC and underlying dipolar SST anomalies over the WNP and tropical IO plays a central role in maintaining this pattern. First, to the southeast of the WNPAC, the SST is cooler (Fig. 3b) because the anomalous northeasterly winds strengthen mean easterlies to the east of the monsoon trough, thereby enhancing evaporation and entrainment. Conversely, the resultant ocean cooling would, in turn, reduce in situ precipitation heating (Fig. 3a), thereby generating descending Rossby waves that reinforce the WNPAC in their westward decaying journey. Numerical experiments with a coupled model demonstrate that an initial SST cooling in the WNP in May can indeed induce and maintain an anomalous WNPAC in the ensuing summer (Fig. 3c).

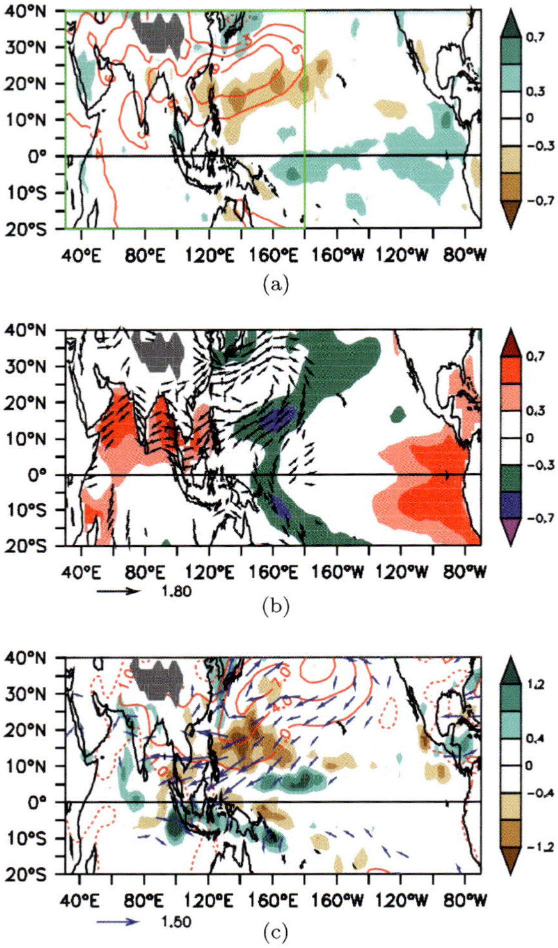

Fig. 3. The first EOF mode derived from summer (JJA) mean 850 hPa geopotential height (H850) in the Asian–Australian monsoon domain (20°S–40°N, 30°–180°E) for 1979–2009. (a) Spatial pattern (contour) and the correlated precipitation (shading) in the Indo-Pacific domain with correlations significant at 90% confidence ($r > 0.3$). (b) The correlated SSTA (shading) and 850 hPa wind anomalies with reference to the principal component of EOF-1. (c) Simulated JJA mean H850 (contour in m), precipitation (shading in mm/day), and 1,000 hPa wind anomaly in response to an initial SST perturbation over the WNP in a coupled model experiment. Adopted from Wang *et al.* (2013).

Second, to the southwest of the enhanced WNPAC, the SCS and northern IO are warm because the easterly anomalies associated with the WNPAC weaken the southwest monsoons (Fig. 3b), thereby reducing surface latent heat flux. Conversely, the northern IO warming would help sustain the WNPAC (Xie *et al.* 2009; Wu *et al.* 2009). Thus, the interplay between the WNPAC and this dipolar SSTA pattern contributes to the maintenance of the WNPAC (Kosaka *et al.* 2013).

2.4. *Central Pacific SSTA forcing theory*

The second leading EOF mode of the summer mean H850 field (Wang *et al.* 2013) features a strong anomalous WNPAC and a weak IO low (Fig. 4a). The anomalous WNPAC ridge and the associated suppressed convection extend from the Philippine Sea southeastward to the equatorial western Pacific. The enhanced WNPAC concurs with equatorial central Pacific cooling (Fig. 4b); the corresponding principal component is negatively correlated with the Niño3.4 index ($r = -0.68$), suggesting that the WNPAC in this mode reflects a developing or a persisting La Niño event.

The enhanced WNPAC associated with the second mode is arguably forced by the central

Pacific cooling that shifts the east–west circulation (Walker) cell with the rising limb shifting westward (Yun *et al.* 2015), thus reducing convection around 160°E and increasing convection over the maritime continent (Fig. 4a). The suppressed convection can directly strengthen the WNPAC by emanation of descending Rossby waves. This assertion was confirmed by numerical experiments with an AGCM forced by the central Pacific SST cooling (Fig. 4c).

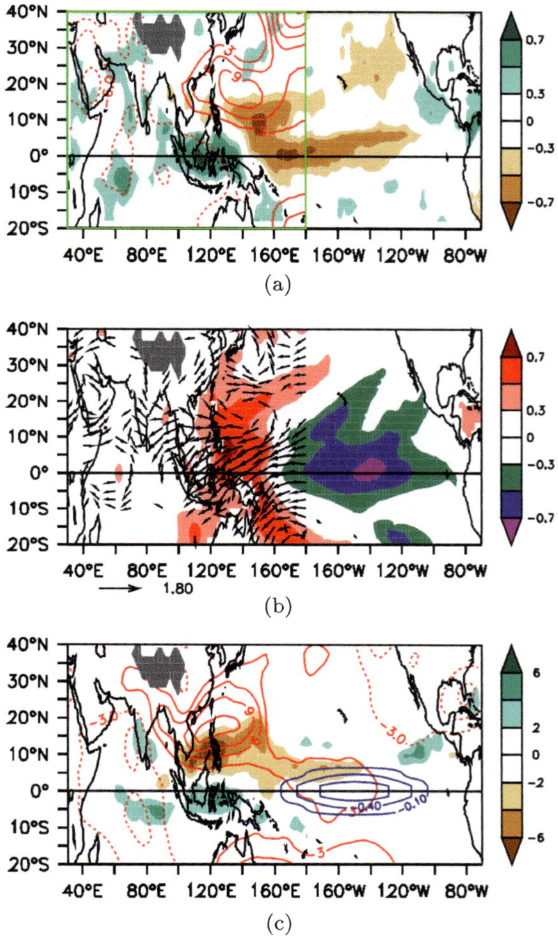

Fig. 4. The second EOF mode derived from summer (JJA) mean 850 hPa geopotential height in the Asian–Australian monsoon domain (20°S–40°N, 30°–180°E) for 1979–2009. (a) Spatial pattern (contour) and the correlated precipitation (shading) in the Indo-Pacific domain with correlations significant at 90% confidence ($r > 0.3$). (b) The correlated SSTA (shading) and 850 hPa wind anomalies with reference to the principal component of the EOF mode. (c) Simulated JJA H850 (contour in m) and precipitation (shading in mm/day) anomalies using an AGCM model forced by the prescribed cold SSTA over the equatorial central Pacific (blue contour in °C). Adopted from Wang *et al.* (2013).

In summary, the central Pacific SSTA forcing may be regarded as an additional mechanism for the development of the WNPAC in the boreal summer. It can occur during either the rapid decaying of an El Niño episode or the developing/persisting phase of a La Niño.

2.5. *A combination mode theory*

Stuecker *et al.* (2015) proposed a "combination mode" (C-mode) theory and attempted to use the theory to explain the formation and maintenance of the WNPAC during the El Niño. The essence of this theory lies in nonlinear interaction between the annual cycle and the ENSO. Mathematically, the product of two time series with an annual period and an approximately 4-year period can lead two "near-annual" periods, roughly 0.8- and 1.2-year periods.

There are two issues related to the C-mode theory. The first issue is whether or not the near-annual periods can be detected in real data. By conducting a power spectrum analysis of observed circulation anomalies at six representative regions across the whole tropical Indo-Pacific domain, Li *et al.* (2016) noted that both the 0.8- and 1.2-year periods are not detectable from the observational data (Fig. 5). The most dominant and significant signals in these regions are at interannual periods. The near-annual period peaks are hardly seen.

In response to this criticism, Stuecker *et al.* (2016) argued that instead of examining whether or not the signal is statistically significant, one needs to pay attention to (a) whether a significant linear correlation exists between the theoretical C-mode time series and the WNPAC indices, and (b) if significant spectral coherence exists between these time series at combination tone frequencies. As shown in Fig. 5, the dominant periodicity for the WNP

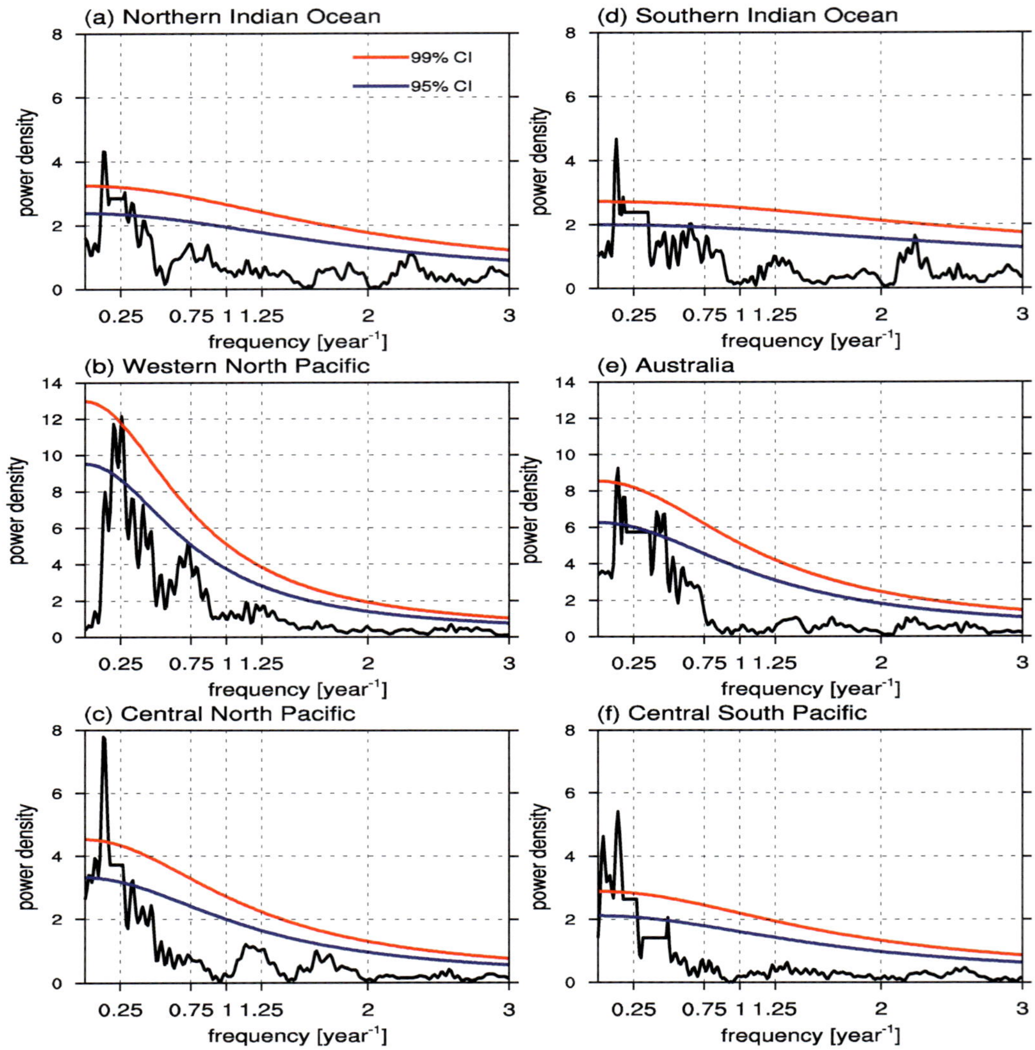

Fig. 5. Power spectra for sea-level pressure (SLP) anomalies averaged in (a) Northern Indian Ocean (5°–20°N, 60°–100°E), (b) Western North Pacific (5°–20°N, 120°–160°E), (c) Central North Pacific (5°–20°N, 180°–140°W), (d) Southern Indian Ocean (20°–5°S, 60°–100°E), (e) Australia (20°–5°S, 120°–160°E), and (f) Central South Pacific (20°–5°S, 180°–140°W). Black lines indicate the power density and colored lines indicate the respective confidence interval (CI) based on a red noise null hypothesis. From Li *et al.* (2016). © Copyright 2016 American Meteorological Society (AMS).

low-level circulation anomaly is at the interannual timescale. Given the wide spectrum of ENSO periodicity, it is likely that ENSO interaction with the annual cycle may still generate an interannual period. In that case, a significant linear correlation between the theoretical C-mode time series and the WNPAC indices does not support the claim that a nearly annual frequency is detected from observed circulation anomalies in the WNP.

To demonstrate how strong the SLP anomaly associated with the C-mode is, Li *et al.* (2016) compared the amplitudes of the near-annual (13–19-month) mode and the interannual mode, and found that the former is about 15% of the latter (see their Fig. 3). Stuecker *et al.*

(2016) disagreed and re-calculated with a different low-pass filter. Figure 6 shows their calculation result. By eye, one may estimate the relative contribution of the two modes. It varies with time and space, ranging from a ratio of 1.5 to 3. The common feature between the two studies is that the interannual component is greater than the near-annual component.

The second issue, which perhaps reflects the most serious limitation of this theory, is that the C-mode does not provide a clear physical interpretation as how the El Niño remotely induces the WNPAC through its nonlinear interaction with the annual cycle. One might wonder why the distinct circulation anomaly appears only in the WNP, not in other regions, and why an anomalous anticyclone, rather than an anomalous cyclone, forms in the region during the El Niño mature winter. Stuecker *et al.* (2015) claimed that the local air-sea interaction is not critical. If so, what atmospheric processes are involved in generating the WNPAC? These questions motivated Wu *et al.* (2017a, b) to address what the relative importance is of the local air-sea interaction and the remote El Niño forcing in generating the WNPAC, and if the atmospheric process alone is able to reproduce the WNPAC, then what internal atmospheric dynamics are involved. The effort leads to the development of a new theory to be discussed in Section 2.6.

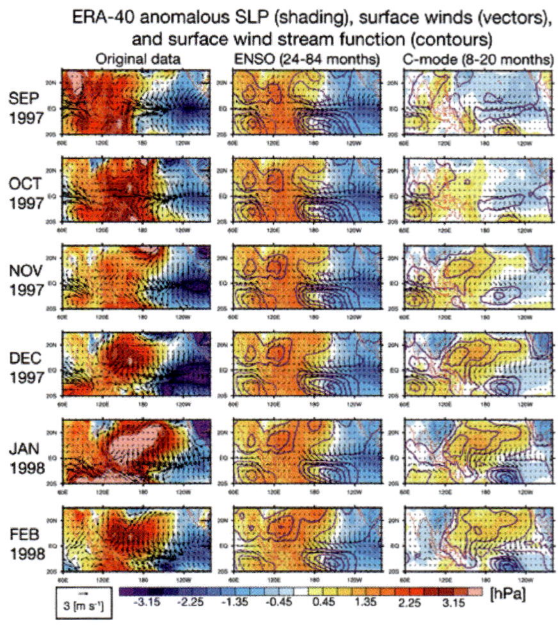

Fig. 6. Monthly anomalous sea-level pressure (shading; hPa) and surface winds (vectors; m s^{-1}) for the development and peak time of the 1997/98 El Niño event (September 1997–February 1998) using the ERA-40 (anomalies relative to the 1958–2001 climatology). Shown are (from left to right) the original data, the low-frequency ENSO component (filtered with a 24–84-month Lanczos bandpass filter; Duchon 1979), and the high-frequency C-mode component (filtered with a 8–20-month Lanczos bandpass filter; Duchon 1979). Additionally, the anomalous surface stream function (purple contours; $0.5 \times 106\,m^2s^{-1}$) is shown for the filtered reanalysis data. From Stuecker *et al.* (2016).

2.6. *Moist enthalpy advection/ Rossby wave modulation theory*

Can the remote El Niño forcing alone "trigger" the WNPAC? To address the questions, Wu *et al.* (2017a, b) conducted two sets of idealized experiments. In the first experiment, the model atmosphere and ocean are coupled in all the regions except in the tropical eastern Pacific (east of the dateline) where the model SST is nudged toward the observed. In the second experiment, the same SST in the tropical eastern Pacific is specified, and in addition climatological monthly SST is nudged over the tropical IO and western Pacific so that the local air-sea interaction is intentionally suppressed.

It is found that both the local air-sea interaction and the remote El Niño forcing contribute equally to the formation of the WNPAC during by El Niño mature winter. To understand how the El Niño generates the WNPAC, Wu *et al.* (2017a, b) conducted vertically integrated moisture and moist static energy (MSE) budget diagnoses using observational data. The results showed that a negative rainfall anomaly, which is the driving source for the WNPAC, results from anomalous vertical advection, which in turn is caused by negative MSE advection of anomalous low-level southerly induced by the El Niño heating in the central equatorial Pacific. Because low-level MSE advection is primarily controlled by moisture and temperature components, this advection process is referred to as moist enthalpy advection.

The reversal of background moisture and vorticity gradients holds a key to the development of the WNPAC in the late fall/early winter. Figure 7 shows schematically how this mechanism works in the presence of a change of the background moisture and vorticity fields. It is noted that the zonal extent of the low-level cyclonic flow anomaly during the El Niño developing summer is much greater than that of the cyclonic anomaly during the El Niño mature winter. The atmospheric Rossby wave response may be determined by an effective beta parameter (β_*), which can be expressed as

$$\beta_* = \beta + \partial_y \overline{\zeta}, \quad (1)$$

where β is the meridional gradient of the Coriolis parameter and $\overline{\zeta}$ is climatological monthly mean relative vorticity.

Climatological maximum relative vorticity (and moisture) center appears over the northern part of the WNP in September and shifts over the southern part of the WNP in November. As a result, the background relative vorticity (and moisture) gradient changes its sign from the northern summer to winter (Figs. 7a, c). A weaker effective beta parameter in the northern winter implies a weaker Rossby wave response to the El Niño heating (Fig. 7c). A smaller cyclonic cell at a low level in Fig. 7c implies an eastward shift of northerly anomalies and thus an eastward shift of negative precipitation anomaly (due to a negative MSE advection anomaly), which would leave space for the WNPAC to develop. Thus, the combined moist enthalpy advection and Rossby wave modulation process holds a key to explaining the distinctive summer–winter contrast of the WNP circulation anomalies from the El Niño developing summer to its mature winter.

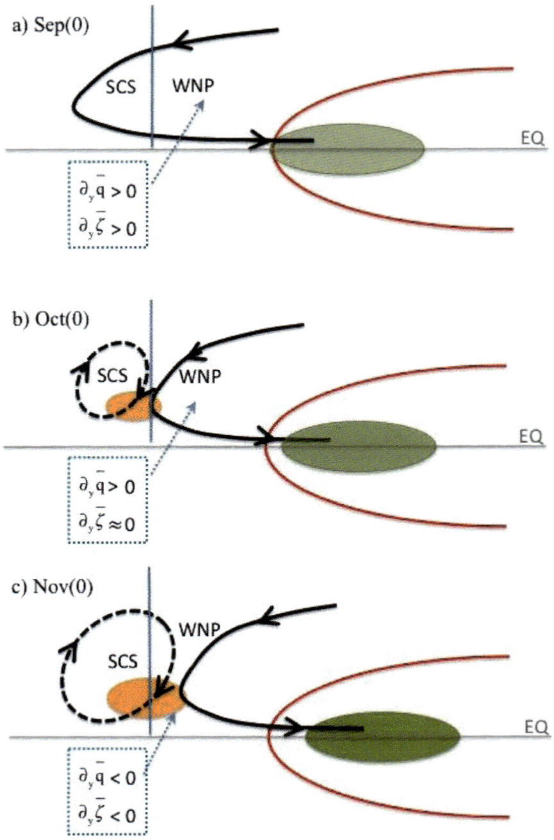

Fig. 7. Schematic of the combined moist enthalpy advection/Rossby wave modulation mechanism. During the El Niño developing summer [say, September (0)], westward stretch of the cyclonic anomalies is enhanced by the positive meridional gradient of the mean relative vorticity ($\partial_y\overline{\zeta} > 0$). With the occurrence of negative $\partial_y\overline{\zeta}$ over the tropical WNP in November (0), the cyclonic anomalies withdraw eastward and leave space for the onset of the WNPAC. Meanwhile, $\partial_y\overline{q}$ changes its sign from positive to negative in the region. The two processes, combined with the effect of underlying cold SSTAs, cause the WNPAC to form in November (0) and maintain throughout the subsequent winter and spring. From Wu *et al.* (2017b).

2.7. *Atlantic capacitor theory*

Recently, a new theory has emerged and has been named as the Atlantic capacitor theory (Li *et al.* 2017). The essence of this theory lies in interbasin teleconnection between tropical Pacific and tropical Atlantic. On the one hand, the El Niño during its mature winter induces significant warm SSTA in the tropical North Atlantic in the boreal spring through decreased northeasterly trades and surface latent heat fluxes (e.g., Klein *et al.* 1999). On the other hand, the Atlantic warming exerts a remote impact on the wind in the Pacific in the subsequent summer.

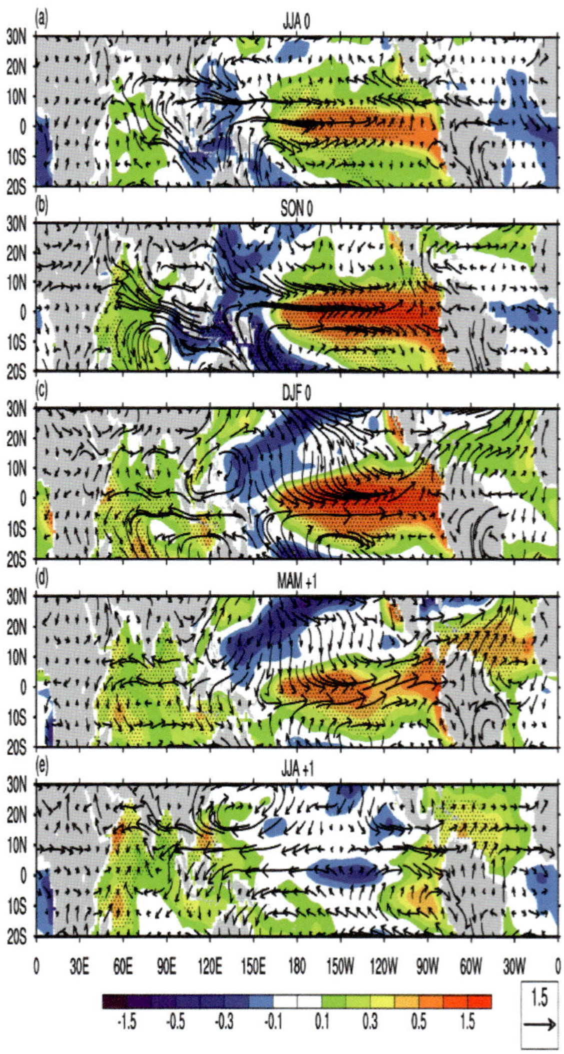

Fig. 8. Anomalous SST (shaded, K) and 850 hPa wind (ms^{-1}) fields regressed against the Nin3.4 index for the period of 1980–2015. Year 0 denotes El Niño developing year, and year +1 denotes El Niño decaying year. Dotted areas indicate that SSTA signals exceed a 95% confidence level. From Li *et al.* (2017).

To clearly illustrate the Atlantic capacitor effect, Li *et al.* (2017) plotted the composite SST evolution patterns from the El Niño developing summer to its decaying summer. Note that a statistically significant positive SSTA signal appears in the tropical Atlantic during the El Niño decaying spring and summer (Fig. 8). In response to the positive Atlantic SSTA, westerly (easterly) anomalies appear over equatorial eastern Pacific (equatorial IO and western Pacific) in the concurrent spring and summer. An anticyclonic center is clearly seen in the WNP during the El Niño decaying summer (Fig. 8e).

Physically, the Atlantic SSTAs may remotely affect the wind anomaly in the tropical Pacific through either a Kelvin wave response to its east or a Rossby wave response to its west. In the former, the anomalous easterly in the equatorial western Pacific could be a direct Kelvin wave response to the Atlantic heat source (Rong *et al.* 2010) or indirect an one through an IO SSTA relaying effect (Yu *et al.* 2016). In the latter, the anomalous wind response in the WNP is a part of the Pacific meridional mode (PMM) interacting with underlying cold SSTA in the extratropical Pacific through Rossby wave-induced evaporation–wind–SST feedback (Ham *et al.* 2013).

3. Concluding Remarks

This chapter summarizes progress made in the past 20 years regarding the understanding of physical mechanisms responsible for the formation and maintenance of the WNPAC during the El Niño. Table 1 lists all relevant mechanisms operating at different phases of ENSO. The main mechanisms responsible for the initial development of the WNPAC prior to and during the El Niño peak winter are the local atmosphere–ocean interaction (Wang *et al.* 2000) and the moist enthalpy advection/Rossby wave modulation (Wu *et al.* 2017a, b). The two mechanisms above are also responsible for the maintenance of the WNPAC from the El Niño

Table 1. Major mechanisms operating in different phases of ENSO.

ENSO phase	Key processes responsible for formation and maintenance of WNPAC
El Niño developing fall SON(0) El Niño mature winter DJF(1) El Niño decaying spring MAM(1)	Local atmosphere-ocean interaction (Wang *et al.* 2000); Moist enthalpy advection/Rossby wave modulation (Wu *et al.* 2017a, b)
El Niño decaying summer JJA(1)	IO Capacitor (Xie *et al.* 2009; Wu *et al.* 2009); Local SSTA forcing (Wu *et al.* 2010); A self-sustained IO-WNP mode (Wang *et al.* 2013) Atlantic capacitor (Li *et al.* 2017)
La Niño developing or persisting summer JJA	Central Pacific SSTA forcing (Wang *et al.* 2013)

mature winter to the subsequent spring. During the El Niño decaying summer, the WNPAC is maintained by local SSTAs in the WNP (Wu *et al.* 2010a), remote SSTAs from the tropical IO (Xie *et al.* 2009; Wu *et al.* 2009), and remote SSTAs from the tropical Atlantic (Li *et al.* 2017). A self-sustained coupled air-sea mode may exist across the entire Indo-Pacific warm pool region during the El Niño decaying summer (Wang *et al.* 2013). The central Pacific cold SSTAs appear important during the summers of the rapid El Niño decaying or La Niña developing/persistence. The near-annual periods predicted by the combination mode theory are hardly detected from the observations, and its contribution to formation of the WNPAC is unclear.

While great progress has been made in the past 20 years in understanding the dynamics of the WNPAC formation and maintenance, some issues remain. For example, the effect of the IO SSTA on the WNPAC remains controversial. On the one hand, the El Niño composite does show a basin-wide warming and associated basin-wide positive precipitation anomalies in the boreal summer over the IO. On the other hand, the EOF analysis based on summertime H850 analysis (Wang *et al.* 2013) showed that the WNPAC is associated with an IO warming but there is no significant rainfall signal associated with the warming (Fig. 3). Besides, no significant SSTA signal is found in the IO at all in the second EOF mode (Fig. 4). This raises a question about the IO SSTA effect on WNPAC. Another issue is about the persistence of the WNPAC from the El Niño mature winter to the subsequent spring. If separating the strong and weak El Niño groups, one may find that the WNPAC does not persist through the spring in the weak El Niño composite, and local cold SSTAs are also quite weak (Bin Wang, personal communication). This raises a question about different processes that might operate in the strong and weak El Niño cases. Therefore, further in-depth observational and modeling studies are needed to understand these issues.

Acknowledgments

This work is supported by NSFC Grant 41630423, NSF Grant AGS-20-06553, and NOAA Grant NA18OAR4310298. This work has an SOEST contribution number of 11114, an IPRC contribution number of 1465, and an ESMC number of 318.

References

Chang, C.-P., Y. Zhang, and T. Li, 2000: Interannual and interdecadal variations of the East Asian summer monsoon and tropical Pacific SSTs. Part I: Roles of the subtropical ridge. *J. Clim.*, **13**, 4310–4325.

Chen, M.-C., T. Li, X.-Y. Shen, and B. Wu, 2016: Relative roles of dynamic and thermodynamic processes in causing evolution asymmetry between El Niño and La Niña. *J. Clim.*, **29**, 2201–2220.

Duchon, C. E., 1979: Lanczos filtering in one and two dimensions. *J. Appl. Meteor.*, **18**, 1016–1022, doi:https://doi.org/10.1175/1520-0450(1979)018<1016:LFIOAT>2.0.CO;2.

Ham, Y.-G., J.-S. Kug, J.-Y. Park, and F.-F. Jin, 2013: Sea surface temperature in the north tropical Atlantic as a trigger for El Niño/Southern Oscillation events. *Nat. Geosci.* **6**, 112–116.

Klein, S. A., B. J. Soden, and N. C. Lau, 1999: Remote sea surface temperature variations during ENSO: Evidence for a tropical atmospheric bridge. *J. Clim.*, **12**, 917–932.

Kosaka, Y., S.-P. Xie, N.-C. Lau, and G. A. Vecchi, 2013: Origin of seasonal predictability for summer climate over the Northwestern Pacific. *Proc. Natl. Acad. Sci. USA*, **110**, 7574–7579.

Li, T., B. Wang, and L. Wang, 2016: Comments on "combination mode dynamics of the anomalous northwest pacific anticyclone". *J. Clim.*, **29**, 4685–4693.

Li, T., B. Wang, B. Wu, T. Zhou, C.-P. Chang, and R. Zhang, 2017: Theories on Formation of an Anomalous Anticyclone in Western North Pacific during El Niño: A review. *J. Meteorol. Res.*, **31**, 987–1006.

Rong, X., R. Zhang, and T. Li, 2010: Impacts of Atlantic SST anomalies on the Indo-East Asian summer monsoon-ENSO relationship. *Chin. Sci. Bull.*, **55**, 1397–1408.

Stuecker, M. F., F.-F. Jin, A. Timmermann, S. McGregor, 2015: Combination mode dynamics of the Anomalous Northwest Pacific anticyclone. *J. Clim.*, **28**, 1093–1111.

Stuecker, M. F., F.-F. Jin, A. Timmermann, and S. McGregor, 2016: Reply to "comments on combination mode dynamics of the Anomalous Northwest Pacific anticyclone." *J. Clim.*, **29**(12), 4695–4706, doi:10.1175/JCLI-D-15-0558.1.

Wang, B., R. G. Wu, and X. H. Fu, 2000: Pacific-East Asian teleconnection: How does ENSO affect East Asian climate? *J. Clim.*, **13**, 1517–1536.

Wang, B., B. Xiang, and J.-Y. Lee, 2013: Subtropical High predictability establishes a promising way for monsoon and tropical storm predictions. *PNAS*, doi: 10.1073/pnas.1212646110.

Wu, B., T. Zhou, and T. Li, 2009: Seasonally evolving dominant interannual variability modes of East Asian climate. *J. Clim.*, **22**, 2992–3005.

Wu, B., T. Li, and T. Zhou, 2010: Relative contributions of the Indian Ocean and local SST anomalies to the maintenance of the western North Pacific anomalous anticyclone during El Niño decaying summer. *J. Clim.*, **23**, 2974–2986.

Wu, B., T. Zhou, and T. Li, 2017a: Atmospheric dynamic and thermodynamic processes driving the western North Pacific anomalous anticyclone during El Niño. Part I: Maintenance mechanisms. *J. Clim.*, **30**, 9621–9635.

Wu, B., T. Zhou, and T. Li, 2017b: Atmospheric dynamic and thermodynamic processes driving the western North Pacific anomalous anticyclone during El Niño. Part II: Formation processes. *J. Clim.*, **30**, 9637–9650.

Xie, S.-P., K. Hu, J. Hafner, H. Tokinaga, Y. Du, G. Huang, and T. Sampe, 2009: Indian Ocean capacitor effect on Indo-western Pacific climate during the summer following El Niño. *J. Clim.*, **22**, 730–747.

Yu, J.-H., T. Li, Z. Tan, and Z. Zhu, 2016: Effects of tropical North Atlantic SST on tropical cyclone genesis in the western North Pacific. *Clim. Dynam.*, **46**(3), 865–877.

Yun, K.-S. *et al.*, 2015: Covariability of western tropical Pacific-North Pacific atmospheric circulation during summer. *Sci. Rep.*, **5**, 16980, doi: 10.1038/srep16980.

Zhang, R. H., A. Sumi, and M. Kimoto, 1996: Impact of El Niño on the East Asian monsoon: A diagnostic study of the '86/87 and '91/92 events.*J. Meteor. Soc. Jpn.*, **74**, 49–62.

Chapter 9

Monsoon Rainfall Prediction Problem in the Western Maritime Continent

Chih-Pei Chang*,§, Tim Li† and Song Yang‡

*Department of Meteorology, Naval Postgraduate School, USA

†Department of Meteorology, University of Hawaii, USA

‡School of Atmospheric Sciences, Sun-yat Sun University, China

§cpchang@nps.edu

Because of strong ENSO influences in the western equatorial Pacific, the seasonal predictability of Maritime Continent rainfall is often perceived to be higher than many places around the world. This notion turned out to be problematic after the Association of South East Asian Nations (ASEAN) Climate Outlook Forum (COF) was established in 2013, when ASEAN national meteorological and hydrological services began to issue consensus seasonal forecasts over the region. The forecasters found that large forecast errors often presented over the western Maritime Continent during the boreal winter, the wet monsoon season. These errors can be seen in many models and are related to the low correlation between ENSO and western Maritime Continent rainfall during the boreal winter monsoon. This Chapter summarizes the research that led Chang *et al.* (2020) to conclude that the forecast difficulty is due to the offsetting between two effects: the anomalous Walker circulation and the anomalous Philippines Sea circulation that interacts with the Sumatra and the Malay Peninsula Mountains.

1. ASEAN Climate Outlook Forums

The Climate Outlook Forum for World Meteorological Organization Region V was established in early December 2013 as the Association of Southeast Asian Nations (ASEAN) Climate Outlook Forum (ASEANCOF). The forum has been held twice a year to produce consensus climate outlooks for boreal winter and boreal summer seasons. The outlooks are derived by consensus from discussions of both climate model forecasts from major world centers and subject forecasts from experienced National Meteorological and Hydrological Service (NMHS) forecasters of individual member nations attending the forum. The results of all six boreal winter outlooks (DJF, 2013/2014–2018/2019) produced as to the writing of this chapter highlight the significant difficulty in forecasting the boreal winter rainfall in the western part of the Maritime Continent, particularly over a local region covering Sumatra, the Malay Peninsula, and western Borneo. The challenges in reaching a consensus rainfall forecast (ASEANCOF-1, -3, -5, -7, -9, -11) are a key reason why this is the only local region to have the highest probability assigned to the middle tercile (Near Normal, NN) in five of the six boreal winters. It was not until the most recent consensus outlook (for DJF 2018/2019) that one of the non-normal terciles, the "Above Normal, AN" category, received a probability that exceeded the NN category.

In the first five DJF outlooks the choice of high probability for the NN tercile did not mean the NMHS forecasters intended to predict a near-normal rainfall season. Instead, iIt reflected the lack of confidence due to the diverging results from different models. This is indicated in the remark in the ASEANCOF-10

The Multiscale Global Monsoon System
Edited by C. P. Chang et al.

Bulletin, which reviewed the outlook of DJF 2017/2018 produced in ASEANCOF-9: *For the regions where predictions favoured near-normal conditions but the season developed into above- or below normal-conditions, the near-normal predictions were considered low-confidence predictions due to the low probability attached to the middle tercile category.* This lack of confidence dominated the first three DJF outlooks (ASEANCOF-1, -3, and -5 for DJF 2013/2014, 2014/2015, and 2015/2016, respectively).

The outlook for DJF 2018/2019 (ASEANCOF-11) was the first time the NN tercile had a lower probability than a non-normal tercile. In the primary region the whole of Sumatra was assigned a 50% probability of Above Normal, indicating a strong confidence that was not seen before. However, Chang *et al.* (2020) showed that against the observed anomaly pattern published by ASEANCOF-12, this high-confidence forecast turned out to also be wrong as the region experienced below-normal rainfall.

Models other than those used in ASEANCOF also faced serious challenges in the seasonal forecast for boreal winter rainfall in the western Maritime Continent. For example, Zhang *et al.* (2016a,b) showed that the NCEP Climate Forecast System version 2 (CFSv2) usually produced skillful rainfall forecast in both dry and wet seasons for the eastern Maritime Continent, but only in the dry season for the western Maritime Continent.

2. Wind-Terrain Interaction and Rainfall

The annual cycle of rainfall in the Southeast Asia Maritime Continent region is mainly driven by the interactions of the local terrain with the prevailing winds from the surrounding oceans (Chang *et al.*, 2005, 2016). Chang *et al.* (2004) suggested that similar effects may give rise to local rainfall–ENSO relationships that vary among different sub-regions during boreal winter, when the region is affected by cold surges off the Asian continent from the north, Indian Ocean zonal wind anomalies from the west, and cross-equatorial flow from the south. These external influences from different directions each have their own dynamics and interannual variations that are usually not well correlated, so that one part inside the domain may have a localized relationship with ENSO that is different from other parts.

The low correlation between ENSO and Maritime Continent rainfall during boreal winter is limited to a western section of the Maritime Continent rather than across the full width of Indonesia (Fig. 1). Outside of this limited sub-region the correlation is significant in most of the central part of the Maritime Continent ("CMC" in Fig. 1) and the southwestern oceanic area ("SWO" in Fig. 1) adjacent to the western coast of Sumatra. The low-correlation sub-region covers mainly Sumatra and the Malay Peninsula ("SMP" in Fig. 1) and extends to Java and the western Borneo coast. It coincides with the region in which the ASEANCOF has difficulty in producing skillful consensus outlooks for the boreal winter monsoon rainfall.

Various names have been designated to this sub-region over Western Indonesia and the Malay Peninsula, including "SMP" by Chang *et al.* (2004), "SMB" by Zhang *et al.*, (2016a), "WMC" by Zhang *et al.* (2016b), and "the Sumatra, Malay Peninsula and western Sarawak/Kalimantan regions" by ASEANCOF. Chang *et al.* (2020) used the name "WIMP" for the region as it implies three meanings. It describes the geographic region of West Indonesia and the Malay Peninsula; it can stand for "Weak Interacting Monsoon Precipitation" to describe the relationship with ENSO; and it reflects operational forecasters' difficulties in reaching a consensus decision.

The one-point correlations between CMAP rainfall and two indices that represent the area-averaged rainfall over SMP and SWO are shown in Fig. 2. A "SMP (WIMP) rainfall mode" may be delineated from the pattern in Fig. 2a, which

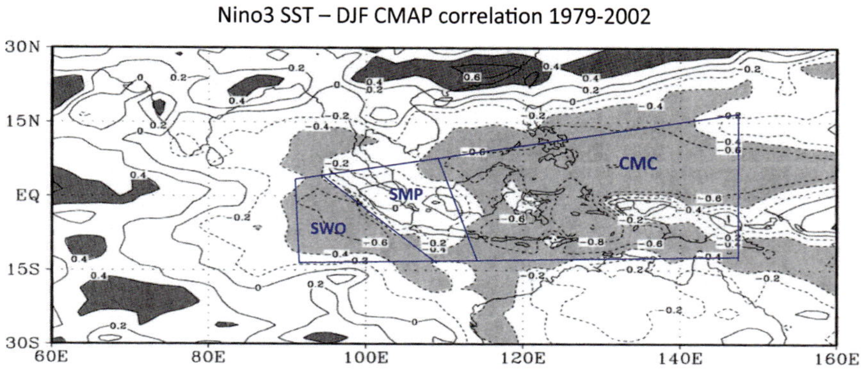

Fig. 1. Correlations of 1979–2002 CMAP rainfall with Nino3 SST. Areas above the 5% significance level are shaded. The rainfall index regions are as follows: SMP: Sumatra–Malay Peninsula; SWO: Southwest oceanic area southwest of Sumatra; and CMC: Central Maritime Continent (from Chang *et al.* 2004, courtesy of American Meteorological Society). *The SMP region is over Western Indonesia–Malay Peninsula and was renamed the WIMP region by Chang et al. (2020).*

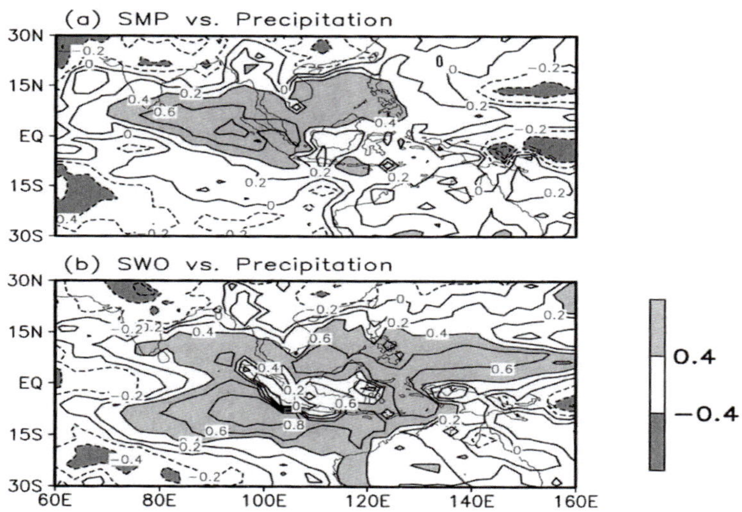

Fig. 2. Correlations of 1979–2002 CMAP rainfall with the (a) SMP rainfall index and (b) SWO rainfall index. Areas above the 5% significance level are shaded (from Chang *et al.* 2004, courtesy of American Meteorological Society).

shows that it is distinctly different from the "ENSO rainfall mode" shown in Fig. 1. On the contrary, a "SWO rainfall mode" (Fig. 2b) is very similar to the ENSO mode, with a conspicuous low-correlation area in the WIMP area and the equatorial region east of 130°E. These different characteristics are confirmed by the correlations among the area-averaged rainfall indices. The correlation between the SMP index and the SWO and CMC indices is 0.41 and 0.34, respectively, while the correlation between the SWO and CMC indices is significantly higher at 0.68.

The SMP mode in Fig. 2a has an east–west-spreading V shape with significant variances that extend into eastern Indian Ocean. As a result, the correlation between the area-averaged SMP index and the individual point rainfalls in the SWO region is higher (average 0.5) than that between the area-averaged SWO

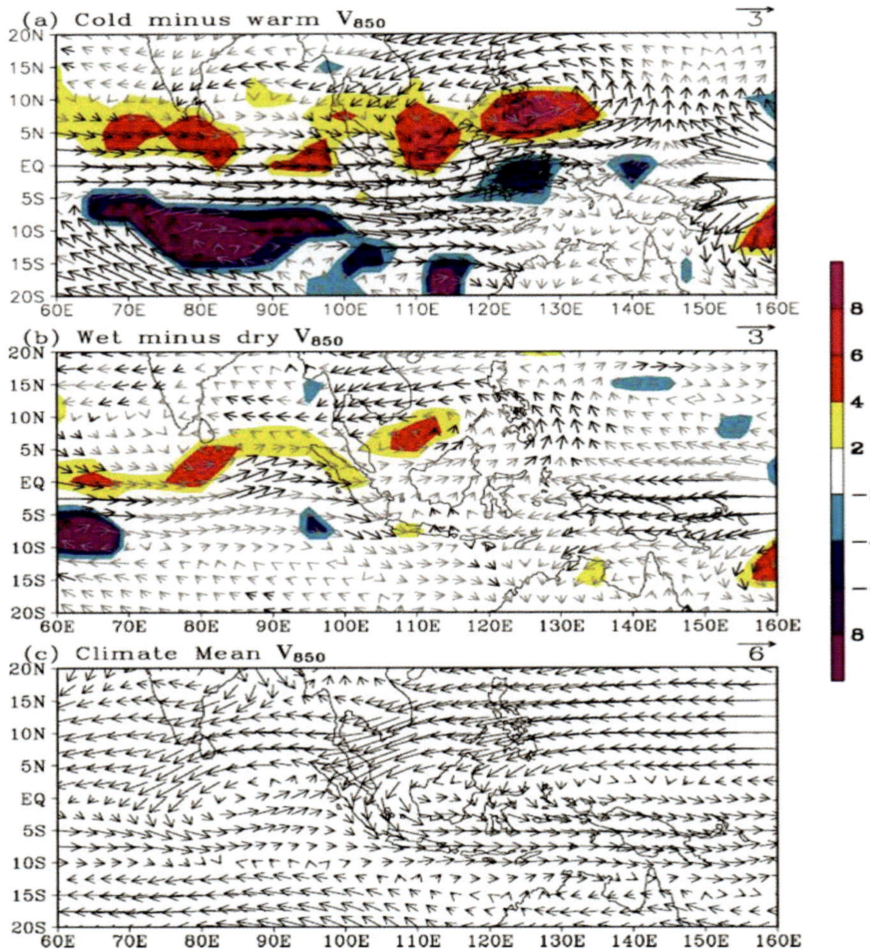

Fig. 3. (a) Composite of Nino3 SST "cold minus warm" 850-hPa wind (heavy arrows indicate differences with at least a 95% confidence level in either the zonal or the meridional component) and vorticity (10^{-6} s^{-1}, only differences at 95% or higher confidence level are plotted) during 1979–2002; (b) same as (a) except for "wet minus dry" in the WIMP rainfall index (adapted from Chang *et al.* 2004, courtesy of American Meteorological Society).

index and the individual point rainfalls in the SMP region (average 0.3).

The composite of the anomalous 850-hPa winds from the 1979–2002 NCEP Reanalysis with respect to the Nino3 SST (Fig. 3a) is consistent with the typical planetary-scale rainfall anomalies over the Maritime Continent regions: wet during La Niña and dry during El Niño. The main difference between the Nino3 *cold* anomaly (*cold-minus-warm*, Fig. 3a) and the WIMP *wet* anomaly (*wet-minus-dry*, Fig. 3b) composites is the equatorial 850-hPa westerly wind, which in the cold anomaly (Fig. 3a) is stronger and extends over a large longitudinal span from the Indian Ocean to the central Maritime Continent. This is a manifestation of two typical circulation responses during La Niña: the Indian Ocean anomalous Walker cell that is often enhanced by the Indian ocean zonal SST gradient (Li *et al.* 2003) and the anomalous cyclonic circulation centered near Philippines which is established and extends southwestward into the equatorial South China Sea. The latter is the counterpart of the Philippine Sea anticyclone that develops in the boreal fall of the El Niño developing phase (Wang *et al.* 2000; Wang and Zhang 2002; Li *et al.* 2016). The convergence with the anomalous easterly wind from the Pacific Walker cell implies a strong rising motion and favors convection in the eastern and central parts of the Maritime Continent. On the contrary, in the WIMP wet anomaly (Fig. 3b) the equatorial westerlies from the Indian Ocean are weaker and attenuate in the WIMP region as they cross the Sumatra Mountains. As a result, the convergence with the low-level easterlies from the Pacific Walker cell occurs mainly in the WIMP region leading to more rainfall there.

Figure 3 may be flipped to consider El Niño events by reversing the anomaly signs. In such a case the circulation centered near the Philippines is anticyclonic and strengthens the equatorial easterlies in the South China Sea, which favors rainfall in the WIMP area as the anomalous easterly winds will encounter mountains in Sumatra and the Malay Peninsula from the east. In either the La Niña or El Niño scenario the rainfall in the WIMP area tends to have an anomaly sign that is opposite to the rest of the Maritime Continent during strong NINO3 events.

For weak NINO3 anomalies the development of the Philippines circulation (and possibly the Indian Ocean SST gradient) will likely be less pronounced. The opposite tendency of rainfall anomalies between WIMP and the other parts of the Maritime Continent may be less pronounced too. In such cases it is conceivable that the rainfall in the WIMP area will be dominated by the planetary-scale Walker cells driven by ENSO, and the correlation pattern between rainfall and NINO3 may be more uniform over the entire Maritime Continent. This is another reason why the sign of the anomalies in the WIMP area is not well correlated with the phase of ENSO.

Jiang and Li (2017) used 1979–2012 data to study the relationship between Maritime Continent rainfall and ENSO, and confirmed that interactions between low-level circulations and the Sumatra mountains is important in separate analyses of El Niño and La Niña events. Figure 4 shows their computation of seasonal evolution of vertical motion and moisture in the WIMP region from boreal summer to boreal

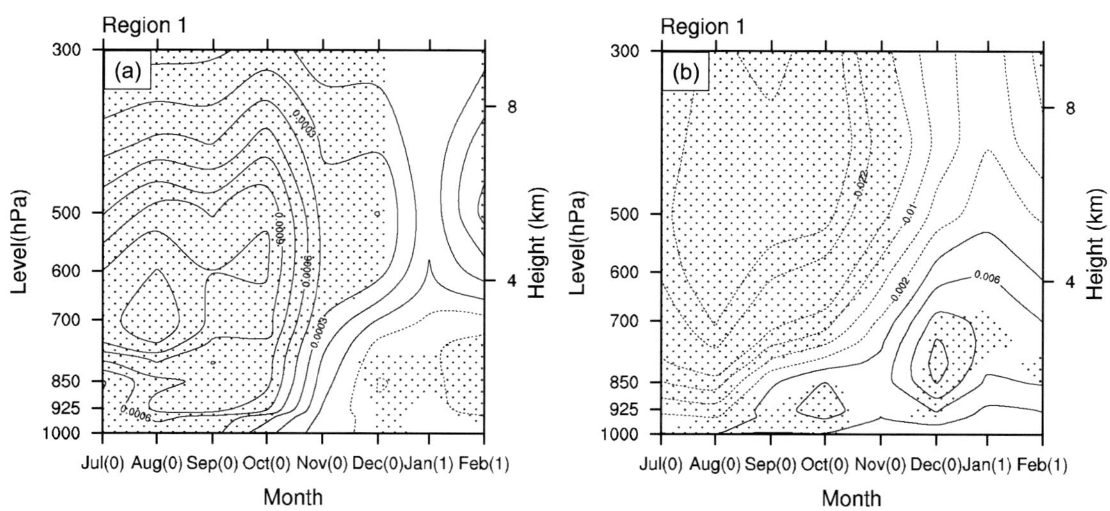

Fig. 4. Time-vertical cross sections of composite of (a) specific humidity (kg.kg^{-1}) and (b) vertical p-velocity anomaly (Pa.s^{-1}) in the WIMP area east of the Sumatra mountains during La Niña from July (0) to February (1) (from Jiang and Li 2019, courtesy of American Meteorological Society).

winter during the La Niña development phase. A deep and moist rising motion dominates the region until early boreal fall, when sinking drier air began to develop in the lee side of the Sumatra mountains as a result of the anomalous westerlies from the Indian Ocean. They found that in El Niño events the effect of the anomalous anticyclone is more prominent, and in La Nina events the effect of the Indian Ocean westerlies is more prominent. The difference in importance of these two circulation systems in the opposite phases of ENSO points to the same important mechanism: in both phases of ENSO the rainfall produced from the lifting of moist air by the Sumatra mountains is the dominant effect of the observed rainfall anomalies in the WIMP region.

Several investigators (Haylock and McBride 2001; Hendon 2003) suggested that the low correlation between boreal winter rainfall over Indonesia and ENSO is due to the effects of the local SST anomalies (SSTA). Jiang and Li (2017) examined the evolution of area-averaged vertical velocity and SSTA during the ENSO development phase. They found that the anomalous ascending motion that peaks in January starts before October, while local warming of the SSTA does not start until two months later. Thus, while the SSTA is still cold, the factor leading to anomalous rainfall in December already starts to gradually build up the moist layer from the orographic lifting.

3. Concluding Remarks

During ENSO the interaction between the low-level anomalous circulations and the Sumatra Mountains tends to offset the effect of the Walker circulation in boreal winter, resulting in low correlation between rainfall in the western Indonesia–Malay Peninsula (WIMP) region and NINO3 SST. This offset is shown in the schematic diagram (Fig. 5) produced by Chang *et al.* (2020). During the La Niña development phase (Fig. 5a) the planetary-scale Walker circulation produces anomalous westerlies around the equator, which generally leads to low-level convergence and anomalous rainfall in the Maritime

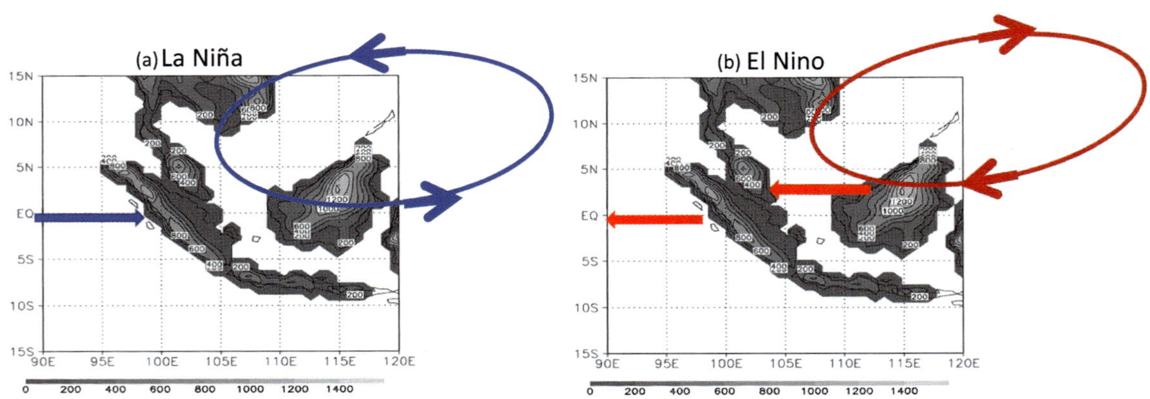

Fig. 5. Schematic diagram of the low-level anomalous circulation over the WIMP area during La Niña (a) and El Niño (b) development phases. For La Niña the circulation centered near Philippines is cyclonic and the equatorial anomalous westerlies are part of the lower branch of the planetary-scale Walker circulation, which produces low-level convergence in the broad Maritime Continent area. However, the westerlies in the eastern Indian Ocean encounter the Sumatra mountains and produce dry desent in the WIMP region, resulting in decreased rainfall. For El Niño the equatorial anomalous easterlies are part of the lower branch of the Walker circulation, which produces low-level divergence over the broad Maritime Continent area, while the easterlies associated with the Philippine Sea anticyclone encounters the Sumatra mountains from the west, producing moist ascent and increased rainfall in the WIMP region (from Chang *et al.* 2020, courtesy of Chinese Meteorological Society).

Continent region, but the westerlies in the equatorial Indian Ocean encounter the Sumatra Mountains causing more rainfall on the windward side and less on the lee side. Thus, the WIMP area experiences less rainfall. During the El Nino development phase (Fig. 5b), the planetary-scale Walker circulation produces anomalous easterlies around the equator, which leads to low-level divergence and rainfall deficits in the broad region. But the circulation of the anomalous anticyclone near Philippines produces upward motion and more rainfall on the east side of the Sumatra and the Malay Peninsula mountains (the WIMP region). So, in both cases the local wind-terrain interaction due to the anomalous horizontal circulation offset the effect of the planetary-scale anomalous Walker circulation. The result is a less deterministic rainfall anomaly in the WIMP region.

Most climate model predictions appear to be rooted in the linkage between SST and rainfall (Zhang *et al.* 2016a, b). Thus, it is not surprising that seasonal forecasting in the western Maritime Continent WIMP area lacks skills. In order to improve forecasts for the WIMP area, climate models need to remove the overdominance of the SST linkage and properly incorporate the local terrain effects that are very important for Maritime Continent monsoon rainfall.

References

ASEAN CLIMATE OUTLOOK FORUM Reports (http://asmc.asean.org/asmc_asean_cof_about/) ASEANCOF-1 Consensus Statement and Meeting Report for DJF 2013–2014, Dec 2013, 9pp. ASEANCOF-3 Consensus Statement and Meeting Report for DJF 2014–2015, Nov 2014, 15pp. ASEANCOF-5 Consensus Bulletin for DJF 2015–
2016, Nov 2015, Singapore, 7pp. ASEANCOF-7 Consensus Bulletin for DJF 2016–2017, Nov 2016, Manila, 10pp. ASEANCOF-8 Consensus Bulletin for JJA 2017, May 2017, Singapore, 8pp. ASEANCOF-9 Consensus Bulletin for DJF 2017–2018, Nov 2017, Hanoi, 14pp. ASEANCOF-10 Consensus Bulletin for JJA 2018, May 2018, Kuala Lumpur, 13pp. ASEANCOF-11 Consensus Bulletin for DJF 2018–2019, 10 Oct–1 Nov 2018, Kuala Lumpur, 14pp. ASEANCOF-12 Consensus Bulletin for JJA 2019, May 2019, Bangkok, 13pp.

Chang, C.-P., Z. Wang, J. Ju, and T. Li, 2004: On the relationship between western Maritime Continent monsoon rainfall and ENSO during northern winter. *J. Clim.*, **17**, 665–672, doi:10.1175/1520-0442(2004)017<0665:OTRBWM>2.0.CO;2.

Chang, C.-P., Z. Wang, J. McBride, and C. H. Liu, 2005: Annual cycle of Southeast Asia — Maritime Continent rainfall and the asymmetric monsoon transition. *J. Clim.*, **18**, 287–301.

Chang, C.-P., M. M. Lu, and H. Lim, 2016: Monsoon convection in maritime continent: interaction of large-scale motion and complex terrain. *In Multiscale Convection-Coupled Systems in the Tropics*, Eds. R. Fovell, and W. Tung, *Meteoro. Monogr.*, 56, American Meteorological Society, doi:10.1175/AMSMONOGRAPHS-D-15-0011.1.

Chang, C.-P., T. Li, and S. Yang, 2020: Boreal winter seasonal prediction in western maritime continent during ENSO. *J, Meteorol. Res.*

Haylock, M., and J. McBride, 2001: Spatial coherence and predictability of Indonesian wet season rainfall. *J. Clim.*, **14**, 3882–3887, doi:10.1175/1520-0442(2001)014<3882:SCAPOI>2.0.CO;2.

Hendon, H. H., 2003: Indonesian rainfall variability: Impacts of ENSO and local air–sea interaction. *J. Clim.*, **16**, 1775–1790, doi:10.1175/1520-0442(2003)016<1775:IRVIOE>2.0.CO;2.

Jiang, L., and T. Li, 2017: Why rainfall response to El Niño over maritime continent is weaker and non-uniform in boreal winter than in boreal summer. *Clim. Dynam.*, **51**, 1465–1483, doi:10.1007/s00382-017-3965-6.

Li, T., B. Wang, C.-P. Chang, Y. Zhang, 2003: A theory for the Indian Ocean dipole–zonal mode. *J. Atmos. Sci.*, **60**, 2119–2135, doi: 10.1175/1520-0469(2003)060<2119:ATFTIO>2.0.CO;2.

Li, T., B. Wang, and L. Wang, 2016: Comments on "combination mode dynamics of the anomalous northwest pacific anticyclone". *J. Clim.*, **29**, 4685–4693, doi: 10.1175/JCLI-D-15-0385.1.

Wang, B., R. Wu, and X. Fu, 2000: Pacific–east asian teleconnection: how does ENSO affect east asian climate? *J. Clim.*, **13**, 1517–1536, doi: 10.1175/1520-0442(2000)013<1517:PEATHD>2.0.CO;2.

Zhang, T., S. Yang, X. Jiang, and P. Zhao, 2016a: Seasonal-interannual variation and prediction of wet and dry season rainfall over the Maritime Continent: Roles of ENSO and monsoon circulation. *J. Clim.*, **29**, 3675–3695.

Zhang, T., S. Yang, X. Jiang, and B. Huang, 2016b: Roles of remote and local forcings in the variation and prediction of regional Maritime Continent rainfall in wet and dry seasons. *J. Clim.*, **29**, 8871–8879.

Part II

Extreme Weather and Convection

Chapter 10

High-resolution Simulations of Heavy Rainfalls in Association with Monsoon Systems and Typhoons Using Cloud-Resolving Models

Kazuhisa Tsuboki*,‡ and Yali Luo†

*Institute for Space-Earth Environmental Research, Nagoya University, Furo-cho, Chikusa-ku, Nagoya, 464-8601 Japan

†Chinese Academy of Meteorological Sciences, Beijing 100081, China

‡tsuboki@nagoya-u.jp

Monsoon systems and typhoons are major causes of heavy rainfall events in East Asia and occasionally give rise to serious disasters. Most heavy rainfall systems are composed of intense convective clouds. Cloud-resolving models are, therefore, indispensable for studies of their mechanisms and processes, and also for accurate and quantitative predictions of heavy rainfall. Advances in computing now make it feasible to use cloud-resolving models for studies and predictions of heavy rainfalls associated with monsoon systems and typhoons. This chapter reviews recent studies of heavy rainfall events using regional cloud-resolving models such as MM5, WRF, JMA-NHM, and CReSS. As an example, the details of the CReSS model are summarized and two simulation experiments of heavy rainfall events are introduced: one caused by the Baiu front in Japan in 2017 and the other accompanying a typhoon in 2004. These high-resolution experiments, which successfully simulated the structure of heavy rainfall systems and provided quantitative predictions of rainfall, were performed with a horizontal resolution of 1 km. These results indicate that cloud-resolving models are very useful and indispensable for mechanism studies and quantitatively accurate predictions of heavy rainfall associated with monsoon systems and typhoons.

1. Introduction

In East Asia, heavy rainfalls occur in association with monsoon systems such as Meiyu, Baiu, Shurin, and Changma. The Meiyu in China and Taiwan, the Baiu in Japan, and the Changma in Korea are quasi-persistent rains or rainy seasons that extend from mid to late spring through early to midsummer. The Shurin (Akisame) is also a similar rainfall or rainy season in Japan and it occurs from late summer through autumn. These systems usually include a quasi-stationary front along a weak baroclinic zone oriented east–west in the lower troposphere, occasionally accompanied by mesoscale convective systems. Typhoons are another major weather system responsible for heavy rainfalls. Combinations of typhoons and monsoon systems often cause long-lasting and unusually heavy rainfall.

Heavy rainfalls occasionally give rise to severe floods and landslides that may result in major disasters. To prevent or mitigate such disasters, quantitative precipitation forecasts are indispensable. Mesoscale weather systems that deliver heavy rainfall are usually composed of intense convective clouds such as cumulonimbus clouds. Realistic simulations of these storms and quantitative predictions of their rainfall require high-resolution and non-hydrostatic numerical models. If the models resolve individual convective clouds and predict time change of water substances in clouds, they are called "cloud-resolving models". One of the most important objectives of regional cloud-resolving models is high-resolution simulations of heavy rainfall

The Multiscale Global Monsoon System
Edited by C. P. Chang et al.

systems. Because the structure of heavy rainfall systems has a wide range of horizontal scales, a large computational domain and a grid with very high resolution to resolve individual classes of the multiscale structure are necessary to simulate accurately the evolution of the weather systems. The recent rapid development in computing enables us to use cloud-resolving models for realistic simulations and quantitative predictions of heavy rainfall although they are still imperfect and under development.

There are some regional non-hydrostatic numerical models such as the Fifth-Generation Penn State/NCAR Mesoscale Model (MM5) (Grell *et al.* 1994), the Weather Research and Forecasting (WRF) (Skamarock *et al.* 2008), the Japan Meteorological Agency Non-Hydrostatic Model (JMA-NHM) (Saito *et al.* 2006, 2007), and the Cloud Resolving Storm Simulator (CReSS) (Tsuboki and Sakakibara 2002). These models have been used for simulation experiments of high-impact mesoscale weather systems with a high resolution of the order of 1 km–100 m because heavy rainfall systems consist of intense cumulonimbus clouds and explicit computation of individual convective clouds is necessary for quantitative simulations of heavy rainfall. Physical processes such as cloud microphysics and planetary boundary layer (PBL) physics are also important for accurate prediction. At a resolution of the order of 1 km–100 m, these processes are parameterized while convections are explicitly calculated.

In general, increase of horizontal resolution is believed to yield better simulations of heavy rainfall. Roberts and Lean (2008) compared the forecast skills of 12-, 4-, and 1-km resolutions using the non-hydrostatic Met Office's Unified Model and showed the biggest improvement for heavier, highly localized rain with a resolution of 1 km. Oizumi *et al.* (2018) successfully simulated a heavy rainfall event on the island of Izu Oshima in October 2013 caused by a typhoon-induced intense rainband using an ultra-high-resolution non-hydrostatic model. They examined four different grid spacings (5 km, 2 km, 500 m, and 250 m) and two PBL schemes and found that a 500-m resolution with Deardorff (1980) PBL scheme gives a quantitatively good precipitation forecast for the heavy rainfall event. Byun *et al.* (2015) performed simulation experiments of a heavy rainfall event that occurred on 21 September 2010 in the Korean Peninsula (maximum total rainfall of 259 mm day^{-1}) using the WRF model with a horizontal grid spacing of 3 km for the innermost domain and showed that the horizontal distribution of the 36-h accumulated precipitation was successfully simulated. They tested 25 combinations of five different PBL schemes and five cloud microphysical schemes and concluded that there was no best combination of the schemes to simulate the heavy rainfall.

To illustrate the importance of horizontal resolution for quantitative prediction of heavy rainfall systems and convective clouds, we show simulation results with different horizontal resolutions using the CReSS model. A detailed description of CReSS will be given in Section 3. Figure 1 compares simulation results of a supercell with two different horizontal resolutions: 1 km and 100 m. Both simulation experiments started from the same initial condition and initial disturbance in a sufficiently large computational domain. The supercell simulated at the 1-km resolution shows a vague hook-shaped structure in the southern part of the cell and the rain intensity is weak. On the contrary, the simulation at the 100-m resolution shows a clear hook-shaped structure at the south end, which is a characteristic of a supercell and the maximum rain mixing ratio is 5 g kg^{-1}. A tornado-like vortex is simulated in the hook-shaped region. Clearly, increased horizontal resolution can bring about both qualitative and quantitative improvements in simulation.

A recent objective for cloud-resolving models was to explicitly simulate dynamic circulations of clouds and their organized systems in a large domain (larger than 1000×1000 km) with

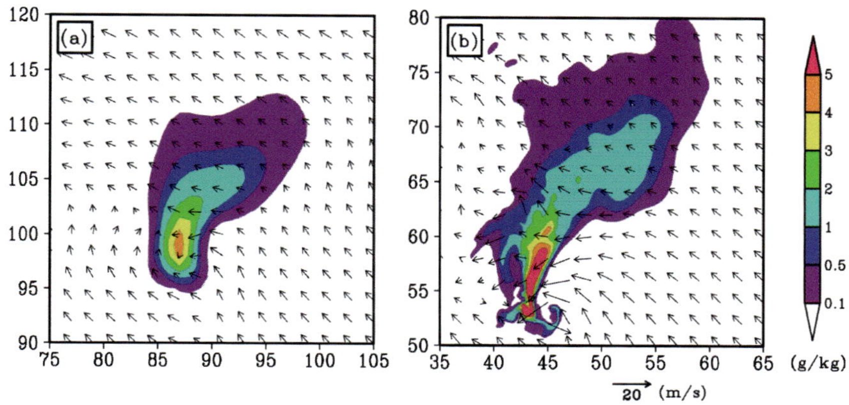

Fig. 1. Simulation experiments of a supercell using CReSS with different horizontal resolutions of (a) 1 km and (b) 100 m. Color shadings are rain mixing ratio ($\mathrm{g\,kg^{-1}}$) and arrows are horizontal velocity vector at a height of 1 km.

resolving individual clouds using a very fine grid system (~1 km). This chapter reviews the current state of numerical simulations and quantitative predictions using cloud-resolving models for heavy rainfall events associated with monsoon systems and typhoons in the East Asian region.

2. Heavy Rainfall Studies Using Cloud-resolving Models

This section reviews some representative studies using cloud-resolving models at high resolutions conducted in East Asia.

2.1. *Japan*

Heavy rainfalls often occur during the warm season in Japan. Since they tend to be localized, they are typically called "localized torrential rain" in Japan. It is also referred to as a heavy rainfall in this chapter. Tsuguchi and Kato (2014) studied statistically synoptic environments of heavy rainfall in Japan. They classified the following six synoptic disturbances as environments of heavy rainfall in Japan: extra-tropical cyclones (including warm fronts), quasi-persistent stationary fronts (Baiu and Shurin fronts), cold fronts, near tropical cyclones (within 500 km), distant tropical cyclones (500–1500 km), and other disturbances. They identified 386 cases of heavy rainfall in the warm season during the period from 1995 to 2009 and showed that 50.3% of the synoptic environments were distant and near tropical cyclones and 21.2% were quasi-persistent stationary fronts. Because quasi-persistent stationary fronts are considered part of a monsoon system, more than 70% of heavy rainfall events in Japan are related to monsoon systems or typhoons.

There are some case studies of heavy rainfall in Japan using high-resolution cloud-resolving models. Oizumi *et al.* (2018) performed ultra-high-resolution numerical weather prediction experiments relating to the Izu Oshima heavy rainfall event using JMA-NHM on the "K computer" at the Riken research institute in Japan. This event occurred on 15–16 October 2013 in Izu Oshima as Typhoon Wipha (2013) approached Tokyo. The JMA surface weather station on the island recorded 24-h rainfall amounts as great as 824 mm and 3-h amounts reaching 328 mm. Simulation experiments used four different grid spacings (5 km, 2 km, 500 m, and 250 m) in a large computational domain. The maximum intensity of heavy rainfall localized over Izu Oshima was simulated quantitatively better at the higher resolutions.

The maximum rainfall intensity in the 500-m experiment using the turbulent closure scheme of Deardorff (1980) was nearly the same as that in the 250-m experiment. The use of finer terrain data (50-m grid spacing digital elevation data) in the 250-m experiment improved the prediction of maximum rainfall intensity. This result indicates that a detailed terrain representation is essential for accurate predictions of localized heavy rainfall.

Tsuboki and Sakakibara (2006) performed a prediction experiment of the heavy rainfall associated with Typhoon Tokage (2004) using CReSS on the Earth Simulator, a huge parallel computer at the Japan Agency for Marine-Earth Science and Technology (JAMSTEC). Typhoon Tokage made landfall in central Japan and caused the largest number of fatalities in western Japan by a typhoon in 2004. The maximum rainfall intensity was more than $50\,\mathrm{mm\,h^{-1}}$ and the total rainfall amount was more than 500 mm. The typhoon track, rainfall distribution, and rainfall intensity were quantitatively simulated by the prediction experiment with a horizontal resolution of 1 km in a large domain. This result indicated that a cloud-resolving model with high resolution is useful and effective for accurate, quantitative predictions of heavy rainfall caused by a typhoon. A more detailed summary of the simulation is given in Section 5.

Monsoon-related heavy rainfalls in Japan have also been investigated using cloud-resolving models. The Baiu front is one of major monsoon systems and heavy rainfalls often occur in the Baiu frontal zone. In particular, serious disasters are occasionally caused by a Baiu frontal heavy rainfall around the late period of the Baiu season. A typical case of heavy rainfall occurred in the Baiu frontal zone on 1 August 1993 in southern Kyushu, Japan. Using the anelastic version of Non-Hydrostatic Model (NHM) developed in the JMA Meteorological Research Institute (MRI) (Ikawa and Saito 1991; Saito 1994), Kato (1998) successfully simulated the back-building process forming the stationary band-shaped torrential rainfall system with a horizontal resolution of 2 km. Another typical back-building formation of a quasi-stationary rainband that lasted for several hours was observed on 4 August 1998 in the Niigata region, Japan. Kato and Goda (2001) simulated a stationary band-shaped rainfall system using the fully compressible version of the MRI-NHM (Saito 1997) with 2-km resolution. They showed that convection-enhanced low-level convergence maintained the quasi-stationary rainband. Kato and Aranami (2005) studied heavy rainfall events which occurred in Niigata–Fukushima on 13 July 2004 and in Fukui, Japan, on 18 July 2004 using the JMA-NHM model with 1.5-km resolution. They attributed the heavy rainfall to the intensification of convective instability in the Baiu frontal zone due to low-level humid inflow and middle-level dry air. According to these simulation experiments, they claimed that accurate data for water vapor over the sea are very important for the improvement of heavy rainfall predictions. These same heavy rainfall events were studied by Yamasaki (2007) and Yamasaki (2008) using a mesoscale convection-resolving model with a resolution of 5 km. These studies concluded that latent instability is important for the formation of these heavy rainfalls and that accuracy of low-level water vapor field is crucial for heavy rainfall predictions. Kato (2006) studied the structure of a band-shaped precipitation system in northern Kyushu on 29 June 1999 by data analysis and numerical simulation using MRI-NHM with a horizontal resolution of 2 km. The study demonstrated the multiscale structure of the rainfall system and the effects of middle-level dry air on the convective activity of the rainfall system.

2.2. *Mainland China*

Heavy rainfall events are occasionally caused by intense mesoscale convective systems (MCSs) and result in severe floods in east China during the Meiyu season. To study multiscale processes involved in triggering and organization of a heavy-rain-producing MCS in east China along the Meiyu front, Zhang and Zhang (2012) used MM5 with the finest grid spacing of 444 m. They showed that an organization of a large MCS was triggered by lifting of moist monsoonal air over convectively generated cold domes. They also emphasized the importance of a high resolution less than 1 km. Luo *et al.* (2014) observed a quasi-linear extreme-rain-producing MCS along a Meiyu front in the Yangtze–Huai rivers' basin in east China and performed a cloud-resolving simulation to study the convective initiation and organization process. Using a one-way triply nesting technique of WRF with a horizontal resolution of 1.11 km for the innermost domain, the observed MCS was successfully simulated. According to the observation and simulation results, they found that a cold dome left by the previous convective activity caused ascension of the moist southwesterly monsoon flow repeatedly. This resulted in a back-building formation of echo training and caused the extreme rainfall. The rainband training was important for the longevity of the rainbands and the formation of the quasi-linear MCS. For improvement of heavy rainfall forecasts, Luo and Chen (2015) studied forecast sensitivity to uncertainties of initial conditions and model physics schemes by ensemble simulations of a quasi-linear extreme-rain-producing MCS along the Meiyu front in the Yangtze–Huai rivers' basin using WRF with a resolution of 1.11 km in the innermost domain. They showed a strong sensitivity of the initial conditions in reproducing the heavy rainfall and that the quasi-linear MCS was formed by the convective back-building and echo-band training processes. They emphasized the importance of the cold dome to lift up the moist air in the southwesterly monsoon airflows.

Southern China is a region where a large amount of heavy rainfall occurs (Ding 1992). However, the quantitative precipitation forecast (QPF) skill for heavy rainfall in the region is not sufficiently improved (Luo 2016). To understand the key process of heavy rainfall and to improve QPF of heavy rainfall in Southern China during the pre-summer rainy season, the Southern China Monsoon Rainfall Experiment (SCMREX) was conducted (Luo 2016; Luo *et al.* 2017). A convection-permitting reginal model for Southern China was developed on the basis of the Global/Reginal Assimilation and Prediction System (GRAPES) (Xue 2004; Xue and Liu 2007), which was named GRAPE-Mesoscale Atmospheric Regional Model for South China with 3 km (MARS3KM) (Zhang *et al.* 2016). Using the model, Zhang *et al.* (2016) assimilated wind profiling radar data and showed improvements of wind and moisture fields and the QPF. Bao *et al.* (2017) examined data assimilation of S-band Doppler radar radial velocity using the WRF-EnKF system for the improvement of forecast of heavy-rainfall-producing MCSs in Southern China. Using four one-way-nested domains with horizontal grid spacings of 27, 9, 3, and 1 km, they successfully predicted the MCSs and found that the shortest time interval of the data assimilation and the longest period of assimilation showed the best performance of the MCSs' prediction. The SCMREX field observations provided valuable radar data of heavy-rainfall-producing MCSs including squall lines in Southern China for improvement of high-resolution models. Qian *et al.* (2018), using WRF with a highest horizontal resolution of 3 km, examined bulk microphysical parametrizations for sensitivity of a squall line developed in a weak synoptic forcing environment and found that microphysical processes and the resulting cold pool strongly change the characteristics of the simulated squall line. Using a local area model (LAM) configuration of the Met Office Unified Model, Furtado *et al.* (2018) examined the sensitivity of subtropical deep convection to cloud microphysical complexity. From a comparison of single- and double-moment scheme experiments with radar observations of SCMREX, they showed that the single-moment scheme gives sufficient results of simulation if it is well tuned.

Heavy rainfalls in Southern China could be influenced by the Tibetan Plateau through the thermodynamic effect on the large-scale atmospheric circulation. Li *et al.* (2014) studied the heating effect of the Tibetan Plateau on the persistent heavy rainfall during 14–24 June 2010. They performed ensemble simulations and sensitivity experiments with five different initial conditions using the 4-km WRF model in a large domain including the Tibetan Plateau and found that the Tibetan Plateau heating significantly affected the temperature distributions and atmospheric circulation, which resulted in the intensification of precipitation in Southern China.

2.3. *Korea*

The monsoon system called the Changma of Korea is similar to the Meiyu of China and the Baiu of Japan. The Changma front is a major factor bringing heavy rainfalls in the warm season to the Korean Peninsula as well as typhoons (Sun and Lee 2002). A quasi-stationary long-lived intense rainband brought a heavy rainfall over the middle Korean Peninsula on 5–6 August 1998. The maximum rainfall was 619 mm in 15 h at the coastal station (Ganghwa). Sun and Lee (2002) studied the formation mechanism of the intense rainband by observational data analysis and numerical simulation using MM5 version 3 (Grell *et al.* 1994) with a two-way interactive nested grid system with 45-, 15-, and 5-km grid sizes. The 5-km simulation showed successive convective cell formation on the upstream side on the west coast to form the intense convective rainband.

On 6–7 August 2002, a heavy rainfall with a maximum rainfall more than 300 mm was caused by several convective rainbands over the middle Korean Peninsula. Shin and Lee (2005) simulated numerically the heavy rainfall using MM5 version 3 with 30- and 10-km grid sizes. Although they showed two formation processes of the rainbands, there were inconsistent points in comparison with observations: one or two rainbands in simulation but several rainbands in observation. Following the previous study, Cho and Lee (2006) conducted a numerical study of the heavy rainfall using the WRF model. They performed many simulations with different horizontal resolutions (30, 10, 3.3, and 3 km), cumulus parameterizations, and cloud microphysics schemes. Multiple convective rainbands were successfully simulated when the horizontal resolution was 3.3 km or 1.1 km without cumulus parameterizations, although some discrepancies were present between the observations and the simulations.

A record-breaking heavy rainfall was caused by a back-building and quasi-stationary mesoscale convective system associated with the Changma front in southeastern South Korea on 7 July 2009. The maximum rainfall amount was 310 mm in less than 12 h in the Busan metropolitan area (Jeong *et al.* 2016). Jeong *et al.* (2016) simulated the heavy rainfall system using CReSS with a horizontal resolution of 2 km and showed that the quasi-stationary, back-building mesoscale convective system was successfully reproduced at approximately the correct location and time. They found that a cold pool produced by evaporative cooling was essential for the stationarity of the heavy rainfall system according to the sensitivity experiment without evaporation. On the contrary, Kim and Lee (2016) showed that a cold pool and the associated outflow were not found around a long, quasi-stationary convective rainband on 1–2 August 2008 over the Yellow Sea, west of the Korean Peninsula. They utilized the WRF model with a grid size of 2 km to simulate the development process of the convective rainband and found that the convective rainband formed along the convergence line within a confluent flow of the trough. According to these studies, there are various formation mechanisms of heavy-rainfall-producing convective rainbands in association with monsoon systems in Korea.

2.4. *Taiwan*

A devastating heavy rainfall occurred in Taiwan in association with Typhoon Morakot (2009). With the landfall of Morakot, a total rainfall amount of 3031.5 mm was recorded and this large amount of rainfall was related to the reduction of moving speed and the influence of monsoonal flows (Liang *et al.* 2011), among other factors. Wang *et al.* (2013b) demonstrated the successful simulation of the total amount, distribution, and timing of the heavy rainfall caused by Morakot in Taiwan about 2 days prior to the main heavy rainfall using a 3-km resolution CReSS model. Wu *et al.* (2011) performed numerical experiments with WRF using three two-way interactive domains of 27-, 9-, and 3-km horizontal resolutions and indicated that interactions of monsoon gyres and Typhoon Morakot reduced the westward movement of the storm and resulted in the unusually long residence of the storm and the large amount of rainfall in southern Taiwan. They also showed that the terrain of Taiwan enhanced the amount of the extreme rainfall. Apart from large-scale influence, Wang *et al.* (2012) found that the slowdown of movement was also related to the asymmetric latent heating structure of Morakot using CReSS with a horizontal resolution of 3 km. Their sensitivity experiment with reduced moisture content showed an increase in the typhoon's transition speed and a decrease in rainfall amount. They emphasized the contribution of asymmetric heating to the typhoon motion and the importance of cloud microphysics for more accurate prediction of typhoon track.

Using the CReSS model with a 3-km resolution, Wang *et al.* (2013a) also successfully simulated the slowdown of the transition speed of Typhoon Fanapi (2010) with the change from a symmetric to an asymmetric structure, which caused heavy rainfall of more than 800 mm in Taiwan. According to the simulation and sensitivity experiments with reduced moisture content, they concluded that the sudden reduction of the transition speed of Fanapi was caused by the asymmetric latent heating and not the environmental flow.

As pointed out by Wu *et al.* (2009), the interaction of a typhoon and the Asian monsoon flow causes heavy rainfall over Taiwan. They conducted simulation and sensitivity experiments of Super-typhoon Babs (1998) and the East Asian winter monsoon using MM5 with three nested domains and a highest resolution of 6.67 km, and showed that the heavy rainfall events were attributed to the monsoon mode and the topographic mode, which are the low-level convergence between the typhoon circulation and the monsoon northeasterly, and interaction of the typhoon circulation and the terrain of Taiwan, respectively. The interaction between a typhoon and monsoon flow often causes a heavy rainfall in Taiwan even when the typhoon is located at a distance. This type of heavy rainfall is referred to as remote rainfall event or indirect rainfall event (Chen and Wu 2016). A typical case of remote heavy rainfall occurred over northeastern Taiwan when Super-typhoon Megi (2010) was located in the South China Sea. Chen and Wu (2016) conducted ensemble simulations using WRF in the four nested domains with grid spacings of 54, 18, 6, and 2 km. They successfully performed a quantitative precipitation forecast of the remote heavy rainfall in northeastern Taiwan and found that the interaction of typhoon circulation and monsoon flow, frontal lifting of moist air, orographic lifting by the terrain of Taiwan, and the direction of the prevailing wind were all important for the remote heavy rainfall.

A heavy-rainfall-producing convective line associated with the Meiyu front observed in Taiwan was studied by Wang *et al.* (2005) using CReSS. They performed a simulation of the convective line and sensitivity experiments of Taiwan's topography with a 2-km grid size and found that the intensity and position of the convective line were highly sensitive to the terrain height. Wang *et al.* (2011) studied the heavy rainfall over central Taiwan on 8 June 2007 using the 2-km grid CReSS model to clarify the triggering mechanism for the deep convection. The simulation showed that preexisting storms produced pauses of cold outflow and gravity waves. The deep convection was triggered by the arrival of the gravity waves ahead of the cold outflow, and this is referred to as the remote trigger of deep convection.

In 2008, a field campaign named the Southwest Monsoon Experiment/Terrain-influenced Monsoon Rainfall Experiment (SoWMEX/TiMREX) was conducted in the Meiyu season to study monsoon rainfall in Taiwan (Jou *et al.* 2011). Wang *et al.* (2014a, b) studied two propagating episodes of mesoscale convection that caused heavy rainfall in Taiwan. They performed a simulation experiment of the eastward- and westward-propagating characteristics of the convective systems using CReSS with a 2.5-km resolution and showed that the simulation strongly agrees with the observation. According to the simulation, they showed that structure, environmental shear, and organization process are different in eastward- and westward-propagating systems. Wang and Huang (2009) performed a high-resolution simulation on a convective line formed off the southeastern coast of Taiwan in the Meiyu season using CReSS with grid sizes of 10, 2, and 0.5 km, and showed the formation process of the convective line.

Using CReSS, Wang *et al.* (2016) developed a new ensemble quantitative precipitation forecast system for Taiwan with a grid spacing of 2.5 km with a large domain and a long forecast period of 8 days. They evaluated the system for six typhoon events (Wang 2014; 2015) and showed that the system provided highly useful quantitative precipitation forecasts with longer lead times that allowed for advanced preparation for disaster prevention.

2.5. *Simulations under climate change conditions*

Recently, cloud-resolving models have been used for downscaling simulations of future projection of typhoon activity with climate change. Tsuboki *et al.* (2015) performed this type of downscaling simulation of typhoons of the western North Pacific in the global warming climate of the late 21st century and showed that the maximum intensity of a super-typhoon will increase and the maximum latitude attained by future super-typhoons will be further north. Kanada *et al.* (2017) studied future enhancement of heavy rainfall associated with a typhoon in the mid-latitude region. Using CReSS coupled with the three-dimensional non-hydrostatic ocean model named the Non-Hydrostatic Ocean model for the Earth Simulator (NHOES) (Aiki *et al.* 2006, 2011, 2015), they conducted a pseudo-global warming experiment of the Typhoon Chanthu (2016) which made landfall over Hokkaido and caused severe floods. Typhoon Chanthu was characterized by a predecessor rain (PRE; e.g., Cote 2007; Galarneau *et al.* 2010). They showed that rainfall in PRE and that of typhoon circulation will be more intense and more concentrated in future warmer climate in the mid-latitude.

3. Cloud-resolving Model: CReSS

To perform high-resolution simulations of weather systems on large parallel computers we have been developing a cloud-resolving model named the Cloud Resolving Storm Simulator (CReSS) at Nagoya University, Japan. The basic equations of CReSS are the non-hydrostatic and compressible equation systems. The horizontal coordinates are of the orthogonal curvilinear type including the orthogonal linear coordinates (Cartesian coordinates) as a special case. The vertical coordinate is a hybrid coordinate that combines a terrain-following coordinate and a height coordinate. Using a stretching technique of the vertical coordinate, the lower levels are calculated with fine grid spacing. A variety of map projections and longitude-latitude coordinates can be adopted. Prognostic variables include the three-dimensional velocity components, perturbations of pressure and potential temperature, water vapor mixing ratio, sub-grid scale turbulent kinetic energy (TKE), and mixing ratios and number concentrations of cloud microphysical variables. A finite difference scheme is used for spatial discretization. The dependent variables are set on staggered grids: the Arakawa-C grid in horizontal and the Lorenz grid in vertical. The mode-splitting technique (Klemp and Wilhelmson 1978) is used for time integration. For parallel processing in calculation, the message passing interface (MPI) and Open MP are used.

Cloud microphysical processes are formulated by bulk parameterization of cold rain, which is based on several sources (Lin *et al.* 1983; Cotton *et al.* 1986; Murakami 1990; Ikawa and Saito 1991; Murakami *et al.* 1994), which consider water vapor, rain, cloud, ice, snow, and graupel. A double-moment option is available for liquid and ice phases. Parameterizations of sub-grid scale eddy motions are either the one-order closure of Smagorinsky (1963) or the 1.5-order closure with TKE (Deardorff 1980). Fluxes of heat, moisture, and momentum at the earth surface are calculated using the scheme of Louis *et al.* (1981) for the land and the scheme of Kondo (1975) for the ocean. A one-dimensional thermal conductivity model is used for calculation of ground temperature (Segami *et al.* 1989). A one-dimensional diffusion model is used for the ocean mixing layer.

More detailed descriptions of CReSS are presented in Tsuboki and Sakakibara (2002, 2007). CReSS has been used for studies of convective systems in the East Asian monsoon regions (e.g., Wang *et al.* 2014a, b; Tsuboki 2008), intense supercells (Shimizu *et al.* 2008; Wang *et al.* 2009; Akter and Tsuboki 2010), winter snow storms (Liu *et al.* 2004, 2006 Maesaka *et al.* 2006), and tropical cyclones (e.g., Wang *et al.* 2012; Akter and Tsuboki 2012; Tsujino *et al.* 2017).

4. CReSS Simulation of Baiu Front Heavy Rainfall

The Baiu front extends zonally for several thousand kilometers, whereas a localized heavy rainfall has a horizontal scale of a few hundred kilometers and consists of intense convective clouds. Various types of multiscale structures are observed along the Baiu front. To clarify the processes in each class of the multiscale system of precipitation along the Baiu front, it is necessary to perform simulations with a grid fine enough to resolve cloud-scale features and with a domain large enough to encompass the whole Baiu front. The explicit representation of cumulonimbus clouds is especially important for quantitatively accurate simulations of localized heavy rainfalls in the Baiu frontal zone.

A typical case of Baiu frontal localized heavy rainfall in western Japan on 5 July 2017 was studied here using CReSS. The JMA radar network observed an intense rainband extending from the southern Korean Peninsula to western Japan along the Baiu front (Fig. 2). The maximum rainfall intensity exceeded 80 mm h^{-1}, and a localized heavy rainfall caused severe flooding in the coastal region of western Japan.

The localized heavy rainfall was simulated using CReSS with a horizontal resolution of 1 km. The initial and boundary conditions were provided by the JMA forecast model output. The 24-h simulation began at 1200 UTC 4 July 2017. The simulated rainband extended from the southern Korean Peninsula to western Japan along the Baiu front, in good agreement with the observed rainband (Fig. 3). A highly humid southwesterly prevailing over the sea formed a strong convergence with the northerly from a mesoscale high over the Sea of Japan. Intense

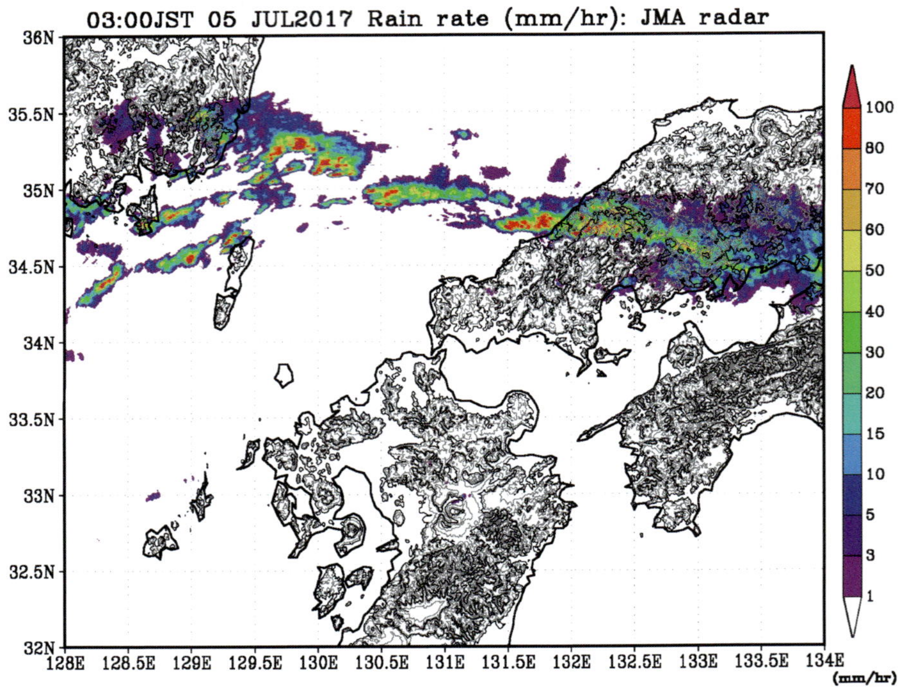

Fig. 2. Rainfall intensity (color shadings; mm h^{-1}) observed by the JMA radar at 0300 JST 5 July 2017 over western Japan. Topography is indicated by contours.

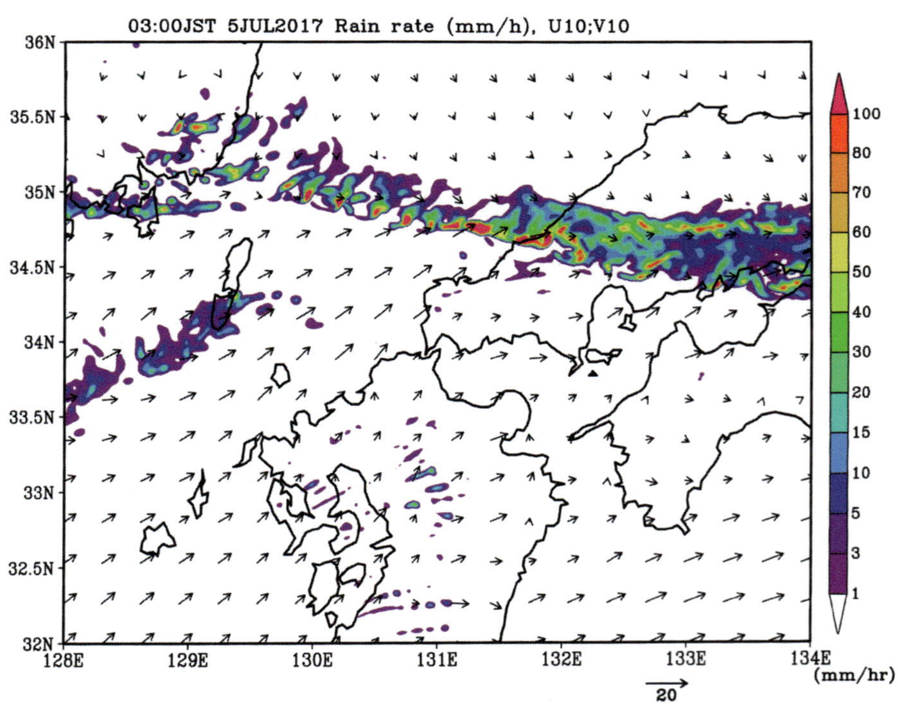

Fig. 3. Rainfall intensity (color shadings; mm h^{-1}) and surface horizontal velocity (arrows) simulated by CReSS at 0300 JST 5 July 2017.

convective cells formed along the southern edge of the rainband with maximum rainfall intensity exceeding 80 mm h^{-1}, roughly the same as the observed rainfall intensity. Most stationary rainbands in Japan form by the back-building process. In this case, however, back-building of convective cells was limited early in the formation stage of the rainband. Intense convective cells formed along the convergence zone and moved eastward to reach the coastal region of western Japan. The simulation showed that successive landfalls of the intense convective cells caused the localized heavy rainfall.

Figure 4 shows the total amount of rainfall observed by the JMA radar between 0000 and 0900 JST (the Japan Standard Time; JST = UTC + 9 h) 5 July 2017. This duration covers almost all of the heavy rainfall in the region during that period. Since the rainband extended from west to east, the total rainfall distribution also extends along the same direction and the maximum rainfall area with more than 300 mm is located near the coast of western Japan. At the JMA surface observation point indicated by the cross mark in Fig. 4, the total rainfall amount of the period (0000–0900 JST, 5 July 2017) is 318 mm. This is roughly the same as that estimated by the JMA radar. The large amount of the total rainfall was successfully simulated by CReSS, as seen in Fig. 5. The simulated heavy rainfall area extends in the west–east direction and the maximum rainfall area is located around the coast although the band-like heavy rainfall area in the simulation is about 10 km south of the observed rainfall band. The rainfall intensity and total rainfall amount are quantitatively simulated using the cloud-resolving model. This result shows that quantitative simulations of monsoon-related heavy rainfall events require a high-resolution model and detailed cloud microphysical processes.

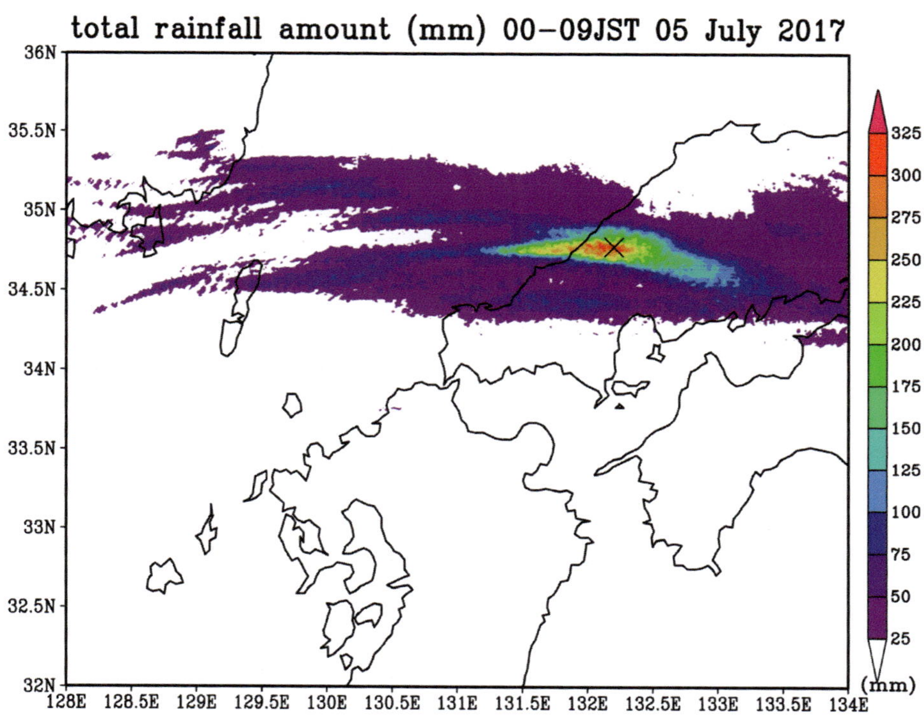

Fig. 4. Total rainfall amount (mm) obtained from the JMA radar for the period from 00 to 09 JST 5 July 2017. The cross mark indicates the location of the JMA surface observation point Haza (132.20 E, 34.78 N).

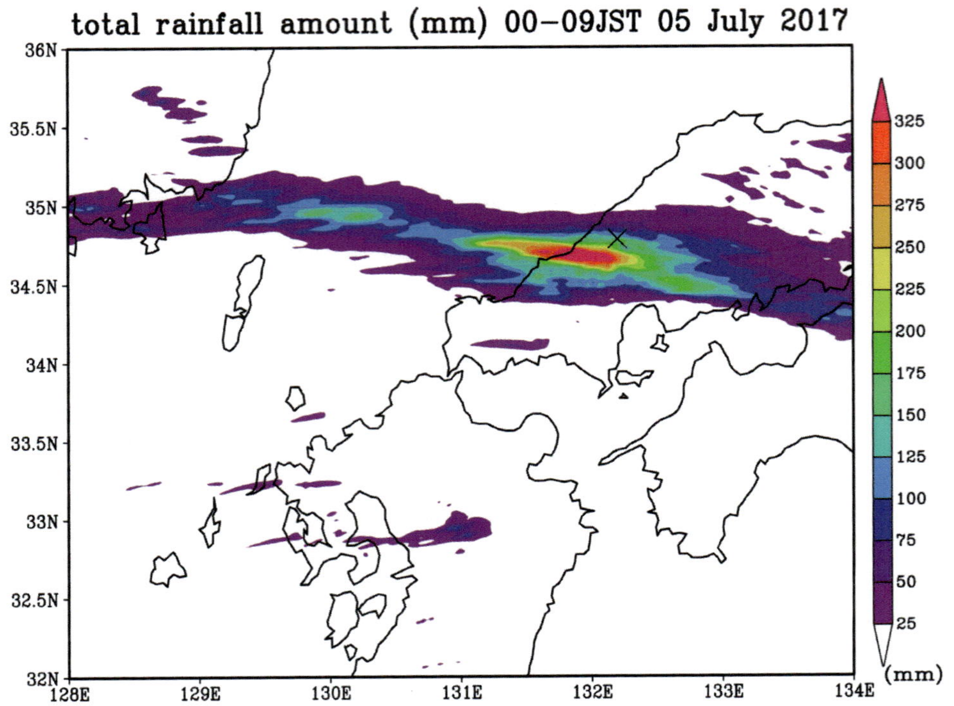

Fig. 5. Total rainfall amount (mm) obtained from the CReSS simulation for the period from 00 to 09 JST 5 July 2017. The cross mark is the same as in Fig. 4.

5. CReSS Simulation of Typhoon-related Heavy Rainfall

Typhoon Tokage (2004) was one of the typical storms that caused serious disasters (Tsuboki and Sakakibara 2006). The typhoon moved northeastward along the Okinawa Islands on 19 October and struck Shikoku Island on 20 October. Heavy rainfall associated with Typhoon Tokage occurred over a wide area of western Japan, causing severe floods and landslides. There were 98 fatalities in the disaster in Japan.

We performed a simulation of Typhoon Tokage using CReSS with a horizontal resolution of 1 km for the 36-h period starting at 12 UTC 19 October 2004. The simulation successfully reproduced the movement of the typhoon and its associated heavy rainfall. The rainfall intensity in western Japan was compared with the JMA surface observations. Figure 6 shows the time series of rainfall intensity at Nobeoka City, Kyushu, as measured by the JMA surface observations and as simulated by CReSS. The maximum intensity (roughly $50\,\mathrm{mm\,h^{-1}}$) and the simulated time series of hourly rainfall intensity from CReSS, including termination of rainfall, matched the observations well. In contrast, a hydrostatic model with a horizontal resolution of 20 km badly underestimated the maximum intensity of the rainfall.

In the simulation, a large northward moisture flux appeared on the east side of the typhoon center, which resulted in heavy rainfall along the Pacific Ocean side. Figure 7 shows the total rainfall amount obtained from the simulation during the period from 1500 UTC 19 to 1500 UTC 20 October 2004. Since the horizontal resolution is 1 km, the terrain heights in the model are almost as high as the real orography, and detailed distribution of heavy rainfall is

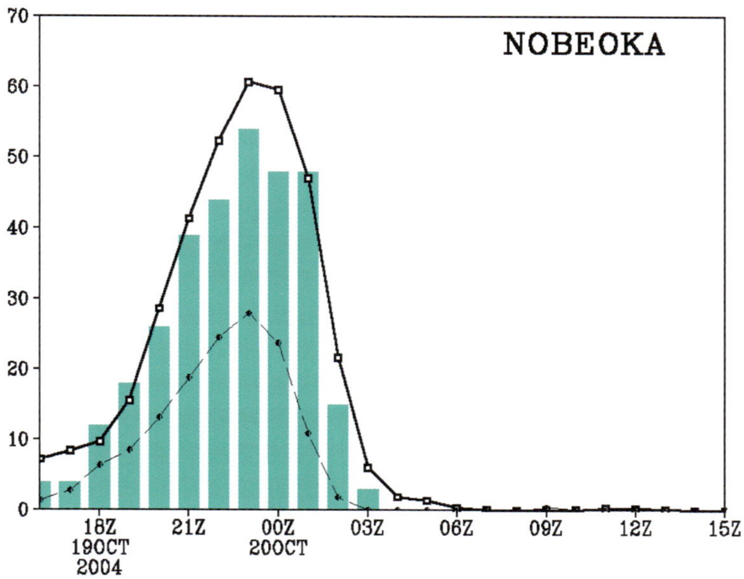

Fig. 6. Time series of rainfall intensity ($\mathrm{mm\,h^{-1}}$) at Nobeoka City in Kyushu, Japan, during the period from 16 UTC 19 October to 15 UTC 20 October 2004: from observations (bars), a CReSS simulation (solid line with open circles), and a 20-km resolution hydrostatic model (dashed line with dots).

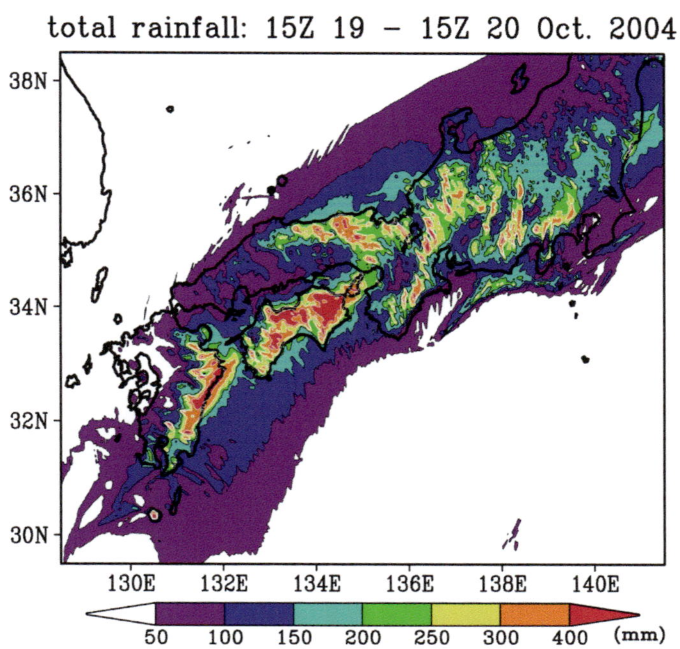

Fig. 7. Distribution of total rainfall amount in the period from 1500 UTC 19 October to 1500 UTC 20 October 2004 obtained from the 1-km resolution simulation using CReSS.

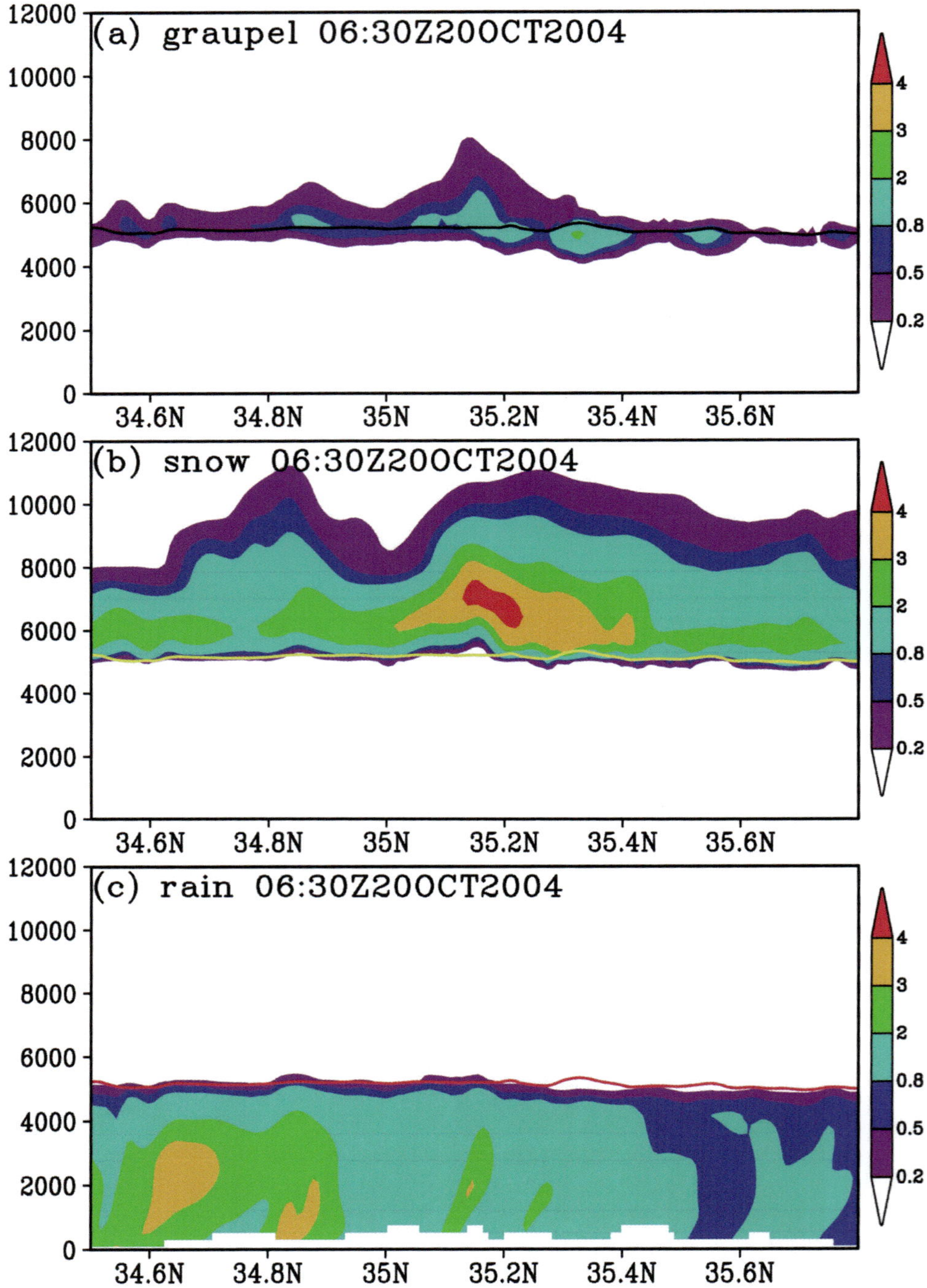

Fig. 8. South–north vertical cross-sections along 135.1°E at 0630 UTC 20 October 2004 of mixing ratios ($g\,kg^{-1}$) of (a) graupel, (b) snow, and (c) rain derived from the CReSS simulation. The solid line in each figure indicates the height of 0°C.

reproduced in the simulation. Large amounts of rainfalls were simulated along the Pacific side of western Japan. Another region of a large amount of simulated rainfall is found on the Japan Sea side of central Japan. The distribution of the total rainfall amount and the rainfall intensity correspond well to those of the JMA rain gauge observation.

The heavy rainfall extended to the region of the Japan Sea side and severe floods occurred in the region. Most heavy rainfalls are usually caused by intense convective clouds. However, the heavy rainfall in the Japan Sea side occurred without intense updrafts in the simulation. The south–north vertical sections crossing the area of the dense mixing ratio of solid particle in the upper level (Fig. 8) show that the solid particles are composed of a large amount of snow particle (Fig. 8b) and a small amount of graupel particle (Fig. 8a). These solid particles melt below the freezing level and produce rain (Fig. 8c). These findings indicate that abundant solid particles in the upper layer were produced by weak upward motions rather than intense convections. The snow and graupel were transported by the upper-level strong southerly to the Japan Sea side, where they produced heavy rainfall. Since CReSS predicts the amounts of each hydrometeor, the detailed process of the formation of heavy rainfall is clarified. These results indicate that a non-hydrostatic cloud-resolving model is useful for mechanism studies of heavy rainfall and for quantitative predictions of rainfall amount and intensity associated with typhoons.

6. Concluding Remarks

Heavy rainfall events in East Asia often occur in association with Baiu (or Meiyu, Shurin, and Changma) as well as typhoons. Quantitative precipitation forecasts of the heavy rainfall events are crucial for disaster prevention and mitigation. Because most heavy rainfalls are caused by intense convective clouds, a cloud-resolving model is necessary for high-resolution quantitative predictions of heavy rainfall. This chapter has summarized recent studies of heavy rainfalls using high-resolution cloud-resolving models and presented the results of two heavy rainfall simulations using CReSS, a cloud-resolving model developed at Nagoya University. Intense rainfalls of 80 mm h^{-1} along the Baiu front and in a typhoon were successfully simulated by CReSS with a horizontal resolution of 1 km.

For more improvement of heavy rainfall prediction, the model physical processes such as cloud microphysics, the planetary boundary layer, and turbulence should be improved. Data assimilations of polarimetric Doppler radars and satellites are also important. Ensemble predictions may be effective for predictions of localized convective rainfall in an unstable atmosphere. Although these topics are beyond the scope of the present chapter, their importance and any promising developments in the future should be emphasized.

Acknowledgments

The simulations and calculations in the present chapter were performed on the Earth Simulator at JAMSTEC and the supercomputers at Nagoya University. The authors thank Professor Chung-Chieh Wang of National Taiwan Normal University for his kindness in reviewing the manuscript. They also thank the anonymous reviewer for valuable comments and suggestions.

References

Aiki, H., K. Takahashi, and T. Yamagata, 2006: The Red Sea outflow regulated by the Indian Monsoon. *Cont. Shelf Res.*, **26**, 1448–1468.

Aiki, H., J. P. Matthews, and K. G. Lamb, 2011: Modeling andenergetics of tidally generated wave trains in the Lombok Strait: Impact of the Indonesian Throughflow. *J. Geophys. Res.*, **116**, C03023, doi:10.1029/2010JC006589.

Aiki, H., M. Yoshioka, M. Kato, A. Morimoto, T. Shinoda, and K. Tsuboki, 2015: A coupled atmosphere-ocean-surface-wave modeling system for understanding air–sea interactions under tropical cyclone conditions. *Bull. Coastal Oceanogr.*, **52**, 139–148 (in Japanese).

Akter, N. and K. Tsuboki, 2010: Characteristics of supercells in the rainband of numerically simulated Cyclone Sidr. *SOLA*, **6A**, Special Edition, 25–28.

Akter, N. and K. Tsuboki, 2012: Numerical simulation of Cyclone Sidr using a Cloud-Resolving Model: Characteristics and formation process of an outer rainband. *Mon. Wea. Rev.*, **40**, 789–810. doi: http://dx.doi.org/10.1175/2011MWR3643.1

Bao, X., Y. Luo, J. Sun, Z. Meng, and J. Yue, 2017. Assimilating doppler radar observations with an ensemble kalman filter for convection-permitting prediction of convective development in a heavy rainfall event during the pre-summer rainy season of South China. *Sci. China Earth Sciences*, **60**, 1866–1885, doi: 10.1007/s11430-017-9076-9.

Byun, U.-Y., J. Hong, S.-Y. Hong, and H. H. Shin, 2015: Numerical simulations of heavy rainfall over central Korea on 21 September 2010 using the WRF model. *Adv. Atmos. Sci.*, **32**, 855–869.

Chen, T.-C. and C.-C. Wu, 2016: The remote effect of Typhoon Megi (2010) on the heavy rainfall over northeastern Taiwan. *Mon. Wea. Rev.*, **144**, 3109–3131, doi:https://doi.org/10.1175/MWR-D-15-0269.1.

Cho, H.-S. and T.-Y. Lee, 2006: A numerical study of multiple convective bands over the Korean Peninsula. *J. Korean Meteor. Soc.*, **42**, 87–105.

Cote, M. R., 2007: Predecessor rain events in advance of tropical cyclones. M.S. thesis, Dept. of Atmospheric and Environmental Sciences, University at Albany, State University of New York, 200 pp. [Available online at http://cstar.cestm.albany.edu/CAP_Projects/Project10/index.htm.]

Cotton, W. R., G. J. Tripoli, R. M. Rauber, and E. A. Mulvihill, 1986: Numerical simulation of the effects of varying ice crystal nucleation rates and aggregation processes on orographic snowfall. *J. Climate Appl. Meteor.*, **25**, 1658–1680.

Deardorff, J. W., 1980: Stratocumulus-capped mixed layers derived from a three-dimensional model. *Bound. — Layer Meteor.*, **18**, 495–527.

Ding, Y. H., 1992: Summer monsoon rainfalls in China. *J. Meteor. Soc. Japan*, **70**, 373–396.

Furtado, K., P. Field, Y.-L. Luo, X. Liu, Z. Guo, T. Zhou, B. Shipway, A. Hill, and J. Wilkinson, 2018: Cloud-microphysical factors affecting simulations of deep convection during the presummer rainy-season in southern China. *J. Geophys. Res. Atmos.*, **123**, https://doi.org/10.1029/2017JD028192.

Galarneau, T. J., L. F. Bosart, and R. S. Schumacher, 2010: Predecessor rain events ahead of tropical cyclones. *Mon. Wea. Rev.*, **138**, 3272–3297, doi:https://doi.org/10.1175/2010MWR3243.1.

Grell, G., J. Dudhia, and D. Stauffer, 1994: A description of the Fifth-Generation Penn State/NCAR Mesoscale Model (MM5), *NCAR Technical Note*, NCAR/TN-398+STR, 117 pp.

Ikawa, M. and K. Saito, 1991: Description of a nonhydrostatic model developed at the Forecast Research Department of the MRI. *Technical Report of the MRI*, **28**, 238 pp.

Jeong, J.-H., D.-I. Lee, and C.-C. Wang, 2016: Impact of the cold pool on mesoscale convective system–produced extreme rainfall over southeastern South Korea: 7 July 2009. *Mon. Wea. Rev.*, **144**, 3985–4006.

Jou, B., J.-D., W.-C. Lee, and R. H. Johnson, 2011: An overview of SoWMEX/TiMREX and its operation. In *The GlobalMonsoon System: Research and Forecast*, Ed. C.-P. Chang, 2nd edition, World Scientific, pp. 303–318.

Kanada, S., K. Tsuboki, H. Aiki, S. Tsujino, and I. Takayabu, 2017: Future enhancement of heavy rainfall events associated with a typhoon in the midlatitude regions. *SOLA*, **13**, 246–251, doi:10.2151/sola.2017-045.

Kato, T., 1998: Numerical simulation of the band-shaped torrential rain observed over Southern Kyushu, Japan on 1 August 1993. *J. Meteor. Soc. Japan*, **76**, 97–128.

Kato, T. and H. Goda, 2001: Formation and maintenance processes of a stationary band-shaped rainfall observed in Niigata on 4 August 1998. *J. Meteor. Soc. Japan*, **79**, 899–924.

Kato, T. and K. Aranami, 2005: Formation factors of 2004 Niigata-Fukushima and Fukui heavy rainfalls and problems in the predictions using a cloud-resolving mode. *SOLA*, **1**, 1–4.

Kato, T., 2006: Structure of the band-shaped precipitation system inducing the heavy rainfall observed over northern Kyushu, Japan on 29 June 1999. *J. Meteor. Soc. Japan*, **84**, 129–153.

Kim, W. and T.-Y. Lee, 2016: A study of the formation mechanism of a long convection band over the Yellow Sea. *Atmos. Research*, **176–177**, 134–147, doi: http://dx.doi.org/10.1016/j.atmosres.2016.02.019.

Klemp, J. B. and R. B. Wilhelmson, 1978: The simulation of three dimensional convective storm dynamics. *J. Atmos. Sci.*, **35**, 1070–1096, https://doi.org/10.1175/1520-0469(1978)035,1070:TSOTDC.2.0.CO;2.

Kondo, J., 1975: Air–sea bulk transfer coefficients in diabatic conditions. *Bound. — Layer Meteor.*, **9**, 91–112, https://doi.org/10.1007/BF00232256.

Li, X., Y. Luo, and Z. Guan, 2014: The persistent heavy rainfall over Southern China in June 2010: Evolution of synoptic systems and the effects of the Tibetan Plateau heating. *J. Meteor. Res.*, **28**, 540–560.

Liang, J., L. Wu, X. Ge, and C.-C. Wu, 2011: Monsoonal Influence on Typhoon Morakot (2009). Part II: Numerical Study. *J. Atmos. Sci.*, **68**, 2222–2235.

Lin, Y. L., R. D. Farley, and H. D. Orville, 1983: Bulk parameterization of the snow field in a cloud model. *J. Climate Appl. Meteor.*, **22**, 1065–1092.

Liu, A. Q., G. W. K. Moore, K. Tsuboki, and I. A. Renfrew, 2004: A high-resolution simulation of convective roll clouds during a cold-air outbreak. *Geophys. Res. Lett.*, **31**, L03101.

Liu, A. Q., G. W. K. Moore, K. Tsuboki, and I. A. Renfrew, 2006: The effect of the sea-ice zone on the development of boundary layer roll clouds during cold air outbreaks. *Boundary-Layer Meteor.*, **118**, 557–581.

Louis, J. F., M. Tiedtke, and J. F. Geleyn, 1981: A short history of the operational PBL parameterization at ECMWF. *Workshop on Planetary Boundary Layer Parameterization*, 25–27 Nov. 1981, pp. 59–79.

Luo, Y., Y. Gong, and D.-L. Zhang, 2014: Initiation and organizational modes of an extreme-rain-producing mesoscale convective system along a Meiyu Front in East China. *Mon. Wea. Rev.*, **142**, 203–221.

Luo, Y. and Y. Chen, 2015: Investigation of the predictability and physical mechanisms of an extreme-rainfall producing mesoscale convective system along the Meiyu front in East China: An ensemble approach. *J. Geophys. Res. Atmos.*, **120**, 10593–10618, doi:10.1002/2015JD023584.

Luo, Y., 2016: Advances in understanding the early-summer heavy rainfall over South China. The Global Monsoon System: III — A WMO Quadrennial Review, Eds. C. P. Chang *et al.*, Vol. 9, World Scientific Series on Asia-Pacific Weather and Climate, pp. 215–226.

Luo, Y., R. Zhang, Q. Wan, B. Wang, W. K. Wong, Z. Hu *et al.*, 2017: The Southern China Monsoon Rainfall Experiment (SCMREX). *Bull. Ame. Meteorol. Soc.*, **98**, 999–1013.

Maesaka, T., G. W. Kent Moore, A. Q. Liu, and K. Tsuboki, 2006: A simulation of a lake effect snowstorm with a cloud resolving numerical model. *Geophys. Res. Lett.*, **33**, L20813.

Murakami, M., 1990: Numerical modeling of dynamical and microphysical evolution of an isolated convective cloud — The 19 July 1981 CCOPE cloud. *J. Meteor. Soc. Jpn.*, **68**, 107–128.

Murakami, M., T. L. Clark, and W. D. Hall 1994: Numerical simulations of convective snow clouds over the Sea of Japan; Two-dimensional simulations of mixed layer development and convective snow cloud formation. *J. Meteor. Soc. Jpn.*, **72**, 43–62.

Oizumi, T., K. Ssaito, J. Ito, T, Kuroda, and L. Duc, 2018: Ultra-high-resolution numerical weather prediction with a large domain using the K Computer: A case study of the Izu Oshima heavy rainfall event on October 15–16, 2013. *J. Meteor. Soc. Jpn.*, **96**, 25–54. Doi: https://doi.org/10.2151/jmsj.2018-006.

Qian, Q., Y. Lin, Y.-L. Luo, X. Zhao, Z.-C. Zhao, Y. Luo, and X. Liu, 2018: Sensitivity of a simulated squall line during SCMREX to parameterization of microphysics. *J. Geophys. Res. Atmos.*, **123**, 4197–4220. https://doi.org/10.1002/2017JD027734.

Roberts, N. M. and H. W. Lean, 2008: Scale-selective verification of rainfall accumulations from high-resolution forecasts of convective events. *Mon. Wea. Rev.*, **136**, 78–97, doi:https://doi.org/10.1175/2007MWR2123.1.

Saito, K., 1994: A numerical study of the local downslope wind "Yamaji-kaze" in Japan. Part 3: Numerical simulation of the 27 September 1991 windstorm with a non-hydrostatic multi-nested model. *J. Meteor. Soc. Jpn.*, **72**, 301–329.

Saito, K., 1997: Semi-implicit fully compressible version of the MRI mesoscale nonhydrostatic model-Forecast experiment of the 6 August 1993 Kagoshima torrential rain-. *Geophys. Mag. Ser. 2*, **2**, 109–137.

Saito, K., T. Fujita, Y. Yamada, J. Ishida, Y. Kumagai, K. Aranami, S. Ohmori, R. Nagasawa, S. Kumagai, C. Muroi, T. Kato, H. Eito, and Y. Yamazaki, 2006: The operational JMA nonhydrostatic model. *Mon. Wea. Rev.*, **134**, 1266–1298.

Saito, K., J. Ishida, K. Aranami, T. Hara, T. Segawa, M. Narita, and Y. Honda, 2007: Nonhydrostatic atmospheric models and operational development at JMA. *J. Meteor. Soc. Jpn.*, **85B**, 271–304.

Segami, A., K. Kurihara, H. Nakamura, M. Ueno, I. Takano, and Y. Tatsumi, 1989: Operational mesoscale weather prediction with Japan Spectral Model. *J. Meteor. Soc. Jpn.*, **67**, 907–924, https://doi.org/10.2151/jmsj1965.67.5_907.

Shimizu, S., H. Uyeda, Q. Moteki, T. Maesaka, Y. Takaya, K. Akaeda, T. Kato, and M. Yoshizaki, 2008: Structure and formation mechanism on the 24 May 2000 supercell-like storm developing in a moist environment over Kanto Plain, Japan. *Mon. Wea. Rev.*, **136**, 2389–2407.

Shin, C.-S. and T.-Y. Lee, 2005: Development mechanisms for the heavy rainfalls of 6–7 August 2002 over the middle of the Korean peninsula. *J. Meteor. Soc. Japan*, **83**, 683–709.

Skamarock, W. C., J. B. Klemp, J. Dudhia, D. Gill, D. Barker, M. Duda, X. Y. Huang, W. Wang, and J. G. Powers, 2008: A description of the advanced research WRF Version 3. *NCAR Technical Note NCAR/TN-475+STR*, Boulder, Colorado, doi:10.5065/D68S4MVH.

Smagorinsky, J., 1963: General circulation experiments with the primitive equations. I. The basic experiment. *Mon. Wea. Rev.*, **91**, 99–164.

Sun, J. H. and T.-Y. Lee, 2002: A numerical study of an intense quasi-stationary convection band over the Korean peninsula. *J. Meteor. Soc. Japan*, **80**, 1221–1245.

Tsuboki, K. and A. Sakakibara, 2002: Large scale parallel computing of Cloud Resolving Storm Simulator. *High Performance Computing*, Eds. H. P. Zima *et al.*, Springer, pp. 243–259.

Tsuboki, K. and A. Sakakibara, 2006: The prediction experiment of a localized heavy rainfall associated with a typhoon using the cloud-resolving model: A case study of the heavy rainfall event in the northern Kinki District associated with the typhoon 0423 on 20 October 2004. *J. Japan Soc. Natural Disaster Sci.*, **25**, 351–373 (in Japanese with English abstract).

Tsuboki, K. and A. Sakakibara, 2007: Numerical prediction of high-impact weather systems. *The Textbook for Seventeenth IHP Training Course in 2007*, Hydrospheric Atmospheric Research Center, Nagoya University, and UNESCO, 273 pp.

Tsuboki, K., 2008: High-resolution simulations of high-impact weather systems using the cloud-resolving model on the Earth Simulator. *High Resolution Numerical Modeling of the Atmosphere and Ocean*, Springer, New York, Kevin Hamilton and Wataru Ohfuchi, pp. 141–156.

Tsuboki, K., M. K. Yoshioka, T. Shinoda, M. Kato, S. Kanada, and A. Kitoh, 2015: Future increase of super-typhoon intensity associated with climate change. *Geophys. Res. Lett.*, **42**, 646–652, doi:10.1002/2014GL061793.

Tsuguchi, H. and T. Kato, 2014: Objective extraction of heavy rainfall events and statistical analysis on their characteristic features. *Tenki*, **61**, 455–469 (in Japanese).

Tsujino, S., K. Tsuboki, and H.-C. Kuo, 2017: Structure and maintenance mechanism of long-lived concentric eyewalls associated with simulated Typhoon Bolaven (2012). *J. Atmos. Sci.*, **74**, 3609–3634, https://doi.org/10.1175/JAS-D-16-0236.1.

Wang, C.-C., G. T.-J. Chen, T.-C. Chen, and K. Tsuboki, 2005: A numerical study on the effects of Taiwan topography on a convective line during the mei-yu season. *Mon. Wea. Rev.*, **133**, 3217–3242.

Wang, C.-C. G. T.-J. Chen, S.-C. Yang, and K. Tsuboki, 2009: Wintertime supercell thunderstorms in a subtropical environment: Numerical simulation. *Mon. Wea. Rev.*, **137**, 2175–2202.

Wang, C.-C., and W.-M. Huang, 2009: High resolution simulation of a nocturnal narrow convective line off the southeastern coast of Taiwan in the Mei-yu season, *Geophys. Res. Lett.*, **36**, L06815, doi:10.1029/2008GL037147.

Wang, C.-C., G. T.-J. Chen, and S.-Y. Huang, 2011: Remote trigger of deep convection by cold outflow over the Taiwan Strait in the mei-yu season: A modeling study of the 8 June 2007 case. *Mon. Wea. Rev.*, **139**, 2854–2875, doi:https://doi.org/10.1175/2011MWR3613.1.

Wang, C.-C., H.-C. Kuo, Y.-H. Chen, H.-L. Huang, C.-H. Chung, and K. Tsuboki, 2012: Effects of asymmetric latent heating on typhoon movement crossing Taiwan: The case of Morakot (2009) with extreme rainfall. *J. Atmos. Sci.*, **69**, 3172–3196.

Wang, C.-C., Y.-H. Chen, H.-C. Kuo, and S.-Y. Huang, 2013a: Sensitivity of typhoon track to asymmetric latent heating/rainfall induced by Taiwan topography: A numerical study of Typhoon Fanapi (2010). *J. Geophys. Res. Atmos.*, **118**, 3292–3308, doi:10.1002/jgrd.50351.

Wang, C.-C., H.-C. Kuo, T.-C. Yeh, C.-H. Chung, Y.-H. Chen, S.-Y. Huang, Y.-W. Wang, and C.-H. Liu, 2013b: High-resolution quantitative precipitation forecasts and simulations by the

Cloud-Resolving Storm Simulator (CReSS) for Typhoon Morakot (2009). *J. Hydrol.*, **506**, 26–41, doi:10.1016/j.jhydrol.2013.02.018.

Wang, C.-C., J. C.-S. Hsu, G. T.-J. Chen, and D.-I. Lee, 2014a: A study of two propagating heavy-rainfall episodes near Taiwan during SoWMEX/TiMREX IOP-8 in June 2008. Part I: Synoptic evolution, episode propagation, and model control simulation. *Mon. Wea. Rev.*, **142**, 2619–2643, https://doi.org/10.1175/MWR-D-13-00331.1.

Wang, C.-C., J. C.-S. Hsu, G. T.-J. Chen, and D.-I. Lee, 2014b: A study of two propagating heavy-rainfall episodes near Taiwan during SoWMEX/TiMREX IOP-8 in June 2008. Part II: Sensitivity tests on the roles of synoptic conditions and topographic effects. *Mon. Wea. Rev.*, **142**, 2644–2664, doi:https://doi.org/10.1175/MWR-D-13-00330.1.

Wang, C.-C., 2014: On the calculation and correction of equitable threat score for model quantitative precipitation forecasts for small verification areas: The example of Taiwan. *Wea. Forecasting*, **29**, 788–798, doi:https://doi.org/10.1175/WAF-D-13-00087.1.

Wang, C.-C., 2015: The more rain, the better the model performs — The dependency of quantitative precipitation forecast skill on rainfall amount for typhoons in Taiwan. *Mon. Wea. Rev.*, **143**, 1723–1748, doi:https://doi.org/10.1175/MWR-D-14-00137.1.

Wang, C.-C., S.-Y. Huang, S.-H. Chen, C.-S. Chang, and K. Tsuboki, 2016: Cloud-resolving typhoon rainfall ensemble forecasts for Taiwan with large domain and extended range through time-lagged approach. *Wea. Forecasting*, **31**, 151–172, doi:https://doi.org/10.1175/WAF-D-15-0045.1.

Wu, C.-C., K. K. W. Cheung, and Y.-Y. Lo, 2009: Numerical study of the rainfall event due to the interaction of Typhoon Babs (1998) and the northeasterly monsoon. *Mon. Wea. Rev.*, **137**, 2049–2064, doi:10.1175/2009MWR2757.1.

Wu, L., J. Liang, and C.-C. Wu, 2011: Monsoonal Influence on Typhoon Morakot (2009). Part I: Observational Analysis. *J. Atmos. Sci.*, **68**, 2208–2221.

Xue, J. S., 2004: Progresses of researches on numerical weather prediction in China: 1999–2002. *Adv. Atmos. Sci.*, **21**, 467–474, doi:10.1007/BF02915573.

Xue, J. S. and Y. Liu, 2007: Numerical weather prediction in China in the new century progress, problems and prospects. *Adv. Atmos. Sci.*, **24**, 1099–1108, doi:10.1007/s00376-007-1099-1.

Yamasaki, M., 2007: A numerical study of rainfall associated with a Baiu front: A case of Niigata-Fukushima heavy rainfall in 2004. *SOLA*, **3**, 57–60.

Yamasaki, M., 2008: A study of Fukui heavy rainfall in 2004. *J. Meteor. Soc. Japan*, **86**, 369–376.

Zhang, M. and D.-L. Zhang, 2012: Subkilometer simulation of a torrential-rain-producing mesoscale convective system in East China. Part I: Model verification and convective organization. *Mon. Wea. Rev.*, **140**, 184–201.

Zhang, X., Y. Luo, Q. Wan, and Coauthors, 2016: Impact of assimilating wind profiling radar observations on convection-permitting quantitative precipitation forecasts during SCMREX. *Wea. Forecast.*, **31**, 1271–1292.

Chapter 11

What is the Most Important Cause of Diurnal Cycles in Meiyu Frontal Precipitation over China?

Ming Xue*,†,§, Xia Luo*,‡, Kefeng Zhu*, Zhengqi Sun* and Jianfang Fei‡

**School of Atmosphere Sciences, Nanjing University, China*
†Center for Analysis and Prediction of Storms, University of Oklahoma, USA
‡College of Meteorology and Oceanography,
National University of Defense Technology, Nanjing, China
§mxue@ou.edu

The most important cause of Meiyu precipitation diurnal cycle over China is investigated using numerical simulations at 4 km grid spacing. Six days from the 2014 Meiyu season are used to produce a 'north' composite rainband over the Yangtze-Huaihe River Basin and another six days used to produce a 'south' composite band. Both rainbands have peak rainfall in the early morning while the south band has a secondary peak in the afternoon. Low-level ageostrophic winds are found to exhibit diurnal cycles with clockwise rotations and their directions and magnitudes depend on the background geostrophic monsoon flows. Net moisture flux into a control volume enclosing each rainband is almost purely due to the ageostrophic winds. For both rainbands, net flux reaches maximum at about 04 LST, about 3–4 hours before morning precipitation peak. For the north band, a prominent minimum occurs at ∼19 LST, 4 hours before the precipitation minimum. The moisture fluxes through the southern control volume boundary make the largest contributions to the net flux and its diurnal variations. The diurnal variations of the ageostrophic winds and their relationship with the background monsoon flows agree very well with the prediction of the Blackadar boundary layer inertial oscillation theory, and the convergence forcing by the ageostrophic winds resulting from the inertial oscillations play a most important role in modulating the diurnal cycles of Meiyu front precipitation, including the creation of early morning peak and evening minimum. Feedback of latent heat release plays only a secondary role. The commonly recognized diurnal monsoon variability can be explained by the Blackadar inertial oscillation theory.

1. Introduction

The Meiyu season is known as a period of persistent rainfall that, over the Yangtze–Huaihe River Basin (YHRB) of China, typically occurs from mid-June to mid-July each year. Meiyu precipitation usually occurs in the form of a rainband, along a quasi-stationary Meiyu front that is a confluent zone between southerly summer monsoon flows and typically weaker northerly flows from the mid-latitudes (Ding 2010, 1992). Persistent precipitation from Meiyu rainfall can cause serious flooding along the river basin (e.g., Luo *et al.* 2014).

The Meiyu rainband has been shown to have large variabilities on timescales from hours to interannual frequency; the latter has a close linkage with the seasonal advancement and retreat of the East Asian summer monsoon (Chang *et al.* 2000; Ding and Chan 2005; Sampe and Xie 2010; Li *et al.* 2017). An understanding of the variabilities of Meiyu precipitation is important for improving heavy rainfall prediction.

Early studies on diurnal variations of precipitation over East Asia often employed satellite data as the proxy for precipitation (e.g., Murakami 1983; Asai and Kodama 1998; Wang *et al.* 2004), and some studies (e.g., Wang *et al.*

The Multiscale Global Monsoon System
Edited by C. P. Chang et al.

2004) had identified diurnal and semi-diurnal signals in precipitation over China in summer months. In more recent years, with the availability of hourly precipitation data, diurnal variations in summer season precipitation over China have received much attention (Zhou *et al.* 2008; Bao *et al.* 2011; Chen *et al.* 2009a, 2012, 2014; Xu *et al.* 2009; Yu and Li 2016; Yu *et al.* 2007a, 2007b, 2014; Yuan *et al.* 2012). It has been shown that the summer precipitation diurnal cycle over China is strongly region dependent, which may result from the combined effects of large-scale atmospheric circulations and inhomogeneous surface conditions over the region (Hirose and Nakamura 2005; Yuan *et al.* 2012; Yu *et al.* 2014). In eastern China, two rainfall peaks with comparable amplitudes appear in the late afternoon and early morning (Yu *et al.* 2007b; He and Zhang 2010; Yuan *et al.* 2012). Over regions with concentrated summer precipitation, such as the upper and middle reaches of the YHRB, most precipitation occurs in the early morning (Yu *et al.* 2007b; Yin *et al.* 2009). The diurnal cycle of the Meiyu rainband, first explored in Geng and Yamada (2007), also attains such a morning peak, suggesting that morning rainfall may dominate these areas in the Meiyu season. Unlike the afternoon rainfall peak, the early morning peak is much less well understood (Luo *et al.* 2014). Some studies have suggested that nighttime cloud top radiative cooling could be the process behind the nocturnal rain in the western part of China (Yin *et al.* 2009).

The synoptic-scale, southwesterly low-level flows coupled with active monsoon systems have long been recognized as crucial factors for maintaining the rainfall along the YRHB (Wang *et al.* 2000; Ding *et al.* 2001). The close relationship between the low-level southwesterly jet and heavy precipitation over the East Asia has been well recognized (Chen and Yu 1988; Chen *et al.* 2005). The synoptic-scale low-level jets (LLJs), which have their peak speed between 850 and 700 hPa levels, do not explain the diurnal variations of rainfall, however.

In contrast, the important role of nocturnal LLJ in affecting the diurnal cycle of rainfall has long been recognized over the Great Plains of the United States (Bonner 1966; Carbone and Tuttle 2008; Wallace 1975; Helfand and Schubert 1995; Tuttle and Davis 2006; Trier *et al.* 2014; Zhong *et al.* 1996). The primary mechanisms of the Great Plains LLJ include the inertial oscillation mechanism of Blackadar (1957) associated with the diurnal variation of surface frictional effect on the boundary layer, and the thermal wind adjustment mechanism in response to diurnal change in the pressure gradient force due to the heating and cooling of sloped terrains (Holton 1967).

Based on the inertial oscillation mechanism of Blackadar, a diurnally oscillating ageostrophic wind vector would rotate clockwise with time at a period of $2\pi/f$, where f is the Coriolis parameter in idealized conditions (Blackadar 1957). More specifically, in the afternoon when surface heating is active, the boundary layer is well mixed and surface friction causes the boundary layer flow to slow down significantly, creating a three-force-balanced (pressure gradient force, Coriolis force, and frictional force) state, the wind vector deviates toward the low-pressure side of the pressure contours, and the wind speed is sub-geostrophic. After sunset in the evening, the boundary layer becomes stable and turbulence mixing dies off, causing the boundary layer to be "decoupled" with the underlying ground, i.e., the boundary layer flow no longer "feels" the surface friction. The effective removal of the frictional force causes the boundary layer flow to accelerate, passing the geostrophic balanced state to become super-geostrophic (and up-pressure gradient), like a pendulum swinging past its gravitationally balanced position. The peak wind speed is usually reached at 2–3 o'clock in the morning, creating a nocturnal boundary layer LLJ. In general, the Blackadar inertial oscillation theory explains better the behaviors of the Great Plains LLJ than the Holton theory, although most recently a unified theory has been proposed that combines the two effects (Shapiro *et al.* 2016), resulting in a better fit of the theoretical prediction to observations. In comparison, the boundary layer LLJ and its diurnal oscillations over China have been studied less, with Du and Zhang (2012) and Du *et al.* (2014, 2015a, 2015b) being few of the examples, suggesting that inertial oscillations play important roles.

During the warm season of China, the onset of southerly or southwesterly summer monsoon sets up strong low-level flows, often in the form of strong synoptic-scale LLJs (Du *et al.* 2014). Such a condition is favorable for strong boundary layer LLJs, as the amplitude of inertial oscillations is proportional to the geostrophic wind speed (Blackadar 1957; Klein *et al.* 2016). In a climatological study examining the relation between rainfall duration and diurnal variation in the warm season precipitation over central eastern China, Yu *et al.* (2009) noticed that the winds at surface stations located on top of mountains (of about 1–1.5-km height MSL) showed distinct diurnal variations, and the winds at those stations were strongest pre-dawn and weakest in the afternoon. The daily wind anomaly also rotated clockwise from late night to late afternoon, "similar to the diurnal variation of the troposphere low-level wind". While the potential linkage of such lower tropospheric wind variations to diurnal precipitation variations was recognized, no clear explanation was offered on the cause of the diurnal variations of wind speed and direction by Yu *et al.* (2009) or in any other related studies (e.g., Chen *et al.* 2009a, 2010). In Chen *et al.* (2009a), the low-level winds over southern China were found to be strongest at late night or in the early morning, and they referred to this diurnal change as diurnal monsoon variability (DMV). They did point out that the DMV was related closely to ageostrophic wind component, and suggested that the clockwise rotation of the wind vector at night was probably a response to the inertial oscillation, with reference to Blackadar (1957). The relationship between DMV and Meiyu frontal rainfall was further documented in Chen *et al.* (2009b), although no clear explanation on the cause of the DMV was given.

In Sun and Zhang (2012), the upward branch of the nighttime thermally driven mountain–plains solenoid (MPS) circulation over the plains east of the Tibetan Plateau (TP) was believed to be primarily responsible for the midnight-to-early-morning rainfall enhancement along the Meiyu front while daytime downward circulation acted to suppress convection over the plains, although an enhanced nocturnal LLJ was also suggested to be a possible important contributor. Sun and Zhang (2012) also suggested that future studies should be conducted to investigate the mechanisms of LLJ diurnal variations and their impacts on the diurnal cycle of precipitation over China. In a more recent study (Chen *et al.* 2017), the role of nocturnal low-level jets (NLLJs) in modulating mesoscale convective systems (MCSs) and the associated heavy rainfall along a Meiyu front was emphasized much more, based on a study of a 5-day period in June 1998 when moist convection repeatedly occurred in the midnight-to-morning hours causing severe flooding. Their study found that "a train of MCSs concurrently developed along a quasi-stationary Meiyu front and coincided with the impact of a monsoon surge on a frontogenetic zone at night. This process was regulated primarily by a NLLJ in the southwesterly monsoon that formed over southern China and extended to central China." The lifting and the destabilization by the convergence at the northern terminus of the LLJ were found to be much stronger than free atmospheric forcing. Their study suggests that the MPS circulation forcing likely played a much smaller role than the LLJ forcing, for their period of study at least.

It is clear that the most important factor responsible for the diurnal cycles of Meiyu precipitation is still a matter of debate. In a recently published study, by Xue *et al.* (2018), forecasts using the WRF model at a 4-km grid spacing over China during a full Meiyu season were used to investigate the primary mechanism of the Meiyu rainfall diurnal oscillations. Specifically, 6 days from the 2014 Meiyu season were used to produce a "north" composite rainband over the YHRB and another 6 days used to produce a "south" composite band. Both rainbands had peak rainfall in the early morning while the south band had a secondary peak in the afternoon. Low-level ageostrophic winds (AGWs) were found to exhibit diurnal cycles with clockwise rotations, and their directions and magnitudes depended on the background geostrophic monsoon flows. Net moisture flux into a control volume enclosing each rainband was found to be almost purely due to the ageostrophic winds. For both rainbands, net flux reached a maximum at ~04 LST, about 3–4 h before the morning precipitation peak. The moisture fluxes through the southern control volume boundary made the largest contributions to the net flux and its diurnal variations. The diurnal variations of the ageostrophic winds and their relationship with the background monsoon flows agreed very well with the prediction of the Blackadar boundary layer inertial oscillation theory, and the convergence forcing by the ageostrophic winds resulting from the inertial oscillations was found to play a paramount role in modulating the diurnal cycles of the Meiyu front precipitation, including the creation of an early morning peak and evening minimum. Feedback of latent heat release was shown to only play a secondary role. This chapter is an attempt to highlight the key findings and views of Xue *et al.* (2018), based on figures reproduced from their paper. Much of the text is also reproduced from their paper.

2. Diurnal Variations of Meiyu Precipitation Bands

The WRF 4-km forecasts used in Xue *et al.* (2018) to study the Meiyu precipitation diurnal cycles were documented in Zhu *et al.* (2018). It was shown that the seasonal migration of the precipitation zones, the spatial distribution and propagation of precipitation systems, and the diurnal cycles of precipitation were well reproduced by the 4-km WRF model in the warm seasons of 2013 and 2014. In Xue *et al.* (2018), 12–36-h forecasts of 2014 during the Meiyu season were used to produce composites of 24-h forecasts to perform diagnostic analyses. The ageostrophic component of winds was separated from the geostrophic component, which was calculated from the smoothed geopotential height field. Low-level moisture fluxes by the total wind and AGW components were examined, in terms of the contribution to the total low-level moisture flux convergence forcing into the Meiyu frontal zone.

The YHRB Meiyu season of 2014 started on 14 June and ended on 11 July. During the season, 12 days had a clearly identifiable Meiyu rainband. Among the 12 days, on June 18, 19, 20, 21, 27, and 28, the Meiyu rainband was at more southern locations, and on June 25, 26, 29, 30, and July 1 and 2, the rainband was located further north in the YHRB region. The two groups of days were used to produce two composite rainbands, referred to as the south and north rainband, respectively. As shown in Xue *et al.* (2018), compared to rainfall observations, the average characteristics of the rainbands, including their diurnal cycles and propagation, were fairly well captured by the model, except for some location and intensity biases. Figure 2 shows the latitude–time and longitude–time Hovmöller diagrams for precipitation within the 20°–35°N and 105°–120°E domains, for observations and model simulations averaged over the 12 days. Generally, even at the latitudes of the primary rainband between 27° and 30°N, the precipitation between 20 and 02 LST is very weak. The precipitation is strongest in the early morning hours around 08 LST (Fig. 2a). There is a hint of secondary maximum in the precipitation in the 23°–27°N latitudinal zone, corresponding to the southern locations of rainband mentioned earlier. In the forecasts (Fig. 2c), the two latitudinal zones of heavier precipitation are reproduced reasonably well, together with the trend of southward propagation in the afternoon, although there are clear biases in the predicted intensity and location of precipitation. In Figs. 2a and 2c, we also notice a precipitation maximum at around 23°N in the afternoon hours between 13 and 18 LST. Such precipitation is clearly due to afternoon surface heating, which tends to trigger convection more easily at the lower latitudes in southern China.

In the longitude–time Hovmöller diagram (Figs. 2b and 3d), the most pronounced feature seen is the precipitation maximum in the early morning hours, which starts earlier shortly after 00 LST and reaches a maximum between 04 and 06 LST in the western part of the domain (between 106°E and 110°E), and occurs later after 04 LST and reaches a maximum around 08 LST over the eastern part (between 112° and 120°E); such an east–west phase difference of diurnal precipitation cycles is still a topic of debate (Chen *et al.* 2010, 2017) and is not the focus of this chapter.

The secondary afternoon peak is also evident in both observations (Fig. 2b) and WRF forecasts (Fig. 2d), with the model peak being stronger and delayed by a couple of hours. Ignoring the afternoon thermal convection, precipitation starts to weaken significantly after 10 LST west of 114°E, and east of 117°E the morning precipitation peak is maintained through the afternoon until after 18 LST. The minimum is found in the evening through midnight, with some phase delay going eastward. Overall, the intensity, location, diurnal cycles of precipitation, the peak and minimum hours, and the phase delay along the Meiyu rainband are all reasonably well reproduced in the model forecasts. In the rest of this chapter, model output will be used to perform diagnostic analyses.

Figure 1 shows the composite (average) hourly precipitation of the south and north rainbands, at 08, 14, and 22 LST and 06 LST (of the second day). At 08 LST (Fig. 1a and e), the Meiyu rainband is best organized and strongest at both southern (Fig. 1a) and northern (Fig. 1e) locations. In the early afternoon (Fig. 1b and f), the rainband becomes broken and splits into somewhat disconnected patches in the western half of the domain while the band over eastern China remains continuous. At 14 LST, scattered afternoon thermal convection is evident over southern China near the coast, contributing to the afternoon secondary maximum. By 22 LST, however, the northern rainband has weakened markedly (Fig. 1g). The southern band (Fig. 1c) is also much weakened at the original location while convective cells reminiscent of afternoon convection remain further south. By 06 LST in the early morning of the next day, both rainbands have regained much of their strength (Figs. 1d and 1h) although they are not as well organized yet, compared to those at 08 LST (Fig. 1a and e).

The mean hourly rainfalls within the parallelograms shown in Fig. 1 for the two rainbands are plotted in Fig. 3. It is clear that the north band has a pronounced minimum at around 23 LST and a maximum around 08 LST. The afternoon peak around 14 LST is hardly identifiable from the curve. For the south band, the evening minimum occurs around 00 LST, and the major morning peak also occurs at 08 LST. The secondary afternoon peak occurs at 17 LST, and it is much more pronounced than that of north band, reflecting more favorable conditions for afternoon thermal convection in the southern part of China.

3. Boundary Layer Inertial Oscillations and Their Relationship with Meiyu Rainfall

3.1. *Diurnal variations of low-level geostrophic and ageostrophic winds*

Figure 4 presents the diurnal variations of composite large-scale fields which include the filtered geopotential height, the geostrophic and

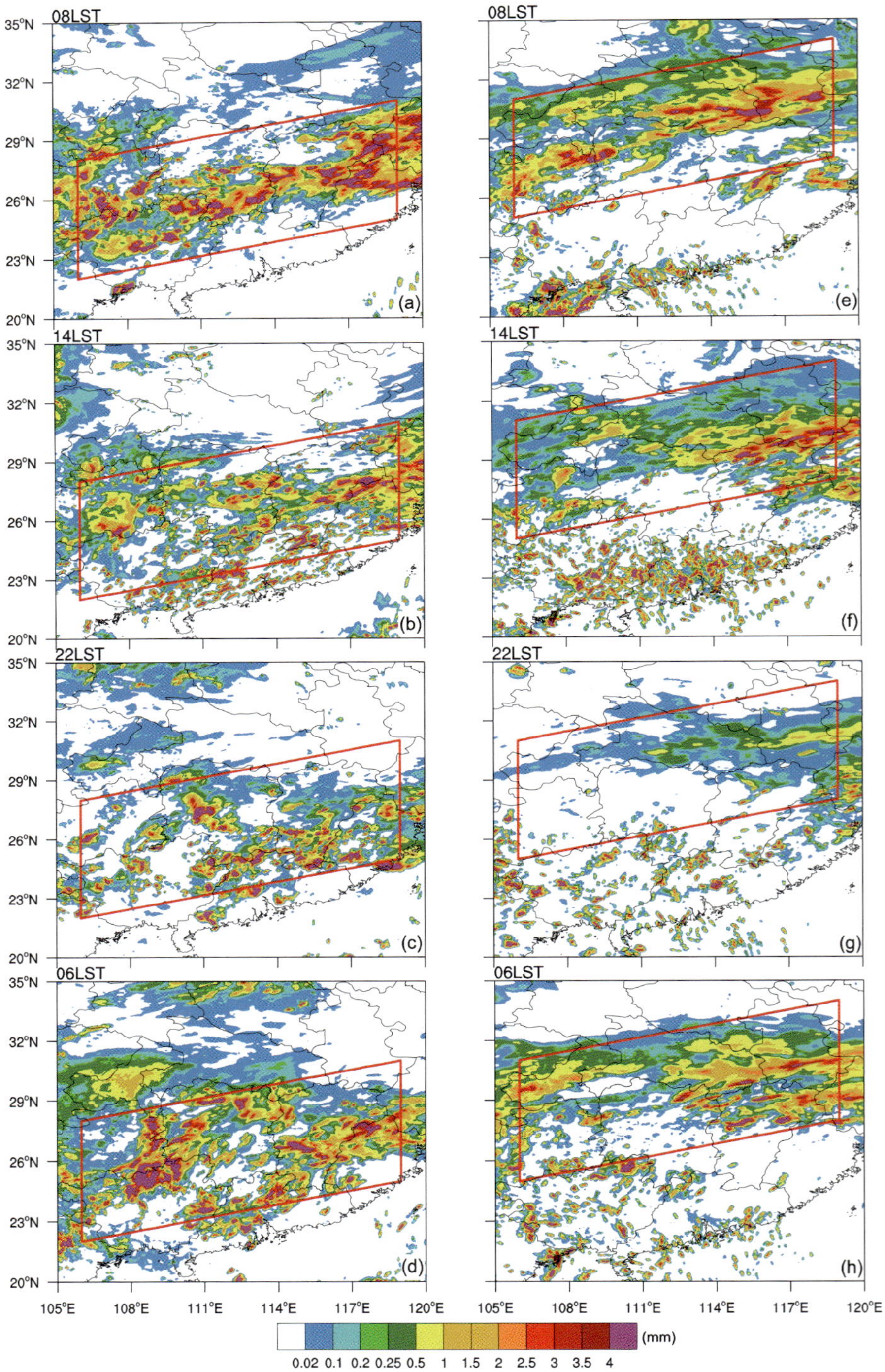

Fig. 1. The composite (average) hourly accumulated Meiyu rainfall of the south rainband cases at (a) 08 LST (LST = UTC + 8 h), (b) 14 LST, (c) 22 LST, and (d) 06 LST. (e)–(h) are the same as Figs. 1a–3d but for the north rainband cases. Red parallelogram in left and right panels denotes the specific south and north rainband regions, respectively (reproduced from Xue *et al.* 2018).

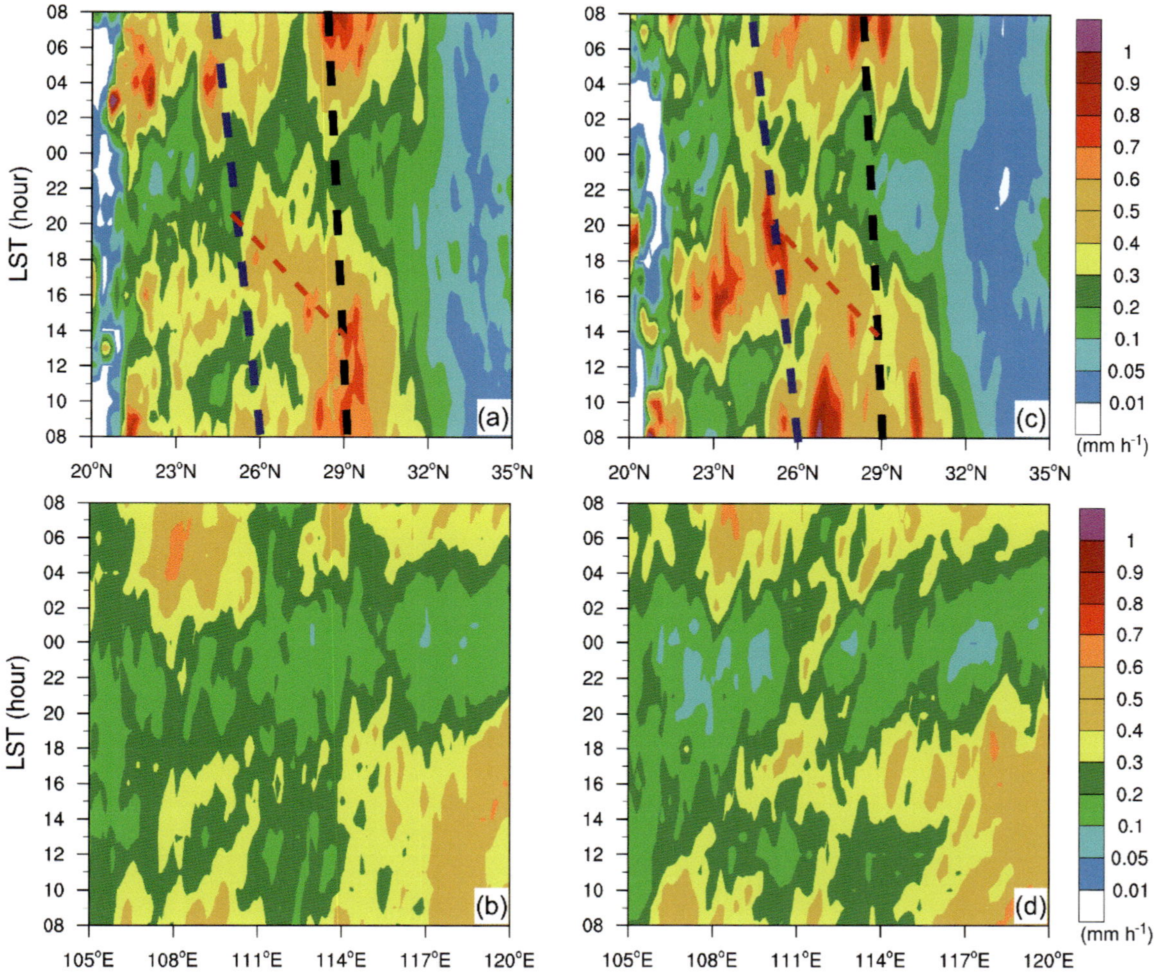

Fig. 2. (a) Latitude–time Hovmöller diagram of rain rate averaged over 105°−120°E (see Fig. 1) and (b) longitude-time Hovmöller diagram of rain rate averaged over 20°–35°N derived from observations, corresponding to all 12 days of WRF forecasts used in this study. (c) and (d) correspond to (a) and (b), respectively, but for WRF simulations. Blue and black dashed lines in (a) and (b) roughly indicate the south and north rainbands, respectively (reproduced from Xue *et al.* 2018).

ageostrophic winds at 925 hPa, and the level where the nocturnal boundary layer LLJ is expected. The geostrophic wind vectors are clearly parallel to the geopotential height contours. The large-scale circulation patterns are maintained throughout the day, although there are some diurnal changes; there is a tendency for the western Pacific subtropical high (WPSH) to retreat toward the ocean in the afternoon and move back toward the interior continent at night. This should be due to daytime continental heating that reduces pressure over land more than that over the ocean.

In comparison, there are very large diurnal changes with the ageostrophic wind vectors (Fig. 4). For the south rainband, the southerly and southwesterly geostrophic flows form a confluent Meiyu frontal zone with the low-pressure trough north of 29°N at 08 LST (Fig. 4a). At this time, the AGW speeds are as strong as those of geostrophic flows, and the winds are at wide angles to the geopotential height contours and

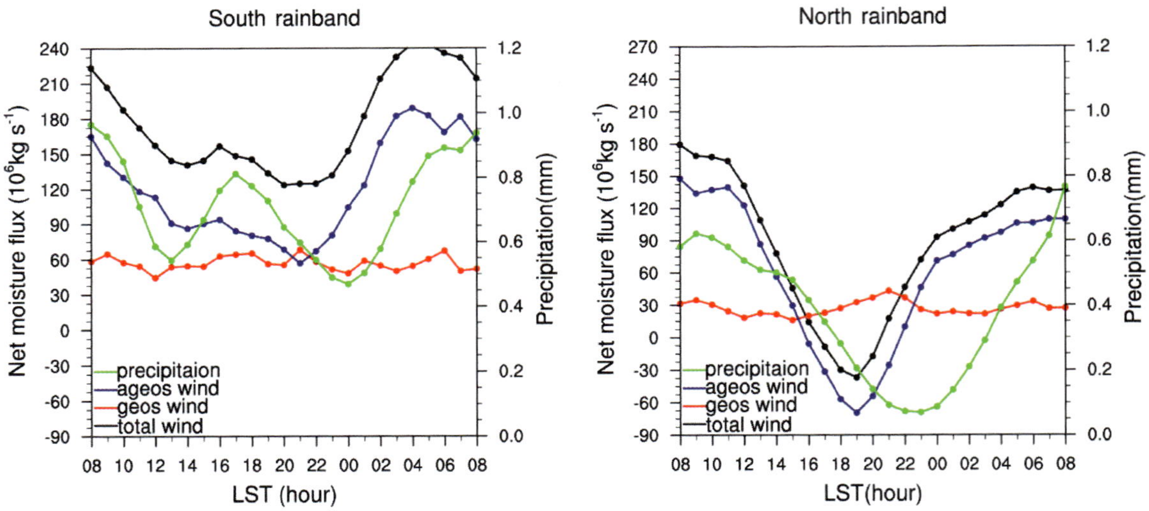

Fig. 3. (a) Diurnal variations of net moisture fluxes (10^6 kg s^{-1}) into the budget region of the south rainband, due to the total wind (black line), geostrophic wind (red line), and ageostrophic wind (blue line) and the mean precipitation rate (green line) within the budget region. The fluxes are integrated from the surface to 700 hPa. (b) is the same as (a) but for the north band.

point toward the convergence zone from both the south and north sides, creating significant low-level convergence forcing. The AGWs on the south side also significantly enhance the northward moisture fluxes into the frontal zone.

In the afternoon at 14 LST (Fig. 4b), the AGWs become much weaker and are mostly parallel to the geostrophic winds on the south side, significantly reducing the convergence forcing in the frontal zone and the moisture transport into the zone. At 20 LST, both geostrophic and ageostrophic winds along the southern boundary of the parallelogram are weak (Fig. 4c) while AGWs across the northern boundary are much stronger. By 02 LST, much of the cross-isobaric AGWs seen at 08 LST has redeveloped near the southern boundary, contributing again significantly to the flow convergence within the frontal zone (Fig. 4d). Over the next 6 h, the WPSH moves inland further, accompanied by the associated northward flow advancement. At 08 LST, the wind directional shear is also strongest in the frontal zone (Fig. 4a). In general, there are clockwise rotations in the AGW vectors throughout the day.

In the later period of the Meiyu season corresponding to most of the north band days, the WPSH advances further inland, and its diurnal position change, while still present, becomes less pronounced (right panels of Fig. 4). This provides a more stable framework to apply the Blackadar inertial oscillation theory that assumes that the background pressure gradient field does not change. The north band cases are more representative of the typical YHRB Meiyu rainband (while the south band days are reminiscent of the late spring south China rainfall season).

In the morning at 08 LST, perhaps partly due to the inland advancement of the WPSH, the higher-latitude low-pressure system is weak (Fig. 4e). The seasonal advancement of summer monsoon concurrently with the northward advancement of WPSH creates a broad region of strong moisture transport from the southern oceans into the frontal zone. On the southern boundary of the parallelogram, the AGWs have significant cross-boundary components, enhancing convergence forcing in the frontal zone (Fig. 4e). By 14 LST, the ageostrophic winds

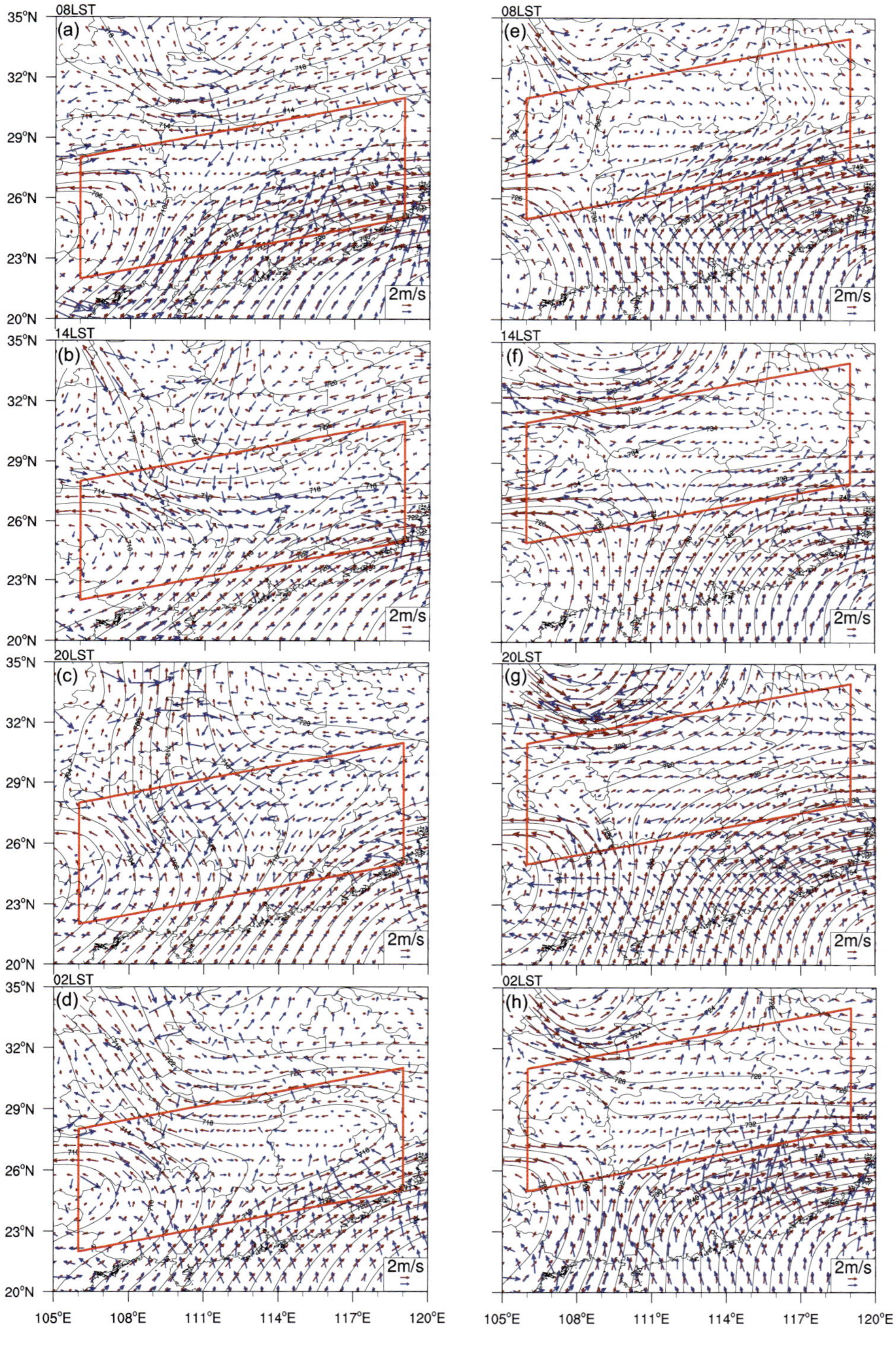

Fig. 4. Composite, spatially filtered geopotential height (black lines, at 2 gpm intervals), geostrophic winds (brown arrows), and AGWs (blue arrows) at 925 hPa of the south rainband days at (a) 08 LST, (b) 14 LST, (c) 20 LST, and (d) 02 LST. (e)–(h) are the same as (a)–(d) but for the north band days (reproduced from Xue *et al.* 2018).

have again rotated mostly into the geostrophic wind directions, removing much of the cross-boundary component. At 20 LST, the situation is still similar, except that the AGWs are now pointing northwestward on the south side. By 02 LST, much of the cross-isobaric AGWs has redeveloped, contributing again to convergence into the frontal zone (Fig. 4h). The above diurnal oscillations of low-level boundary layer flows, especially in the ageostrophic component, create diurnal variations in low-level convergence forcing and southerly moisture fluxes into the frontal zone, which are consistent with the early morning precipitation peak and evening minimum observed with both south and north bands.

3.2. *Diurnal variations of moisture fluxes into frontal zone and their relationship with Meiyu rainfall*

The moisture fluxes through the four boundaries of the parallelogram shown in Fig. 4 up to 700 hPa are calculated, and the net moisture flux (water mass per second) into the control volume can be calculated from the four fluxes. The net moisture fluxes into the two control volumes due to total, geostrophic, and ageostrophic winds are plotted in Fig. 3 as a function of the time in a day. The net fluxes due to geostrophic winds do not change much throughout the 24 h and therefore should not contribute much to the diurnal cycle of precipitation. They are positive mainly because moisture content is higher in the flow into the control volume from the south, even though the geostrophic winds should be more or less non-divergent; the geostrophic winds themselves provide little dynamic convergence forcing to convection, but they do bring net moisture into the frontal zone. The net ageostrophic moisture fluxes are, however, much larger in magnitude, and their diurnal variations closely follow those of total winds (Fig. 3); in fact, almost all of the diurnal variations come from the ageostrophic fluxes. For the south band, the net flux has a maximum at 04 LST, the same as that of AGWs, and the flux maximum is 4 h earlier than the precipitation maximum at 08 LST (Fig. 3a), suggesting that the net flux peak is the cause rather than the result of the precipitation peak. The minimum net fluxes occur around 21 LST, about 3 h ahead of the precipitation minimum, again suggesting that the flux minimum leads to a precipitation minimum in the evening. For the afternoon precipitation peaks around 17 LST, the total flux peak appears almost at the same time, consistent with the mutual support between the two peaks for deep thermal convection. Throughout the day, the net moisture flux remains positive (Fig. 3a), supporting sustained though diurnally varying rainfall in the south rainband region.

For the northern band, a more pronounced net flux minimum is seen that precedes an equally pronounced precipitation minimum, by 3–4 h again. In fact, after 16 LST, the net total wind flux turns negative, which means moisture is being drawn out of the control volume; as a result, existing precipitation rapidly decreases until almost zero at 22–23 LST. After 19 LST, the net flux rapidly increases but it is not until 20 LST that the flux turns positive again. After 23 LST, the total precipitation starts to increase again (Fig. 3b), and reaches the maximum strength around 08 LST. The sequence of events shown in Fig. 3 provides rather convincing evidence that the net moisture flux variations to a very large extent cause diurnal variations in precipitation in the Meiyu rainband, and the net moisture fluxes come almost exclusively from the AGW component. We note that there was some inaccuracy with the total flux calculations in Fig. 9 of Xue *et al.* (2018), which somewhat underestimate the total net fluxes due to geostrophic winds. The inaccuracy does not affect the conclusions, and the accurate version is show here in Fig. 3.

Xue *et al.* (2018) further calculated the time-lagged correlations between the two series for both bands. For the north band, the correlation is maximized with a 2–3 h time lead in the net flux while for the south band, the maximum occurs at a 2-h lead time. At a phase lead of 2–3 h, the correlation coefficient between the two time series is 0.77 and 0.86. The results suggested that the diurnal changes in the net moisture fluxes are possibly the cause of the precipitation diurnal changes.

3.3. *Ageostrophic moisture fluxes into frontal zone and the role of boundary layer inertial oscillations*

Hourly variations of the mean vector moisture fluxes at each boundary of the parallelogram (per unit horizontal length) are shown in Fig. 5 to give a more direct view of flux direction changes. In the figure, the tips of the mean vector moisture fluxes are sequentially connected, and it can be seen that at all boundaries for both bands, both total and AGW flux vectors, show clockwise rotation, with the ageostrophic fluxes forming nearly closed circles while those of total fluxes tend to remain open due to diurnal changes in geostrophic winds. In about half of the cases, the ageostrophic circles enclose the coordinate origin (ageostrophic wind vector rotates around the origin as predicted by the Blackadar inertial oscillation theory) while in other cases the circles do not. The non-stationarity of the geostrophic flow around which the AGW vector is supposed to rotate should be part of the reason. Underestimation of the geostrophic component of the winds is another possible reason, given that the filter procedure used likely leads to underestimation of the horizontal pressure gradient. Still, the clockwise rotational behaviors indicate that the Blackadar boundary layer inertial oscillation theory should explain to a large extent the systematic diurnal wind speed and direct changes seen here. With the theory, the amplitude of the AGW is proportional to the speed of the geostrophic wind; hence, the amplitudes of oscillations are the largest at the southern and western boundaries where the southerly and southwesterly monsoon flows are the strongest. This is consistent with the observation by Chen *et al.* (2013) that diurnal variations of monsoon are stronger on days when the summer monsoon is active. On the northern boundary of the north band, the oscillation amplitude is the smallest (Fig. 5h). The Blackadar theory also predicts strongest winds in the form of boundary layer LLJs around 3 LST, and the weakest winds in the afternoon when boundary layer mixing is the strongest.

The moisture fluxes through individual boundaries (per unit horizontal length in $\mathrm{kg\,m^{-1}\,s^{-1}}$) over the depth of 1,000–700 hPa are plotted in Fig. 6. As mentioned, geostrophic winds and associated moisture fluxes exhibit small diurnal variations; we can see an increase in the geostrophic fluxes in the afternoon through evening, which can at least be partially attributed to the large-scale mountain–plains and land-ocean solenoids (during the a day, a stronger pressure gradient can develop between the WPSH and lower pressure over the continent). The fluxes at the east and west boundaries tend to be in phase, and those at the north and south boundaries also tend to be in phase; therefore, their contributions to the net fluxes tend to cancel each other, leading to small contributions to the net fluxes overall by the geostrophic fluxes.

At the east and west boundaries, the ageostrophic fluxes are also generally in phase, and the fluxes are generally positive from midnight through early afternoon, and negative in the evening, consistent with the fluxes seen in Fig. 5. Such fluxes point mostly eastward at 14 LST and westward at 20 LST (not shown), and are consistent with clockwise rotation due to inertial oscillations (Fig. 5). At the west boundary, the diurnal variation of westerly inflow

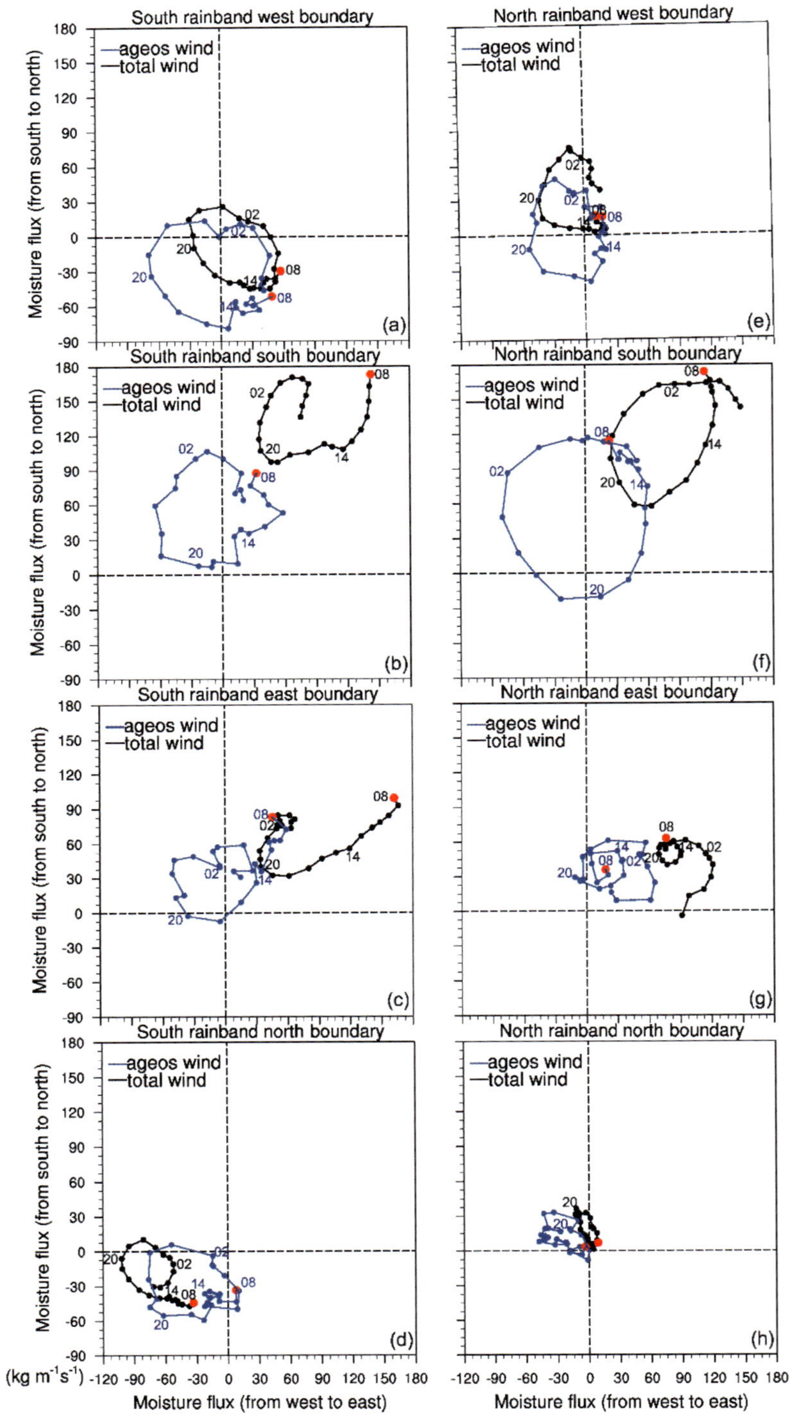

Fig. 5. Plots of vector moisture fluxes (per unit horizontal length) by total wind (black line) and AGW (blue line) through the (a) west, (b) south, (c) east, and (d) north boundaries of the south rainband control volume. (e)–(h) are the corresponding plots for the north rainband region. The dots on the curves correspond to the tips of the flux vectors and the red dot on each curve marks the start time at 08 LST (reproduced from Xue *et al.* 2018).

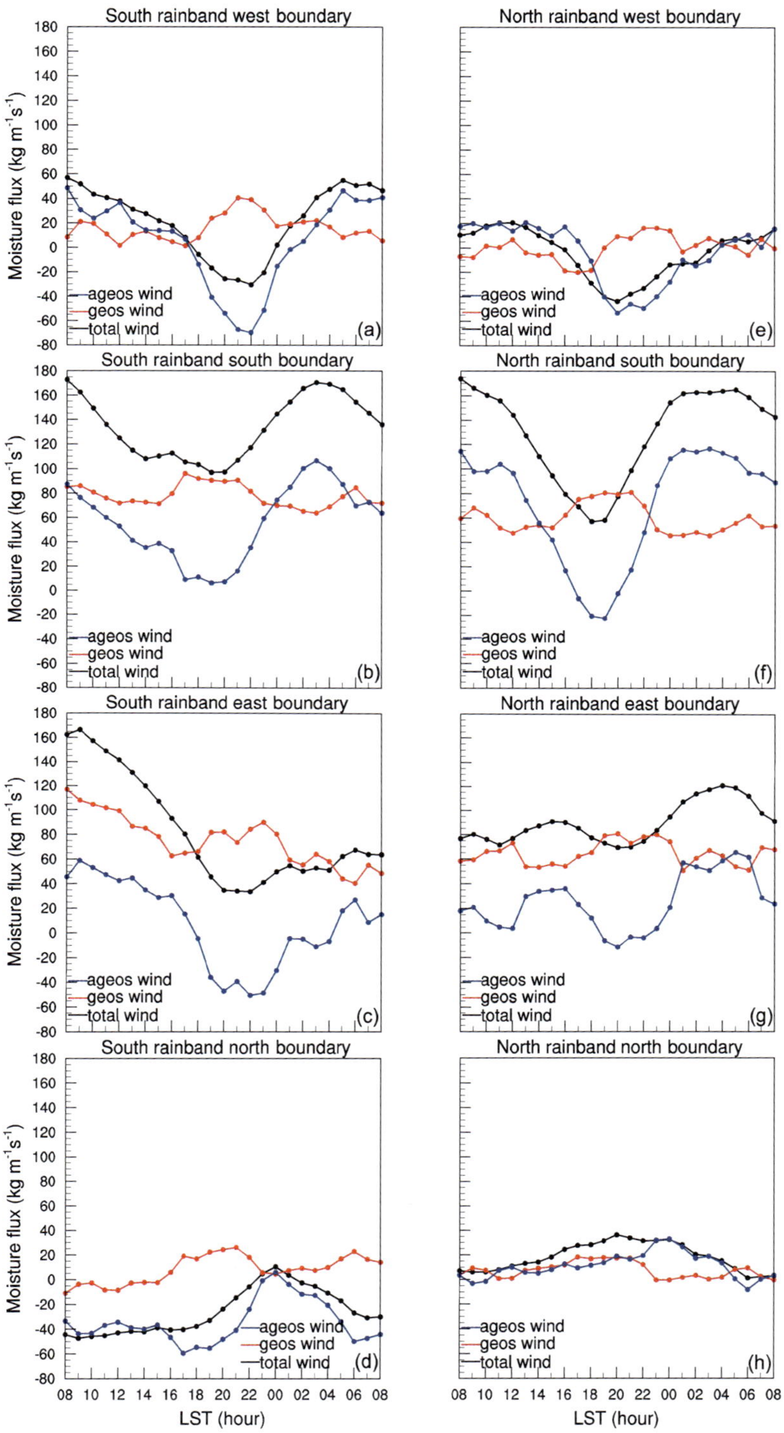

Fig. 6. Diurnal variations of moisture fluxes (per unit length) by total winds (black line), geostrophic winds (red line), and AGWs (blue line) at (a) west, (b) south, (c) east, and (d) north boundary of the south rainband region. In (e)–(h) are the corresponding fluxes for the north rainband region (reproduced from Xue *et al.* 2018).

for the south band is more distinct, probably because it is more affected by the mountain–plains circulations. At the northern boundary, the diurnal variations of fluxes of the north rainband are negligible (Fig. 6h), a result of weak geostrophic flows there.

The relatively fast phase rotation for the north rainband seems to be more evident in Fig. 6. For instance, at the south boundary, for the south rainband, the morning peak of southwesterly inflow occurs at 03 LST (Fig. 6b), which is 2 h later than the peak time for the north rainband (Fig. 6f). The larger Coriolis parameter may have played a role (for faster rotation due to inertial oscillation), as may the geostrophic wind direction differences.

3.4. *Relative role of latent heat feedback in frontal precipitation*

Given the heavy and persistent precipitation involved, latent heating can be a significant contributor to increasing and maintaining Meiyu frontogenesis and the associated frontal circulations, thereby enhancing the lower tropospheric convergence and rainband development (Ding and Chan 2005; Kuo and Anthes 1982). For this reason, Xue *et al.* (2018) asked the following question: "What kind of role does the release of latent heating play in the Meiyu rainfall diurnal cycle and what is the consequence of the latent heat release?" To help answer this question, they reran the forecast of 12 UTC June 29, 2014 with the effect of latent heating turned off 10 h into the model simulation. In the experiment, condensation was allowed to occur within the microphysics but the latent heat was discarded, disabling the latent heating feedback from the precipitation. Their results showed that in spite of the latent heating feedback, the peak value in the low-level moisture flux through the southern boundary of the control volume at around 04 LST is little changed, in both magnitude and timing, indicating that any positive feedback of the latent heat release associated with the enhanced nighttime precipitation plays at most a secondary role in modulating the precipitation diurnal cycles associated with the Meiyu rainband.

4. Conceptual Model and Summary

Based on the previous results, Xue *et al.* (2018) proposed a conceptual model for Meiyu rainfall diurnal cycles, which highlights the key role of boundary layer inertial oscillations (Fig. 7). According to Xue *et al.* (2018), in the typical Meiyu season, strong southerly monsoon flows between the WPSH and a low-pressure system over the Yunnan–Guizhou and Tibetan Plateaus to the west create a large-scale confluence zone with mid-latitude flows from the north and a quasi-stationary Meiyu front between south China and YHRB regions. The geostrophic components of such flows are more or less non-divergent and hence do not contribute much to the low-level convergence forcing at the front. In other words, such flows form a roughly classic deformation field that is non-divergent (though it can be frontogenetic through the development of ageostrophic circulations but the temperature gradient across the Meiyu front is typically weak).

In the afternoon, when active surface heating exists, surface friction and boundary layer mixing slow down the boundary layer flow significantly, causing the total wind vector to deviate to the left of the isobars and the geostrophic wind vector, creating a three-force-balanced state, and making the total wind speed sub-geostrophic (see Fig. 7b). After sunset, in the evening, the boundary layer is stabilized and surface friction ceases to exert a drag on the boundary layer flow, and the three-force balance in the afternoon is broken. The net force between the pressure gradient and Coriolis force then causes the flow to accelerate and according to the Blackadar inertial oscillation theory, the tip of ageostrophic wind vector will draw a full circle, clockwise, around

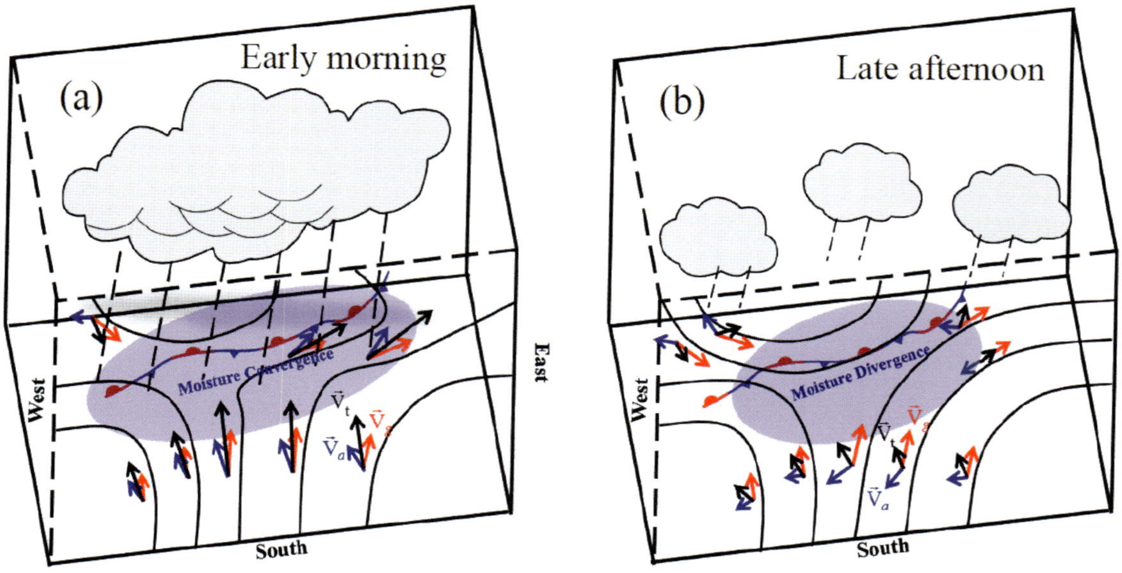

Fig. 7. Conceptual diagram of diurnal variations of boundary layer wind fields, including the total wind $\vec{V}_t$ (black arrows), geostrophic wind $\vec{V}_g$ (red arrows), and AGW $\vec{V}_a$ (blue arrows) over the Meiyu rainband region in the (a) early morning and (b) late afternoon. The geopotential height contours also shown with black lines and the quasi-stationary Meiyu front is marked by standard symbols (adopted from Xue *et al.* 2018).

the tip of the geostrophic wind vector, assuming that the background pressure gradient force does not change, and the surface friction of the next day does not interfere. The time to complete the full circle is $2\pi/f$, where f is the Coriolis parameter. This creates the inertial oscillations of the boundary layer flow. The magnitude of the AGW is proportional to the background geostrophic wind, hence, the stronger the background (southerly monsoon) flow, the stronger the ageostrophic inertial oscillation.

Usually, between midnight and early morning, the AGW vector points in a similar direction as the geostrophic wind, making the total wind super-geostrophic (see Fig. 7a). It is the AGW that creates low-level flow convergence and hence low-level lifting forcing. Strong, sustained super-geostrophic flows from the south also transport more moisture into the frontal zone. The two processes create a precipitation maximum in the early morning, a few hours after the net moisture flux into the frontal zone reaches its maximum. In the afternoon, the AGWs are directed in roughly the opposite direction as the geostrophic winds, effectively drawing air out of the frontal zone, creating a net moisture flux out of the frontal zone, thereby causing the frontal precipitation to weaken quickly in the late afternoon and evening. As the southerly monsoon flows weaken in the afternoon cold pool and gust front forming from the morning precipitation in the Meiyu rainband region can push ahead toward the south, producing propagating squall lines that eventually detach themselves from the main confluence zone and decay.

Summarizing the key results of Xue *et al.* (2018), the diurnal variation of the boundary layer LLJs as caused by the boundary layer inertial oscillations and in the form of AGW vectors rotating around the tip of geostrophic wind vectors is primarily responsible for the net low-level moisture flux maximum at 03–04 LST into the front zone, and such a flux convergence maximum is the main forcing mechanism for the climatological early morning rainfall maximum in the Meiyu frontal zone. The latent heating feedback from precipitation plays at most a secondary role in modulating the precipitation diurnal cycles. Solenoidal circulations between the Tibeter Plateau and the central and eastern China Plains also likely play only secondary roles.

References

Asai, T., S. Ke, and Y.-M. Kodama, 1998: Diurnal variability of cloudiness over east asia and the western pacific ocean as revealed by gms during the warm season. *J. Meteor. Soc. Jpn.*, **76**, 675–684.

Bao, X., F. Zhang, and J. Sun, 2011: Diurnal variations of warm-season precipitation east of the tibetan plateau over china. *Mon. Wea. Rev.*, **139**, 2790–2810.

Blackadar, A. K., 1957: Boundary layer wind maxima and their significance for the growth of nocturnal inversions. *Bull. Amer. Meteor. Soc.*, **38**, 283–290.

Bonner, W. D., 1966: Case study of thunderstorm activity in relation to the low-level jet. *Mon. Wea. Rev.*, **94**, 167–178.

Carbone, R. E. and J. D. Tuttle, 2008: Rainfall occurrence in the united states warm season: The diurnal cycle. *J. Clim.*, **21**, 4132–4136.

Chang, C. P., Y. Zhang, and T. Li, 2000: Interannual and interdecadal variations of the east asian summer monsoon and tropical pacific ssts. Part i: Roles of the subtropical ridge. *J. Clim.*, **13**, 4310–4325.

Chen, G.-X., W. Sha, and T. Iwasaki, 2009a: Diurnal variation of precipitation over southeastern china: Spatial distribution and its seasonality. *J. Geophy. Res.*, **114**, D13.

Chen, G.-X., W. Sha, and T. Iwasaki, 2009b: Diurnal variation of precipitation over southeastern china: 2. Impact of the diurnal monsoon variability. *J. Geophys. Res.*, **114**, D21105.

Chen, G.-X., W. Sha, T. Iwasaki, and K. Ueno, 2012: Diurnal variation of rainfall in the yangtze river valley during the spring-summer transition from trmm measurements. *J. Geophys. Res.*, **117**, D06106.

Chen, G.-X., W. Sha, M. Sawada, and T. Iwasaki, 2013: Influence of summer monsoon diurnal cycle on moisture transport and precipitation over eastern china. *J. Geophys. Res.*, **118**, 3163–3177.

Chen, G.-X., W. Sha, T. Iwasaki, and Z. Wen, 2017: Diurnal cycle of a heavy rainfall corridor over east asia. *Mon. Wea. Rev.*, **45**, 3365–3389.

Chen, G. T.-J. and C. C. Yu, 1988: Study of low-level jet and extremely heavy rainfall over northern taiwan in the meiyu season. *Mon. Wea. Rev.*, **116**, 884–891.

Chen, G. T.-J., C.-C. Wang, and D. T.-W. Lin, 2005: Characteristics of low-level jets over northern taiwan in meiyu season and their relationship to heavy rain events. *Mon. Wea. Rev.*, **133**, 20–43.

Chen, H., R. Yu, J. Li, W. Yuan, and T. Zhou, 2010: Why nocturnal long-duration rainfall presents an eastward-delayed diurnal phase of rainfall down the yangtze river valley. *J. Clim.*, **23**, 905–917.

Chen, X.-C., K. Zhao, and M. Xue, 2014: Spatial and temporal characteristics of warm season convection over pearl river delta region, china, based on 3 years of operational radar data. *J. Geophys. Res. Atmos.*, **119**, 12447–12465.

Ding, Y., 1992: Summer monsoon rainfalls in china. *J. Meteorol. Soc. Jpn.*, **70**, 373–396.

Ding, Y., 2010: *Monsoons Over China*. Springer, 420 pp.

Ding, Y. and J. C. L. Chan, 2005: The east asian summer monsoon: An overview. *Meteor. Atmos. Phys.*, **89**, 117–142.

Ding, Y., Y. Zhang, Q. Ma, and G. Hu, 2001: Analysis of the large-scale circulation features and synoptic systems in east asia during the intensive observation period of game/hubex. *J. Meteor. Soc. Jpn.*, **79**, 277–300.

Du, Y. and Q. Zhang, 2012: Characteristics of low-level jets in shanghai during the 2008 and 2009 warm seasons as inferred from wind profiler radar data. *J. Meteor. Soc. Jpn.*, **90**, 891–903.

Du, Y., R. Rotunno, and Q. Zhang, 2015a: Analysis of wrf-simulated diurnal boundary layer winds in eastern china using a simple 1d model. *J. Atmos. Sci.*, **72**, 714–727.

Du, Y., Y.-L. Chen, and Q. Zhang, 2015b: Numerical simulations of the boundary layer jet off the southeastern coast of china. *Mon. Wea. Rev.*, **143**, 1212–1231.

Du, Y., Q. Zhang, Y.-L. Chen, Y. Zhao, and X. Wang, 2014: Numerical simulations of spatial distributions and diurnal variations of low-level jets in china during early summer. *J. Clim.*, **27**, 5747–5767.

Geng, B. and H. Yamada, 2007: Diurnal variations of the meiyu/baiu rain belt. *SOLA*, **3**, 61–64.

He, H. and F. Zhang, 2010: Diurnal variations of warm-season precipitation over northern china. *Mon. Wea. Rev.*, **138**, 1017–1025.

Helfand, H. M. and S. D. Schubert, 1995: Climatology of the simulated great plains low-level jet and its contribution to the continental moisture budget of the united states. *J. Clim.*, **8**, 784–806.

Hirose, M. and K. Nakamura, 2005: Spatial and diurnal variation of precipitation systems over asia observed by the trmm precipitation radar. *J. Geophys. Res.*, **110**, D05106.

Holton, J. R., 1967: The diurnal boundary layer wind oscillation above sloping terrain. *Tellus*, **19**, 199–205.

Klein, P. M., X.-M. Hu, A. Shapiro, and M. Xue, 2016: Linkages between boundary-layer structure and the development of nocturnal low-level jets in central oklahoma. *Bound. Layer Meteor.*, **158**, 383–408.

Kuo, Y. H. and R. A. Anthes, 1982: Numerical simulation of a meiyu system over southeastern asia. *Pap. Meteor. Res.*, **5**, 15–36.

Li, T., B. Wang, B. Wu, T. Zhou, C.-P. Chang, and R. Zhang, 2017: Theories on formation of an anomalous anticyclone in western north pacific during el niño: A review. *J. Meteor. Res.*, **31**, 987–1006.

Luo, Y., Y. Gong, and D.-L. Zhang, 2014: Initiation and organizational modes of an extreme-rain-producing mesoscale convective system along a mei-yu front in east china. *Mon. Wea. Rev.*, **142**, 203–221.

Murakami, M., 1983: Analysis of the deep convective activity over the western pacific and southeast asia. Part i: Diurnal cycle. *J. Meteor. Soc. Jpn.*, **61**, 60–75.

Sampe, T. and S. Xie, 2010: Large-scale dynamics of the meiyu-baiu rainband: Environmental forcing by the westerly jet. *J. Clim.*, **23**, 113–134.

Shapiro, A., E. Fedorovich, and S. Rahimi, 2016: A unified theory for the great plains nocturnal low-level jet. *J. Atmos. Sci.*, **73**, 3037–3057.

Sun, J. and F. Zhang, 2012: Impacts of mountain–plains solenoid on diurnal variations of rainfalls along the mei-yu front over the east china plains. *Mon. Wea. Rev.*, **140**, 379–397.

Trier, S. B., C. A. Davis, and R. E. Carbone, 2014: Mechanisms governing the persistence and diurnal cycle of a heavy rainfall corridor. *J. Atmos. Sci.*, **71**, 4102–4126.

Tuttle, J. D. and C. A. Davis, 2006: Corridors of warm season precipitation in the central united states. *Mon. Wea. Rev.*, **134**, 2297–2317.

Wallace, J. M., 1975: Diurnal variations in precipitation and thunderstorm frequency over the conterminous united states. *Mon. Wea. Rev.*, **103**, 406–419.

Wang, C.-C., G. T.-J. Chen, and R. E. Carbone, 2004: A climatology of warm-season cloud patterns over east asia based on gms infrared brightness temperature observations. *Mon. Wea. Rev.*, **132**, 1606–1629.

Wang, W.-C., W. Gong, and H. Wei, 2000: A regional model simulation of the 1991 severe precipitation event over the yangtze huai river valley. Part i: Precipitation and circulation statistics. *J. Clim.*, **13**, 74–92.

Xu, W., E. J. Zipser, and C. Liu, 2009: Rainfall characteristics and convective properties of meiyu precipitation systems over south china, taiwan, and the south china sea. Part i: Trmm observations. *Mon. Wea. Rev.*, **137**, 4261–4275.

Xue, M., X. Luo, K. Zhu, Z. Sun, and J. Fei, 2018: The controlling role of boundary layer inertial oscillations in meiyu frontal precipitation and its diurnal cycles over china. *J. Geophy. Res.*, **123**, 5090–5115, https://doi.org/10.1029/2018JD028368.

Yin, S., D. Chen, and Y. Xie, 2009: Diurnal variations of precipitation during the warm season over china. *Int. J. Climatol.*, **29**, 1154–1170.

Yu, R. and J. Li, 2016: Regional characteristics of diurnal peak phases of precipitation over contiguous china (in chinese with english abstract). *Acta. Meteor. Sinica.*, **74**, 18–30.

Yu, R., J. Li, and H. Chen, 2009: Diurnal variation of surface wind over central eastern china. *Clim. Dynam.*, **33**, 1089–1097.

Yu, R., Y. Xu, T. Zhou, and J. Li, 2007a: Relation between rainfall duration and diurnal variation in the warm season precipitation over central eastern china. *Geophys. Res. Lett.*, **34**, L13703.

Yu, R., J. Li, H. Chen, and W. Yuan, 2014: Progress in studies of the precipitation diurnal variation over contiguous china (in chinese with english abstract). *Acta. Meteor. Sinica.*, **72**, 948–968.

Yu, R., T. Zhou, A. Xiong, Y. Zhu, and J. Li, 2007b: Diurnal variations of summer precipitation over contiguous china. *Geophys. Res. Lett.*, **34**, L01704.

Yuan, W., R. Yu, M. Zhang, W. Lin, H. Chen, and J. Li, 2012: Regimes of diurnal variation of summer rainfall over subtropical east asia. *J. Clim.*, **25**, 3307–3320.

Zhong, S., J. D. Fast, and X. Bian, 1996: A case study of the great plains low-level jet using wind profiler network data and a high-resolution mesoscale model. *Mon. Wea. Rev.*, **124**, 785–806.

Zhou, T., R. Yu, H. Chen, A. Dai, and Y. Pan, 2008: Summer precipitation frequency, intensity, and diurnal cycle over china: A comparison of satellite data with raingauge observations. *J. Clim.*, **21**, 3997–4010.

Zhu, K., M. Xue, B. Zhou, K. Zhao, Z. Sun, P. Fu, Y. Zheng, X. Zhang, and Q. Meng, 2018: Evaluation of real-time convection-permitting precipitation forecasts in china during the 2013–2014 summer season. *J. Geophy. Res.*, **123**, 1037–1064.

Chapter 12

Mesoscale and High-Impact Weather in the South American Monsoon

Leila M. V. Carvalho*,‡ and Maria A. F. Silva Dias†

**Department of Geography & Earth Research Institute, University of California, Santa Barbara, USA*

†Department of Atmospheric Sciences, University of Sao Paulo, Sao Paulo, Brazil

‡leila@eri.ucsb.edu

The South American Monsoon System (SAMS) is the most important climatic feature in the continent. Millions of people rely on the monsoonal rainfall for agriculture, hydropower, and fresh water supply, and unique ecosystems depend on the seasonal rains for their survival. The intensity, amount, and regularity of the monsoon precipitation are controlled by complex coupled mechanisms on multiple spatiotemporal scales. Convection organized in mesoscale convective systems (MCSs) during the warm season is often associated with vertically developed thunderstorms capable of producing extreme precipitation and high-impact events such as hail, destructive winds, and lightning. The occurrence of MCSs depends on moisture supply and convective instability. The South American low-level jet (SALLJ) is responsible for moisture transport across South America, and variations in the characteristics of the LLJ influence the initiation and development of MCSs. Additionally, active and break phases of the SAMS accompany the variations in circulation and thermodynamics that affect the diurnal cycles of the MCSs and their geographic location. The rapid rate of tropospheric warming in recent decades has altered circulations and thermodynamic conditions with impacts on convection and precipitation that are still understudied. This is particularly critical in large tropical areas of the Amazon, Central Brazil, and in the Andes, where sparse, intermittent, and notably short records limit the ability to assess climatic changes in these regions. This chapter provides an overview of the characteristics and temporal variability of the SAMS and discusses recent studies on trends in extreme precipitation. Mechanisms driving the spatiotemporal characteristics of MCSs are also reviewed. Two combined field campaigns designed to advance scientific understanding of cloud processes over tropical South America (GoAmazon and CHUVA) are presented here, demonstrating the importance of enhancing observations to evaluate critical processes responsible for high-impact weather in South America.

1. Introduction

The South American Monsoon system (SAMS) is defined by the onset of deep convection and precipitation that accompany pronounced seasonal variations in moisture flux and heat content throughout the troposphere (Zhou and Lau 1998, 2001). Precipitation associated with the SAMS peaks between December and February (DJF). The South Atlantic Convergence Zone (SACZ), which is the SAMS's main feature, is characterized by persistent precipitation and convection driven by enhanced moisture transport from the tropics toward subtropical southwest Atlantic (Fig. 1a) (Kodama 1992; Carvalho *et al.* 2002a, 2004). Precipitation bands along the SACZ provide water supply for millions of people and are often related to high-impact weather in populated areas (Liebmann *et al.* 2001; Carvalho *et al.* 2002a). Additionally, the SAMS and the SACZ are characterized by organized mesoscale convective systems (MCSs) that bring substantial rainfall and severe weather to vast regions over tropical

The Multiscale Global Monsoon System
Edited by C. P. Chang et al.

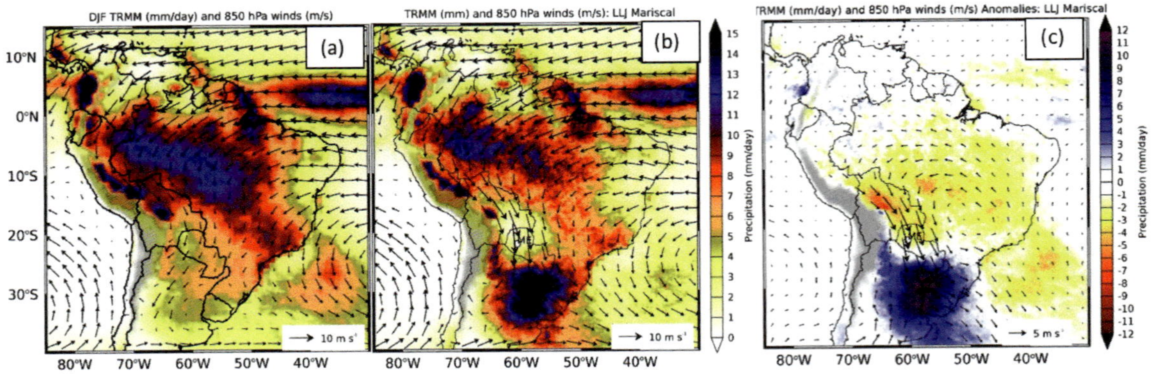

Fig. 1. Composites of TRMM rainfall (mm/day) and 850 hPa (m/s) winds (1998–2013) from the Climate Forecast System Reanalysis (CFSR): (a) DJF climatology, (b) during SALLJ observed in Mariscal Estigarriba–Paraguay, and (c) differences between DJF climatology and SALLJ events (right).

South America, including the Andes (Garreau 1999; Poveda *et al.* 2005; Durkee *et al.* 2009; Boers *et al.* 2015). The South American low-level jet (SALLJ) east of the Andes (Marengo *et al.* 2004; Vera *et al.* 2006) plays an important role in transporting moisture from the tropics to the subtropics during the wet season (Salio *et al.* 2007; Nascimento *et al.* 2016; Montini *et al.* 2019; Jones and Carvalho 2018).

The intensity, duration, spatial distribution, and propagation characteristics of MCSs depend on dynamic and thermodynamic conditions resulting from the interactions between synoptic systems and the local environment in which MCSs evolve (Mohr and Zipser 1996; Silva Dias *et al.* 2002; Halverson *et al.* 2002; Carvalho *et al.* 2002b; Rickenbach *et al.* 2002; 2004; Laurent *et al.* 2002; Salio *et al.* 2007; Anabor *et al.* 2008; Durkee and Mote 2010; Alcantara *et al.* 2011). Additionally, convection in the SAMS and the SACZ is modulated by coupled modes of variability on intraseasonal (Paegle *et al.* 2000; Jones and Carvalho 2002; Carvalho *et al.* 2002b; 2011; Cunningham and Cavalcanti 2006; Muza *et al.* 2009), interannual (Zhou and Lau 2001; Grimm and Zilli 2009; Bombardi *et al.* 2013), decadal (Grimm and Saboia 2015; Apaéstegui *et al.* 2014; Garcia and Kayano 2015), and multidecadal timescales (Jones and Carvalho 2018).

Tropical South America (SA) has experienced a large warming rate in the lower troposphere in recent decades with possible consequences for the distribution of precipitation, particularly extreme rainfall (Carvalho and Jones 2013; Cavalcanti 2012).This chapter presents the most recent studies on the observed variations and changes in precipitation in the SAMS and the SACZ, with particular focus on extreme events. An overview of the characteristics of MCSs over tropical SA and the recent advancement in observing and modeling deep convection in the Amazon are also discussed.

2. Trends in Extreme Rainfall

There is observational evidence that extreme rainfall events have increased over tropical SA in the past few decades (Haylock *et al.* 2006; Marengo *et al.* 2010; Skansi *et al.* 2013; Heidinger *et al.* 2017, Carvalho 2019), which includes populated areas of south and southeastern Brazil (Dufek and Ambrizzi 2008; Teixeira and Satyamurty 2011; Dereczynski *et al.* 2013; Sugahara *et al.* 2009; Marengo *et al.* 2010, 2013; Silva Dias *et al.* 2013; Valverde and Marengo 2014; Pedron *et al.* 2016; Zilli *et al.* 2016). Conversely, some studies have shown an increase in the number of consecutive dry days and a decrease in the number of light rainy days (less than 5mm/day) in densely urbanized areas (Xavier *et al.* 1994; Marengo *et al.* 2013; Raimundo *et al.* 2014). Zilli *et al.* (2016) examined 70 years of daily precipitation data over coastal southeastern Brazil and concluded that the frequency of both rainy days and extreme precipitation events has increased southward, whereas fewer rainy days and heavier precipitation are observed north of the region. Zilli *et al.* (2019) showed that these trends are likely associated with the weakening of the northerly winds at the northern margin of the SACZ, which resulted in the poleward shift of the SACZ in recent decades. The remarkable warming of the Equatorial Atlantic and tropical SA over the same period, combined with teleconnections with coupled modes of variability on the decadal to multi-decadal timescales, may have influenced the observed changes in the SAMS and the SACZ, including the frequency of extreme rainfall events (Silva Dias *et al.* 2013; Grimm and Saboia 2015).

3. MCSs and High-impact Weather

One of the most important atmospheric influences on the development of deep convection and heavy precipitation in SA is the SALLJ (Fig. 1b,c). Although the SALLJ occurs year-round, the poleward transport of moisture and heat associated with the jet is greatest in summer (Sun and Barros 2015; Montini *et al.* 2019), creating the necessary synoptic conditions for the development of mesoscale convective complexes (MCCs; Durkee and Mote 2009; Durkee *et al.* 2009) that often deliver severe weather in vast regions (Salio *et al.* 2007). Composites of the Tropical Rainfall Measurement Mission (TRMM) daily rainfall in DJF (1998–2013) during SALLJ events identified in Montini *et al.* (2019) showed remarkable enhancement of rainfall over Northeastern Argentina, Uruguay, and Southern Brazil (Fig. 1b,c). Moreover, SALLJ events are associated with an anticyclonic anomaly over central Brazil and suppression of convection over the SAMS and the SACZ (Fig. 1c).

Montini *et al.* (2019) examined 36-yr climatology of the SALLJ (1979–2015) and showed that the frequency of SALLJ events has decreased during autumn and summer, whereas the annual average intensity of events has increased. These trends likely result from the combination of internal variability of the climate system and tropospheric warming. Among the most important factors influencing these trends are the transition in phase of the Atlantic Multi-decadal Oscillation (AMO) (Jones and Carvalho 2018) and the broadening of the Hadley cell caused by poleward shift of the subtropical jet in the South American sector resulting from global warming (Ding *et al.* 2011; Fu *et al.* 2013). These changes can influence the distribution of moisture and the frequency of extreme events over tropical and subtropical SA.

The SAMS exhibits alternating active and break phases on synoptic to subseasonal timescales (e.g., Paegle *et al.* 2000; Jones and Carvalho 2002). Modifications in circulation, moisture transport, moist static energy, and convective available potential energy that accompany these phases alter the strength and spatiotemporal characteristics of MCSs (Silva Dias *et al.* 2002a,b; Halverson *et al.* 2002; Carvalho *et al.* 2002b). Contrasting precipitation and 850 hPa circulation patterns between SAMS active and break phases (Fig. 2) indicate that enhanced (weakened) SAMS and SACZ are associated with low-tropospheric westerly (easterly) wind anomalies over tropical SA, cyclonic (anticyclonic) anomalies over subtropical western Atlantic, and suppressed (enhanced) convection over northwest Amazon. These anomaly patterns strongly influence convection east of the tropical Andes (Fig. 2). Atmospheric conditions during SAMS's active phase are conducive to the development of MCSs along the SACZ responsible for persistent heavy

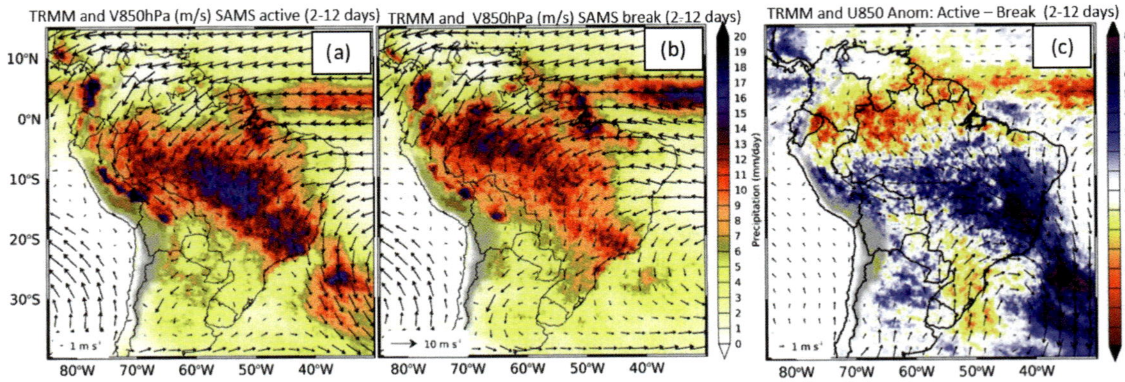

Fig. 2. Same as Fig. 1: (a) SAMS active phase, (b) SAMS break phase, and (c) differences between active and break phases .Phases were identified with the large-scale Index for the South American monsoon (LISAM) (Carvalho *et al.* 2011) band filtered between 2 and 12 days.

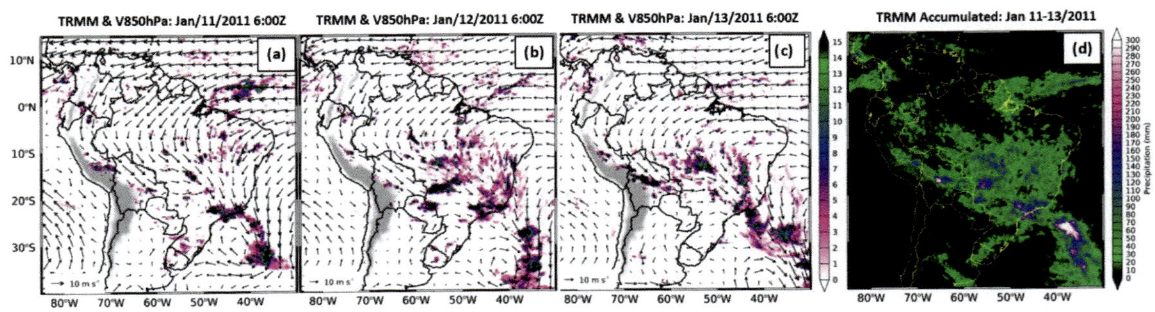

Fig. 3. TRMM-3B42 precipitation (mm/h) at 25° and CFSR 850 hPa winds (0.5°) at 6:00 UTC (a) January 11, 2011; (b) January 12, 2011; (c) January 13, 2011; (d) accumulated precipitation (mm) from January 11–13, 2011.

rainfall. For example, persistent westerly wind anomalies transporting moisture from the tropical Amazon combined with the convergence of winds induced by an extratropical cyclone near the southeast coast of Brazil resulted in numerous floods and landslides in January 2011 (Fig. 3). As a result, some locations over Southeast Brazil accumulated over 150 mm between January 11 and 13. In some areas these records represented more than the mean rainfall during January.

MCSs may be observed across Tropical South America, north of 20°S all year round but with greater frequency during the austral spring (Jaramillo *et al.*, 2017). Severe precipitation features as detected by the TRMM satellite may also be found with greater frequency in the pre-monsoon season (Nunes *et al.*, 2016).

4. GoAmazon and CHUVA Experiments

To advance the scientific understanding of cloud processes over Tropical South America, specifically over the Amazon region, combined field campaigns of GoAmazon (Martin *et al.* 2016) and ACRIDICON-CHUVA (Machado *et al.* 2018; Wendisch *et al.* 2016) have been conducted with the aid of airplane instrumentation and intensive surface observations. Figure 4 shows a schematic of the airplane flights.

GoAmazon has resulted in the quantification of the effects of the polluted plume of a large urban area, the city of Manaus with its 2.5 million inhabitants, on the downwind region and specifically on cloud properties (Cechini *et al.* 2016; Giangrande *et al.* 2017).

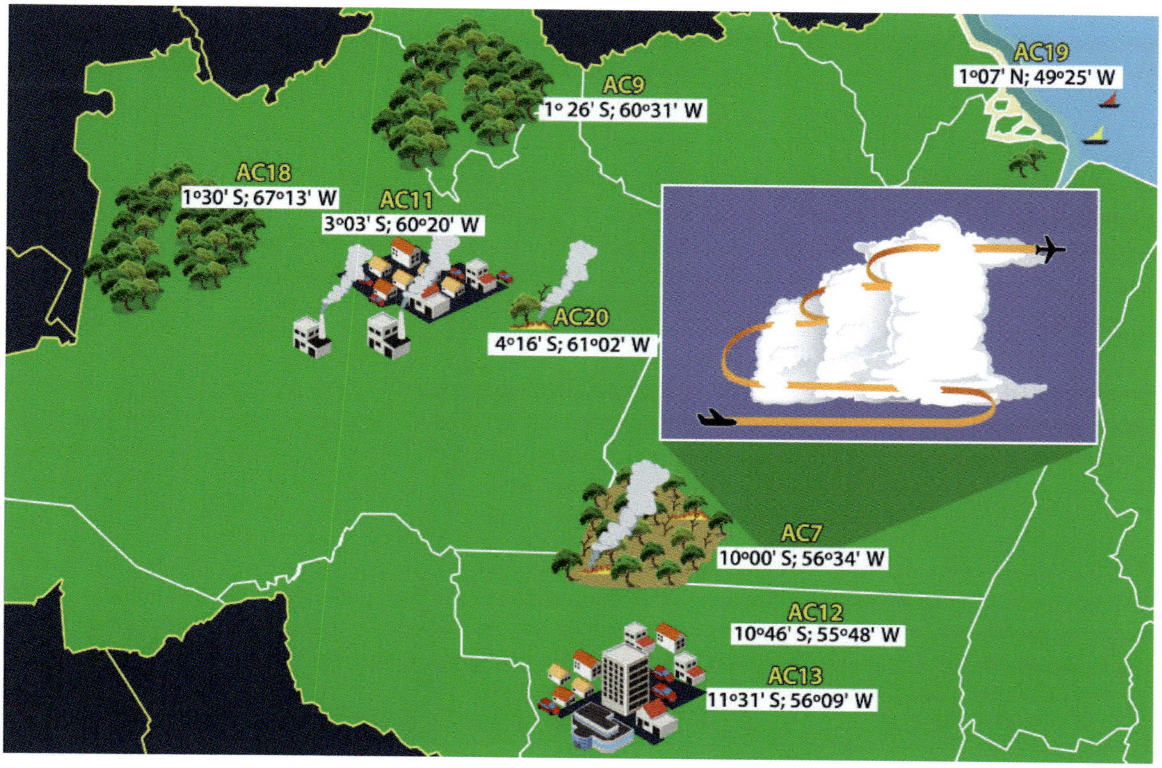

Fig. 4. Locations where cloud profiles were collected for different flights during the ACRIDICON-CHUVA campaign. Clouds formed over southern Amazonia and in the Manaus region are subject to higher aerosol loadings due to the presence of the deforestation arc and urban emissions. Clouds formed over the northern and northwestern Amazon are driven by background conditions with low aerosol concentration (from Cechini *et al.* 2016; Wendisch *et al.* 2016).

The ACRIDICON-CHUVA campaign flew a high-altitude airplane and was able to monitor the outflow of deep convective systems and detect the formation of new ultrafine particles in the upper troposphere (Wang *et al.* 2016, Andreae *et al.* 2018). These ultrafine particles undergo downward mixing and transport and upon reaching the boundary layer may increase in size by acquiring biogenic material, thus becoming a source of cloud condensation nuclei.

A particularly comprehensive study of cloud microphysical properties and updraft speeds in convective clouds evolving in the Amazon was presented by Cechini *et al.* (2016), which was based on the ACRIDICON-CHUVA flight paths. They show that the main driver of droplet size distributions is the boundary layer aerosol concentration. Updraft speeds in relation to different land covers, forests, deforested regions, and agriculture areas mainly affect the total number of droplets and the liquid water content. The flights cover pristine areas, and moderately and heavily polluted clouds, thus giving a whole range of conditions associated to warm-phase microphysics. The results provide a framework for testing cloud physics parameterizations in numerical models with a focus on tropical rainfall. Oceanic and continental behaviors, typical of wet and dry season clouds, respectively, during GoAmazon 2014/2015 of discussed by Wang *et al.* (2018).

Convective scale updrafts and downdrafts and also mass fluxes and convective area fraction were observed during GoAmazon 2014/2015 by a radar wind profiler, with results presented by Giangrande *et al.* (2016). Convective profiles are stratified by environmental conditions and by season, thus providing an overall view. Results indicate more vigorous convection in transitional months when compared with the wet season and for break periods as opposed to monsoon rains as already discussed by previous authors (Williams *et al.* 2002; Albrecht *et al.* 2011).

The variety of convective strength and convective area fraction documented over the two-year sampling indicates that wet season conditions are associated with a substantially larger area fraction and mass flux. The environmental conditions and radiative fluxes associated with convection during the whole GoAmazon 2014/2015 campaign were summarized by Giangrande *et al.* (2017) from a multi-sensor perspective. Diurnal and seasonal variations and cloud types, along with their properties, are also presented. These data and discussions provide a comprehensive framework for comparisons with modeling results over the Amazon, particularly for the role of shallow convection, congestus, and deep convective development, their diurnal and seasonal cycles.

The relationship among column water vapor, local circulations, and convective development has been studied in the Amazon Basin. Dos Santos *et al.* (2014), Adams *et al.* (2016), Fitzjarrald *et al.* (2008), and Silva Dias *et al.* (2004) indicated that the very large Amazon rivers have an effect on local convection in all seasons. Adams *et al.* (2013) showed that typical timescales derived from column water vapor of intense change during about 4 h are associated with the transition from shallow to deep convection in the Amazon. New research based on GoAmazon 2014/15 has a particular focus on the transition from shallow to deep convection.

Machado *et al.* (2014) showed that greater rainfall rates and lower season accumulations are observed during the dry season, as opposed to the wet season. That study also indicated that typical MCSs are 10 times larger and have lifetimes approximately 3 times that of individual rain cells. MCS can have a 300-km radius and lifetimes eventually longer than 24 h. The behavior of overall convection, isolated as opposed to organized in MCS, indicates that the latter represents less than 1% of the whole population (Jaramillo *et al.*, 2017). Using GoAmazon 2014/2015 data Schiro and Neelin (2018) showed that MCSs induce changes in boundary layer moisture that take longer to recover than those for isolated convection. These authors also showed that column-integrated water vapor increases are larger preceding MCS occurrence.

These new results suggest a complex interaction between aerosol and cloud life cycles with implications on rainfall prediction during SAMS. The modeling of such interaction remains a scientific challenge.

Acknowledgments

LMV Carvalho is thankful for the support provided by NSF- 1505198 and MAF Silva Dias for the support provided by FAPESP 2017/17047-0, 2015/14497-0, and 2013/05014-0.

References

Adams David K., R. M. S. Fernandes, K. L. Holub, S. I. Gutman, and L. M. S. Tanaka, 2016: The Amazon Dense GNSS Meteorological Network: A new approach for examining water vapor and deep convection interactions in the tropics. *Bull. Amer. Met. Soc.*, https://doi.org/10.1175/BAMS-D-13-00171.1

Adams, D. K., S. I. Gutman, K. L. Holub, and D. S. Pereira, 2013: GNSS observations of deep convective time scales in the Amazon. *Geophys. Res. Lett.*, **40**, 2818–2823, doi:https://doi.org/10.1002/grl.50573.

Alcântara, C. R., M. A. Silva Dias, E. P. Souza, and J. C. Cohen, 2011: Verification of the role of the low level jets in Amazon squall lines. *Atmosph. Res.*, **100**(1), 36–44.

Anabor, V., D. J. Stensrud, and O. L. de Moraes, 2008: Serial upstream-propagating mesoscale convective system events over southeastern South America. *Mon. Wea. Rev.*, **36**, 3087–3105.

Andreae, M. O. *et al.*, 2018: Aerosol characteristics and particle production in the upper troposphere over the Amazon Basin, Atmos. *Chem. Phys.*, **18**, 921–961, https://doi.org/10.5194/acp-18-921-2018.

Apaéstegui, J. *et al.*, 2014: Hydroclimate variability of the South American Monsoon System during the last 1600 yr inferred from speleothem isotope records of the north-eastern Andes foothills in Peru. *Clim. Past Discuss.*, **10**, 533–561.

Boers, N. *et al.*, 2015: Extreme rainfall of the South American monsoon system: A dataset comparison using complex networks. *J. Clim.*, **28**(3), 1031–1056.

Bombardi R. J., L. M. Carvalho, C. Jones, and M. S. Reboita, 2014: Precipitation over eastern South America and the South Atlantic Sea surface temperature during neutral ENSO periods, *Clim. Dynam.*, **42**, 1553–1568.

Carvalho, L. M. V., C. Jones, and B. Liebmann, 2002a: Extreme precipitation events in southeastern South America and large-scale convective patterns in the South Atlantic convergence zone. *J. Clim.*, **15**(17), 2377–2394.

Carvalho, L. M. V., C. Jones, and M. A. Silva Dias, 2002b: Intraseasonal large-scale circulations and mesoscale convective activity in tropical South America during the TRMM-LBA campaign. *J. Geophys. Res.: Atmosph.*, **107**(D20).

Carvalho, L. M. V., C. Jones, and B. Liebmann, 2004: The South Atlantic Convergence Zone: intensity, form, persistence and relationships with intraseasonal to interannual activity and extreme rainfall. *J. Clim.*, **17**, 88–108.

Carvalho, L. M. V. *et al.*, 2011: Moisture transport and intraseasonal variability in the South America Monsoon System. Climate Dynamics (DOI 10.1007/s00382-010-0806-2).

Carvalho, L. M. V. and C. Jones, 2013: CMIP5 simulations of low-level tropospheric temperature and moisture over tropical Americas. *J. Clim.*, **26**, 6257–6286, doi: 10.1175/JCLI-D-12-00532.1.

Carvalho, L. M. V. 2019: Assessing precipitation trends in the Americas with historical data: a review. Climatic Change, https://doi.org/10.1002/wcc.627.

Cavalcanti, I. F. A., 2012: Large scale and synoptic features associated with extreme precipitation over South America: A review and case studies for the first decade of the 21st century. *Atmos. Res.*, **118**, 27–40.

Cechini, M. *et al.*, 2016: Impacts of the Manaus pollution plume on the microphysical properties of Amazonian warm-phase clouds in the wet season. *Atmos. Chem. Phys.* (Online), **16**, 7029–7041.

Cunningham, C. A. and I. F. A. Cavalcanti, 2006: Intraseasonal modes of variability affecting the South Atlantic Convergence Zone. *Int. J. Climatol.*, **26**(9), 1165–1180.

Dereczynski, C. P., W. Luiz Silva, and J. A. Marengo, 2013: Detection and projections of climate change in Rio de Janeiro, Brazil. *Am. J. Clim. Change*, **2**, 25–33.

Ding, Q., E. J. Steig, D. S. Battisti, and M. Küttel, 2011: Winter warming in West Antarctica caused by central tropical Pacific warming. *Nat. Geosci.* **4**(6), 398.

Dos Santos, M. J., M. A. F. Silva Dias, and E. D. Freitas, 2014: Influence of local circulations on wind, moisture, and precipitation close to Manaus City, Amazon Region, Brazil. *J. Geophys. Res. — Atmosph.* DOI: 10.1002/2014JD021969.

Dufek, A. S. and T. Ambrizzi 2008: Precipitation variability in São Paulo state, Brazil. *Theor. Appl. Climatol.*, **93**, 167–178, doi: 10.1007/s00704-007-0348-7.

Durkee, J. D. and T. L. Mote, 2019: A climatology of warm-season mesoscale convective complexes in subtropical South America. *Int. J. Climatol.*, **30**(3), pp. 418–431.

Durkee, J. D., T. L. Mote, and J. M. Shepherd, 2009: The contribution of mesoscale convective complexes to rainfall across subtropical South America. *J. Clim.*, **22**(17), pp.4590–4605.

Espinoza, J. C., S. Chavez, J. Ronchail, C. Junquas, K. Takahashi, and W. Lavado, 2015: Rainfall hotspots over the southern tropical Andes: Spatial distribution, rainfall intensity, and relations with large-scale atmospheric circulation. *Water Resour. Res.*, **51**, 3459–3475.

Fitzjarrald, D. R. *et al.*, 2008: Spatial and temporal rainfall variability near the Amazon-Tapajós confluence, *J. Geophys. Res.*, **113**, G00B11, doi:10.1029/2007JG000596.

Fu, R. *et al.*, 2013: Increased dry-season length over southern Amazonia in recent decades and its implication for future climate projection. *Proc. Nat. Acad. Sci.*, **18**, 02584.

Garcia, S. R. and M. T. Kayano, 2015: Multidecadal variability of moisture and heat budgets of the South American Monsoon system. *Theoret. Appl. Climatol.*, **121**(3–4), 557–570.

Garreaud, R., 1999: Multiscale analysis of the summertime precipitation over the central Andes. *Mon. Wea. Rev.*, **127**(5), 901–921.

Giangrande, S. E. *et al.*, 2017: Cloud characteristics, thermodynamic controls and radiative impacts during the observations and modeling of the green ocean Amazon (GoAmazon2014/5) Experiment, *Atmosph. Chem. Phys.*, **17**, 14-519–14-541.

Giangrande, S. E., T. Toto, M. P. Jensen, M. J. Bartholomew, Z. Feng, A. Protat, C. R. Williams, C. Schumacher, and L. Machado, 2016: Convective cloud vertical velocity and mass-flux characteristics from radar wind profiler observations during GoAmazon2014/5, *J. Geophys.* 35 *Res. Atmos.*, **121**(12), 891–12,913, https://doi.org/10.1002/2016JD025303.

Grimm, A. M. and J. P. Saboia, 2015: Interdecadal variability of the South American precipitation in the monsoon season. *J. Clim.*, **28**(2), 755–775.

Grimm, A. M. and M. T. Zilli, 2009: Interannual variability and seasonal evolution of summer monsoon rainfall in South America. *J. Clim.*, **22**(9), 2257–2275.

Halverson, J. B., T. Rickenbach, B. Roy, H. Pierce, and E. Williams, 2002: Environmental characteristics of convective systems during TRMM-LBA. *Mon. Wea. Rev.*, **130**, 1493–1509.

Haylock, M. R. *et al.*, 2006. Trends in total and extremes South American rainfall in 1960–2000 and links with sea surface temperature. *J. Clim.*, **19**, 1490–1512, doi: 10.1175/JCLI3695.1.

Heidinger, H., L. M. V. Carvalho, C. Jones, A. Posadas, and R. Quiroz, 2017: A new assessment in total and extreme rainfall trends over central and southern Peruvian Andes during 1965–2010. *Int. J. Climatol.* doi: 10.1002/joc.5427

Jaramillo, L., G. Poveda, and J. F. Mejíab, 2017: Mesoscale convective systems and other precipitation features over the tropical Americas and surrounding seas as seen by TRMM. *Int. J. Climatol.*, doi: 10.1002/joc.5009.

Jones, C. and L. M. Carvalho, 2002: Active and break phases in the South American monsoon system. *J. Clim.*, **15**(8), 905–914.

Jones, C. and L. M. Carvalho, 2018: The influence of the Atlantic Multidecadal Oscillation on the eastern Andes low-level jet and precipitation in South America. Nature Climate and Atmospheric Science, doi:10.1038/s41612-018-0050-8.

Kodama, Y.-M. 1992: Large-scale common features of subtropical precipitation zones (the Baiu Frontal Zone, the SPCZ and the SACZ) part I: Characteristics of subtropical frontal zones. *J. Meteorol. Soc. Jpn.*, **70**, 813–835.

Laurent, H., L. A. Machado, C. A. Morales, and L. Durieux, 2002: Characteristics of the Amazonian mesoscale convective systems observed from satellite and radar during the WETAMC/LBA experiment. *J. Geophys. Res. Atmos.*, **107**(D20).

Liebmann, B., C. Jones, and L. M. V. Carvalho, 2001: Interannual variability of daily extreme precipitation events in the state of Sao Paulo, Brazil. *J. Clim.*, **14**, 208–218.

Liebmann, B. *et al.*, 2004: An observed trend in central South American precipitation. *J. Clim.*, **17**, 4357–4367.

Machado, L. A. *et al.*, 2014: The CHUVA project: How does convection vary across Brazil? *Bull. Amer. Meteor.*, **95**(9), 1365–1380.

Machado, L. A., A. J. Calheiros, T. Biscaro, S. Giangrande, M. A. Silva Dias, M. A. Cecchini, R. Albrecht, M. O. Andreae, W. F. Araujo, P. Artaxo, and S. Borrmann, 2018: Overview: Precipitation characteristics and sensitivities to environmental conditions during GoAmazon2014/5 and ACRIDICON-CHUVA. *Atmos. Chem. Phys.*, 7, **18**(9), 6461–6482.

Marengo, J. A., W. R. Soares, C. Saulo, and M. Nicolini, 2004: Climatology of the low-level jet east of the Andes as derived from the NCEP–NCAR reanalyses: Characteristics and temporal variability. *J. Clim.*, **17**(12), 2261–2280.

Marengo, J. A., M. Rusticucci, O. Penalba, and M. Renom, 2010: An intercomparison of observed and simulated extreme rainfall and temperature events during the last half of the twentieth century: Part 2: historical trends, *Clim. Change*, **98**, 509–529, doi: 10.1007/s10584-009-9743-7.

Marengo, J. A., M. C. Valverde, and G. O. Obregon, 2013: Observed and projected changes in rainfall extremes in the Metropolitan Area of São Paulo. *Clim. Res.*, **57**, 61–72.

Marengo, J. A., G. F. Fisch, L. M. Alves, N. V. Sousa, R. Fu, and Y. Zhuang, 2017: Meteorological context of the onset and end of the rainy season in Central Amazonia during the GoAmazon2014/5. *Atmos. Chem. Phys.*, **17**(12), 7671–7681.

Martin, S. T. *et al.*, 2016: The Green Ocean Amazon Experiment (GoAmazon2014/5) Observes Pollution Affecting Gases, Aerosols, Clouds, and Rainfall over the Rain Forest. *Bulletin of the Am. Meteorol. Soc.*, **97**, doi:10.1175/BAMS-D-.

Mohr, K. I. and E. J. Zipser, 1996. Mesoscale convective systems defined by their 85-GHz ice scattering signature: Size and intensity comparison over tropical oceans and continents. *Mon. Wea. Rev.*, **124**(11), pp. 2417–2437.

Montini, T., C. Jones, and L. M. V. Carvalho, 2018: The South America Low-level Jet: New climatology, variability and changes. *J. Geophys. Res.* (accepted).

Montini, T. L., C. Jones, and L. M. V. Carvalho, 2019: The South American low-level jet: a new climatology, variability, and changes. *J. Geophys. Res.: Atmos.*, **124**(3), 1200–1218.

Muza, M. N., L. M. V. Carvalho, C. Jones, B. Liebmann, 2009: Intraseasonal and interannual variability of extreme dry and wet events over southeastern South America and the subtropical Atlantic during austral summer. *J. Clim.*, **22**, 1682–1699, doi: 10.1175/2008JCLI2257.1.

Nascimento, M. G., D. L. Herdies, and D. O. Souza, 2016: The South American Water Balance: The Influence of Low-Level Jets. *J. Clim.*, **29**(4), 1429–1449.

Paegle, J. N., L. A. Byerle, and K. C. Mo, 2000. Intraseasonal modulation of South American summer precipitation. *Mon. Wea. Rev.*, **128**(3), 837–850.

Poveda, G. *et al.*, 2005: The diurnal cycle of precipitation in the tropical Andes of Colombia. *Mon. Wea. Rev.*, **133**, 228–240.

Raimundo, C. C., C. A. Sansigolo, and L. C. B. Molion, 2014: Trends of rainfall classes in the metropolitan region of Sao Paulo. *Rev. Bras. Meteor.* **29**, 397–408, doi: 10.1590/s0102-778620130655.

Rickenbach, T. M., R. N. Ferreira, J. B. Halverson, D. L. Herdies, and M. A. Silva Dias, 2002. Modulation of convection in the southwestern Amazon basin by extratropical stationary fronts. *J. Geophys. Res.: Atmos.*, **107**(D20).

Rickenbach, T. M., 2004: Nocturnal cloud systems and the diurnal variation of clouds and rainfall in southwestern Amazonia. *Mon. Wea. Rev.*, **132**, 1201–1219.

Salio, P., M. Nicolini, and E. J. Zipser, 2007. Mesoscale convective systems over southeastern South America and their relationship with the South American low-level jet. *Mon. Wea. Rev.*, **135**(4), 1290–1309.

Schiro, K. A. and D., Neelin, 2018: Tropical continental downdraft characteristics: mesoscale systems versus unorganized convection Atmos. *Chem. Phys.*, **18**, 1997–2010, https://doi.org/10.5194/acp-18-1997-2018

Silva Dias, M. A. *et al.*, 2002a: A case study of convective organization into precipitating lines in the Southwest Amazon during the WETAMC and TRMM-LBA. *J. Geophys. Res.*, **27**, 107(D20).

Silva Dias, M. A., J. Dias, L. Carvalho, E. Freitas, and P. Silva Dias, 2013: Changes in extreme daily rainfall for Sao Paulo, Brazil. *Clim. Change*, **116**, 705–722, doi: 10.1007/s10584-012-0504-7.

Silva Dias, M. A. *et al.*, 2002b. Cloud and rain processes in a biosphere-atmosphere interaction context in the Amazon region. *J. Geophys. Res., Atmos.*, **107**(D20).

Silva Dias, M. A. F., P. L. Silva-Dias, M. Longo, D. R. Fitzjarrald, and A. S. Denning, 2004: River breeze circulation in eastern Amazonia: Observations and modelling results. *Theor. Appl. Climatol.*, **78**, 111–121, doi:https://doi.org/10.1007/s00704-004-0047-6.

Skansi, M. M. *et al.*, 2013: Warming and wetting signals emerging from analysis of changes in climate extreme indices over South America. *Global and Planetary Change*, **100**, 295–307, doi: 10.1016/j.gloplacha.2012.11.004.

Sugahara, S., R. P. Rocha, and R. Silveira, 2009: Non-stationary frequency analysis of extreme daily rainfall in Sao Paulo, Brazil. *Int. J. Climatol.*, **29**, 1339–1349, doi: 10.1002/joc.1760.

Sun, X. M. and A. P. Barros 2015: Isolating the role of surface evapotranspiration on moist convection along the Eastern Flanks of the tropical Andes using a quasi-idealized approach, *J. Atmos. Sci.*, **72**(1), 243–261.

Vera, C. *et al.*, 2006: The South American low-level jet experiment. *Bull. Amer. Meteor.*, **87**, 63–77.

Wang, D. *et al.*, 2016: Amazon boundary layer aerosol concentration sustained by vertical transport during rainfall. *Nature*, **539**, 416–419, doi:10.1038/nature19819.

Wang, D., S. E. Giangrande, M. J. Bartholomew, J. Hardin , Z. Feng, R. Thalman, and L. A. T. Machado, 2018: The Green Ocean: Precipitation Insights from the GoAmazon2014/5 Experiment. *Atmos. Chem. Phys. Discuss.*, https://doi.org/10.5194/acp-2018-101.

Wendisch, M. *et al.*, 2016: ACRIDICON-CHUVA Campaign: Studying Tropical Deep Convective Clouds and Precipitation over Amazonia Using the New German Research Aircraft

HALO, B. *Am. Meteorol. Soc.*, **97**, 1885–1908, doi:10.1175/bams-d-14-00255.1.

Williams, E., D. Rosenfeld, N. Madden, J. Gerlach, N. Gears, L. Atkinson, N. Dunnermann, G. Frostom, M. Antonio, B. Biazon, R. Camargo, H. Franca, A. M. Gomes, and M. A. Lima, Contrasting convective regimes over the Amazon: Implications for cloud electrification. *J. Geophys. Res.*, **107**, 8082, https://doi.org/10.1029/2001JD000380, 2002.

Zhou, J. and K. M. Lau, 2001. Principal modes of interannual and decadal variability of summer rainfall over South America. *Int. J. Climatol.*, **21**(13), 1623–1644.

Zhou, J. and K. M. Lau, 1998: Does a monsoon climate exist over South America? *J. Clim.*, **11**(5), 1020–1040.

Zilli, M. T., L. M. V. Carvalho, and B. Lintner, 2017: The poleward shift of South Atlantic Convergence Zone in recent decades. *Clim. Dynam.*, 1–19.

Zilli, M. T., L. M. V. Carvalho, and B. R. Lintner, 2019: The poleward shift of South Atlantic Convergence Zone in recent decades. *Clim. Dynam.*, **52**(5–6), 2545–2563.

Zilli, M. T., L. M. V. Carvalho, B. Liebmann, and M. A. F. Silva Dias, 2016: A comprehensive analysis of trends in extreme precipitation over southeastern coast of Brazil. *Int. J. Climatol.*, doi: 10.1002/joc.4840.

Zipser, E., P. Salio, and M. Nicolini, 2004. Mesoscale convective systems activity during SALLJEX and the relationship with SALLJ events. *CLIVAR Exchanges*, **9**(1), 14–18.

Chapter 13

An Overview of the Weather Conditions and Climate Trends of Extreme Rainfall in the Sahel

Jean-Philippe Lafore*, Philippe Peyrillé, Romain Roehrig and Florent Beucher
Centre National de Recherches Météorologiques, Météo-France & CNRS, Toulouse, France
**jeanphilippe.lafore2020@gmail.com*

The West African Monsoon (WAM) has undergone drastic changes in the recent past, which has dramatic impacts on the population. After 30 years of devastating droughts, the last two decades experienced a growing number of damaging floods concurrent with ongoing episodes of water shortages and famines. In the past few years, a general emphasis has been placed on the study of extreme rainfall over the Sahel, motivated by the need to understand the physics and the evolution of these high-impact weather systems.

This chapter gives an overview of the current understanding of the occurrence of rainfall extremes in the Sahel. A recent characterization of the rainfall regime based on rain gauge data has aided in obtaining data pertaining to a long time series of extreme events and shows that the contribution of extreme rainfall to the seasonal rainfall amount has increased during the last decade. This intensification of Sahelian rainfall has been recently shown to be consistent with the signal of climate change/global warming.

The weather conditions associated with the occurrence of an extreme rainfall event have also been documented. A case study of a dramatic event — 263 mm of rain recorded in ten in Ouagadougou, Burkina Faso, on September 1, 2009 — shows that a range of different scales are involved in the occurrence of such an event. The development of intense convection is favored by the simultaneous actions of a Kelvin wave and an Equatorial Rossby wave together with a train of three successive African Easterly waves (AEWs). A moist anomaly of regional scale and a stronger than usual monsoon flow helped a deep moist vortex to develop within which the convective system is initiated and favored since dry air is not brought into the convective system. This picture is confirmed by the analysis of 20 extreme storms in Ouagadougou and several cases in other countries documented in the recent "Forecaster's Handbook over West Africa".

1. Introduction

The West African Monsoon (WAM) has undergone drastic changes in the recent past, which has had dramatic impacts on the population. After 30 years of devastating droughts, the last two decades experienced a growing number of damaging floods concurrent with ongoing episodes of water shortages and famines. In the past few years, a general emphasis has been placed on the study of extreme rainfall over the Sahel, motivated by the need to understand the physical processes associated with these high-impact weather (HIW) systems and their recent and future evolution.

2. Recent Trend of Extreme Rainfall Events

Within this climatic context, Panthou *et al.* (2014) used a 65-year record of rain gauge data to analyze the contribution of extreme rainfall events to the annual rainfall regime and their trends during the past decades. Using a regional approach of the Generalized Extreme Value

The Multiscale Global Monsoon System
Edited by C. P. Chang et al.

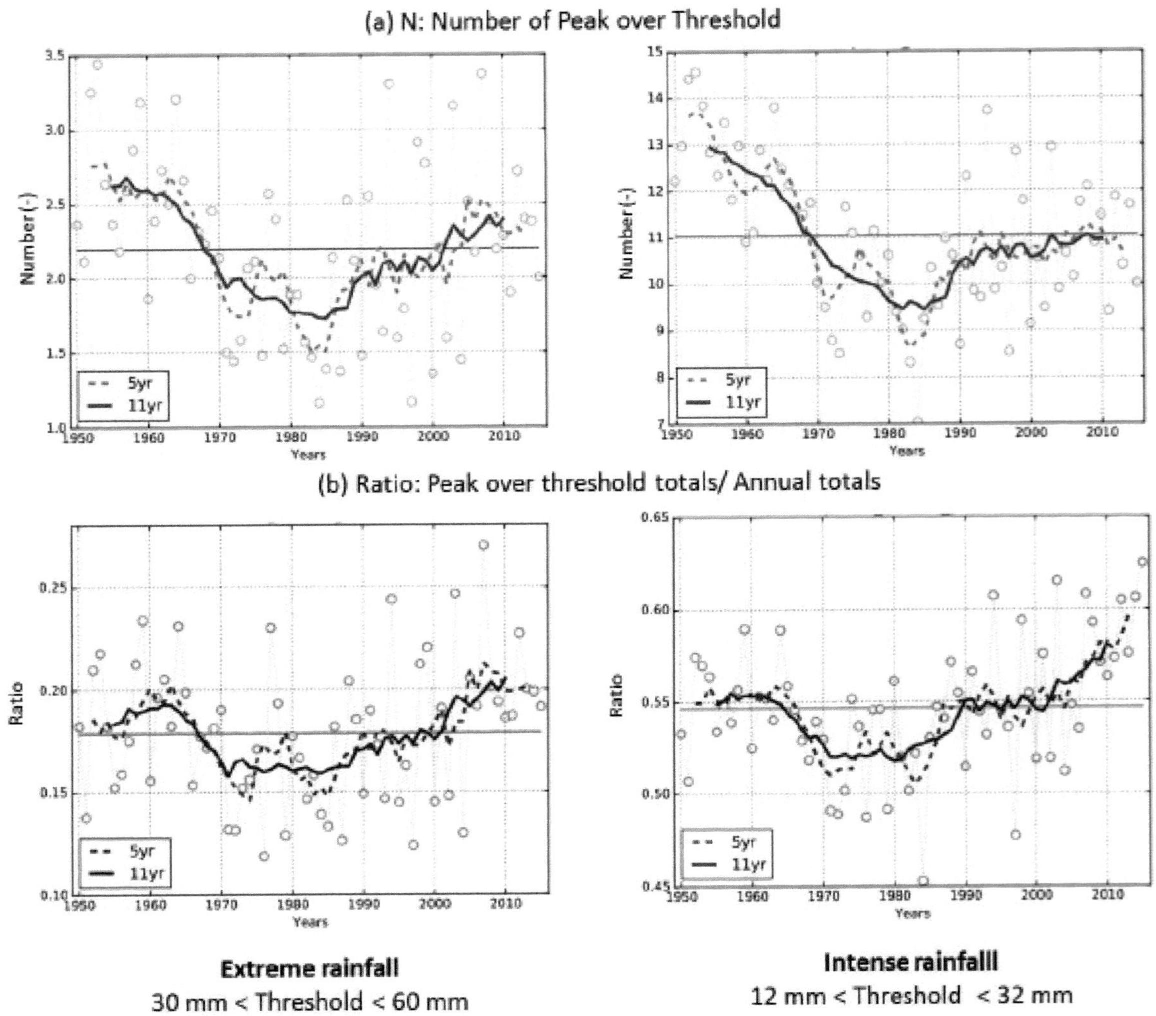

Fig. 1. (a) Evolution of the number of extreme (left) and intense (right) daily rainfall events, and (b) of their contribution to the total rainfall expressed as a ratio for each year (circle) and filtered over 5 and 11 years (dotted and solid lines, respectively). *Source*: Vischel *et al.* (2019), updated after Panthou *et al.* (2014a).

theory, which helps in the sampling of such rare events, they defined a space-dependent threshold to detect intense or extreme precipitating events (e.g., nearly 45 mm of daily rainfall for Ouagadougou for extreme events and 24 mm for intense events). Figure 1 illustrates the changes over the period 1950–2015 in mean annual number of extreme and intense events (Fig. 1a) and their relative contribution to annual rainfall amounts (Fig. 1b). During the last decade, the number of extreme events and their contribution to annual rainfall amounts have increased leading to a more intense hydrological cycle in the Sahel.

The frequency of occurrence of extreme events has significantly increased over the last fifteen years, reaching in 2015 a level close to that of the wet period before 1970. In contrast, the frequency of occurrence of intense events has remained well below those that the regions experienced in the 1950s and 1960s. The relative contribution of both extreme and intense events to the annual totals has sharply increased since 2000 to unprecedented levels: since 1950 and despite the occurrence of two very wet decades in the 1950s and 1960s, the importance of extreme events in the rainfall regime has never been as high as in the last 15 years.

The consistent increase of extreme rainfall without any clear trend in annual mean rainfall confirms the findings of Panthou *et al.* (2014): an important change in the rainfall regime occurred at the turn of this century, with extreme events becoming more pronounced.

More recently Taylor *et al.* (2017) have shown that the more frequent extreme rainfall events over the Sahel are correlated with the signal of global warming and likely reflects its effect on the meridional temperature gradient and subsequently on the African Easterly Jet (AEJ) magnitude. A stronger AEJ corresponds to a stronger vertical wind shear and more structured mesoscale convective systems (MCSs). They noted, however, that the environment of these intense MCSs is not significantly more moist before the arrival of the HIW.

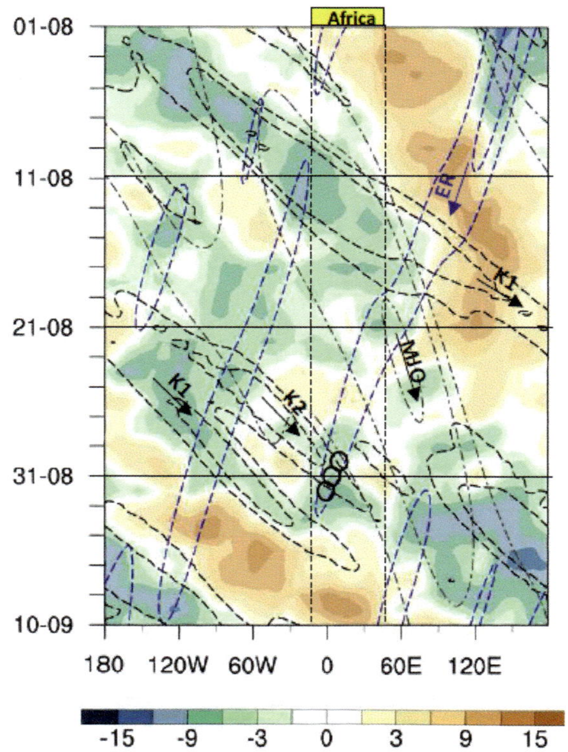

Fig. 2. Longitude–time diagrams from August 1st to September 10th of daily anomalies of velocity potential ($10^6\,m^2\,s^{-1}$) at 200 hPa. Anomalies are averaged in the 5°–15°N band and are relative to the ERA-Interim over the 1979–2014 period. Favorable phases for convection are superimposed for MJO (black dash–dot isolines), Kelvin (black dashes), and ER waves (blue dashes), which are detected owing to the use of the Roundy and Frank (2004) filtering. Isolines are plotted at −1.5, −3, and −6 $10^6\,m^2s^{-1}$. The three black circles mark the location of the trough T2 on 30th, 31st August and 1st September. Africa is delimited by vertical dashed lines (between 15°W and 45°E). *Source*: Lafore *et al.* (2017).

3. The Ouagadougou Extreme Event in 2009

In order to better understand the physical processes involved in the development of these HIW systems, Lafore *et al.* (2017) performed a multiscale analysis of an extreme rain event that occurred in Burkina Faso on 1st September 1, 2009 with an absolute record of 263 mm of rainfall observed at Ouagadougou.

This HIW system results from the combination of several favorable elements at different scales. The sea surface temperature anomaly patterns in July–August 2009 of the Atlantic cold tongue, the Tropical Atlantic Dipole, and the warmer Mediterranean Sea were favorable to the northward penetration of the monsoon flow over the Sahel (for details, see Fig. 2 of Lafore *et al.* 2017). This extreme event occurred during the establishment of a strong El Niño event from June 2009 to April 2010, whereas the Atlantic Ocean was warmer in the Northern Hemisphere due to the warm phase of the Atlantic Multi-decadal oscillation (AMO). The impact of El Niño on the West African Monsoon is unclear. Nevertheless, the Atlantic warm anomaly in the Northern Hemisphere due to the warm phase of the AMO appears to favor Sahel precipitation during El Niño as suggested by Rodriguez-Fonseca *et al.* (2011).

The intense convective activity of the last 10-day period in August 2009 was associated with the crossing of a convectively coupled Kelvin Wave (K2) and of an Equatorial Rossby (ER) wave as shown in Fig. 2. As suggested by Ventrice and Thorncroft (2013), the AEW activity increases during and after the passage of the convectively coupled Kelvin Wave K2 as indicated in Fig. 2. On the contrary, the ER wave contributes to the building up of a large moist anomaly of precipitable water (PW) over East Africa (blue area in Fig. 3a). The combination of these two waves generated favorable key conditions for the occurrence of this extreme event, i.e., enhancement of the AEW activity, and the formation of a large wet spell over the eastern and central Sahel that advected later toward the western Sahel.

At the synoptic scale, this event corresponds to the passage of a train of three AEWs with increasing magnitude (troughs overlaid in yellow in Fig. 3a). Behind the 1st AEW trough axis T1, an intense and deep southerly monsoon burst develops (Fig. 3a). It contributes to the amplification of the 2nd AEW T2 and its breaking as associated with the formation of an intense meso-vortex on the southern flank of the AEJ.

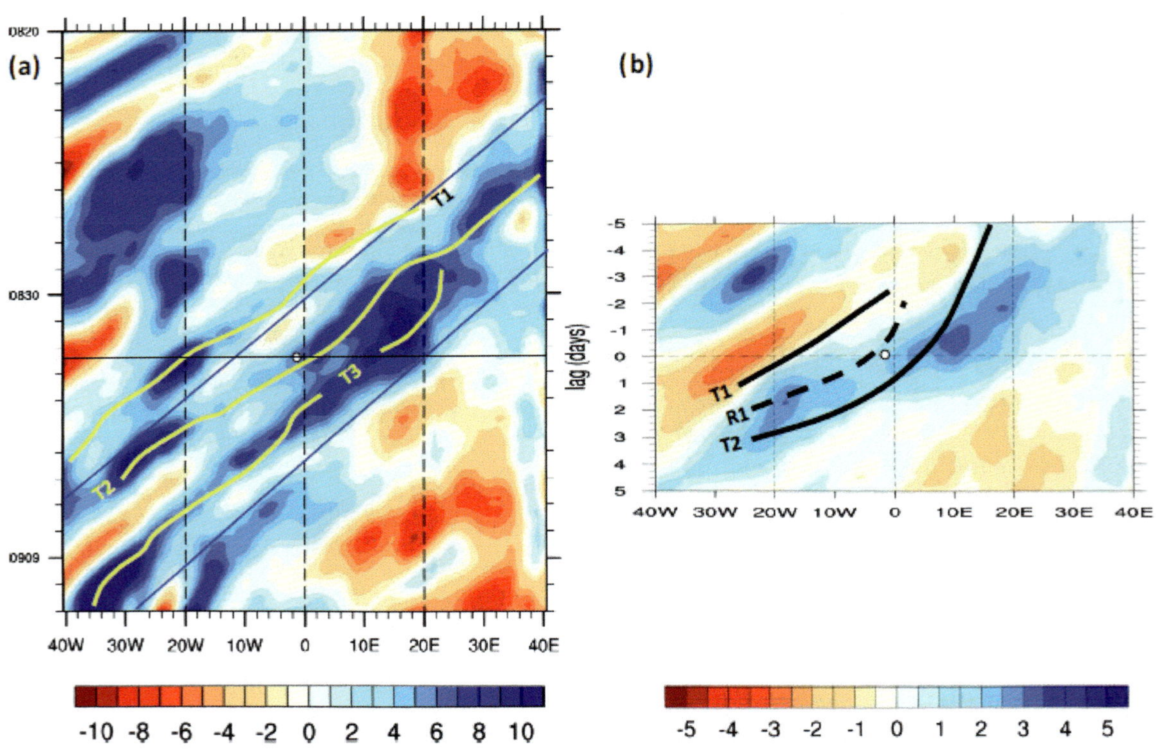

Fig. 3. (a) Longitude–time diagrams between 40°W and 40°E from 20th August to 10th September 2009 for the precipitable water anomaly PW (mm), averaged in the 10–18°N band. (b) Longitude–time lag composite structure of 20 extreme events of PW anomaly (mm) averaged in the 10–18°N band, between 40°W and 40°E, between 5 days prior and after the extreme rainfall event date. The trough axis (solid lines numbered T1, T2, and T3) of the AEW train is overlaid on both panels. The ridge axis R1 is also overlaid on panel (b) with a shaded black line. The white dot indicates the location of the Ouagadougou event. Data are provided by ERA-interim reanalysis (Dee *et al.* 2011). *Source*: Lafore *et al.* (2017) and Vischel *et al.* (2019) for (a) and (b), respectively.

4. A Statistical View of Extreme Rainfall Events

To assess the representativeness of the above results based on a single extreme event, a statistical analysis was performed by Vischel *et al.* (2019). First, daily rainfall observed in Ouagadougou from 1998 allowed the detection of 20 extreme events following the approach of Panthou *et al.* (2014) (above 48.5 mm). The 2009 event is by far the most intense event in these two decades.

The composite structure of these extreme events is then computed as the average of the atmospheric state associated with each of the 20 events occurring over Ouagadougou. A composite sequence is also computed from 5 days prior to the events (Lag -5) to the date of the events (Lag 0). To better emphasize the dynamical and thermodynamical structures associated with these events, and especially how they can depart from the slowly varying state of the WAM, we compute composites of both raw and anomalous atmospheric variables. Anomalies are defined as departure from the mean 1998–2010 annual cycle (see, Poan *et al.* 2013 for further details).

Figure 3b shows the resulting composite for the anomaly of PW to be compared with the same evolution for the Ouagadougou event in 2009 (Fig. 3a). The formation over the eastern Sahel of a large and intense wet spell (zonal extension more than 20°) about 5 days prior to the extreme event is a common feature, except that due to the dispersion between the 20 events, its magnitude is less in the composite than for the Ouagadougou 2009 single event (~4 mm instead ~8 mm). Also, in both studies, the wet spell is preceded by a large dry anomaly (red shading in Fig. 3). Another key common feature is the location of the moist anomaly ahead (west) of the AEW trough (T2) within the northerly flow, instead of being located behind the AEW trough, in the monsoon flow (Poan *et al.* 2014; Cornforth *et al.* 2017, see their AEW schematic in Fig. 2.10). The extreme rain events occur at this location, a few degrees west of the trough within the wet anomaly (~2° and ~4° for the Ouagadougou event and the composite event, respectively). The growing stage of the trough T2 in the lee of the previous trough T1 corresponds to a slower westward propagation than T1, while the ridge R1 reinforces it. As a result, T2 becomes, after about 2 days, independent of its mother wave trough. For most extreme events such as in 2009, the trough T2 becomes a close circulation associated with the breaking of the AEW. This statistical analysis over the last 20 years of the most extreme rainfall events in Ouagadougou shows that this time sequence of their arrival is robust and rather systematic among the events. Besides, it occurs at least once a year (Roehrig *et al.* 2017). Further statistical studies are now needed to assess the role of the equatorial waves in the development of such extreme rainfall events, as suggested by the 2009 case study (Lafore *et al.* 2017), in particular the crossing of convectively coupled Kelvin waves and Equatorial Rossby waves.

5. Conclusion — Towards Applications

To explore the potential forecast skill brought about by the advanced understanding of the intraseasonal variability (ISV) over West Africa gained from the international program AMMA (African Monsoon Multi-disciplinary Analysis, Redelsperger *et al.*, 2006), several initiatives have been launched to monitor the tropical ISV in general and more recently with a focus on Africa (Climate Prediction Center of the NOAA, State University of New York). Among them, the MISVA[1] collaborative project

[1]Real-time monitoring and forecast of IntraSeasonal Variability over Africa http://misva.sedoo.fr/.

between
Senegal and France, started in 2011, aims to monitor and forecast the ISV over Africa in real time. This project has provided a series of diagnostics and indices during the WAM season and the experience has shown that dry and moist intense events can be detected several days in advance, even as Numerical Weather Prediction models have almost no skill in these scales, at least for precipitation. This fruitful collaboration between forecasters and researchers allowed the collection and analyses of numerous cases of intense events. In general these case studies confirm the results of this chapter. In particular, intense events appear as the result of the interactions between different equatorial waves.

The physical understanding of the WAM discussed above helped to improve forecasting methods over West Africa and led to the writing of the "Forecaster's Handbook for West Africa" published recently (Parker and Diop-Kane, 2017), in close collaboration between forecasters and scientific international communities, under the umbrella of the World Meteorological Organization WMO.

References

Cornforth R., Z. Mumba, D. J. Parker, G. Berry, N. Chapelon, K. Diakaria, M. Diop-Kane, V. Ermert, A. H. Fink, P. Knippertz, J-P. Lafore, A. Laing, S. Lepape, R. Maidment, J. Methven, B. N. Orji, D. Osika, E. Poan, R. Roca, S. Rowell, R. Smith, T. Spengler, C. Taylor, C. Thorncroft, J-C. Vincendon, and C. York, 2017: Synoptic Systems, in *Meteorology of Tropical West Africa: The Forecasters' Handbook*, Eds. D. J. Parker and M. Diop-Kane, John Wiley & Sons, Ltd, Chichester, UK, doi: 10.1002/9781118391297.ch2.

Dee, D. P., S. M. Uppala, A. J. Simmons *et al.*, The ERA-Interim reanalysis: Configuration and performance of the data assimilation system. *Q. J. R. Meteorol. Soc.*, **125**, 553–597, doi: 10.1002/qj.828.

Diaz, M. L. and A. Aiyyer. 2013. Energy dispersion in African easterly waves. *J. Atmos. Sci.*, **70**, 130–145, doi: 10.1175/JAS-D-12-019.1.

Lafore, J.-P., F. Beucher, P. Peyrillé, A. Diongue, D. Bouniol, N. Chapelon, F. Ferry, E. Poan, and R. Roehrig, xxxx: The extreme rain event of Ougadougou in 2009 Part I: Multi-scale analysis and conceptual model. *Q. J. R. Meteorol. Soc.*, **141**, doi:10.1002/qj.3165.

Panthou, G., T. Vischel, and T. Lebel, 2014: Recent trends in the regime of extreme rainfall in the Central Sahel. *Int. J. Climatol.*, **34**, 3998–4006.

Parker, D. J. and M. Diop-Kane, 2017: *Meteorology of tropical West Africa: The Forecasters' Handbook*, Wiley-Blackwell, ISBN: 978-1-118-39130-3, 496.

Poan, E., R. Roehrig, F. Couvreux, and J.-P. Lafore, 2013: West Africa intraseasonal variability: A precipitable water perspective. *J. Atmosph. Sci.*, **70**(4), 1035–1052, doi: http://dx.doi.org/10.1175/JAS-D-12-087.1.

Poan, D. E., J.-P. Lafore, R. Roehrig, and F. Couvreux, 2014: Internal processes within African Easterly wave system. *Q. J. R. Meteorol. Soc.*, doi: 10.1002/qj.2420.

Raymond, D. J. and H. Jiang. 1990: A theory for long-lived mesoscale convective systems. *J. Atmos. Sci.*, **47**(24), doi: http://dx.doi.org/10.1175/1520-0469(1990)047<3067:ATFLLM>2.0.CO;2.

Redelsperger, J.-L., C. D. Thorncroft, A. Diedhiou, *et al.*, 2006. African Monsoon multidisciplinary analysis: an international research project and field campaign. *Bull. Am. Meteorol. Soc.*, **87**, 1739–1746.

Rodriguez-Fonseca, B., S. Janicot, E. Mohino, T. Losada, J. Bader, C. Caminade, F. Chauvin, B. Fontaine, J. García-Serrano, S. Gervois, M. Joly, I. Polo, P. Ruti, P. Roucou, and A. Voldoire, 2011. Interannual and decadal SST-forced responses of the West African monsoon. *Atmos. Sci. Lett.*, **12**, 67–74. https://doi.org/10.1002/asl.308.

Roundy, P. E. and W. M. Frank. 2004. A climatology of waves in the equatorial region. *J. Atmosph. Sci.*, **61**, 2105–2132.

Taylor, C., D. Belušić, F. Guichard, D. J. Parker, T. Vischel, O. Bock, P. P. Harris, S. Janicot, C. Klein, and G. Panthou, 2017: Frequency of extreme Sahelian storms tripled since 1982 in satellite observations. *Nature*, doi:10.1038/nature22069.

Ventrice, M. J. and C. D. Thorncroft. 2013. The Role of Convectively Coupled Atmospheric Kelvin Waves on African Easterly Wave Activity. *Mon. Wea. Rev.*, **141**, 1910–1924, doi:10.1175/MWR-D-12-00147.1.

Vischel, T., G. Panthou, P. Peyrillé, R. Roehrig, G. Quantin, T. Lebel, C. Wilcox, F. Beucher, and M. Budiarti: Chapter 4 — Precipitation Extremes in the West African Sahel: Recent Evolution and Physical Mechanisms, Editor(s): V. Venugopal, J. Sukhatme, R. Murtugudde, R. Roca, 2019: *Tropical Extremes, Elsevier*, pp. 95–138, ISBN 9780128092484, https://doi.org/10.1016/B978-0-12-809248-4.00004-2.

Chapter 14

Role of Monsoon Variability in Heat Wave Activity over South China and Indochina

Ming Luo*,†,§ and Ngar-Cheung Lau†,‡,¶

**School of Geography and Planning, Sun Yat-sen University, Guangzhou 510275, China*
†Institute of Environment, Energy and Sustainability, The Chinese University of Hong Kong, Sha Tin, N.T., Hong Kong, China
‡Department of Geography and Resource Management, The Chinese University of Hong Kong, Sha Tin, N.T., Hong Kong, China
§luom38@mail.sysu.edu.cn
¶gabriel.lau@cuhk.edu.hk

Heat waves (HWs) exert serious impacts on human health and the natural environment. With global warming, HW activities in many parts of the world have been intensifying during the past decades and are projected to pose increasingly severe threats to society in the future. In order to assess the global and regional impacts of climate change, it is therefore important to broaden our understanding of HWs. In this chapter, we first review recent advances in the study of HWs in various regions by focusing on the synoptic characteristics and atmospheric circulations associated with such phenomena. Then we examine the behavior of HWs over South China and Indochina, where the local climate is strongly influenced by monsoon circulations. Results are presented on how the HW occurrences in these regions are affected by the intensity and spatial pattern of the local monsoon flow.

1. Introduction

Episodes of extremely high near-surface air temperatures lasting several days or more, often referred to as heat waves (HWs), exert notable impacts on public health as well as the natural environment (Easterling *et al.* 2000; Forzieri *et al.* 2017; Kovats and Kristie 2006; McMichael *et al.* 2006). Such events occur in many parts of the globe, including temperate, subtropical, and tropical zones. Under global warming, HW activities have been increasing noticeably during the past few decades, and are projected to intensify in the future (Cowan *et al.* 2014; Kunkel *et al.* 2010). The WMO (2013) reported that the number of deaths due to HWs during the most recent decade 2001–2010 (136,000) has increased markedly (by 23 times) compared to the number of mortality cases (6,000) in 1991–2000. The synoptic characteristics associated with the development of HWs in various regions have been examined by many investigators using both observational and model-generated data. These studies indicate the strong geographical dependence of the atmospheric processes contributing to HW formation and maintenance. Here, a survey is offered on the distinctive meteorological features and the associated mechanisms of HWs identified in different regions.

At the synoptic scale, HWs in many regions are accompanied by increased surface pressure, suppressed precipitation, and sinking motion (Loikith and Broccoli 2012). Clear sky with increased solar radiation also prevails during HW episodes. The occurrence of atmospheric blocking, which is characterized by quasi-stationary pressure ridges in mid-latitudes, is an important contributing factor for HWs

The Multiscale Global Monsoon System
Edited by C. P. Chang et al.

(Matsueda 2011; Sillmann and Croci-Maspoli 2009). Up to 80% of hot extremes in northern Europe are associated with such blocking episodes (Pfahl and Wernli 2012). The blocking ridges alter the path of mid-latitude jets and impede the propagation of synoptic-scale weather disturbances. The enhanced adiabatic warming within the subsiding air column and clear-sky radiative heating also favor warm temperature extremes during blocking events (Brunner *et al.* 2017; Pfahl *et al.* 2015; Pfahl and Wernli 2012).

In addition to atmospheric effects, HWs are also influenced by land–atmosphere feedbacks. With rising temperature, soil moisture decreases. The resulting reduction in latent heat flux causes temperatures to warm further (Garcia-Herrera *et al.* 2010; Lorenz *et al.* 2010; Miralles *et al.* 2014). Other factors such as sea surface temperature (SST) are also important for strengthening the land–atmosphere feedback process (Della-Marta *et al.* 2007). Further studies on HW characteristics over North America and Europe have been reviewed by Grotjahn *et al.* (2016) and Ward *et al.* (2016), respectively.

Besides the North American and European sectors, HWs have frequently been observed in Asian monsoon regions and have caused even more severe upheavals to human society and the ecosystem due to the dense population and social vulnerability to natural hazards in the latter zones (Chen and Lu 2015; Ding *et al.* 2010; Kothawale *et al.* 2010; Revadekar *et al.* 2013). These regions have also exhibited significant warming trends in recent decades, as reported by many observational studies (Ding *et al.* 2007; Qi and Wang 2012; Zhai and Pan 2003). In China, the intensifying trends in HW activities are especially strong in the North China plain, and the Yangtze and Pearl River Deltas, possibly due to rapid urbanization in these areas (Zhai and Pan 2003; Zhou and Ren 2011; Luo and Lau 2017).

The HWs in many parts of China are accompanied by increases in surface pressure and solar radiation, and decreases in humidity, cloud cover, and precipitation (Ding *et al.* 2010; Lu and Chen 2016; Luo and Lau 2017). During HW events, these areas are also under the influence of subsidence and adiabatic warming. HWs are associated with the westward movement of the western North Pacific subtropical high (WNPSH) system (Ding *et al.* 2010; Luo and Lau 2017). Hsu *et al.* (2017) showed that HWs in Asia are modulated by boreal summer intraseasonal oscillations.

The year-to-year variations of HWs are related to the East Asian jet stream (EAJS) and the circumglobal teleconnection (CGT) pattern, and are modulated by the El Niño–Southern Oscillation (ENSO) (Chen *et al.* 2018; Hu *et al.* 2012, 2013; Luo and Lau, 2018a; Wang *et al.* 2013, 2014). Hu *et al.* (2012) reported that the frequency of HWs occurring south of the Yangtze River valley is influenced by SST anomalies in the Indian Ocean.

The HW-related patterns are closely associated with the Asian monsoon circulation. For instance, a recent investigation classified extreme HWs in eastern China into three categories, namely, typical, monsoonal, and foehn patterns, among which the monsoonal pattern is characterized by anomalous subsidence and low-level monsoonal southwesterlies (Chen and Lu 2015). Wei and Chen (2011) suggested that the abrupt increase in hot extremes over northern China and Mongolia since the mid-1990s is partially due to the weakening in the East Asian summer monsoon (EASM) circulation.

Remote forcings also contribute to hot weather episodes in monsoon Asia. A recent study based on CMIP5 simulations suggested that SST anomalies in the North Pacific and the North Atlantic play an important role in the variability of HW activity in monsoon Asia, via the modification of WNPSH (Chen and Zhou 2017). HW events in northern China are influenced by the North Atlantic oscillation via a zonal wave train along the westerly jet stream (Sun 2012).

HWs in different sectors exhibit both common and region-specific characteristics. Understanding the meteorological features and the governing mechanisms of HWs in various regions would facilitate the design of early warning systems for such extreme events in these locations. In the following sections, we devote particular attention to the behavior of HWs over South China and Indochina, where the local climate is strongly influenced by monsoon circulations. Results are presented on how the HW occurrences in these regions are affected by the variability in the intensity and spatial pattern of the local monsoon flow. The possible influences of climate change on HW behavior are assessed by analyzing the long-term trends of various indicators of intensity, frequency of occurrence, and duration of HWs in the two regions of interest.

Most of the results presented here are based on the recent studies on HWs over South China (Luo and Lau 2017) and Indochina (Luo and Lau 2018b). The primary dataset used in these investigations is the Climate Forecast System Reanalysis (CFSR) produced by the US National Centers for Environmental Prediction for the 1979–2010 period. Results for South China have been obtained using data for the warm season from May through September. As explained fully in Section 3, results for Indochina have been computed separately for the wet season (typically from May to October) and the dry season (typically from November to April).

2. Heat Waves over Southern China

Spatial averages of the daily maximum air temperature ($T_{\max}$) were taken over the Guangdong Province in southern China (GDPC; see the border of this region in Figs. 2 and 3). HW events were identified on the basis of this time series by applying the set of criteria adopted by Meehl and Tebaldi (2004) and Lau and Nath (2012). Specifically, an HW episode needs to satisfy three conditions: exceedance of the

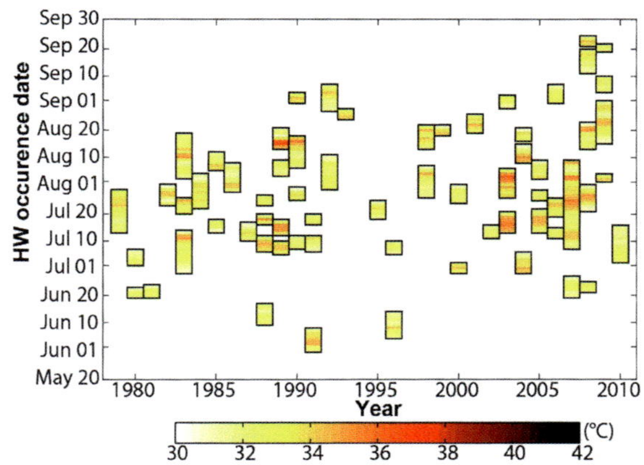

Fig. 1. Occurrences of HWs in South China. Each black box denotes an HW event. Shading indicates the spatial average of $T_{\max}$ during HWs.

95th percentile value of $T_{\max}(T_1)$ for more than three consecutive days, exceedance of T_1 by the event-averaged $T_{\max}$, and exceedance of the 75th percentile value by $T_{\max}$ for each day of the event.

The HWs in the summers of individual years within the 1979–2010 period are shown in Fig. 1. This diagram indicates prominent decadal-scale changes in HW activity. Specifically, HWs mainly occurred in late July and early August in the 1980s, whereas HW occurrences since the 1990s are more frequent in late June, as well as the latter half of August through September.

The patterns of $T_{\max}$, precipitation, horizontal wind vector near the surface, and sea level pressure (SLP), as obtained by compositing the anomaly fields over all identified HW days, are shown in Fig. 2. Warm temperature anomalies prevail throughout GDPC and neighboring regions, including much of southern China and some parts of Indochina (Fig. 2a). Dry conditions are observed over southern China (Fig. 2b). The anomalous near-surface wind and SLP patterns in this region are dominated by an anticyclonic flow around a high-pressure center near Hainan Island (Fig. 2c). This anticyclone is accompanied by an offshore flow toward the South China Sea (SCS), which would reduce the

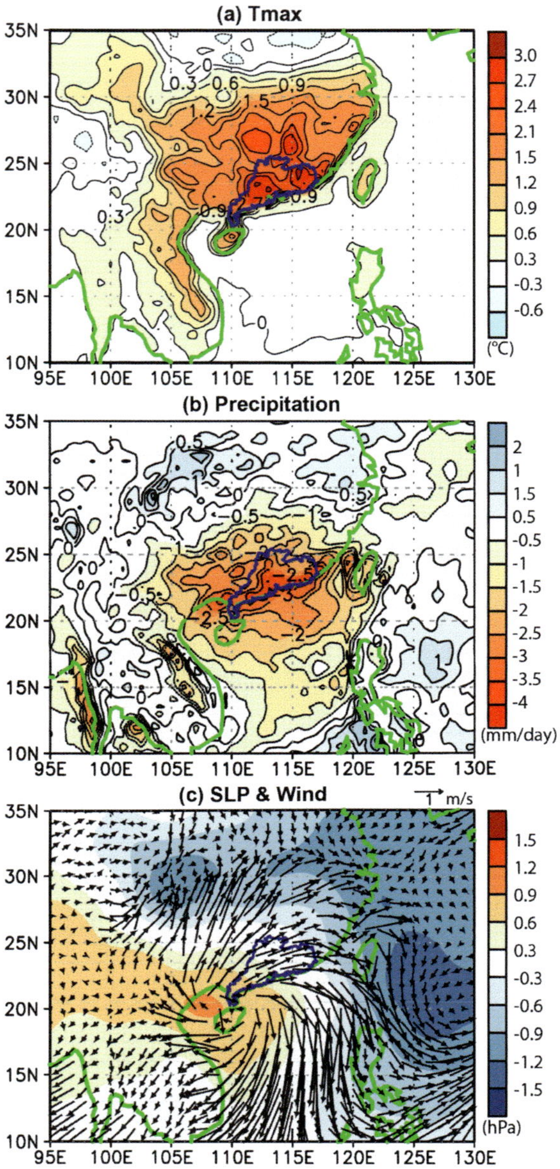

Fig. 2. Composite charts of anomalies of (a) $T_{\max}$, (b) precipitation, and (c) SLP (shading) and near-surface wind (vector) for HWs in South China. Blue boundary denotes the study region, i.e., Guangdong Province of China.

advection of moist and cold air from ocean to land, thus contributing to the dry and hot conditions over southern China. On the contrary, the anomalous southwesterlies on the northwestern flank of the anticyclone lead to enhanced moisture transport and flow convergence toward central China, and may partially account for the increased precipitation over the latter region (Fig. 2b).

The spatio-temporal development of the synoptic features associated with HWs in southern China is illustrated in Fig. 3. These charts show the composite anomalies of 500-mb height (left column) and $T_{\max}$ and near-surface wind vector (right column) for 3, 2, 1, and 0 days prior to the onset of HWs. A prominent positive anomaly center at 500-mb height is discernible over the western Pacific three days before the HW onset (i.e., day –3). This anomaly amplifies as it migrates westward to eastern and southern China in the next three days. These results are indicative of the intensification and westward extension of WNPSH, which is a dominant climatological feature of the monsoonal circulation in this region. This evolution of the WNPSH is seen to be accompanied by westward migration of warm temperature and anticyclonic flow anomalies over southeastern China.

Secular changes in different aspects of HW activity may be investigated by examining the long-term changes in various HW indicators, all computed for individual years. These indicators include number of HW events (HWN) and HW days (HWF), duration of the longest event (HWD), warmest day of hottest event (HWA), mean duration of all events in the year (HWL), mean amplitude of all events in the year (HWM), starting date of the first event (HWO), and ending date of the last event (HWE). The temporal evolution of these yearly measures of HW activity as well as the space–time average of $T_{\max}$ over GDPC and the May–September (MJJAS) season are presented in Fig. 4. The trends of various indicators, as obtained using the Mann–Kendall trend analysis (Sen 1968), are also shown in this diagram.

The results in Fig. 4 indicate prominent trends for the areal-averaged $T_{\max}$ as well as HW activity during the 1979–2010 period. In particular, $T_{\max}$ exhibits a significant upward trend,

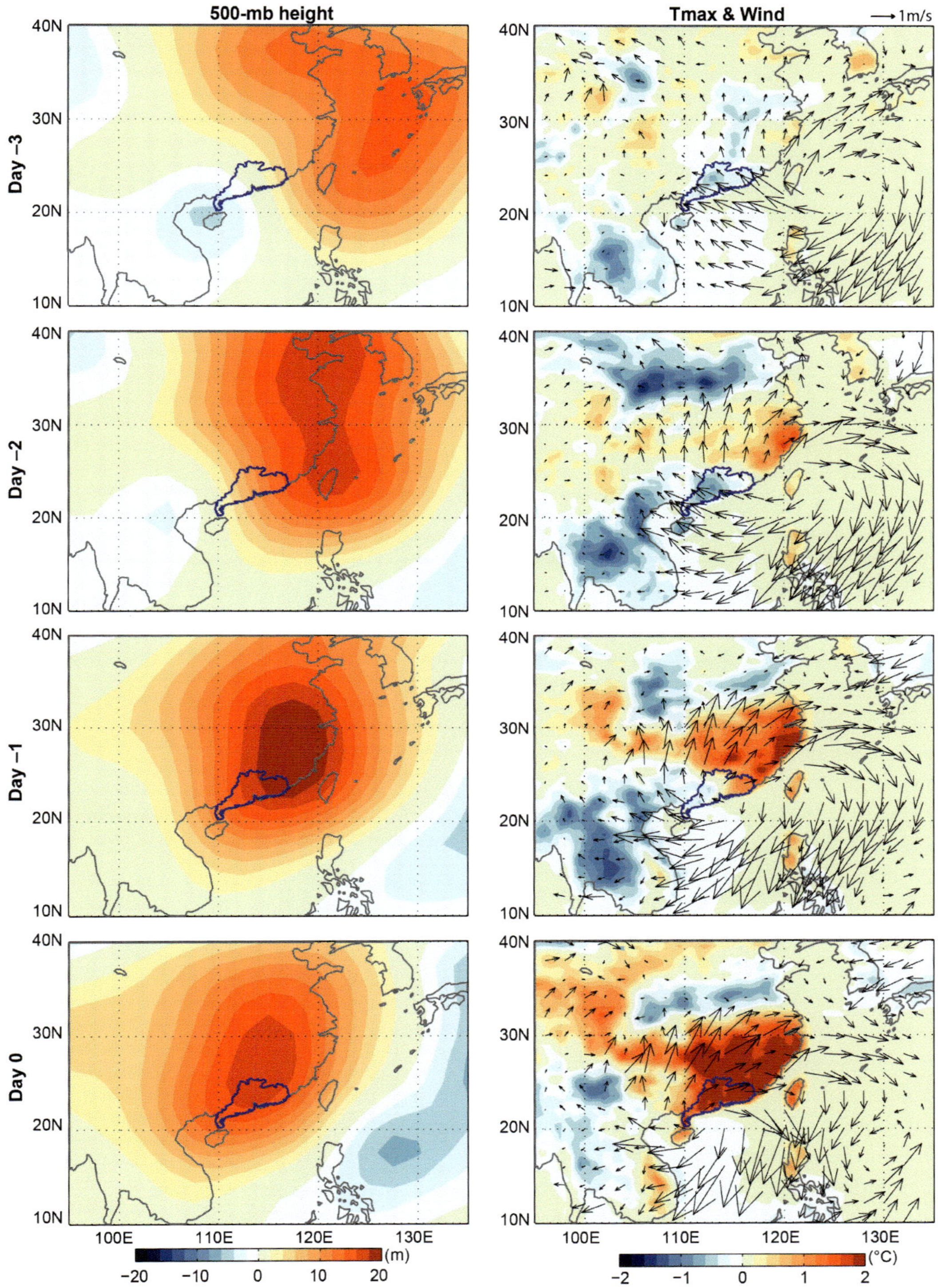

Fig. 3. Composite charts of the evolution of (left) 500-mb geopotential height and (right) near-surface $T_{\max}$ (shading) and wind (vector) prior to the occurrence of HW in South China. Day 0 denotes the day when HW begins, and day −1 denotes the day before day 0, etc. Blue boundary denotes the location of Guangdong Province, South China.

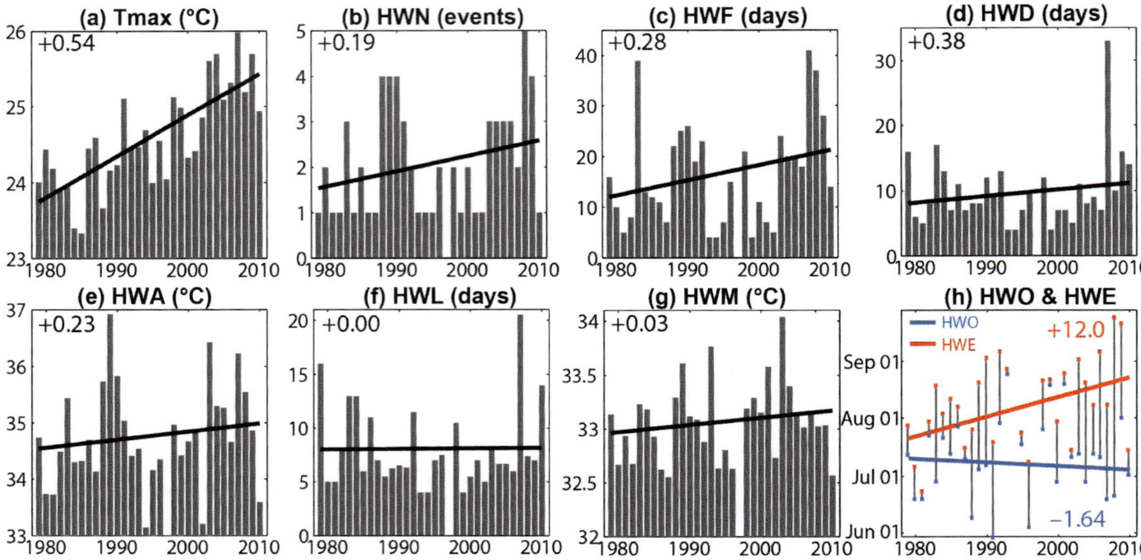

Fig. 4. Time series of (a) summer $T_{\max}$, (b) HWN, (c) HWF, (d) HWD, (e) HWA, (f) HWL, (g) HWM, and (h) HWO and HWE in South China. Straight lines denote their corresponding trends. Numbers indicate the trends (per decade).

and HWs in southern China have become more frequent (HWN, HWF), long-lasting (HWD), and severe (HWA, HWM) in the past three decades. Also noteworthy in Fig. 4h are the downward (upward) trends in HWO (HWE), indicating that the HWs commenced earlier and terminated later within the warm season in more recent years.

3. Heat Waves over Indochina

The Indochinese Peninsula is straddled by the SCS/western Pacific and the Bay of Bengal (BOB)/Indian Ocean. Its climate is linked to the conditions of the monsoon circulation systems in both East Asia and South Asia. This region is also characterized by a unique climatological seasonal cycle. Rather wet and generally cool conditions prevail from mid-May through October. The remaining half of the annual cycle is dry and is marked by rather high temperatures in the period from February through mid-May. Generally speaking, the climatological flow pattern in the vicinity of Indochina in the dry season is dominated by near-surface northeasterlies associated with the East Asia winter monsoon (EAWM), whereas the circulation in the wet season is reversed to a southwesterly pattern driven by the South Asian summer monsoon (SASM). In view of this special basic state, the HW characteristics in the wet and dry seasons have been considered separately.

The procedure for identifying HWs in Indochina is analogous to that used in our study for southern China. The exceedance criteria (see the previous section) are applied to spatial averages of $T_{\max}$ taken over those gridpoints in the central portion of the Indochinese Peninsula where the climatological value of the annually averaged $T_{\max}$ is higher than 30°C.

The occurrence of the identified HWs in individual years is illustrated in Fig. 5. These charts indicate that, for both dry and wet seasons, HW events have become more frequent in the latter half of the 1979–2010 period. HWs started earlier and ended later in both dry and wet seasons during more recent years. In particular, intense HWs in the dry season (with $T_{\max}$ amplitudes

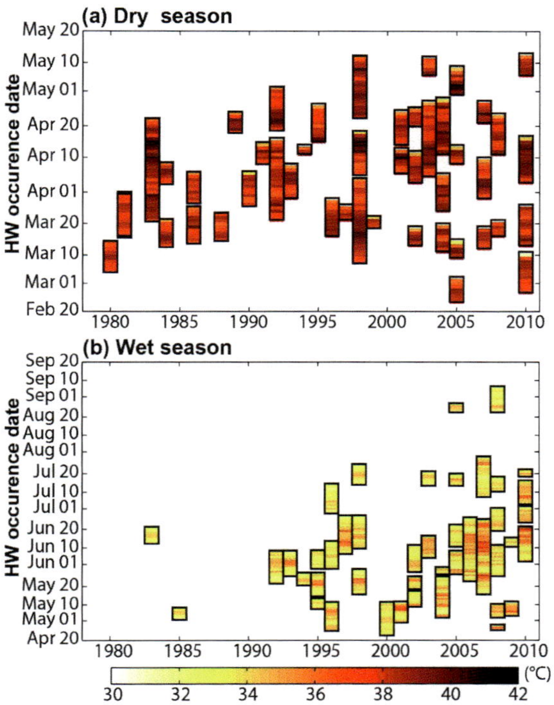

Fig. 5. Same as Fig. 1 but for HWs in the (top) dry and (bottom) wet seasons of Indochina.

reaching as high as 36–40°C) mostly occurred in mid-March through April in 1980–1995, whereas more HWs are seen in early March and early May since 1995. It is also evident from Fig. 5b that HWs were more widespread throughout the wet season after 1995 as compared with the earlier years.

The distributions of anomalous patterns averaged over the course of the HWs are displayed in Fig. 6 for the dry (left column) and wet (right column) seasons. Results are shown for $T_{\max}$, precipitation, and 925-mb wind vector and SLP. HWs in the dry season are seen to be accompanied by warm anomalies in Indochina and southwestern China (Fig. 6a), reduced precipitation over Indochina, SCS, and northeastern Indian Ocean (Fig. 6b), as well as anomalous cyclonic near-surface circulation and low SLP (Fig. 6c) over Northeast Asia. These patterns suggest that HWs in the dry season are associated with a warm-core low structure. The prevalent southwesterly wind anomalies over East Asia and the surrounding waters (Fig. 6c) are directed against the climatological wintertime monsoon flow. This result hence implies that HWs in Indochina during the dry season are coincident with a weakening of the EAWM circulation.

During the wet season (right column of Fig. 6), HWs in Indochina occur in conjunction with warm anomalies in Indochina and central-northern India (Fig. 6d), generally dry conditions over Indochina, BOB, and parts of western India (Fig. 6e), as well as anticyclonic wind and positive SLP anomalies in the northeastern sector of BOB and northern SCS (Fig. 6f). These results suggest that HWs in the wet season are characterized by a warm-core high structure. The easterly wind anomalies on the southern flank (between 10°N and 20°N) of the anticyclonic centers are directed opposite to the local climatological westerly monsoon flow. This anomalous wind pattern hence signifies that the SASM is typically weakened during HW events in Indochina.

The typical evolution of the anomalous $T_{\max}$ and 950-mb wind fields on 3, 2, 1, and 0 days before HW onset over Indochina is depicted in Fig. 7 using composite data for the dry (left column) and wet (right column) seasons. Three days prior to HW onset in the dry season (day –3, see Fig. 7a), a high-temperature anomaly is formed over the northern portion of East Asia. This feature strengthens and migrates equatorward on days –2 and –1 (Fig. 7b and c), and eventually covers most parts of East Asia as well as northeastern Indochina. On day 0, high temperatures cover most parts of the Indochinese Peninsula, with the strongest anomaly being seen in the northern part of that region. Warming is also discernible over southeastern China, Korea, and southern Japan.

The southward movement of the warming center from northern East Asia coincides with the equatorward extension of the anomalous southerly/southwesterly wind. This circulation

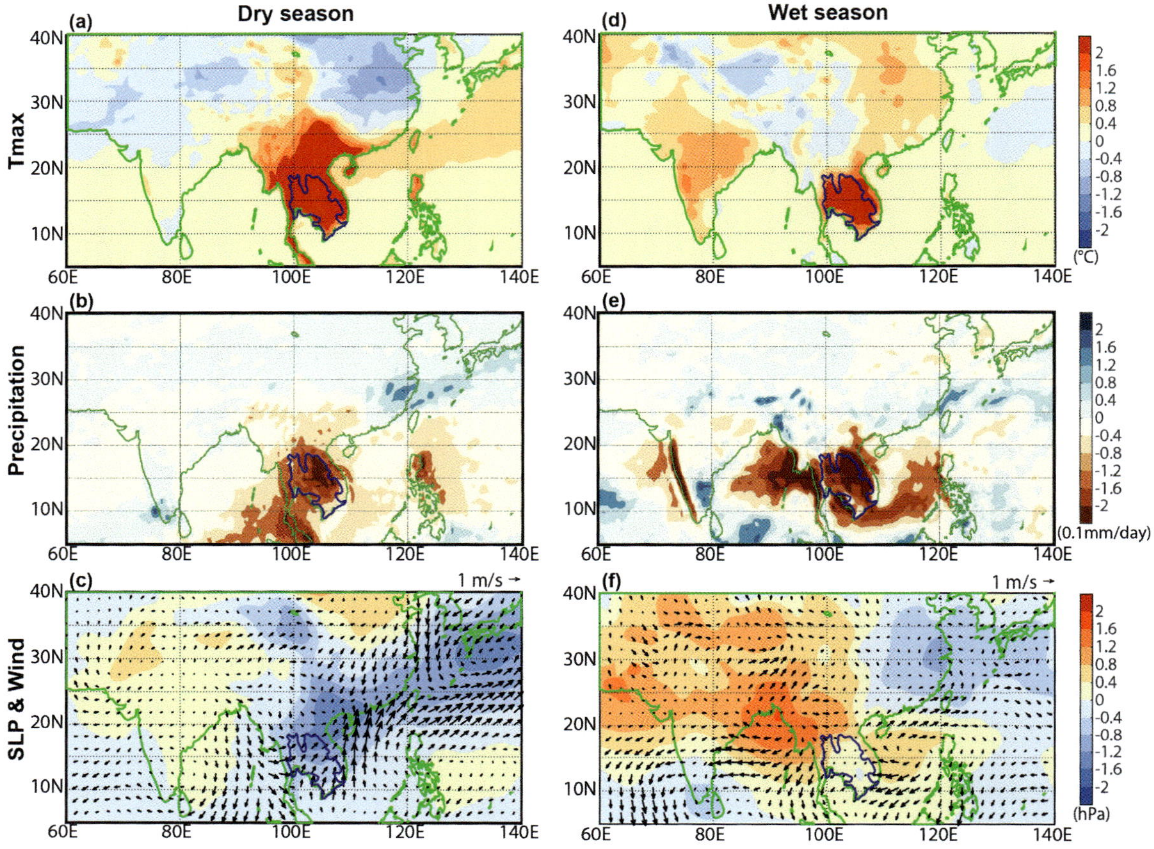

Fig. 6. Same as Fig. 2 but for HWs in the (left) dry and (right) wet seasons of Indochina. Blue boundary denotes the study region in Indochina.

change is indicative of the weakening of the northeasterly EAWM flow. The impacts of this change in the EAWM intensity first appear in the northern interior of Asia. The associated wind and temperature signals migrate progressively along the northeasterly path of the climatological monsoon system. They arrive at Indochina several days later and contribute to the onset of HW events in that region.

The development of HWs in the wet season is quite different from that in the dry season. On day –3, in the wet season (Fig. 7e), warm temperature anomalies are mostly found over west Asia and western India. These anomalies grow in amplitude and spatial extent on days –2 and –1 (Fig. 7f and g), and a prominent warm center is established over Indochina on day 0 (Fig. 7h).

The evolution of temperature changes as described above is accompanied by the prevalence of easterly wind anomalies within the tropical belt between 5°N and 20°N. The easterly wind anomalies that initially appear over the Arabian Sea (day −3, Fig. 7e) are seen to spread eastward with time, and their influences are felt over BOB and Indochina on day 0. These results again support the notion that weakening of the westerly climatological SASM plays a significant role in the occurrence of HW events over Indochina during the wet season.

The variations of different characteristics of HW activity over Indochina during the study

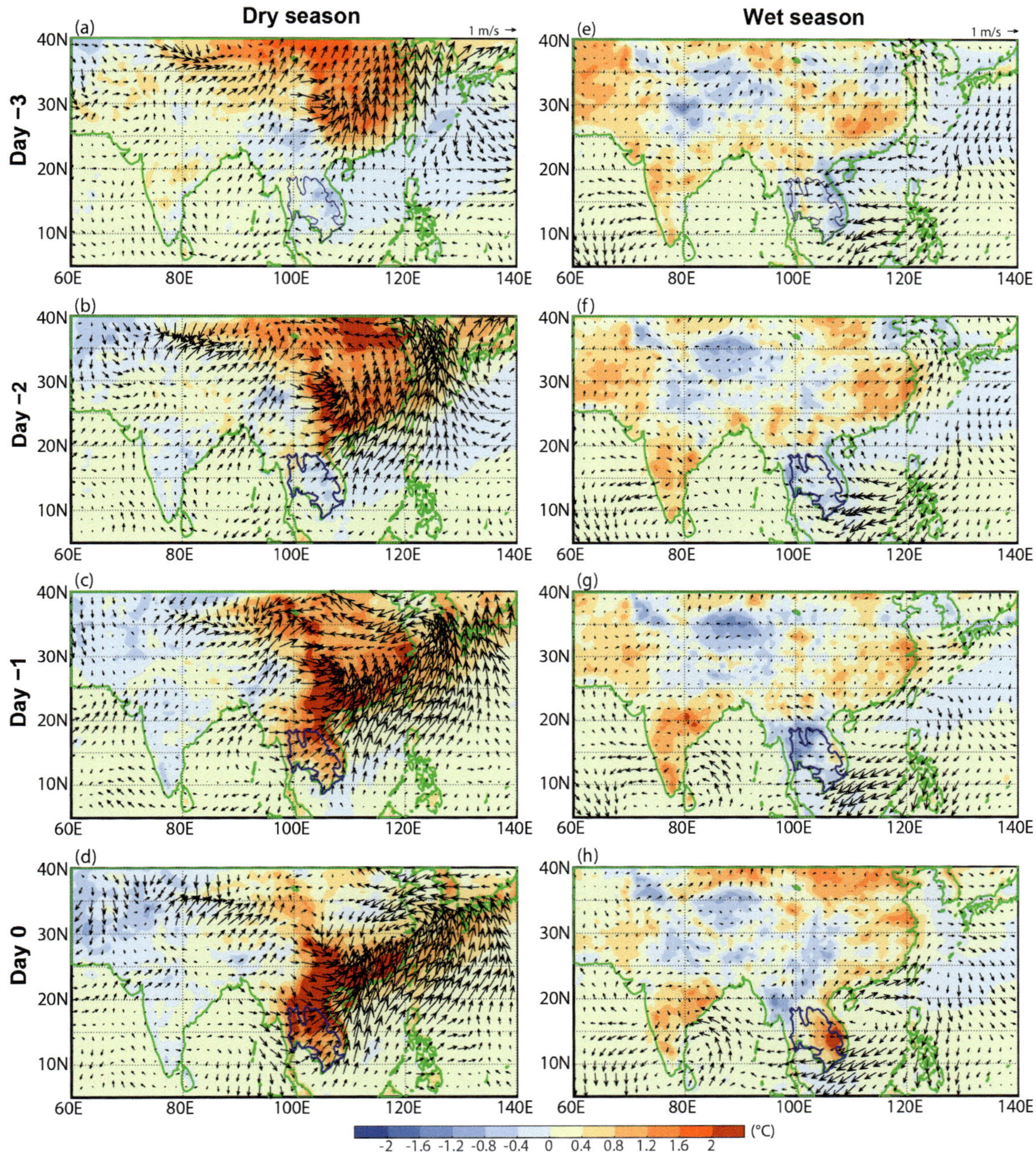

Fig. 7. Composite charts of the evolution of near-surface $T_{\max}$ (shading) and wind (vector) prior to the occurrence of HW in the (left) dry and (right) wet seasons of Indochina. Day 0 denotes the day when HW begins, and day −1 denotes the day before day 0, etc. Blue boundary denotes study region in Indochina.

period may be discerned from Fig. 8, which shows the time series and long-term trends of individual HW indices (see definitions in the previous section). Results are presented for the dry (panels in the upper half) and wet (lower half) seasons. It is seen that $T_{\max}$, as well as the frequency (HWN, HWF), duration (HWD), and severity (HWA, HWM) of the HW events in the

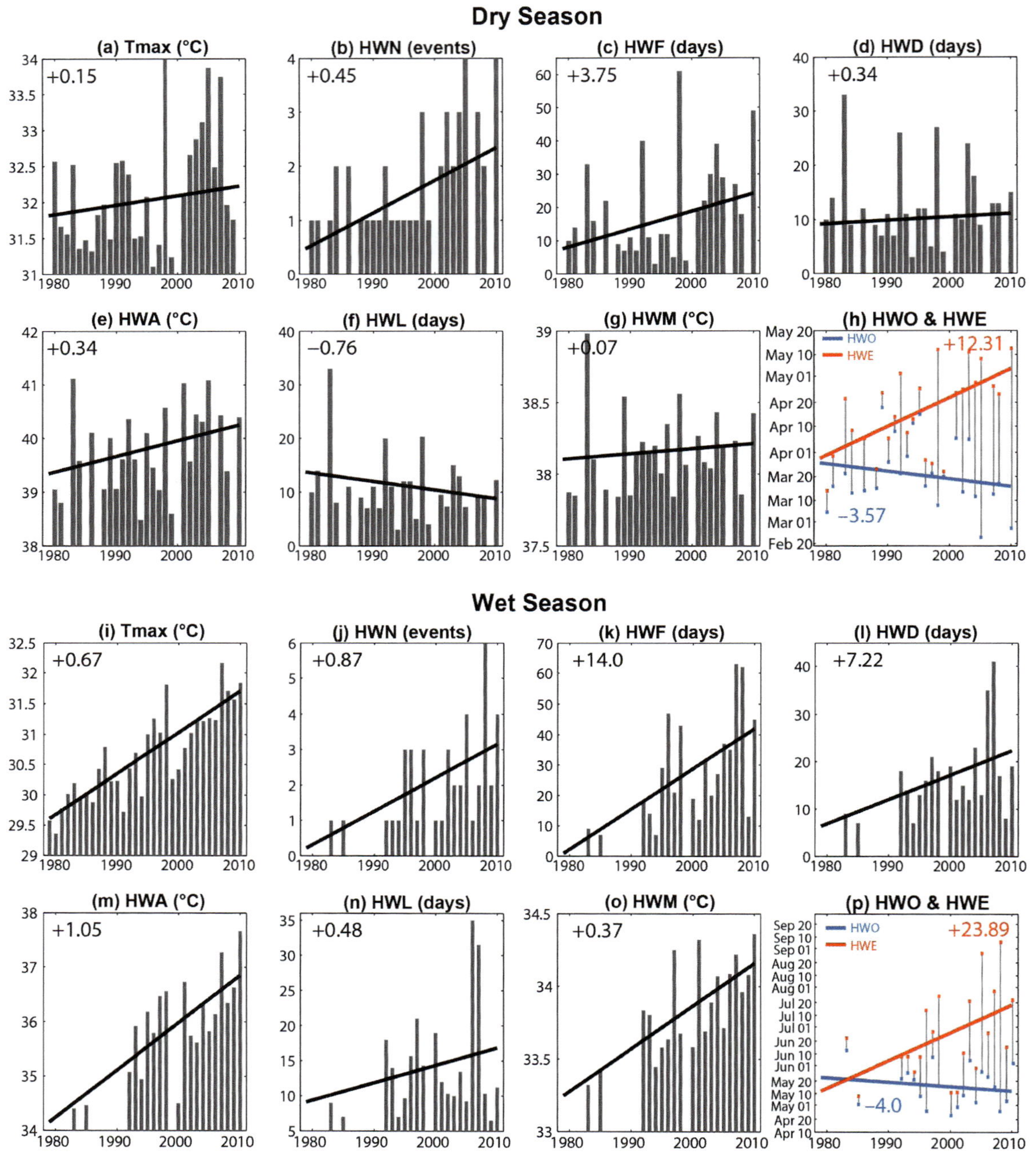

Fig. 8. Same as Fig. 4 but for HWs in the (a)–(h) dry and (i)–(p) wet seasons of Indochina.

dry season exhibit salient upward trends. The results for HWO and HWE (Fig. 8h) indicate that the HWs commence progressively earlier, and terminate later, within the dry season.

The secular trends of various HW indices for the wet season (lower half of Fig. 8) are generally 2–3 times stronger than those for the dry season. The finding based on HWO and HWE (Fig. 8p) signifies that HW events also tended to start earlier and end later in the wet season during more recent years.

4. Summary and Discussion

In this chapter, we diagnose the behavior of HWs over South China and Indochina and examine the possible role of local monsoon flows. All aspects of HWs, in terms of frequency, duration, magnitude, and timing, over both regions exhibit a profound intensifying trend during the past three decades. However, HWs in different areas are accompanied by distinct circulation patterns. The development of HWs over South China is closely associated with the intensification and westward movement of the WNPSH, a dominant climatological feature of the monsoonal circulation in East Asia, whereas HWs in the dry and wet seasons of Indochina are characterized by the weakening in the EAWM and SASM circulations, respectively.

It is noted that, during HWs over South China, positive temperature anomalies extend to many parts of Indochina (Fig. 2a) and anomalous low-level easterlies appear over Indochina, which are similar to the circulation feature associated with the HWs in the dry season of Indochina. These results suggest that HWs over South China may be related to the weakening of the SASM circulation as well. The diagrams in Figs. 1 and 4 display an interdecadal-scale change in HW activity over South China, with a gap between two peak subperiods: 1979–1990 and 1995–2010. Such a change is in accord with the decadal eastward retreat and northward shift of the WNPSH in the early 1990s, possibly caused by the tropical Indian Ocean warming (Lin *et al.* 2017).

Previous studies reported that the circulation associated with both EAWM and SASM weakened after the 1980s (Fan *et al.* 2010; Wang and Chen 2010; Wang *et al.* 2009). These weakening trends in monsoon circulation are in accord with the intensifying trends of HW activities in Indochina during 1979–2010. Considering that the strength of both EAWM and SASM circulations is projected to decrease in the future (Hong *et al.* 2017; IPCC 2013), it is expected that HW activities will become more frequent and stronger in the coming decades.

Our examinations mainly focus on the monsoon flows associated with HWs. It should be noted that other processes, such as the local hydroclimate condition and remote forcing of SST, may also contribute to the occurrence and maintenance of HWs. It is, therefore, of interest and significance to examine the roles of these factors. Moreover, our results are based on reanalysis data; it is also interesting to assess the robustness of these results. These assessments can be conducted by comparing observational/reanalysis data with the output from model simulations.

Acknowledgments

This study is partially supported by the National Natural Science Foundation of China (No. 41871029). The appointment of N.-C. Lau at The Chinese University of Hong Kong is partially supported by the AXA Research Fund.

References

Brunner, L., G. C. Hegerl, and A. K. Steiner, 2017: Connecting atmospheric blocking to European temperature extremes in spring. *J. Clim.*, **30**, 585–594.

Chen, R. and R. Lu, 2015: Comparisons of the circulation anomalies associated with extreme heat in different regions of eastern China. *J. Clim.*, **28**, 5830–5844.

Chen, R., Z. Wen, and R. Lu, 2018: Large-scale circulation anomalies and intraseasonal oscillations associated with long-lived extreme heat events in South China. *J. Clim.*, **31**(1), 213–232.

Chen, X. and T. Zhou, 2017: Relative contributions of external SST forcing and internal atmospheric variability to July–August heat waves over the

Yangtze River valley. *Clim. Dynam.*, **51**, 4403–4419.

Cowan, T., A. Purich, S. Perkins, A. Pezza, G. Boschat, and K. Sadler, 2014: More frequent, longer, and hotter heat waves for Australia in the twenty-first century. *J. Clim.*, **27**, 5851–5871.

Della-Marta, P. M., J. Luterbacher, H. von Weissenfluh, E. Xoplaki, M. Brunet, and H. Wanner, 2007: Summer heat waves over western Europe 1880–2003, their relationship to large-scale forcings and predictability. *Clim. Dynam.*, **29**, 251–275.

Ding, T., W. Qian, and Z. Yan, 2010: Changes in hot days and heat waves in China during 1961–2007. *Int. J. Climatol.*, **30**, 1452–1462.

Ding, Y., G. Ren, G. Shi, P. Gong, X. Zheng, P. Zhai, D. e. Zhang, Z. Zhao, S. Wang, H. Wang, Y. Luo, D. Chen, X. Gao, and X. Dai, 2007: China's national assessment report on climate change (I): Climate change in China and the future trend (in Chinese). *Adv. Clim. Change Res.*, **3**, 1–5.

Easterling, D. R., G. A. Meehl, C. Parmesan, S. A. Changnon, T. R. Karl, and L. O. Mearns, 2000: Climate extremes: observations, modeling, and impacts. *Science*, **289**, 2068–2074.

Fan, F., M. E. Mann, S. Lee, and J. L. Evans, 2010: Observed and modeled changes in the South Asian summer monsoon over the historical period. *J. Clim.*, **23**, 5193–5205.

Forzieri, G., A. Cescatti, F. B. Silva, and L. Feyen, 2017: Increasing risk over time of weather-related hazards to the European population: A data-driven prognostic study. *Lancet Planet. Health*, **1**, e200–e208.

Garcia-Herrera, R., J. Diaz, R. M. Trigo, J. Luterbacher, and E. M. Fischer, 2010: A review of the European summer heat wave of 2003. *Crit. Rev. Environ. Sci. Tech.*, **40**, 267–306.

Grotjahn, R., R. Black, R. Leung, M. F. Wehner, M. Barlow, M. Bosilovich, A. Gershunov, W. J. Gutowski, J. R. Gyakum, R. W. Katz, Y.-Y. Lee, Y.-K. Lim, and Prabhat, 2016: North American extreme temperature events and related large scale meteorological patterns: A review of statistical methods, dynamics, modeling, and trends. *Clim. Dynam.*, **46**, 1151–1184.

Hong, J.-Y., J.-B. Ahn, and J.-G. Jhun, 2017: Winter climate changes over East Asian region under RCP scenarios using East Asian winter monsoon indices. *Clim. Dynam.*, **48**, 577–595.

Hsu, P.-C., J.-Y. Lee, K.-J. Ha, and C.-H. Tsou, 2017: Influences of boreal summer intraseasonal oscillation on heat waves in monsoon Asia. *J. Clim.*, **30**, 7191–7211.

Hu, K., G. Huang, X. Qu, and R. Huang, 2012. The impact of Indian Ocean variability on high temperature extremes across the southern Yangtze River valley in late summer. *Adv. Atmos. Sci.*, **29**, 91–100.

Hu, K., G. Huang, and R. Wu, 2013: A strengthened influence of ENSO on August high temperature extremes over the southern Yangtze River valley since the late 1980s. *J. Clim.*, **26**, 2205–2221.

IPCC, 2013. Climate change 2013: The physical science basis. Contribution of Working Group I to the Fifth Assessment Report of the *Intergovernmental Panel on Climate Change*. Cambridge University Press, Cambridge, United Kingdom and New York, NY, USA pp. 1535.

Kothawale, D., J. Revadekar, and K. R. Kumar, 2010: Recent trends in pre-monsoon daily temperature extremes over India. *J. Earth Syst. Sci.*, **119**, 51–65.

Kovats, R. S. and L. E. Kristie, 2006: Heatwaves and public health in Europe. *Eur. J. Pub. Health*, **16**, 592–599.

Kunkel, K. E., X.-Z. Liang, and J. Zhu, 2010: Regional climate model projections and uncertainties of US summer heat waves. *J. Clim.*, **23**, 4447–4458.

Lau, N.-C. and M. J. Nath, 2012: A model study of heat waves over North America: Meteorological aspects and projections for the twenty-first century. *J. Clim.*, **25**, 4761–4784.

Lin, R., J. Zhu, and F. Zheng, 2017: Decadal shifts of East Asian summer monsoon in a climate model free of explicit GHGs and aerosols. *Sci. Rep.*, **6**, 38546.

Loikith, P. C. and A. J. Broccoli, 2012: Characteristics of observed atmospheric circulation patterns associated with temperature extremes over North America. *J. Clim.*, **25**, 7266–7281.

Lorenz, R., E. B. Jaeger, and S. I. Seneviratne, 2010: Persistence of heat waves and its link to soil moisture memory. *Geophys. Res. Lett.*, **37**, L09703.

Lu, R.-Y. and R.-D. Chen, 2016: A review of recent studies on extreme heat in China. *Atmos. Ocean. Sci. Lett.*, **9**, 114–121.

Luo, M. and N.-C. Lau, 2017: Heat waves in southern China: Synoptic behavior, long-term change, and urbanization effects. *J. Clim.*, **30**, 703–720.

———, 2018a: Amplifying effect of ENSO on heat waves in China. *Clim. Dynam.*, **54**, 631–648.

———, 2018b: Synoptic characteristics, atmospheric controls, and long-term changes of heat waves over the Indochina Peninsula. *Clim. Dynam.*, **51**, 2707–2723.

Matsueda, M., 2011: Predictability of Euro-Russian blocking in summer of 2010. *Geophys. Res. Lett.*, **38**, L06801.

McMichael, A. J., R. E. Woodruff, and S. Hales, 2006: Climate change and human health: present and future risks. *Lancet*, **367**, 859–869.

Meehl, G. A. and C. Tebaldi, 2004: More intense, more frequent, and longer lasting heat waves in the 21st century. *Science*, **305**, 994–997.

Miralles, D. G., A. J. Teuling, C. C. Van Heerwaarden, and J.V.-G. de Arellano, 2014: Mega-heatwave temperatures due to combined soil desiccation and atmospheric heat accumulation. *Nat. Geosci.*, **7**, 345–349.

Pfahl, S., C. Schwierz, M. Croci-Maspoli, C. M. Grams, and H. Wernli, 2015: Importance of latent heat release in ascending air streams for atmospheric blocking. *Nat. Geosci.*, **8**, 610–614.

Pfahl, S. and H. Wernli, 2012: Quantifying the relevance of atmospheric blocking for co-located temperature extremes in the Northern Hemisphere on (sub-) daily time scales. *Geophys. Res. Lett.*, **39**, L12807.

Qi, L. and Y. Wang, 2012: Changes in the observed trends in extreme temperatures over China around 1990. *J. Clim.*, **25**, 5208–5222.

Revadekar, J., S. Hameed, D. Collins, M. Manton, M. Sheikh, H. Borgaonkar, D. Kothawale, M. Adnan, A. Ahmed, and J. Ashraf, 2013: Impact of altitude and latitude on changes in temperature extremes over South Asia during 1971–2000. *Int. J. Climatol.*, **33**, 199–209.

Sillmann, J. and M. Croci-Maspoli, 2009: Present and future atmospheric blocking and its impact on European mean and extreme climate. *Geophys. Res. Lett.*, **36**, L10702.

Sun, J.-Q., 2012: Possible impact of the summer North Atlantic oscillation on extreme hot events in China. *Atmos. Ocean. Sci. Lett.*, **5**, 231–234.

Wang, L. and W. Chen, 2010: How well do existing indices measure the strength of the East Asian winter monsoon? *Adv. Atmos. Sci.*, **27**, 855–870.

Wang, L., R. Huang, L. Gu, W. Chen, and L. Kang, 2009: Interdecadal variations of the East Asian winter monsoon and their association with quasi-stationary planetary wave activity. *J. Clim.*, **22**, 4860–4872.

Wang, W., W. Zhou, and D. Chen, 2014: Summer high temperature extremes in southeast China: Bonding with the El Niño–Southern Oscillation and East Asian summer monsoon coupled system. *J. Clim.*, **27**, 4122–4138.

Wang, W., W. Zhou, X. Wang, S. K. Fong, and K. C. Leong, 2013: Summer high temperature extremes in southeast China associated with the East Asian jet stream and circumglobal teleconnection. *J. Geophys. Res. Amos.*, **118**, 8306–8319.

Ward, K., S. Lauf, B. Kleinschmit, and W. Endlicher, 2016: Heat waves and urban heat islands in Europe: A review of relevant drivers. *Sci. Total Environ.*, **569–570**, 527–539.

Wei, K. and W. Chen, 2011: An abrupt increase in the summer high temperature extreme days across China in the mid-1990s. *Adv. Atmos. Sci.*, **28**, 1023–1029.

World Meteorological Organization (WMO), 2013: The Global Climate 2001–2010: A decade of climate extremes summary report. *WMO-No. 1119*, World Meteorological Organization, Geneva, Switzerland, pp. 119. [Available online at https://library.wmo.int/pmb_ged/wmo_1119_en.pdf].

Zhai, P. and X. Pan, 2003: Trends in temperature extremes during 1951–1999 in China. *Geophys. Res. Lett.*, **30**, 1913.

Zhou, Y. and G. Ren, 2011: Change in extreme temperature event frequency over mainland China, 1961–2008. *Clim. Res.*, **50**, 125–139.

Chapter 15

Diurnally Forced Convection in the Australian Monsoon

Todd P. Lane*,§,¶, Claire L. Vincent*,§, Martin Jucker†,§, Martin Bergemann*,§ and Sugata Narsey‡,§

*School of Earth Sciences, The University of Melbourne, Australia
†Climate Change Research Centre, University of New South Wales, Australia
‡School of Earth, Atmosphere and Environment, Monash University, Australia
§ARC Centre of Excellence for Climate Extremes, Australia
¶tplane@unimelb.edu.au

The Australian monsoon features diverse multiscalar interactions that manifest as a variety of convective cloud features and modes of organization. Darwin is arguably the best studied and understood region affected by the Australian monsoon, being the focus of numerous international field programs and observational deployments. The monsoon near Darwin is affected considerably by intraseasonal variability, with five distinct regimes that have recently been identified through objective clustering methods. These regimes differ in the direction of the low-level wind, total column moisture, the latitudinal extent of the monsoon trough, the cloud cover, and the diurnal cycle of wind and precipitation. Two of these regimes are closely linked to the traditional categories of "active monsoon" and "break" periods. The break periods near Darwin feature intense diurnally forced organized convective systems that are localized and often initiated or organized by coastal processes such as the sea/land breeze system. In contrast, the active periods have more widespread cloud cover and a weaker diurnal cycle of wind and precipitation.

In this chapter, we present a summary of recent work that has examined diurnally forced convection near Darwin and its variation with the monsoon regime. We present examples using radar observations, satellite precipitation estimates, and high-resolution modeling. We also summarize some of the challenges in realistically simulating convection in the region using high-resolution models.

1. The Australian Monsoon and Its Regimes

The tropical northern regions of Australia can be categorized into two distinct periods: the wet season (~October to ~April) and the dry season. More than 90% of the rainfall occurs during the wet season, leading to marked differences between the two periods of the year and also having meteorological impact on the community, agriculture, and tourism. The climate of Darwin is broadly characterized by these two distinct periods and the behavior of convection in the Darwin wet season has been the focus of many studies and field experiments. Examples of the studies include the Australian Monsoon Experiment (AMEX; Holland *et al.* 1986), the Maritime Continent Thunderstorm Experiment (MCTEX; Keenan *et al.* 2000), and the Tropical Warm Pool — International Cloud Experiment (TWP-ICE; May *et al.* 2008). Although many instruments have been deployed at Darwin (both temporary and long-term installations), a key platform is the Gunn Point C-band polarimetric radar (C-POL; Keenan *et al.* 1998), which has been used for many studies of convection and rainfall processes in the region (see Fig. 1 for its location). Darwin's coastal

The Multiscale Global Monsoon System
Edited by C. P. Chang et al.

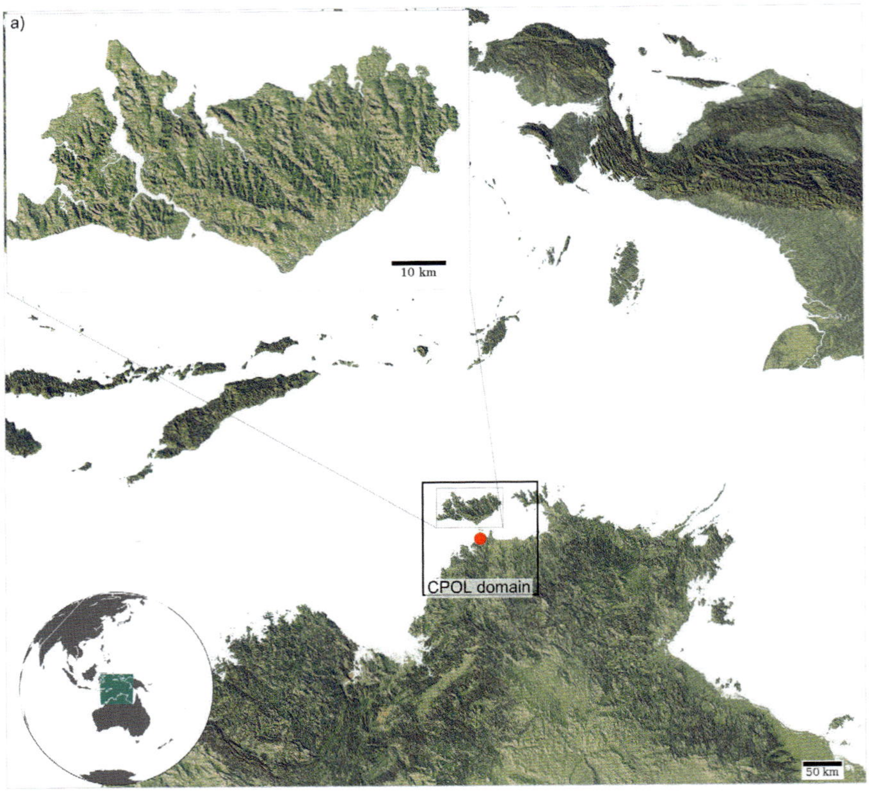

Fig. 1. Region of interest. The maps show Northern Australia and the surrounding islands, with the red dot indicating the location of the CPOL Darwin radar and its observational domain denoted by the black square. The inset shows the Tiwi Islands (Bathurst Island to the west and Melville Island to the east).

location, complex coastline, and the nearby Tiwi Islands make it an ideal location to study monsoonal convection and its multiscale variability.

The Darwin wet season experiences significant intraseasonal variability that manifests as a range of distinct regimes. Much of this variability is thought to be the result of midlatitude influences (e.g., Hung and Yanai 2004, Davidson *et al.* 2007, Berry and Reeder 2016, and Narsey *et al.* 2017). Figure 2 presents an analysis of the mean wet season, and an example season (the 2005/2006 season) to illustrate the rainfall variability, which is closely linked to the monsoonal regimes. These regimes include the traditional descriptions of the *monsoon build-up* period, and following *monsoon onset* there are *active monsoon* and *break* periods. Each of these terms has been defined in a range of ways, but a commonly used method to distinguish active and break periods based on the zonal winds at Darwin is found in Drosdowsky (1996). This method defines the active monsoon (sometimes referred to as a monsoon "burst", e.g., Troup 1961) as when the mean zonal wind from the surface to 500 hPa is westerly, and the mean zonal wind between 300 hPa and 100 hPa is easterly. The monsoon period is defined by the onset (the first burst) and the retreat (the last burst), with intermediate periods (of at least three days) that do not satisfy the active monsoon criteria defined as break periods.

More recently, Pope *et al.* (2009) performed a cluster analysis on wind and thermodynamic radiosonde profiles from Darwin to create objective definitions of the intraseasonal regimes of the Darwin wet season. The study identified

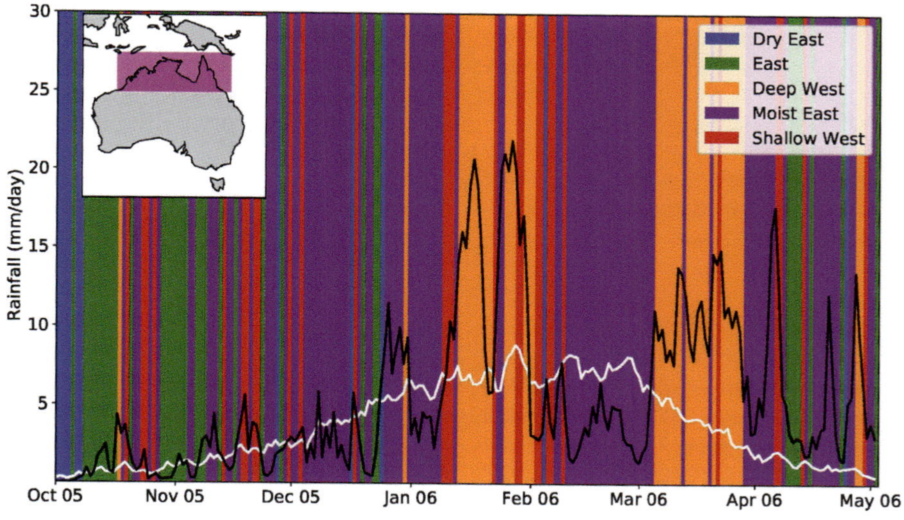

Fig. 2. Rainfall during the Australian monsoon, averaged over the region shown in the inset map. Solid white line is the average rainfall from 1979 to 2010. The solid black line is the rainfall for the 2005/2006 season. Colored shading represents the Pope regime categories (discussed in the text) for the 2005/2006 season. Rainfall is from the Australian Water Availability Project (AWAP, Jones *et al.* 2009) dataset.

five regimes, now often referred to as the "Pope regimes", with distinct wind and moisture profiles: Dry Easterly (DE), Deep Westerly (DW), Easterly (E), Shallow Westerly (SW), and Moist Easterly (ME). Of these regimes there is close correspondence between the DW regime and wind-based definitions of the active monsoon, and also close correspondence between the ME regime and wind-based definitions of the break. The SW regime has features similar to the DW regime, though with less moisture. The SW regime is consistent with previous discussions of the "suppressed monsoon" (e.g., May *et al.* 2008), which can feature a more active monsoon trough over the eastern part of Australia. The DE regime is a dry trade wind dominated flow, occurring mostly before monsoon onset and after monsoon retreat. The E regime is more moist, corresponding to the "build-up". Of all five regimes, the ME and DW regimes are the most moist, as is demonstrated by Fig. 2c of Kumar *et al.* (2013).

Examples of the qualitative differences between the Darwin rainfall patterns during active (DW) and break (ME) periods of the Australian monsoon are shown in Fig. 3, alongside a 2-month average of the daily rainfall. The differences in the meteorological conditions can be summarized as follows: active monsoon periods feature a monsoon trough encroaching to the south of Darwin, low-level westerly winds, large-scale forcing that leads to widespread clouds and convection, and the greatest area-averaged rainfall of any of the other wet season regimes. In contrast, the break periods feature low-level easterly winds and suppressed large-scale forcing that leads to more localized and intense convection, especially over land. Break periods possess some similarities to monsoon build-up regimes, which also contain intense locally forced convection over land, albeit in a drier environment. Of interest, the strong "Hector" thunderstorms that form over the Tiwi Islands (e.g., Keenan *et al.* 1990) are prevalent in the build-up (E regime) and break (ME) regimes of the wet season.

Drosdowsky's (1996) analysis identified the mean length of active monsoon periods to be about two weeks, with a mean return period of about a month. In both estimates the standard

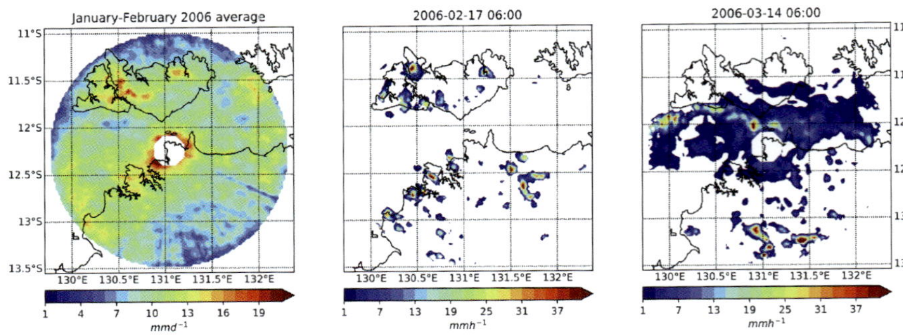

Fig. 3. Radar-derived rainfall from the Gunn Point polarimetric radar (CPOL). The left image is the average daily rainfall from January to February 2006. The center and right images are examples of instantaneous rain rate from the break (ME regime) at 0600 UTC 17 February 2006 (center) and the active monsoon (DW regime) at 0600 UTC 14 March 2006 (right). The coast is shown as a solid line and the nominal 150-km range of the radar is evident in the left image. Note the different color scales between the average rainfall image and the instantaneous images.

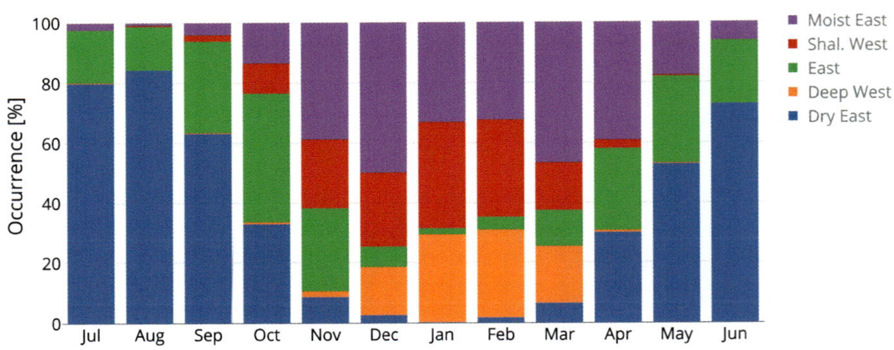

Fig. 4. Frequency of occurrence of the Pope regime categories by month for 1957–2017.

deviation was approximately two weeks, suggesting significant variability in length and recurrence interval. Some of this variability, and monsoon onset, can be linked to the Madden Julian oscillation (MJO; see Wheeler and McBride 2011 for a detailed discussion). For example, Pope *et al.* (2009) showed that the ME (break) regime showed a slight increase in its occurrence during phases 3–4 of the MJO, as measured by the Wheeler and Hendon (2004) index.

Pope *et al.* (2009) showed that the ME regime (break) actually has the highest probability of occurrence (40–50%) of all the regimes between December and March, followed by the SW regime and then the DW regime. However, if the two westerly regimes (DW and SW) were combined together, they would form the most frequent regime. This analysis is updated for the period 1957–2017 in Fig. 4.

Kajikawa *et al.* (2010) created an Australian monsoon index, defined as the spatial average (over the area 5°–15°S, 110–130°E) of the 850-hPa zonal wind. This index allows assessment of the monsoon onset, and intraseasonal, interannual, and interdecadal variability of the Australian monsoon, over a broader region than just Darwin. Using this index, Kajikawa *et al.* (2010) found that the monsoon onset had noteworthy interannual variability, but the monsoon retreat did not as it was more closely linked to the annual cycle. Moreover, they showed that early monsoon onset was related to a strong monsoon season, and vice versa.

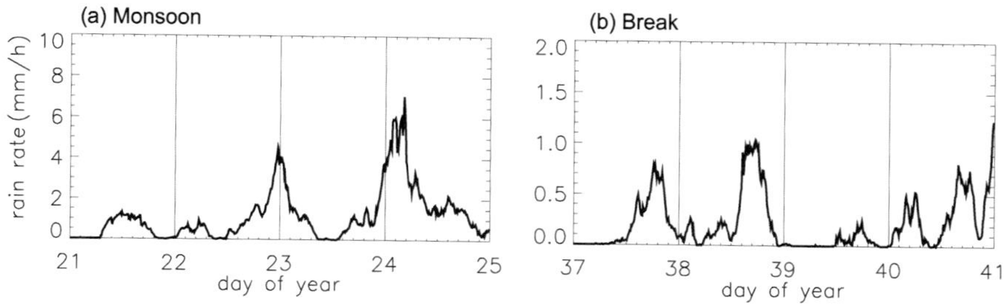

Fig. 5. Rain rate derived from the Gunn Point polarimetric radar, averaged over the range of the radar identified in Fig. 3. Each panel shows four days representative of (a) the active monsoon (DW) and (b) the break (ME). The abscissa is in local time starting 1 January 2006, and (a) and (b) have different scales on the ordinate (adapted from Wapler *et al.* 2010).

2. Diurnally Forced Convection

The regimes of the monsoon are particularly relevant when considering the occurrence of localized storms, which often occur as part of the diurnally forced convective initiation that is dominant in the break. Examples of Darwin rainfall for an active and break period are shown in Fig. 5. Figure 5b illustrates a well-defined diurnal cycle with peak rainfall activity in the late afternoon/early evening on most days, as well as a secondary peak in the early morning on some days. In contrast, even though there is a diurnal signal in rainfall in the active monsoon (Fig. 5a), with afternoon rainfall occurring regularly, the timing of the daily peaks in rainfall is variable and more closely related to large-scale forcing than pure diurnal forcing. Moreover, although the area-averaged rain rate is larger in the active monsoon, the localized and intense nature of the break convection can lead to more intense local rain rates, which is consistent with Fig. 3, and the localized break storms often deepen to penetrate the tropical tropopause layer and lower stratosphere (e.g., Hassim *et al.* 2014).

In terms of the intensity of extreme rainfall, Pope *et al.* (2009, their Fig. 5) demonstrated that the largest 24-h rainfall accumulations at the Darwin Airport gauge occur in the ME regime, followed by the SW and then the DW regimes. However, for 24-h accumulations less than 200 mm, the DW regime has by far the largest frequency of occurrence. Of these three regimes, Kumar *et al.* (2013, their Fig. 4) identified that the SW regime has the deepest convective cells, followed by the ME and then DW regimes. Both the SW and ME regimes have notable lightning activity.

To examine the variation of the diurnal cycle with the monsoon regime, composite diurnal cycles for the five Pope regimes are shown in Fig. 6 for the 2014/2015 wet season for two satellite-derived rainfall products: CMORPH (Climate Prediction Center, 2011) and TRMM 3B42 (Huffman *et al.*, 2007; Goddard Space Flight Center 1998). These cycles are separated into rainfall over land and sea, which captures the mean diurnal cycles over the land/sea areas but loses information about the intensity of localized rainfall events. As can be seen in Fig. 6, all regimes (except the DE regime) have a notable diurnal cycle over land. In the SW and DW regimes this cycle is superimposed on substantial daily background rainfall (consistent with the variable timing in daily maxima seen in Fig. 5a); in these regimes the sea points have a similar background rainfall with only a small diurnal cycle. For the E and ME regimes over land the diurnal peaks occur after a near-zero rainfall over land in the early morning; similarly, over the sea there is minimal rainfall with the exception of the early morning peak. The DE regime has little rainfall.

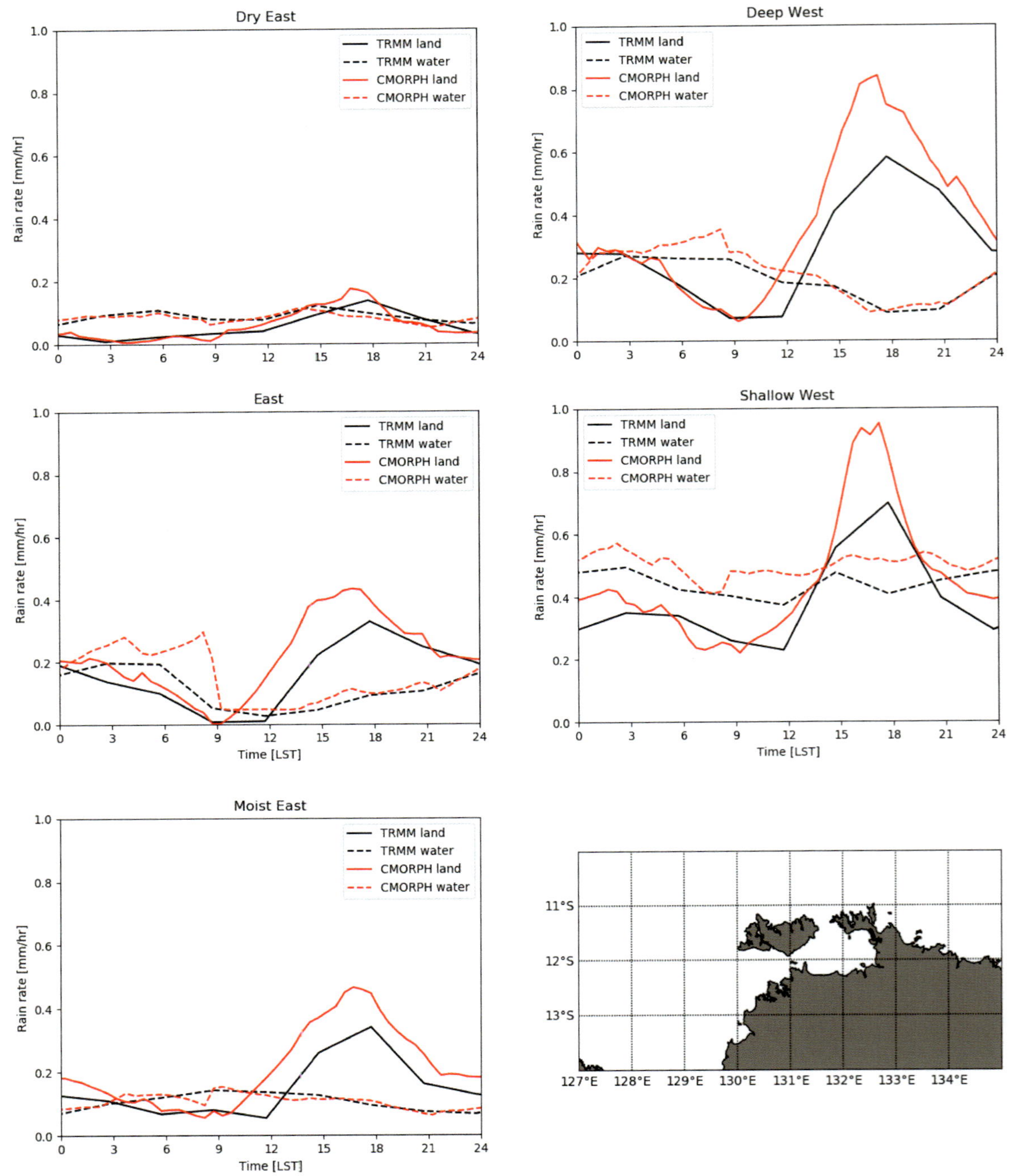

Fig. 6. Composite diurnal cycles of hourly rainfall (mm/hr) separated into the five monsoon regimes. Rain is calculated over land points (solid) and sea points (dashed) for the area shown in the bottom right panel. Lines correspond to rainfall derived from WRF simulations (at 4-km grid spacing), TRMM, and CMORPH for November 2014 to April 2015.

It is worth noting that Fig. 6 also highlights the differences in rainfall derived from the two satellite precipitation products. Not only are there differences in rainfall totals but also differences in the timing of the diurnal cycle. Some of these uncertainties are alleviated by using radar and gauge data. However, the rainfall uncertainties over Darwin (and elsewhere in the Maritime Continent) pose significant difficulties for model evaluation in the region.

During the break (ME) and build-up periods (E) the dominant forcing of convection is associated with topographic effects and coastal processes. The afternoon peak of convective activity occurs predominantly over land (e.g., Fig. 3, center panel) and can be linked to ascent from surface heating, sea breeze initiation, and convergence of sea breezes over the narrow peninsulas and the Tiwi Islands. For example, the regularly occurring Hector storms are formed by converging sea breezes (e.g., Carbone *et al.* 2000). The early morning peak in rainfall (seen in Fig. 6 and on a few days in Fig. 5b) normally occurs offshore and can be related to land breeze circulations and offshore convergence associated with complex coastlines (e.g., Wapler and Lane 2012). These processes governing offshore rainfall near the coast work in concert with radiative effects that enhance nocturnal rainfall over the oceans (e.g., Gray and Jacobson 1973, Randall *et al.* 1991). However, consistent with Vincent and Lane (2016), density currents and gravity waves are likely the dominant mechanisms near coasts.

During the westerly monsoon regimes (DW and SW) the background forcing associated with the monsoon trough leads to substantial daily mean rainfall, albeit with a strong diurnal cycle superimposed upon it by land-based convection. The amplitude of the diurnal cycle during these regimes does not appear suppressed, as might be expected from the mean cloudiness during those periods.

It is clear that many of the processes that determine the diurnal cycle of intense convection near Darwin, especially during the break periods, are similar to those that occur elsewhere in the Maritime Continent (MC), especially during suppressed periods of the MJO. For example, Vincent and Lane (2017) examined aspects of the diurnal cycle of rainfall and convection over the entire MC and documented similar behavior to Darwin, with afternoon peaks in convective activity over land, which then transition to offshore maxima in the early morning. However, the offshore signal is stronger in other parts of the Maritime Continent than Darwin. Bergemann *et al.* (2015) analyzed satellite rainfall observations and identified that a significant proportion of coastal rainfall in the MC is aligned with coastlines, further highlighting the importance of coastal processes. In agreement with the comments above, although the proportion of coastally aligned rainfall over land during the day near Darwin is similar to the other islands in the MC, the offshore coastally aligned rainfall is weaker than the large MC islands. Presumably these geographic differences are related to the height of topography over the large MC islands compared to the relatively flat terrain near Darwin, with both katabatic flows and the sea/land breeze circulation contributing to the coupling between land and oceanic convection.

Recent work by Brown *et al.* (2017) has examined the sea/land breeze system near Darwin using scatterometer wind measurements. This work demonstrated that the Darwin region does have very well-defined sea and land breezes, and the scatterometer measurements documented that the sea/land breeze extends a few hundred kilometers offshore. This offshore extent and the intensity of the sea/land breeze wind perturbations are modulated considerably by the phase of the monsoon. Stronger wind perturbations extend farther offshore during the break compared to the active periods. These differences are related to the reduced cloud cover during break periods, leading to stronger coastal circulations that ultimately initiate intense and localized convection.

3. Convection-permitting Modeling

A number of studies have examined convection processes over Darwin using cloud-resolving and/or convection-permitting models. Wapler *et al.* (2010) evaluated the performance of convection-permitting models during parts of TWP-ICE. They found that the models compared better to observations during break periods when localized forcing is dominant. During active regimes of the monsoon the regional convection-permitting models are reliant on the accuracy of large-scale forcing. Consequently, during active periods the models had more difficulty reproducing the spatial and temporal characteristics of the convection. Caine *et al.* (2013) studied the simulated convective cell characteristics (in a 1.25-km resolution model) during a break period and compared those characteristics to objects identified from radar data. Their study identified a number of significant biases in convective cell depth and size. The positive bias in cell depth is likely a consequence of simulated convective updrafts that are too intense, owing to an ineffective treatment of entrainment at convection-permitting resolutions. Some aspects of these biases were alleviated with higher resolution (420-m horizontal grid spacing).

Other convection-permitting model errors, identified by simulations over Darwin and elsewhere in the MC (e.g., Vincent and Lane 2017), are that the rainfall over land is too intense and occurs a few hours too early. These biases may be linked to the inability of convection-permitting models to properly resolve shallow convection, which also limits their ability to realistically represent the transition from shallow to deep convection. Inadequacies in cloud microphysics are also important sources of systematic errors. Of interest, comparisons between models and satellite observations made by Brown *et al.* (2017) suggest that the sea/land breeze can be well represented by convection-permitting models, adding further confidence in the explanations above.

As an example of the variations in model performance with regimes of the Australian monsoon, Fig. 6 also shows composite simulated rainfall diurnal cycles. These data are from a model simulation of the 2014/2015 wet season using the Weather Research and Forecasting model (WRF, Skamarock *et al.* 2008) with 4-km horizontal grid spacing — the full details of the simulation are described in Brown *et al.* (2017). Consistent with the discussion above, in the E, ME, and DW regimes the convection-permitting WRF simulation produces a rainfall peak that is too large. As an example of this, in the SW regime the model produces a rainfall peak that has some agreement with the CMORPH estimates. The simulations for the DW, SW, and ME regimes are all a few hours too early in the timing of their convective development over land (with agreement for the E regime). These results are consistent with the inability of convection-permitting models to properly simulate the transition from shallow to deep convection in a realistic manner, which can lead to deep convection occurring too early.

4. Summary

This chapter has summarized some recent work that has characterized the behavior of the Australian monsoon, its regimes, and the occurrence of diurnally forced convection in those regimes. The traditional categories of "active monsoon" and "break" have been expanded upon with the creation of the "Pope regimes" that, among other things, introduce the distinct shallow westerly regime, which has been also referred to as a "suppressed monsoon" in the past.

Of interest, with the exception of the Dry East (pre- and post-monsoon trade wind) regime, all of the regimes feature a distinct diurnal cycle near Darwin, especially over the land. However, the most intense rainfall and deepest convection occur during the Moist East (ME) and Shallow West (SW) regimes. The ME (break) regime has received considerable attention in the past, but the SW regime has received far less attention.

Convection-permitting models are now commonly used for prediction and process studies over the region of the Australian monsoon and broader Maritime Continent. These models have notable systematic biases in the timing and intensity of convection and further efforts to improve them, through both improved physical processes and model resolution, are required.

Acknowledgments

This work was supported by the Australian Research Council's Centres of Excellence Program (CE170100023). Simulations were conducted using resources from the National Computational Infrastructure (NCI) facility in Canberra. The authors thank Valentin Louf (Bureau of Meteorology) for supplying the radar data and Alain Protat (Bureau of Meteorology) for supplying the Pope regimes data.

References

Bergemann, M., C. Jakob, and T. P. Lane, 2015: Global detection and analysis of coastline associated rainfall using objective pattern recognition techniques. *J. Clim.*, **28**, 7225–7236.

Berry, G. J. and M. J. Reeder, 2016: The dynamics of Australian monsoon bursts. *J. Atmos. Sci.*, **73**, 55–59.

Brown, A. L., C. L. Vincent *et al.*, 2017: Scatterometer estimates of the tropical sea breeze circulation near Darwin, with comparison to regional models. *Quart. J. Roy. Meteor. Soc.*, in press.

Caine, S., T. P. Lane *et al.*, 2013: Statistical assessment of tropical convection-permitting model simulations using a cell-tracking algorithm. *Mon. Wea. Rev.*, **141**, 557–581.

Carbone, R. E., J. W. Wilson *et al.*, 2000: Tropical island convection in the absence of significant topography. Part I: Life cycle of diurnally forced convection. *Mon. Wea. Rev.*, **128**, 3459–3480.

Climate Prediction Center, 2011: NOAA CPC morphing technique (CMORPH) global precipitation analyses 30 minute estimates at 8 km resolution. [Available online at http://ftp.cpc.ncep.noaa.gov/precip/CMORPH_V1.0/RAW/8km-30min/.]

Davidson, N. E., K. J. Tory, M. J. Reeder, and W. L. Drosdowsky, 2007: Extratropical-tropical interaction during onset of the Australian monsoon: Reanalysis diagnostics and idealized dry simulations. *J. Atmos. Sci.*, **64**, 3475–3498.

Drosdowsky, W., 1996: Variability of the Australian Summer Monsoon at Darwin: 1957–1992. *J. Clim.*, **9**, 85–96.

Goddard Space Flight Center, 1998: TRMM (Tropical Rainfall Measurement Mission) project merged HQ-infrared precipitation estimate (3B42), version 7. [Available online at ftp://trmmopen.gsfc.nasa.gov/pub/merged/3B42RT/]

Gray, W. M. and R. W. Jacobson, 1977: Diurnal variation of deep cumulus convection. *Mon. Wea. Rev.*, **105**, 1171–1188.

Hassim, M. E. E., T. P. Lane, and P. T. May, 2014: Ground-based observations of overshooting convection during the tropical warm pool-international cloud experiment. *J. Geophys. Res. Atmos.*, **119**, 880–905.

Holland, G. J., J. L. McBride *et al.*, 1986: The BMRC Australian Monsoon Experiment: AMEX. *Bull. Amer. Meteor. Soc.*, **67**, 1466–1472.

Huffman, G. J., D. T. Bolvin, E. J. Nelkin, D. B. Wolff, R. F. Adler, G. Gu, E. J. Nelkin, K. P. Bowman, Y. Hong, E. F. Stocker, and D. B. Wolff, 2007: The TRMM Multisatellite Precipitation Analysis (TMPA): Quasi-global, multiyear, combined-sensor precipitation estimates at fine scales. *J. Hydrometeor.*, **8**, 38–55.

Hung, C.-W. and M. Yanai, 2004: Factors contributing to the onset of the Australian summer monsoon. *Quart. J. Roy. Meteor. Soc.*, **130**, 739–758.

Jones, D. A., W. Wang, and R. Fawcett, 2009: High-quality spatial climate datasets for Australia. *Aust. Meteor. Oceanogr. J.*, **58**, 233–248.

Kajikawa, Y., B. Wang, and J. Yang, 2010: A multi-time scale Australian monsoon index. *Int. J. Climatology*, **30**, 1114–1120.

Keenan, T. D., B. R. Morton, X. S. Zhang, and K. Nyguen, 1990: Some characteristics of thunderstorms over Bathurst and Melville Islands near Darwin. *Australia. Quart. J. Roy. Meteor. Soc.*, **116**, 1153–1172.

Keenan, T., K. Glasson *et al.*, 1998: The BMRC/NCAR C-band polarimetric (C-POL) radar system. *J. Atmos. Oceanic Technol.*, **15**, 871–886.

Keenan, T., P. May *et al.*, 2000: The Maritime Continent thunderstorm experiment (MCTEX): Overview and some results. *Bull. Amer. Meteor. Soc.*, **81**, 2433–2455.

Kumar, V. V., A. Protat, C. Jakob, G. Penide, S. Kumar, and L. Davies, 2013: On the effects of large-scale environment and surface types on convective cloud characteristics over Darwin, Australia. *Mon. Wea. Rev.*, **141**, 1358–1374.

May, P. T., J. H. Mather *et al.*, 2008: The tropical warm pool international cloud experiment. *Bull. Amer. Meteor. Soc.*, **89**, 629–645.

Narsey, S., M. J. Reeder, D. Ackerley, and C. Jakob, 2017: A midlatitude influence on Australian monsoon bursts. *J. Clim.*, **30**, 5377–5393.

Pope, M., C. Jakob, and M. J. Reeder, 2009: Regimes of the North Australian wet season. *J. Clim.*, **22**, 6699–6715.

Randall, D. A., Harshvardhan, and D. A. Dazlich, 1991: Diurnal variability of the hydrologic cycle in a general circulation model. *J. Atmos. Sci.*, **48**, 40–62.

Skamarock, W. C. and Coauthors, 2008: A description of the advanced research WRF version 3. NCAR Tech. Note NCAR/TN-475+STR, 113 pp.

Troup, A. J., 1961: Variations in upper tropospheric flow associated with the onset of the Australian summer monsoon. Indian *J. Meteor. Geophys.*, **12**, 217–230.

Vincent, C. L. and T. P. Lane, 2016: Evolution of the diurnal precipitation cycle with the passage of a Madden Julian Oscillation event through the Maritime Continent. *Mon. Wea. Rev.*, **144**, 1983–2005.

Vincent, C. L. and T. P. Lane, 2017: A 10-year Austral summer climatology of observed and modeled intraseasonal, mesoscale and diurnal variations over the Maritime Continent. *J. Clim.*, **30**, 3807–3828.

Wapler, K., T. P. Lane *et al.*, 2010: Cloud-system-resolving model simulations of tropical cloud systems observed during the tropical warm pool-international cloud experiment. *Mon. Wea. Rev.*, **138**, 55–73.

Wapler, K. and T. P. Lane, 2012: A case of offshore convective initiation by interacting land breezes near Darwin, Australia. *Meteor. Atmos. Phys.*, **115**, 123–137.

Wheeler, M. C. and H. H. Hendon, 2004: An All-season real-time multivariate MJO index: Development of an index for monitoring and prediction. *Mon. Wea. Rev.*, **132**, 1917–1932.

Wheeler, M. C. and J. L. McBride, 2011: Australasian monsoon. In Intraseasonal Variability in the Atmosphere–Ocean Climate System Eds. Lau, W. K. M and D. E. Waliser, Chapter 5, 2nd edition Springer-Verlag.

Chapter 16

Warm Season Convective Variability in Snow Mountain Range and Heavy Rains in Taipei

Ben Jong-Dao Jou*,§, Hiroshi Uyeda†, Ultimate Chi-June Jung*, Radian Rong-Kuan Hsiu* and Kevin Shou-Liang Yu*,‡

*Department of Atmospheric Science, National Taiwan University, Taipei, Taiwan
†Nagoya University, Nagoya, Japan
‡Civil Aeronautical Administration, Taipei, Taiwan
§jouben43@gmail.com

Convective activity around metropolitan Taipei City is greatly affected by the Snow Mountain Range (SMR), especially in the warm season with weak synoptic conditions. Pronounced convective activity is frequently observed around the mountain range and causes severe flash floods in the basin of Taipei. The formation mechanisms of a daytime and a nighttime extremely heavy rain-flash flood event are investigated in this study. The results show the daytime heavy rain event was associated with severe air-mass afternoon thunderstorms with a complex structure. Storm initiation was due to insolation-produced sea breeze and upslope flows. The short-duration torrential heavy rain occurred during the merger of convective cells. Radar analysis identifies that cell merging was associated with enhanced low-level convergence induced by the sea breeze and prior storm-produced outflows. On the contrary, the heavy rain at night highlighted the importance of locally enhanced baroclinicity due to intrusion of cold air. Elevated convection was triggered by the lifting of the SW warmer air above the shallow cold surface to form stagnant convection in the basin. Convective cells that formed over the coastal region along the baroclinic zone propagated eastward and merged with the convective cells in the basin. The intensified convection produced heavy rain. Signatures of polarimetric radar parameters are also described to examine the possible differences in the characteristics of bulk cloud microphysics between the daytime and the nighttime heavy rain cases.

1. Introduction

Convective activity over metropolitan Taipei City is greatly affected by the Snow Mountain Range (SMR), especially in the warm season with weak synoptic conditions (Jou 1994; Chen *et al.* 2007; Lin *et al.* 2011). SMR is a mountain range located in northern Taiwan, east of Taipei City, with a width of ~25 km, a length of ~100 km, and an orientation of northeast–southwest (Fig. 1a). The SMR extends in elevation from near sea level to ~3 km with an average height of 500–1000 m. During the summer monsoon season, the prevailing southwesterly flow around the island interacting with local circulations produces local convergence zones in northern Taiwan. Pronounced convective activity is frequently observed along the convergence zones and causes severe flash floods in Taipei.

Occurrence frequency of strong radar reflectivity at low levels (greater than 40 dBZ indicating strong convection) in northern Taiwan from May and June 2008 is given in Fig. 1b (Jou *et al.* 2011). In Fig. 1b, the W–E (S–N) Hovmoller diagrams show the occurrence frequency along the S–N (W–E) direction within the box indicated in Fig. 1a. Pronounced diurnal variability of strong convection in the Taipei basin (west of SMR) is indicated. A single peak of strong convection occurrence in the afternoon is shown. This feature is quite different from what was

The Multiscale Global Monsoon System
Edited by C. P. Chang et al.

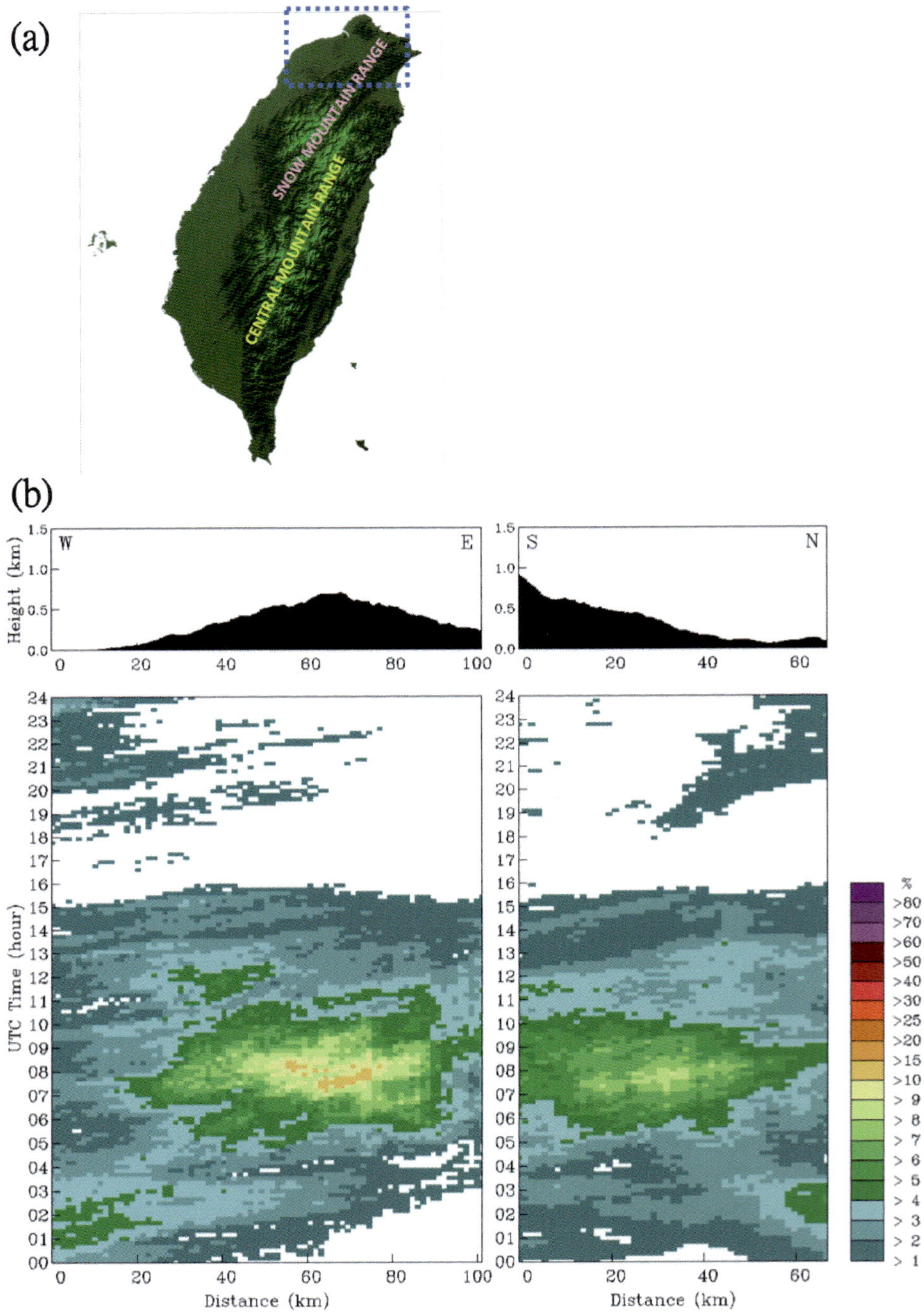

Fig. 1. (a) Major terrain features in Taiwan; the bright green indicates that terrain heights are above 3000 m. The Central Mountain Range and the Snow Mountain Range are highlighted. The rectangle marked by the dash blue line is the area for radar reflectivity study. (b) Diurnal variation (in UTC, +8 the LST) of occurrence frequency in percentage (%) of large reflectivity (>= 40 dBZ) in the Taipei area for May and June 2008. The averaged terrain heights for west–east and south–north of the rectangle are shown in the top figures. It is noted that storms initiated around 1200–1300 LST and reached the maximum extent at around 1600 LST and then decayed rapidly. The south portion of the area can last till 1800 LST. The signal of convection initiated at the sloping area with terrain height around 500 m, from south to north, from west to east. However, starting at 1500 LST, there is a signal that propagates from east to west with a much slower speed. This signal indicates there were storms propagated into the basin from the sloping area.

observed in southern Taiwan in the same period (Davis and Lee 2012; Ruppert *et al.* 2013). In addition to an afternoon peak, a secondary peak existed in the early morning in southern Taiwan. In the Taipei area, the storms initiated around 0500 UTC (+8 for LST, 1300 LST) and occurrence frequency of strong radar reflectivity reached its maximum at around 1500–1600 LST and then decayed rapidly. It is noted that strong reflectivity dissipated earlier in the northern portion of the Taipei basin and lasted longer till 1600 LST in the southern portion.

In northern Taiwan, convection initiated at the sloping area with terrain height around 500 m, starting from south to north and from west to east with an average speed of $\sim 6\,\mathrm{m\,s^{-1}}$. However, at around 1500 LST, there was another signal but with weaker occurrence frequency that propagated from east to west with a much slower speed, i.e., $\sim 2\,\mathrm{m\,s^{-1}}$. This signal draws special attention because the daytime heavy rain storms in the Taipei basin did go through similar tracks. In summary, at the beginning stage, convective storms that initiated at the sloping region propagated along the prevailing southwesterly wind. Nevertheless, some of the storms changed their courses toward the northwest into the basin after maturing. These storms are the ones producing heavy rainfall within the basin (Jou 1994). In this study, one case of this type (daytime heavy rain case) is presented for discussion.

On the contrary, nighttime heavy rain cases are relatively rare in the Taipei basin. Most of these lower-frequency events are associated with southward-propagating frontal systems. In this study, one nighttime extremely heavy rain case is also chosen for comparison. A mesoscale surface network and S-band dual-polarized radar (upgraded from WSR-88D) data from an operational site (Wu–Feng–Shan) are utilized. The kinematic flow structure at low levels and the dynamical and microphysical characteristics of the storms will be described, and the formation and development mechanisms of these extremely heavy rain events in the Taipei City will be discussed.

2. Data

Twenty five surface stations in northern Taiwan with hourly surface wind, temperature, and humidity measurements are used for analysis. Twice-daily upper air soundings at Banciao (south of the Taipei basin) are used for stability and vertical wind shear analysis. Two Doppler radars, one being a Wu–Feng–Shan S-band (WFS/CWB) and the other being a C-band (at Taoyuan airport, CKS/CAA), are used for dual-Doppler wind synthesis every 15 min. The newly upgraded polarimetric radar at WFS is also used for bulk microphysical analysis.

3. Daytime Heavy Rain Case

On the afternoon of 14 June 2015, the southern part of Taipei City experienced an unexpected flash flood with the 60-min accumulated rainfall amount exceeding 130 mm. The heavy rain-induced flash flood was caused by the local afternoon thunderstorm activities. The flash flood storm was characterized by short duration but extreme rainfall intensity and will be shown to correspond to the convective cell merge. The merger of convective cells produces an enlarged precipitation area and extends a strong echo to a much higher altitude. Figure 2 shows radar reflectivity images at a 3.5-km height for (a) 1418, (b) 1429, and (c) 1441 LST on 14 June 2015 indicating merging of several small convective cells into a larger thunderstorm complex. The surface rainfall station Gongkwan (near the NTU campus) observed a 60-min accumulated rainfall event exceeding 130 mm and two consecutive 10-min accumulated rainfall events greater than 30 mm at 1440 and 1450 LST, respectively, and with pronounced lightning activity. Figure 3 shows vertical cross-sections of radar reflectivity covering the time of the afternoon flash flood event. The cells merged around 1441 LST and

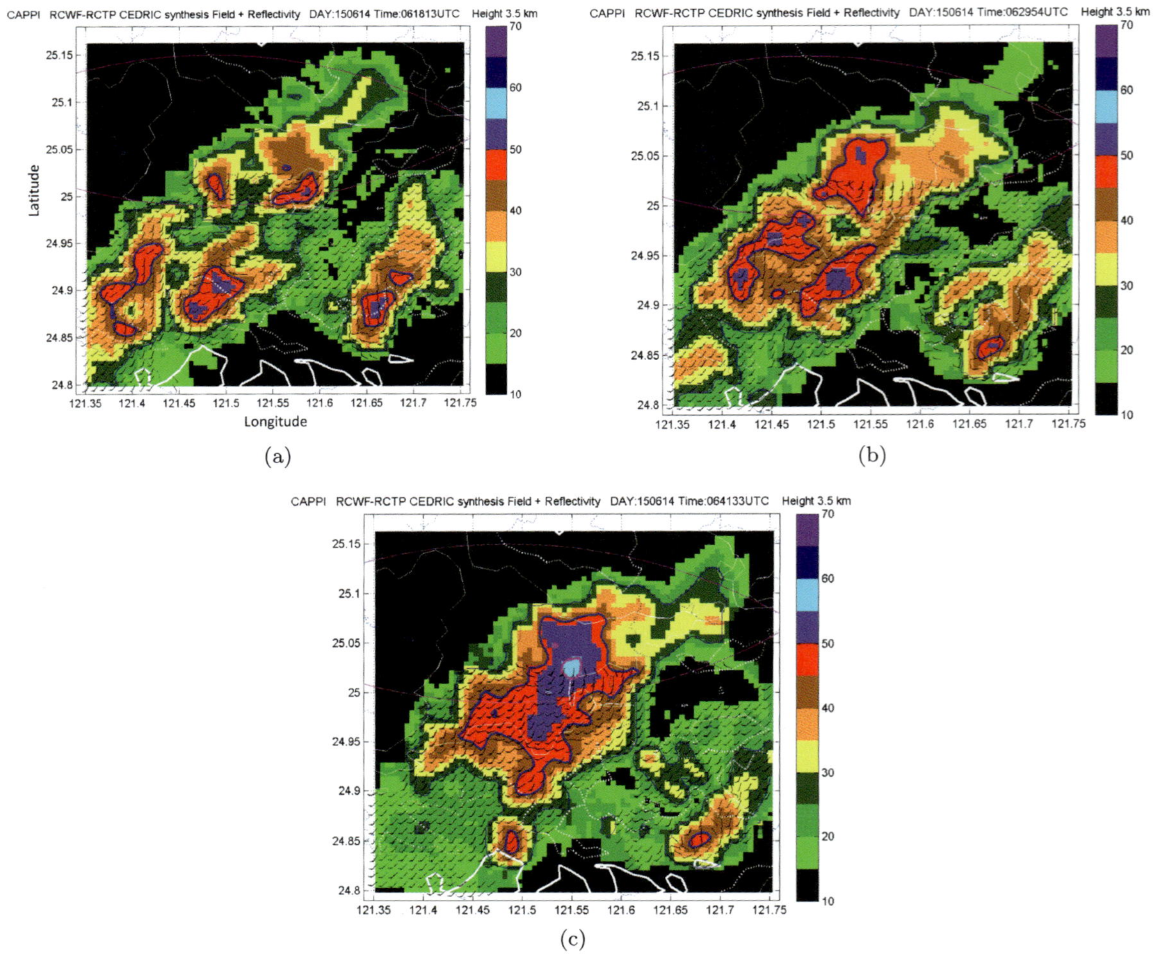

Fig. 2. The radar reflectivity images (color shading, in dBZ) in the Taipei basin at 3.5-km height for (a) 1418, (b) 1429, and (c) 1441 LST, 14 June 2015, indicating merging of several small convective cells into a giant thunderstorm complex. The bold blue contour indicates the area with reflectivity larger than 45 dBZ. In this region, 60-min rainfall larger than 100 mm was observed and associated with pronounced lightning activities. The arrows are the horizontal winds (in $m\,s^{-1}$) synthesized from WFS/CWB and CKS/CAA radars using NCAR/CEDRIC algorithm. Winds within the pink arc lines are on the baseline of dual-Doppler domain.

deeper and wider convective cells with increased reflectivity were observed. Results from dual-Doppler analysis also indicated stronger vertical velocity after the cell merge consistent with the greater vertical development of the reflectivity.

In order to explore the environmental conditions associated with this cell merger, surface temperatures and winds at 1400 LST 14 June 2015 are given in Fig. 4. A low-level convergence zone in the Taipei basin produced by the sea breeze from the west coast and the cooler outflows (blue color indicating the lower-temperature area) from the surrounding mountains in the east provided favorable conditions for cells to merge. The cooler air was the result of evaporation of rainfall from earlier storms that developed over the mountain region. The additional moisture brought by sea breeze also played an important role in producing a stronger

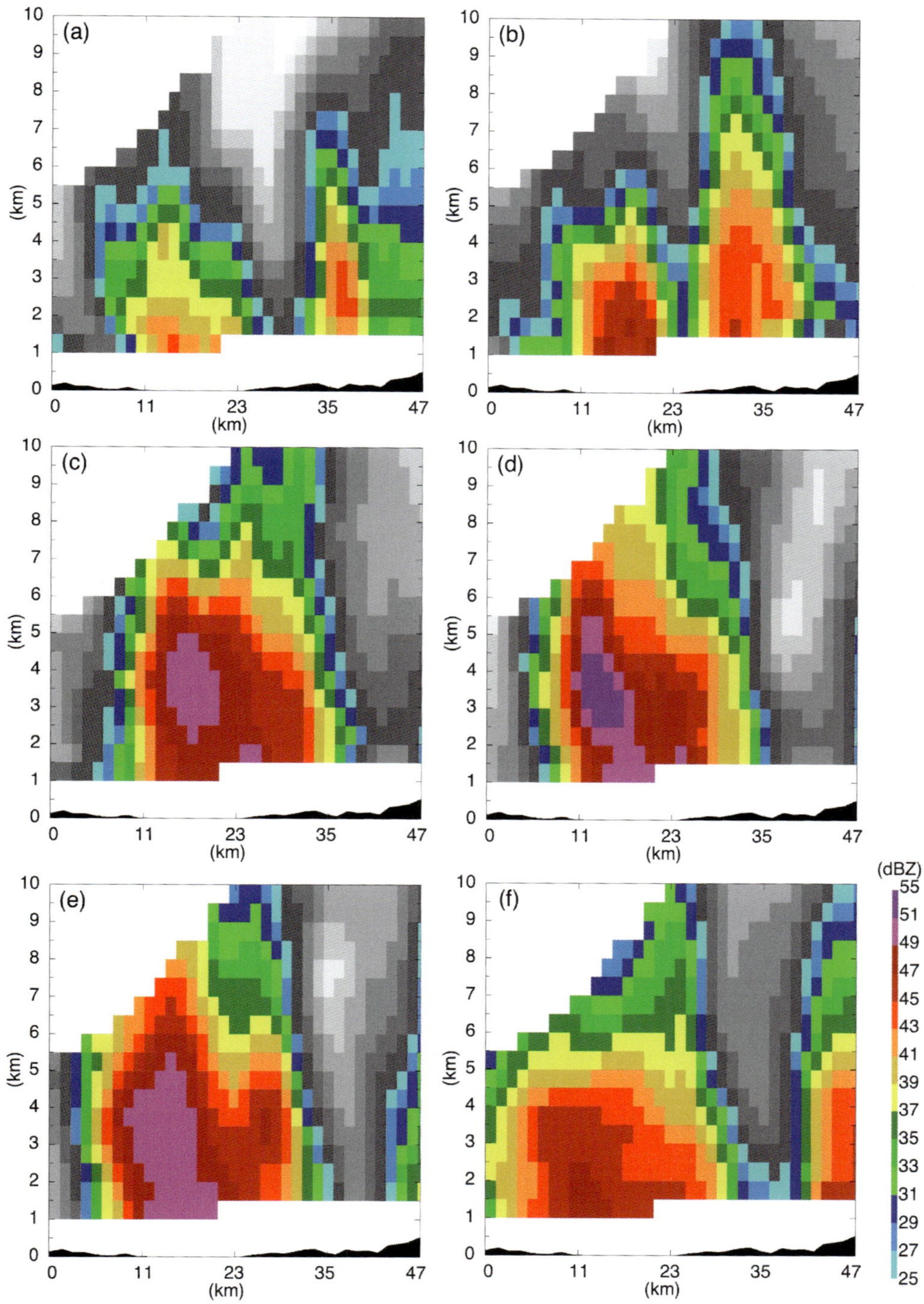

Fig. 3. Vertical cross-sections of radar reflectivity images along the mean southwesterly flow (SW on the left) showing the cell merge for the afternoon flash flood thunderstorm over the Taipei basin. (a) 1418, (b) 1429, (c) 1441, (d) 1452, (e) 1503, and (f) 1521 LST 14 June 2015. Color shadings are radar reflectivity in dBZ. 50-dBZ radar reflectivity reached 6-km height after the cell merge indicating a very strong vertical motion associated with the storm development.

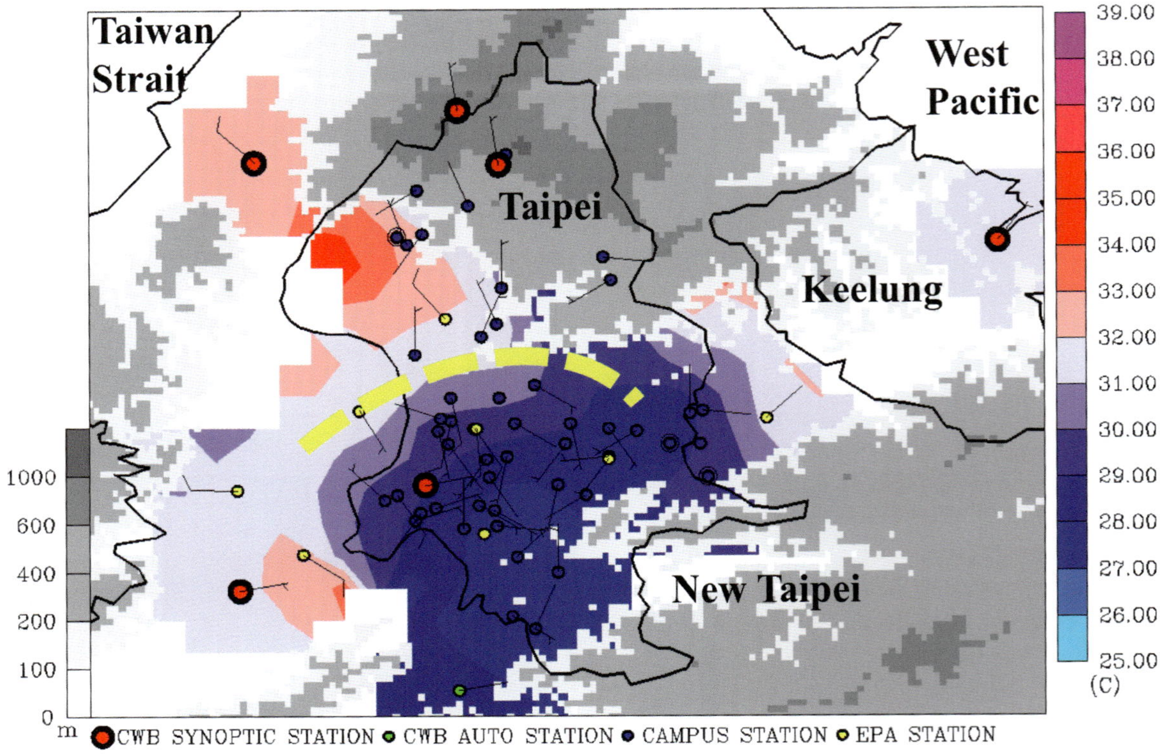

Fig. 4. Surface map at 1400 LST 14 June 2015 showing the low-level convergence zone (marked by yellow dash line) in the Taipei basin. Color shadings are surface temperatures composited from various surface stations. Red dots are surface stations with staff of CWB, light blue are automatic stations of CWB, dark blue are from elementary school campus light-duty weather stations operated by Taipei City Education Department, and white dots are from Environmental Protection Agency (EPA). The gray shading indicates the terrain height and the black lines are administrative divisions, Taipei City in the center surrounded by New Taipei City and with Keelung City on the NE. It is noted that the enhanced horizontal convergence zone over the Taipei basin which is favorable for the cell merge is produced by sea breeze from the west coast and cooler outflows from the surrounding mountains in the east. The additional moisture brought by sea breeze may also play a significant role in producing a stronger and deeper storm after the cell merge.

and deeper storm after the cell merge (Jou *et al.* 2016).

Figure 5 gives the radial velocity at 0.53^0 elevation angle (PPI) observed by WFS radar (located at lower right corner) at (a) 1418 LST and (b) 1435 LST, 14 June 2015. The gray/black color area is the weak wind zone with radial velocity less than $+/-2\,\mathrm{m\,s^{-1}}$. At 1435 LST, the dark red indicates wind speed closer to $10\,\mathrm{m\,s^{-1}}$ receding from the radar. The dark red areas were associated with very strong radar reflectivity (larger than 50 dBZ) indicating occurrence of pronounced precipitation after the cell merge. The intensification of near-surface winds associated with strong reflectivity indicates the outflow signature of the merged storm system. For comparison, in Fig. 6, horizontal wind and divergence/convergence at 3 km calculated using the dual-Doppler wind synthesized algorithm WISDDOM developed by Liou *et al.* (2014) at (a) 1420 LST and (b) 1435 LST, 14 June 2015, are given. The intensified horizontal convergence associated with the merged cell related to strong outflows induced by the pronounced precipitation as indicated in Fig. 5 was retrieved successfully.

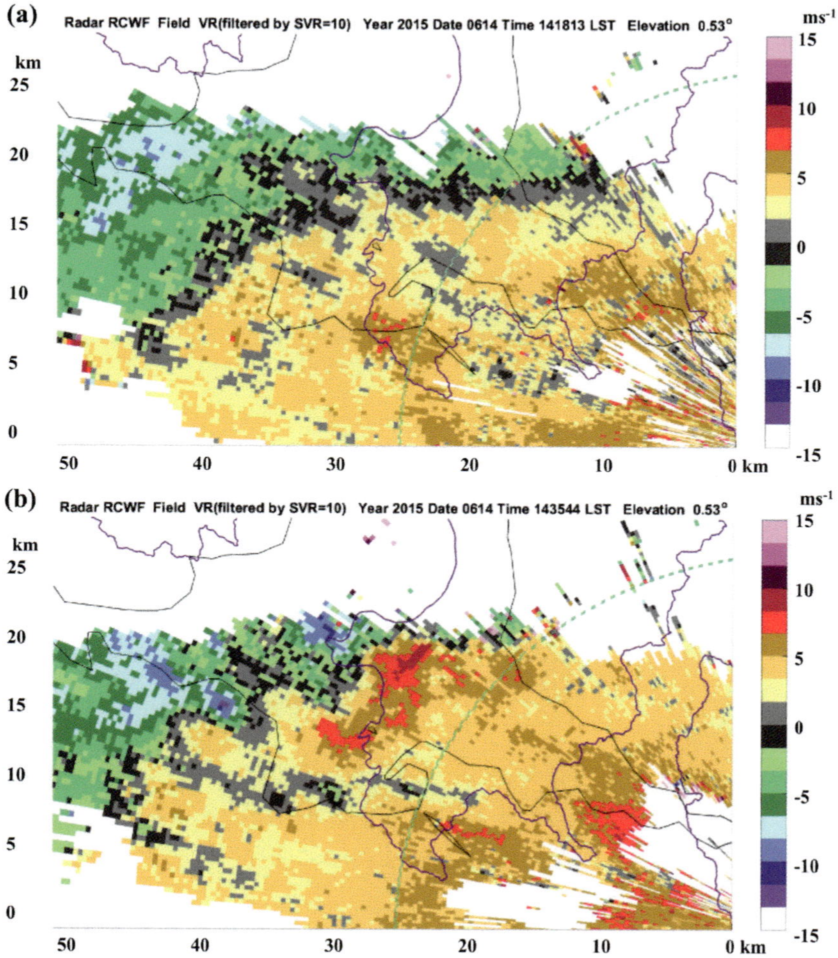

Fig. 5. Plan-position indicator (PPI) of radial velocity at 0.53° elevation angle observed by WFS/CWB radar (located at lower right corner) at (a) 1418 LST and (b) 1435 LST 14 June 2015. The gray/black color area is the weak wind zone with radial velocity less than $+/-2\,\mathrm{m\,s^{-1}}$. At 1435 LST, the dark red (light blue) indicates wind speed closer to $10\,\mathrm{m\,s^{-1}}$ receding from (approaching to) the radar and these areas were associated with very strong low-level winds. These areas also coincide with the enhanced surface convergence zone indicated in Fig. 4. The light green dash line marks the 1-km height circle. The purple line is for administrative division and light black line is for terrain contour of 500-m height.

From the above analysis, the results suggest that enhanced horizontal convergence induced by the sea breeze and downslope flow over the sloping area of SMR east of Taipei City plays an important role in facilitating the cell merge in the basin. The downslope cooler air is the result of the light rainfall from earlier weaker convections in the mountain area. It was suggested by Jou (1994) that the weak convection that developed over SMR was initiated by differential heating-induced upslope flow.

Next, we investigate possible significant variations of the structure and distribution of polarimetric variables before and after the cell merge and the information that could be helpful for short-term storm nowcast. The data of CWB operational S-band polarimetric radar at WFS are used for the microphysical analysis.

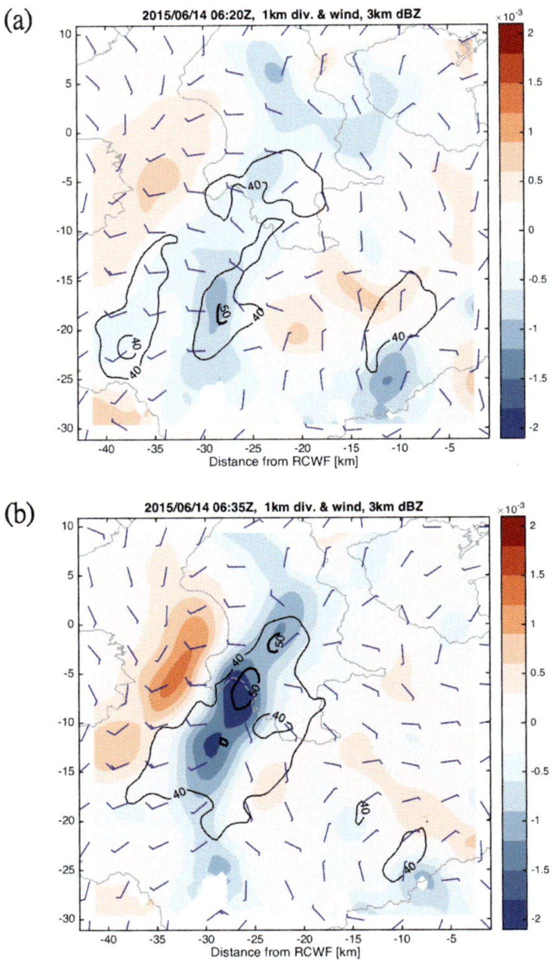

Fig. 6. Horizontal winds (in vector), radar reflectivity larger than 40 dBZ (black contour), and horizontal divergence (in color shadings $\times 10^{-3}\,s^{-1}$) at 3 km calculated using dual-Doppler wind synthesized algorithm WISDDOM developed by Liou *et al.* (2014) at (a) 1420 LST and (b) 1435 LST 14 June 2015. The domain of calculation is 50 km × 40 km in the center of Taipei. The intensified horizontal convergence is closely related to the strong outflows induced by the enhanced precipitation from the merged cell.

A description of principles and application of polarimetric observables can be found in Kumjian (2013). Figure 7 shows (a) 0.5^0 elevation angle radar reflectivity PPI display, and vertical cross-sections along line A–B of (b) radar reflectivity in dBZ, (c) differential reflectivity (Z_{DR}) in dB, and (d) specific differential phase (K_{DP}) in deg. Km^{-1} for the daytime heavy rain in the Taipei basin at 1354 LST, 14 June 2015, before the cell merge. Figure 7e–h is the same as Fig. 7a–d but for 1441 LST after the cell merge.

The comparison between Fig. 7a–d and Fig. 7e–h is summarized.

(1) Radar reflectivity images show that individual thunderstorm cells have clustered together into a giant thunderstorm complex after the cell merge (Fig. 7e). The vertical cross-section of reflectivity indicates that intense reflectivity (>40 dBZ) reached above 10 km and with a much larger horizontal extent (Fig. 7f).

(2) Before the cell merge, the differential reflectivity (Z_{DR}) exhibited a layered structure with large values near the surface, decreasing with height (Fig. 7c). This feature indicates a particle distribution with larger sizes at lower levels and smaller sizes at upper levels, a consequence of particle size sorting (Dawson *et al.* 2014). However, after the cells merged (Fig. 7g), the structure changed. Column features of Z_{DR} were observed. According to Snyder *et al.* (2015), the appearance of Z_{DR} column in a convective system indicates that strong updraft and larger particles can be maintained in the upper level. This result is consistent with what was observed by Z_{HH} (Fig. 7f) that a larger extent and stronger reflectivity appeared after the cell merge.

(3) Before the cell merge, K_{DP} distribution showed a layered structure with large K_{DP} near the surface, decreasing with height. After the cell merge, large K_{DP} was observed above the melting layer (Fig. 7h). The maximum K_{DP} value increased from $1.75°km^{-1}$ to $5.5°km^{-1}$, at least 3 times larger. Large K_{DP} was found well above melting level which was not observed before the cell merge. The collocation of high Z_{HH},

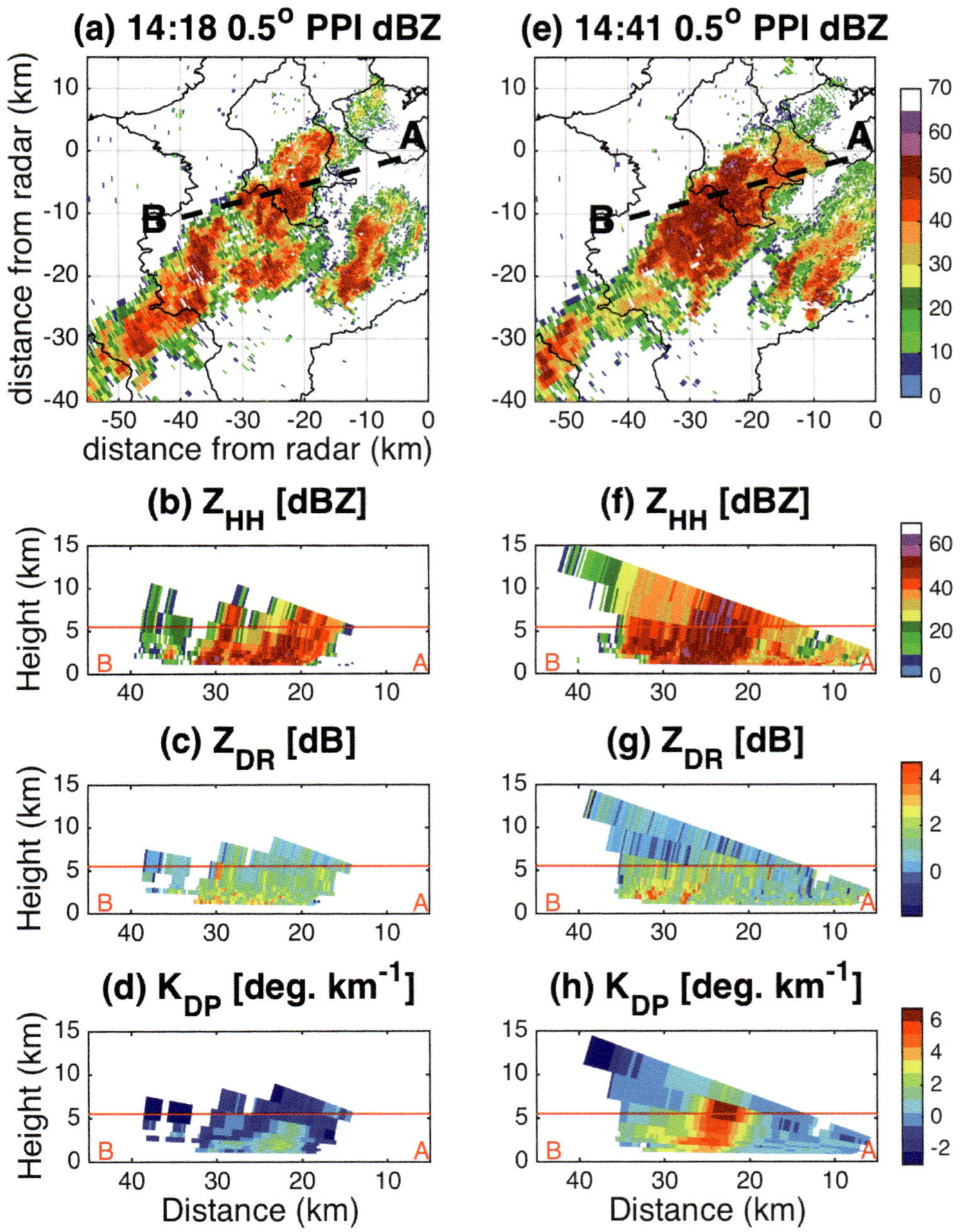

Fig. 7. (a) The reflectivity (dBZ) of 0.5° elevated angle observed by WFS/CWB S-band polarimetric radar at 1354 LST 14 June 2015, and (b) the vertical cross-section indicated by the black dashed line from point A to B. The cross-sections of (c) differential reflectivity (Z_{DR}) in dB and (d) specific differential phase (K_{DP}) in deg. Km^{-1} are also shown. The thin red line at about 5.5-km height in cross-sections represents the height of the melting level. Respectively, the (e), (f), (g), and (h) are the same as (a), (b), (c), and (d) but for 1441 LST after the cell merge.

high Z_{DR} and K_{DP} suggests that the dominant hydrometer types after the cell merge are mixed-phase hydrometeors, e.g., hail with rain and/or graupel with rain (Straka *et al.* 2000; Ryzhkov *et al.* 2005). The distribution of large Z_{DR} and K_{DP} above melting

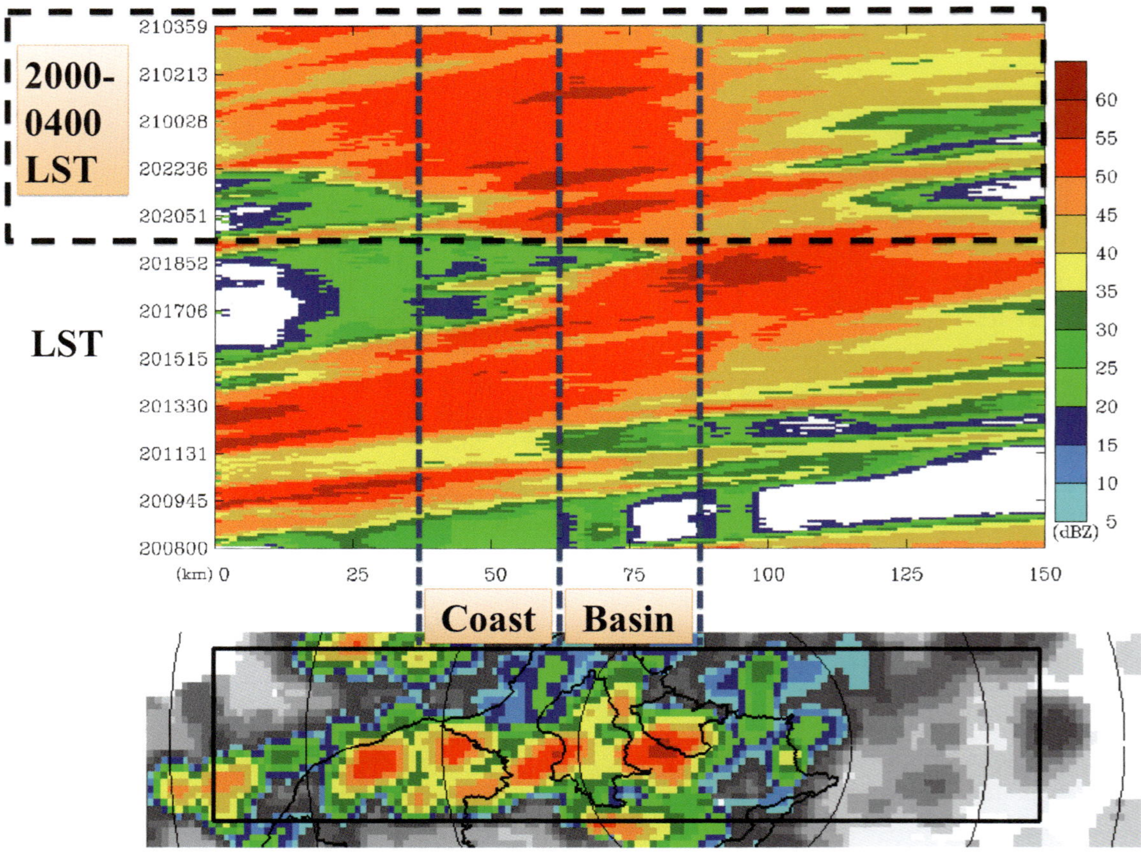

Fig. 8. The Hovmöller diagram of radar reflectivity (ZHH) over the Taipei basin with a west–east orientation. The Taipei basin is located at around 63–87 km (see bottom figure for comparison). The time is from 0800 LST 20 May to 0400 LST 21 May 2014. The period between 20 LST 20 May and 0400 LST 21 May 2014 marked by the black dash rectangle is the period with intensive heavy rainfall in the basin. Convective cells propagated eastward from the west coast into the basin and were intensified while entering the basin. From sequential radar images (Fig. 9), it is suggested that the intensification of the convection was majorly related to the cell merge between the eastward-propagating cell and cell originally initiated within the basin.

layer supports the existence of strong vertical motion, and large drops could be found at the area above freezing level.

4. Nighttime Heavy Rain Case Associated with Weak Surface Front

From late evening on 20 May (2100 LST) to early morning (0200 LST) on 21 May 2014, more than 200 mm of rainfall accumulated in just 6 h as measured by the rain gauges in southern Taipei City. The heavy rain event occurred following the passage of a 500-hPa trough accompanied by intensifying surface low pressure. There was no pronounced upper-level synoptic support for the development of the event. Over the coast of SE China, a quasi-stationary surface front moved slowly eastward to the west coast of Taiwan. A quasi-stationary convective line formed along with it. From surface temperature analysis, cold air intruded from the Keelung river mouth (NE of Taipei) into the southern part of the Taipei basin. From Banciao sounding (not shown), the near-surface atmosphere was absolutely stable. The atmosphere

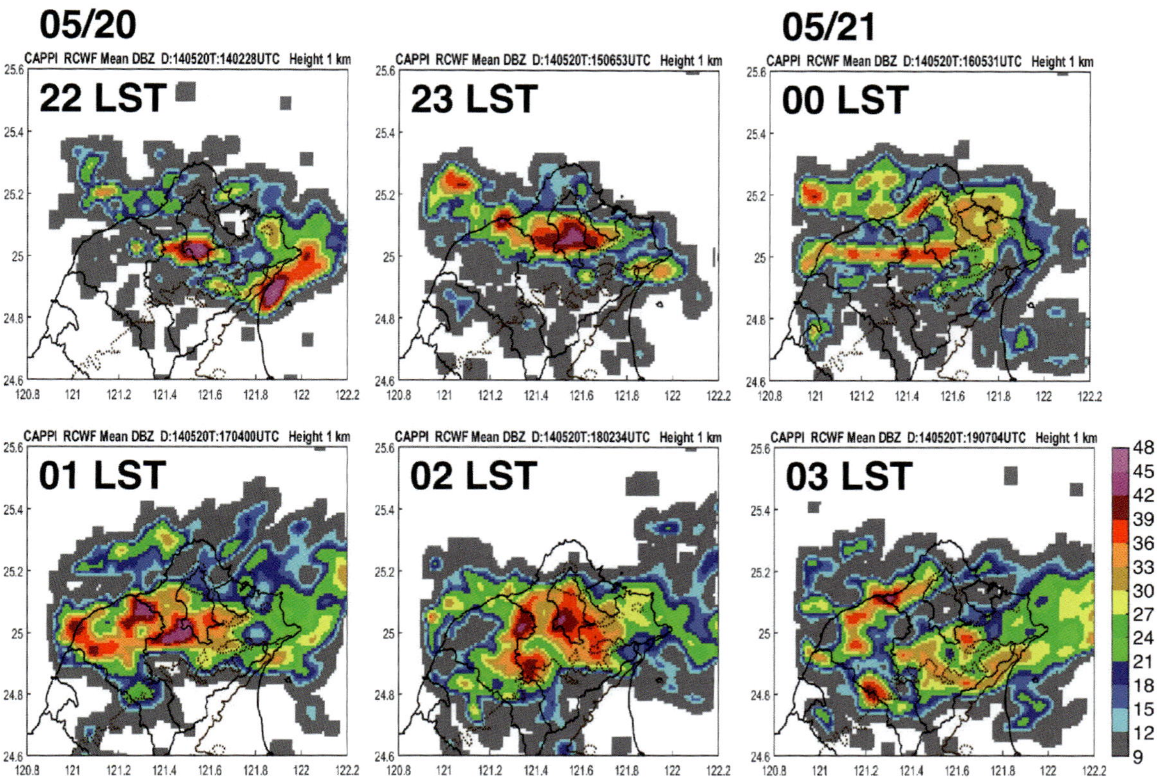

Fig. 9. Sequential hourly radar reflectivity maps (in dBZ) at 1-km height from 2200 LST 20 May to 0300 LST 21 May 2014 taken from WFS radar. The domain is 120.8–122.2°E and 24.6–25.6°N. Black lines are for coast and administrative divisions and the light brown lines are for terrain height above 500 m.

below 1-km height in the Taipei basin was acting like a cold dome. The warmer and moist air from the south was lifted by the cold dome to become unstable above. Convection that developed under these conditions is best described as elevated convection and has been the focus of the recent PECAN field campaign in the US Plains (Geerts *et al.* 2017).

Figure 8 shows the Hovmoller diagram of radar reflectivity over the Taipei basin, revealing two distinct modes of convection development at 0900 LST and 2000 LST, 20 May 2014. At an earlier time (0900–1800 LST), a fast-eastward-propagating mode existed. The propagating speed was around 8–9 $\mathrm{m\,s^{-1}}$ which was consistent with the prevailing westerly flow. Between 2000 LST 20 May and 0400 LST 21 May, a slow-eastward (quasi-stationary) mode existed. Convection was limited within the Taipei basin, indicating a cell-training characteristic (Wang *et al.* 2014).

Before 2000 LST 20 May, the convection initiated off the coast and propagated eastward into the basin. The intensity of the convection decayed rapidly and the heaviest rainfall occurred near the coast. Nevertheless, some stronger ones attained their maximum intensity at the eastern part of the city in the late afternoon (around 1700–1900 LST).

Starting 2000 LST, some new convective cells developed in the basin and became quasi-stationary. These cells merged with eastward-propagating cells which were initiated over the coast to form an east–west convective line (at 2300 LST 20 May, Fig. 9). The line was quasi-stationary and persisted for about 1 h

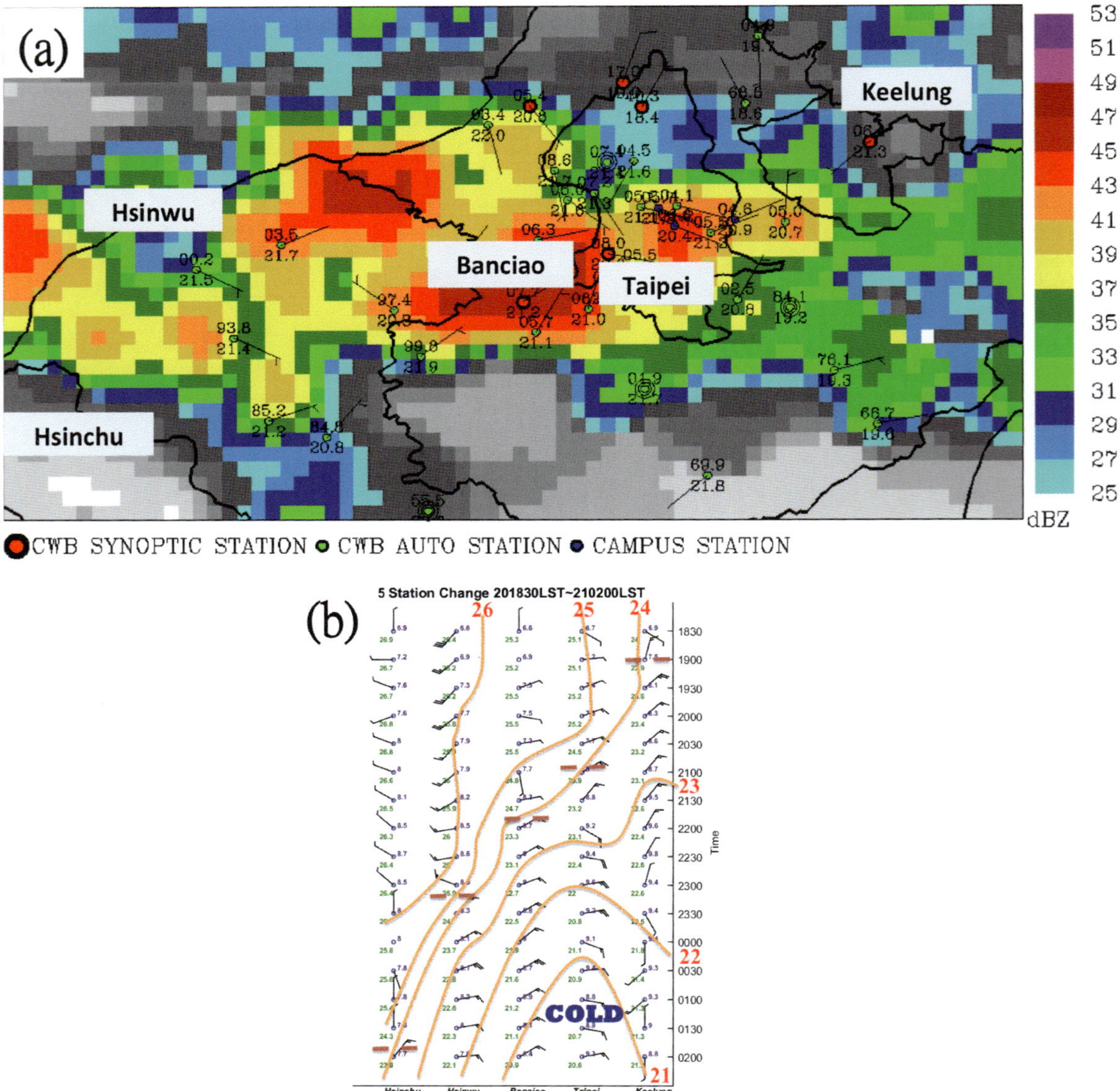

Fig. 10. (a) Reflectivity image at 1-km height at 0100 LST 21 May 2014. The surface winds, temperatures, and dew points are also presented. (b) Time–space section of surface parameters from 1830 LST 20 May to 0200 LST 21 May 2014. For each station, hourly variation of temperatures (in green), station surface pressure (in purple, +1000 hPa), and surface winds (in knots) are given. Contours are isotherms. The intrusion of colder air into the Taipei basin is clearly indicated.

and then drifted northward and dissipated. At the same time, new cells developed to the south of the old line and produced another east–west convective line and propagated northward into the basin. This phenomenon lasted for about 6 h starting from 2100 LST 20 May to 0200 LST 21 May 2014 (Fig. 9). This is the period when the Taipei basin received most of the heavy rainfall.

Figure 10a shows the surface observations overlaid with radar reflectivity at 0100 LST

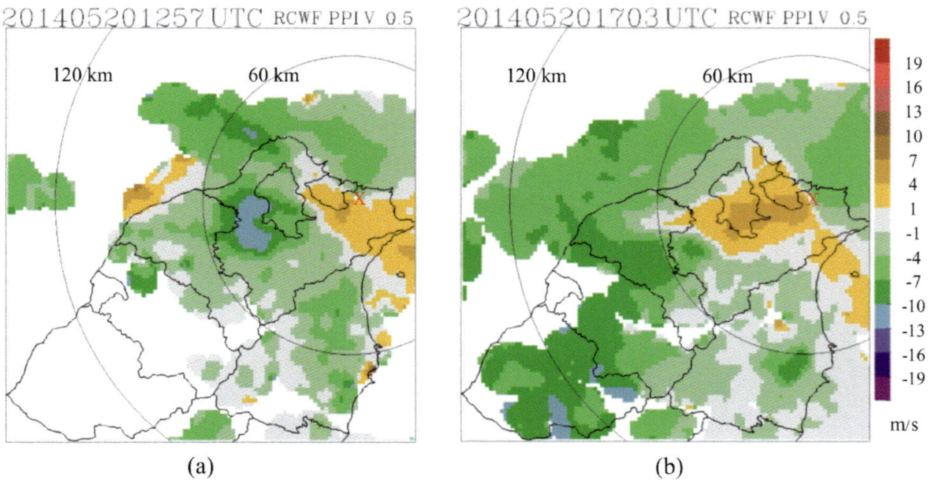

Fig. 11. Plan-position indicator (PPI) displays of 0.5° elevation angle radial velocity observed by WFS radar, (a) for 2057 LST 20 May and (b) for 0103 LST 21 May 2014. Red cross is the radar location, and warm color indicates flows receding away from the radar and the cool color indicates flows approaching the radar.

21 May 2014. Station positions for Keelung, Taipei, Banciao, Hsinwu, and Hsinchu are given for further reference and the total distance is about 100 km. Time–space sections of surface parameters from Keelung to Hsinchu and from 1830 LST 20 May to 0200 LST 21 May 2014 are given in Fig. 10b. Contours are isotherms and winds are shown with arrows. Cold air moved into the basin at around 2100 LST and then stagnated. At Taipei station, the temperature dropped from 25°C to less than 21°C within 5 h (from 2100 LST 20 May to 0100 LST 21 May). However, the coastal stations such as Hsinwu and Hsinchu did not experience such a rapid temperature drop. It is noted that very intense radar reflectivity (>45 dBZ) between Banciao and Taipei was collocated with the coldest air as indicated in Fig. 10b. The observation suggests that intrusion of cold air played an important role in formation of strong convection in the Taipei basin.

The 0.5° elevation angle radial velocity observed by the WFS radar (located at east side of the city) is given in Fig. 11. The yellow colors are areas with winds blowing away from the radar and greens and blues are winds blowing toward the radar. At 2100 LST 20 May (Fig. 11a), most of the winds were westerly within the Taipei basin. However, at 0100 LST 21 May (Fig. 11b), the winds changed direction to become easterly. Since the WFS radar station is about 750 m above sea level, the 0.5° elevation angle observations are at around a 1-km height. As seen from Figs. 10 and 11, winds at the surface and 1-km height all became easterly after 0100 LST 21 May. The intrusion of cold air with a depth near 1 km from the northeast of Taipei is captured by the radar. It is noted that the leading edge of the cold air tongue is collocated with the formation of quasi-stationary line convection in the basin. These observations further facilitate that warmer air from the SW was lifted by the cold air from the NE and initiated convection over the Taipei basin.

Figure 12 shows the WFS-observed polarimetric variables of the nighttime heavy rain system for 20–21 May 2014. The melting layer was located at around 4.5 km which is about 1 km lower than that in the daytime case. Above the melting layer, no large values of Z_{DR} or K_{DP} were observed in this case. This is the major difference between the daytime heavy rain case and the nighttime case. In the nighttime case, most of the time, large K_{DP} showed a layered

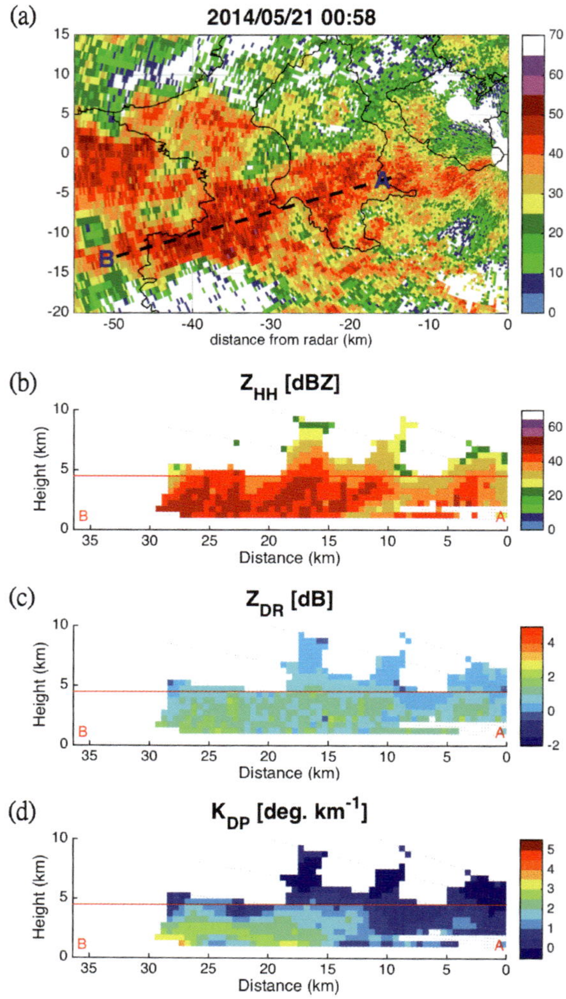

Fig. 12. Polarimetric observations of $Z_{\rm HH}$, $Z_{\rm DR}$, and $K_{\rm DP}$ at 0058 LST 21 May 2014. This figure is the same as Fig. 7(a)–(d) except the red line indicates the melting layer which is at about 4.5-km height, 1 km lower than that for the daytime heavy rain case. No pronounced $Z_{\rm DR}$ and $K_{\rm DP}$ columns were detected during the whole period of heavy rainfall. Most of the time, $K_{\rm DP}$ is less than $2°\ {\rm km}^{-1}$.

structure at low levels and was weaker than $2°{\rm km}^{-1}$. From these observations, we speculate that weaker upward motions were associated with the system compared with the daytime case and the warm rain process possibly dominates the precipitation formation in this situation.

5. Discussion and Conclusion

In this study, two heavy rain-flash flood cases over Taipei, one in the daytime and the other in the nighttime, are analyzed and the differences in formation mechanisms and the bulk microphysics are investigated. It is found that the daytime heavy rain case is associated with a complex afternoon thunderstorm system with notable a cell merge process. Strong vertical motions with significant mixed-phase cloud microphysics were suggested by the appearance of $Z_{\rm DR}$ and $K_{\rm DP}$ columns. The formation of strong vertical motions is produced by the low-level convergence zone which is induced by the sea breeze circulation from the west and downslope winds from previously existent convection over the surrounding mountain slopes in the east.

Nighttime heavy rain is relatively rare in this area. The mesoscale surface network and dual-polarized radar data are utilized to reveal the kinematic flow structure at low levels and the dynamic and microphysical characteristics of the storms. Intrusion of cold air from the ocean interacted with the pre-existent cold pool produced by the afternoon storms helping to trigger convection in the Taipei basin. Convection formed along the coastal area in the west of the city then propagated along the prevailing westerly flow into the basin and merged with the convection in the basin. This process not only intensified the convection but also forced the convection to slow down its propagation speed. The polarimetric measurements indicate that the nighttime system was lacking Z_{DR} and K_{DP} columns, the bulk microphysics was more dominated by the warm rain process, and the rainfall intensity is not as much as the daytime heavy rain system but with a much longer time span.

The WRF model (CWB of Taiwan) predicted the nighttime storm rainfall well at the beginning stage while the rainfall was associated with the approaching surface front. However, the rainfall prediction was poor at a later stage. From radar observations, it is noted that new cells formed at the coastal region, west of the city, and propagated eastward into the basin. A training feature such as back-building type or parallel stratiform MCS (Luo *et al.* 2014; Wang *et al.* 2014; Parker and Johnson 2000) is characterized as the cause for this nighttime event.

Due to the geometric limitation, the synthesized dual-Doppler winds and horizontal convergence fields from WFS/CWB and CKS/CCA are poor near the base line. The direct impact is the retrieved vertical motion field. It is suspected that the upward/downward motions retrieved are underestimated. We are expecting that this difficulty can be improved in the PRECIP 2020 field program as well as through more heavy rain events for comparison in northern Taiwan in May–August 2020 (Bell 2019).

Acknowledgments

Support for this study was provided by the Ministry of Science and Technology (MOST), Fire Department, Taipei City Government (TCG), and Central Weather Bureau (CWB) of Taiwan. The first author would like to take this opportunity to thank fellow students Yu-Cheng Kao, Pin-Fang Lin, and Pao-Liang Chang for their contribution on data processing and diagram plotting. Many constructive suggestions provided by Dr. Angela Rowe and Prof. Richard Johnson are greatly appreciated. Prof. Uyeda's visit to National Taiwan University from September 2017–June 2018 was supported by MOST of Taiwan.

References

Bell, M., 2019: Science Plan for PRECIP 2020. Workshop of PRECIP2020/TAHOPE at National Taiwan University, Taipei, 16 October 2019.

Davis, C. A. and W.-C. Lee. 2012: Mesoscale analysis of heavy rainfall episodes from SoWMEX/TiMREX. *J. Atmos. Sci.*, **69**, 521–537.

Dawson, D. T., E. R. Mansell, Y. Jung, L. J. Wicker, M. R. Kumjian, and M. Xue, 2014: Low-level ZDR signatures in supercell forward flanks: The role of size sorting and melting of hail. *J. Atmos. Sci.*, **71**, 276–299.

Geerts, B., and Coauthors. 2017: The 2015 Plains Elevated Convection at Night (PECAN) field project. *Bull. Amer. Meteor. Soc.*, 767–786, doi:10.1175/BAMS-D-15-00257.1.

Jou, B. J.-D., 1994: Mountain-originated mesoscale precipitation system in northern Taiwan: A case study 21 June 1991. *TAO*, **5**, 169–197.

Jou, B. J.-D., W.-C. Lee, and R. H. Johnson, 2011: An overview of SoWMEX/TiMREX. *The Global Monsoon System: Research and Forecasts* (2nd edition). Eds., C.-P. Chang *et al.*, World Scientific, pp. 303–318.

Jou, B. J.-D., Y.-C. Kao, R.-G. R. Hsiu, C.-J. U. Jung, J. R. Lee and H. C. Kuo, 2016: Observational characteristics and forecast challenge of Taipei flash flood afternoon thunderstorm: Case study of 14 June 2015. *Atmos. Sci. (in Chinese)*, **44**(1), 57–82.

Kumjian, M. R., 2013: Principles and applications of dual-polarization weather radar. Part I: Description of the polarimetric radar variables. *J. Oper. Meteor.*, **1**, 226–242.

Lin, P. F., P. L. Chang, B. J.-D. Jou, J. Wilson, and R. Roberts, 2011: Warm season afternoon thunderstorm characteristics under weak synoptic-scale forcing over Taiwan Island. *Weather and Forecasting*, **26**, 44–60.

Liou, Y. C., J. L. Chiou, W. H. Chen, and H. Y. Yu, 2014: Improving the model convective storm quantitative precipitation nowcasting by assimilating state variables retrieved from multiple-doppler radar observations. *Mon. Wea. Rev.* **142**, 4017–4035.

Luo, Y., Y. Gong, and D. L. Zhang, 2014: Initiation and organizational modes of an extreme-rain-producing mesoscale convective system along a Mei-Yu Front in east China. *Mon. Wea. Rev.*, **142**, 203–221.

Parker, M. D., and R. H. Johnson, 2000: Organizational modes of midlatitude mesoscale convective systems. *Mon. Wea. Rev.*, **218**, 3413–3436.

Ruppert, Jr. J. H., R. H. Johnson, and A. K. Rowe, 2013: Diurnal circulations and rainfall in Taiwan during SoWMEX/TiMREX (2008). *Mon. Wea. Rev.*, **141**, 3851–3872.

Ryzhkov, A. V., T. J. Schuur, D. W., Burgess, P. L. Heinselman, S. E. Giangrande, and D. S. Zrnic, 2005: The joint polarization experiment: Polarimetric rainfall measurements and hydrometeor classification. *Bull. Amer. Meteor. Soc.*, **86**, 809–824.

Snyder, J. C., A. V. Ryzhkov, M. R. Kumjian, A. P. Khan, and J. Picca, 2015: A Z_{DR} column detection algorithm to examine convective storm updrafts. *Weather Forecast*, **30**, 1819–1844.

Straka, J. M., D. S. Zrnic, and A. V. Ryzhkov, 2000: Bulk hydrometeor classification and quantification using polarimetric radar data: Synthesis of relations. *J. Appl. Meteor.*, **39**, 1341–1372.

Tao, W. K. and J. Simpson, 1989: A further study of cumulus interactions and mergers: Three-dimensional simulations with trajectory analyses. *J. Atmos. Sci.*, **46**, 2974–3004.

Wang, C.-C., J. C. Hsu, G. T. Chen, and D. I. Lee, 2014: A study of two propagating heavy-rainfall episodes near Taiwan during SoWMEX/TiMREX IOP-8 in June 2008. Part II: Sensitivity tests on the roles of synoptic conditions and topographic effects. *Mon. Wea. Rev.*, **142**, 2644–2664.

Wang, H., Y. Luo, and B. J.-D. Jou, 2014: Initiation, maintenance, and properties of convection in an extreme rainfall event during SCMREX: Observational analysis. *J. Geophys. Rev.*, **119**, 2100–2123.

Chapter 17

Minimal Model Studies of Stratosphere–Troposphere Two-way Dynamical Coupling in the Tropics through Organizations of Moist Convective Systems

Shigeo Yoden*,§, Hoang-Hai Bui† and Eriko Nishimoto‡

**Department of Geophysics, Kyoto University, 606-8502 Kyoto, Japan,*
†Geophysical Institute, University of Bergen, N-5020 Bergen, Norway,
‡Japan Agency for Marine-Earth Science and Technology, 236-0001 Yokohama, Japan
§yoden@kugi.kyoto-u.ac.jp

In association with recent research progress on stratosphere–troposphere two-way dynamical coupling in the tropics, we have conducted a series of numerical model studies on the dynamical analog of the quasi-biennial oscillation (QBO) in a two-dimensional minimal model framework of a non-hydrostatic regional-scale cloud-system-resolving model. The QBO-like oscillation in the zonal mean zonal wind is a robust feature as a result of wave–mean flow interactions by upward-propagating gravity waves generated by moist convection in the troposphere, not sensitive to the choice of model configuration or boundary conditions. A three-dimensional minimal model run with a rectangular domain confirms a similar QBO-like oscillation in both the domain-averaged zonal and meridional winds in the stratosphere.

The downward influence of the QBO-like oscillation on moist convection is clear in the two-dimensional minimal model framework. Squall-line type and back-building type of precipitation patterns appear alternately, depending on the vertical shears or profiles of the zonal mean zonal wind in the upper troposphere and lower stratosphere (UTLS) and near the surface. The relative importance of the vertical shears in the UTLS and near the surface was investigated by a series of low-level nudging experiments. Large values of the shear near the surface enhance the longevity and intensity of the moist convective systems by separating the updraft and downdraft. On the other hand, large values of shear near the cloud top tend to disrupt the convective structure and lead to weakening moist convection, although this mechanism seems to be secondary to the surface shear effect. Quantitative diagnosis of vertical momentum transports associated with moist convection and upward- and downward-propagating gravity waves also shows their periodic variations with the QBO-like oscillation according to the modulation of tilted convective systems by the vertical shear of the mean zonal wind and the spectral features of gravity wave momentum transports.

1. Introduction

In the last decade or two, the stratosphere–troposphere (ST) dynamical coupling in middle and high latitudes has attracted much attention, such as the annular mode variability in both hemispheres and two-way dynamical coupling during extreme events like stratospheric sudden warming (SSW) (e.g., Baldwin and Dunkerton 2001; Shepherd 2002; Yoden *et al.* 2002; Baldwin *et al.* 2018). The standard paradigms for interpreting and explaining the ST dynamical coupling in the extratropics and polar regions have been based on balanced dynamics: the conservation law of potential vorticity and non-local aspects of potential vorticity inversion, with multifaceted applications for planetary wave propagation, wave–mean flow interactions,

The Multiscale Global Monsoon System
Edited by C. P. Chang et al.

nonlinear wave breaking, wave-induced mean circulations, and so on, in both the troposphere and stratosphere.

Recently, we started a new international research activity on Stratospheric And Tropospheric Influences on Tropical Convective Systems (SATIO-TCS) under the project Stratosphere–troposphere Processes And their Role in Climate (SPARC) of the World Climate Research Programme (WCRP) (https://www.sparc-climate.org/activities/satio-tcs; Geller *et al.* 2017; Anstey *et al.* 2018; Osprey *et al.* 2018). SATIO-TCS focuses on ST dynamical coupling in the tropics, where no comparable interpretive paradigm exists such as in the extratropics. Some observational data analyses and global to regional numerical model studies have pointed out important stratospheric influences on tropical moist convection and organized convective systems, as compactly reviewed below. Multiscale interactions of these systems are likely to play a vital role in determining the influence of stratospheric variations on tropical convective systems, as summarized in Fig. 1 for this research activity (Hitchcock *et al.* 2019).

There is some observational evidence that spatio-temporal variations in the lower stratosphere and around the tropopause do influence tropospheric variability in the tropics in the form of moist convection and its large-scale organization into meso-to-planetary-scale systems, in association with SSW events (e.g., Eguchi and Kodera 2010; Kodera *et al.* 2015; Albers *et al.* 2016; Bal *et al.* 2017), stratospheric Kelvin waves (Zhang *et al.* 2018), the equatorial quasi-biennial oscillation (QBO) (e.g., Collimore *et al.* 2003; Liess and Geller 2012), the 11-year solar cycle (Hood 2017), or the anthropogenic stratospheric cooling trend (e.g., Emanuel *et al.* 2013). The organized moist convective systems and their spatio-temporal modulations include the phenomena shown in Fig. 1, such as the cloud clusters, their diurnal variations, tropical cyclones, the Madden–Julian oscillation (MJO), annually varying monsoon systems, interannual timescale El Niño–Southern oscillations (ENSOs), and features in climate change timescales.

Recently, the importance of multiscale interactions of moist convection has been recognized, as it is much more complex than a simple variation of convection with another timescale, such as the influence of the stratospheric QBO on the boreal wintertime MJO (Yoo and Son 2016; Nishimoto and Yoden 2017). The MJO is the dominant component of intraseasonal variability in the tropics, consisting of smaller-scale equatorial waves and mesoscale convective systems. Figure 2 shows that the composite outgoing longwave radiation (OLR) anomaly (i.e., a measure of the intensity of organized moist convective systems) associated with the MJO events has a larger negative value and slower eastward propagation with a prolonged period of active convection in the easterly phase of the QBO than in the westerly phase. Statistically significant differences of the MJO activities between the QBO phases also exist with dynamical consistency in the divergence of the horizontal wind, the vertical wind, the moisture, the precipitation, and the 100-hPa temperature.

There are some numerical model studies on the influence of the QBO on tropical moist convection with regional-scale cloud (or cloud-system)-resolving models. Yoden *et al.* (2014) reexamined the internally generated oscillation of the zonal mean zonal wind in dynamical analogy to the equatorial QBO, which was firstly obtained by Held *et al.* (1993), and found that the domain-averaged low-pass-filtered precipitation also varies in accordance with the QBO-like oscillation. Nie and Sobel (2015) showed temperature anomalies associated with the QBO influence on tropical deep convection through interactions between convection, radiation, and large-scale vertical motion; the precipitation rate showed slight increases in response to the temperature perturbation of an easterly QBO phase over small SST anomalies and strong decreases over large SST anomalies.

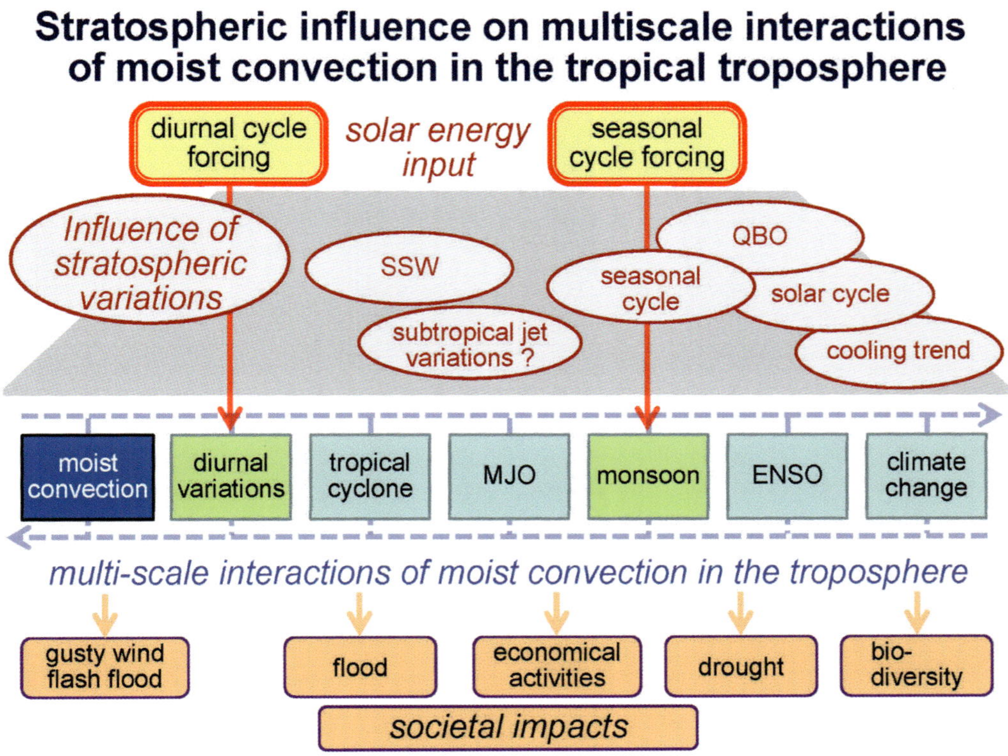

Fig. 1. Schematic diagram of the influence of stratospheric variations on multiscale interactions of moist convection in the tropical troposphere and their societal impacts; the diagram was originally drawn to explain the main purpose of the WCRP/SPARC/SATIO-TCS activity (Hitchcock *et al.* 2019).

Marshall *et al.* (2017) showed improved predictions of the MJO in boreal winter during the easterly QBO phase using a 30-year dataset of subseasonal retrospective forecasts with a global atmosphere–ocean coupled model. These improvements in forecast skill result not just from the initial condition with stronger MJO events but also from the more persistent behavior during easterly QBO phases. On the contrary, investigating the multi-decadal simulation with the Global Ocean Mixed Layer configuration of the Met Office Unified Model, Lee and Klingaman (2018) reported a weak and insignificant correlation between the QBO winds and mean MJO amplitude in boreal winter, in contrast to the significant anti-correlation in recent observational studies. At present, there are no numerical model simulations as yet to show the influence of the stratospheric QBO on the boreal wintertime MJO, and such modeling studies are still in a rather preliminary state.

In this paper, we summarize our recent research progress on ST two-way dynamical coupling in the tropics with a two-dimensional minimal model framework of a non-hydrostatic limited-domain cloud-system-resolving model (Yoden *et al.* 2014; Nishimoto *et al.* 2016; Bui *et al.* 2017), focusing on a fundamental process of the stratospheric and tropospheric influences on tropical convective systems, in particular through the vertical profile or shears of the zonal mean zonal wind in the troposphere. The minimal model used in the present study has internally generated time variations in the vertical profile of zonal mean temperature, but not controlled in any experiments, in contrast with the experimental design by Nie and Sobel (2015). On the contrary, the relative importance

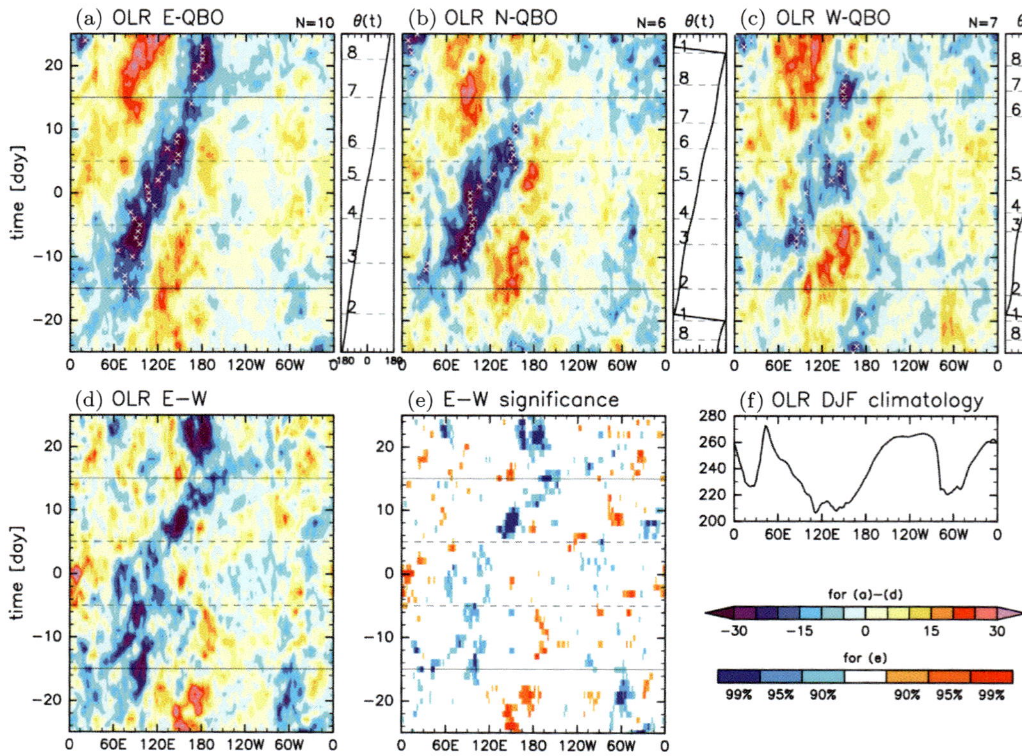

Fig. 2. Longitude–time sections of the composite OLR anomaly (Wm^{-2}) of MJO events averaged over 10S–10N for (a) the easterly phase of the QBO, (b) the neutral phase, and (c) the westerly phase, and for (d) the composite difference of OLR between the easterly and westerly phases; (e) the statistically significance of the composite difference; and (f) the climatological mean state of OLR during DJF in the neutral ENSO period. The MJO convective center is marked with white crosses in (a)–(c). Time series of the OMI phase $\theta(t)$ and the phase number I are shown to the right of (a)–(c). Taken from Nishimoto and Yoden (2017).

of vertical shears in the UTLS and near the surface with regard to their influence on deep moist convection is investigated in detail by a series of nudging experiments.

2. A Minimal Model of QBO-like Oscillation in ST Coupled System

Yoden *et al.* (2014) obtained internally generated QBO-like oscillations as a radiative-moist convective quasi-equilibrium state in a highly idealized minimal model framework with explicit moist convection under a periodic lateral boundary condition without Coriolis effects. Figure 3 is an example of such a QBO-like oscillation with a period of 134.8 days obtained for an experiment with the model top around 40 km. The QBO-like oscillation is a robust feature, not sensitive to the choice of model configuration such as domain size and horizontal resolution, or boundary conditions like sea surface temperature and prescribed zonal wind at the top boundary.

Figure 3 shows clear oscillations both in the stratosphere and the troposphere with a kink around the tropopause. As easily seen by the zero-wind lines, downward propagation of the oscillation from the bottom of the Rayleigh damping layer (~30 km) to the tropopause (~13 km) has a descending time of

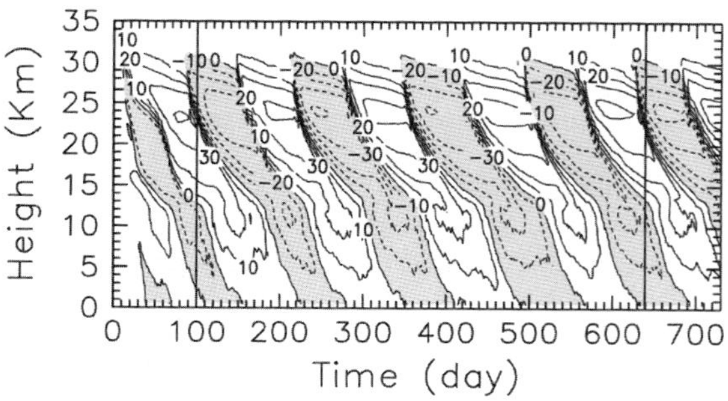

Fig. 3. Time−height section of the zonal mean zonal wind (ms^{-1}) for a QBO-like oscillation, taken from Yoden *et al.* (2014), Fig. 2c High-top case. Negative values are shaded. A pair of vertical lines in the plot indicates the interval of four full cycles of the oscillation.

about 100 days, whereas it is about 50 days from the tropopause to the surface. The downward propagation of the zonal mean signals in the stratosphere reveals similar fundamental dynamical features as seen in observations and driven by wave–mean flow interactions predicted by theory.

In the troposphere, unlike the observed equatorial QBO, there are clear signals of synchronized oscillations not only in the zonal mean zonal wind but also in temperature and precipitation, though day-to-day fluctuation of the latter is very large compared to the long-term oscillation (as seen in Fig. 4b). The possible importance of synchronized modulation of organized moist convective systems was studied further in subsequent papers, pertaining to convective momentum transport due to tilted convective structures by the vertical shear of the zonal mean zonal wind.

3. Downward Influence of QBO-like Oscillation on Moist Convection

Nishimoto *et al.* (2016) investigated the influence of the QBO-like oscillation on moist convection, by analyzing high time resolution output data obtained in the two-dimensional minimal model framework as those corresponding to Fig. 3. They found periodic variations of organized features of moist convective systems, which are characterized as alternate appearance of squall-line (SL) type and back-building (BB) type of precipitation patterns. The vertical shear of the zonal mean zonal wind in the UTLS is small during the SL-type periods compared with during the BB-type periods, consistent with the hypothesis of Collimore *et al.* (2003) on the influence of the mean zonal wind shear around the UTLS on moist convection. At the same time, the precipitation pattern is of the SL type when the vertical shear of the mean zonal wind near the surface is large, whereas it is of the BB type when the vertical shear is small. These are consistent with the numerical experiment by Liu and Moncrieff (2001) who used a two-dimensional cloud-system-resolving model by changing the vertical shear of the initial mean zonal wind near the surface.

Bui *et al.* (2017) further examined the relative importance of the vertical shears or profiles of the zonal mean zonal wind in the UTLS and near the surface in the same two-dimensional minimal model framework with two series of parameter sweep experiments: *model top experiments* with varied model top heights, and *low-level nudging experiments* with reduced zonal mean zonal wind toward zero from the surface to a certain level.

The QBO-like oscillation in the zonal mean zonal wind is a robust feature obtained in all the *model top experiments* with the model top from 40 km to 15 km. Further, in the low-top cases, only the tropospheric portion of the oscillatory features is retained. In all the *model top experiments*, the zonal mean precipitation is modulated with a half period of the mean zonal wind oscillation and a positive correlation exists between the precipitation intensity and the magnitude of vertical shear of the mean zonal wind near the surface (Figs. 4a, b for Control run). On the contrary, the relation of the vertical shear in the upper troposphere with the precipitation is not so systematic among the experiments.

In the *low-level nudging experiments*, the precipitation modulation with the QBO-like

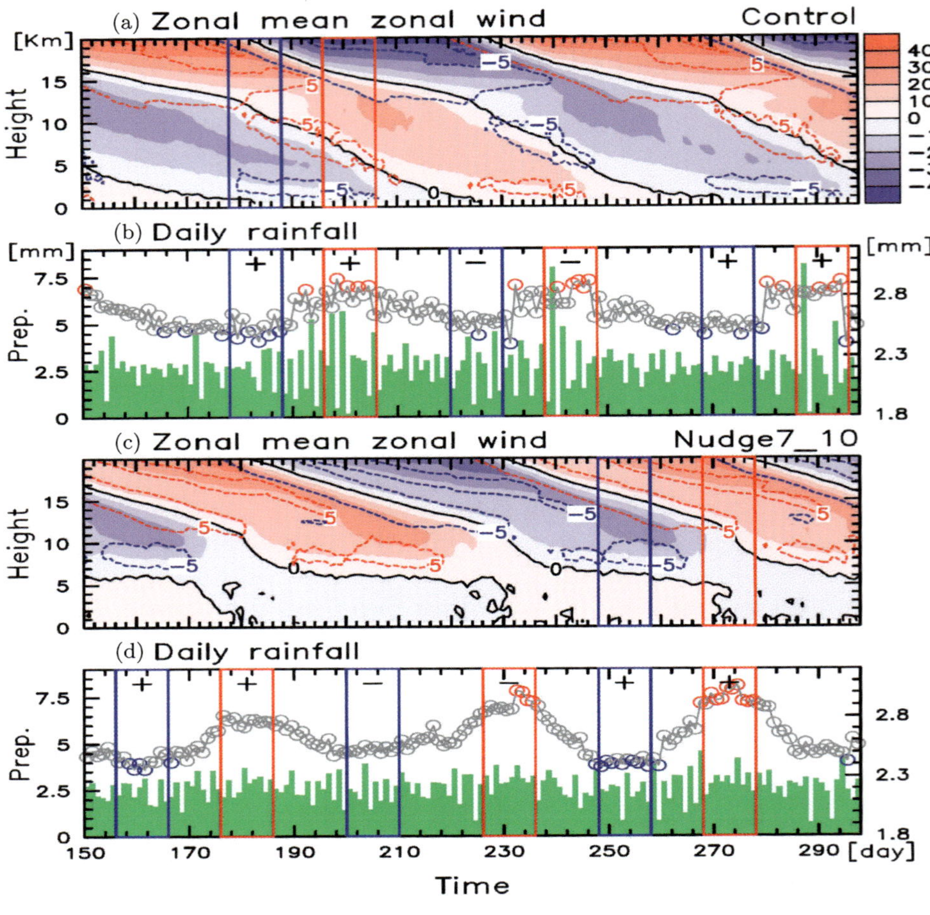

Fig. 4. QBO-like oscillations obtained for (a) and (b) the control experiment with model top at 40 km, and for (c) and (d) the low-level nudging experiment from the surface to 10 km, taken from Bui *et al.* (2017). (a) and (c) Zonal mean zonal wind (red and blue colors shaded with a black contour line for 0; unit is ms^{-1}) and its vertical shear (red and blue dashed contour lines; $\times 10^{-3}\mathrm{s}^{-1}$). (b) and (d) Domain-averaged 1-day precipitation (green bars) and 15-day running mean smoothed precipitation (gray line with circles). Smoothed precipitation over the 90th (under 10th) percentiles is colored red (blue). The red (blue) rectangles indicate time windows for heavy (light) precipitation conditions. Plus (minus) sign in each box in (b) and (d) indicates that it was used for composite analysis in the original paper.

oscillation is weakened by nudging the low-level mean zonal wind toward zero (not shown here), but appears again in the cases with the low-level nudging in the middle and lower troposphere (Figs. 4c, d). There is a negative correlation between the precipitation intensity and the magnitude of vertical shear of the mean zonal wind in the upper troposphere. These results suggest that the QBO-like oscillation modulates the moist convection via two mechanisms related to the vertical shear. Large values of the shear near the surface enhance the longevity and intensity of the moist convective systems by separating the updraft and downdraft (c.f., Liu and Moncrieff 2001). On the contrary, large values of the vertical shear near the cloud top tend to disrupt the convective structure and lead to weakening of moist convection (c.f., Collimore *et al.* 2003), although this mechanism seems to be secondary to the surface shear effect.

4. Vertical Momentum Transports with Moist Convection and Gravity Waves

Nishimoto *et al.* (2016) analyzed details of the momentum budget variation to further study the self-sustained QBO-like oscillation both in the stratosphere and troposphere in a radiative-moist convective quasi-equilibrium state in the present minimal model of the ST coupled system.

The vertical flux of horizontal momentum is separated into three contributions from convective momentum transport (CMT) and upward and downward gravity wave momentum transports (GWMTs). The mean-zonal-flow driving is entirely due to convective momentum transport plus upward/downward-propagating gravity waves generated from the convective events in the present minimal model framework. The time–height variations of each contribution are evaluated quantitatively as shown in Fig. 5. This method is an extension of that introduced by Shaw and Lane (2013) based on the linear theory of gravity waves that separates upward- and non-upward-propagating contributions. Here the phase speed spectra of the total cloud mixing ratio are used to identify the CMT contribution.

The acceleration of the zonal mean zonal wind in the stratosphere occurs for a limited time interval in a confined layer around zero-wind lines as shown in Fig. 3, and is basically due to the convergence of the upward GWMT, consistent with the picture of wave–mean flow interaction theory of the QBO (Figs. 5a, d, g). Switching of the zonal mean zonal wind at the bottom of the Rayleigh damping layer (~30 km) takes place in association with the sudden commencement of large upward GWMTs, which are generated near the tropopause. The generation process of gravity waves that cause the large upward GWMTs near the tropopause is still unclear and needs further diagnostic studies. The CMT and downward GWMT are confined to the troposphere, and the former is the most dominant contribution to the acceleration

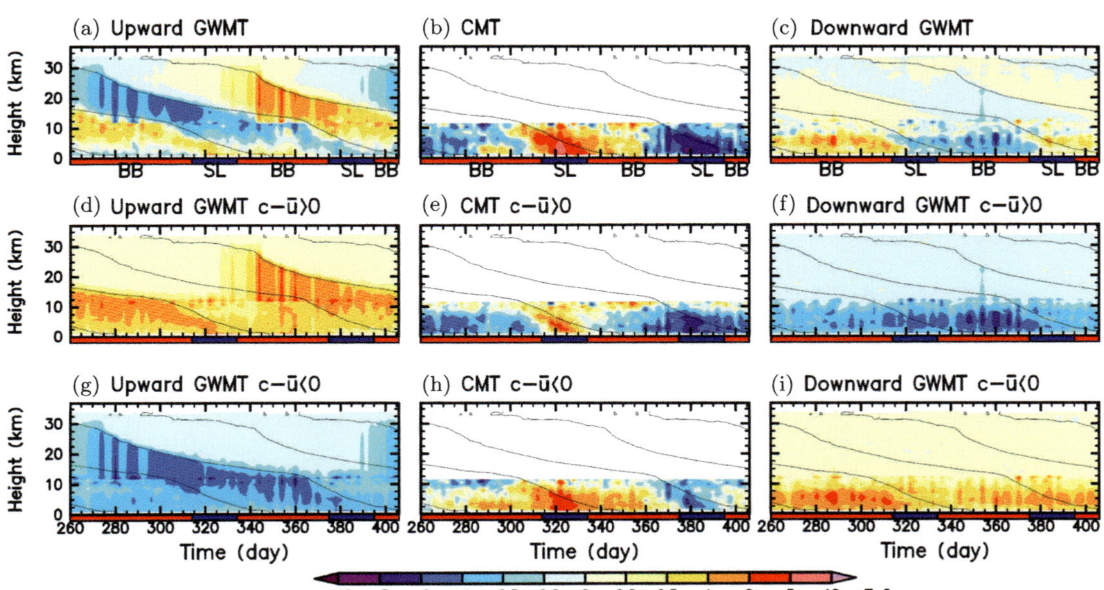

Fig. 5. Time–height sections of momentum transports for (left) upward GWMT, (center) CMT, and (right) downward GWMT, taken from Nishimoto *et al.* (2016). (top) Total value and values integrated over (middle) positive ($c > U$), and (bottom) negative ($c < U$) intrinsic phase speeds. The zero-wind lines are overlaid with black contours. Red and blue bars at the bottom of each plot denote BB- and SL-type periods, respectively.

of the zonal mean zonal wind in the troposphere (Figs. 5b, e, h). The QBO-like oscillations of the mean zonal wind modulate the organization of moist convective systems with the alternate appearance of SL-type and BB-type precipitation patterns through the variation of vertical shear of the mean zonal wind as described in the previous section. According to the modulation of tilted convective systems, the spectral features of every momentum transport also vary periodically.

5. Concluding Remarks

The momentum budget analysis in the present minimal model framework (Nishimoto *et al.* 2016) is a step forward in our understanding of the dynamical coupling processes in the highly idealized two-dimensional model associated with the QBO-like oscillations including tropospheric moist convective systems. The method introduced here can be used to quantitatively evaluate the space–time variations of momentum transport processes in a hierarchy of numerical models that include both contributions of convection and gravity waves of upward- and downward-propagating components.

In this summary paper, we focused on our numerical model studies with a two-dimensional minimal model framework of a non-hydrostatic regional-scale cloud-system-resolving model (Yoden *et al.* 2014; Nishimoto *et al.* 2016; Bui *et al.* 2017). Two-dimensionality in a limited computational domain is the most idealized assumption to investigate the fundamental dynamics of self-sustained variations of organized moist convective systems by securing enough computational time for studying radiative-convective quasi-equilibrium states in wide parameter sweep experiments. The only possible precipitation organization that can be realized in this two-dimensional framework is a SL-type or BB-type pattern. Thus, we have made a three-dimensional minimal model run with a rectangular domain of 640 km × 160 km for a 730-day period and obtained an oscillation with a period of about 300 days in both the domain-averaged zonal and meridional winds in the stratosphere (Bui *et al.* 2019). There is a synchronization of the mean zonal and meridional winds which features a counterclockwise rotation with time of a counterclockwise spiral of wind vectors with height in hodograph. There are modulations of domain-averaged low-pass-filtered tropospheric temperature anomaly and precipitation with irregular periods of about 100 days in which heavy precipitation is associated with positive temperature anomaly. There are indications of more independent variations in the generation process of gravity waves associated with moist convection systems in the three-dimensional model.

In association with the rapid growth in computing powers over the last decade or so, there has been an increase of numerical model studies on self-aggregation of deep moist convection in idealized cloud-resolving simulations (e.g., Bretherton *et al.* 2005; Muller and Held 2012; Wing and Emanuel 2014; Wing *et al.* 2017). However, several differences still exist between these studies of self-aggregation of moist convection and our recent three-dimensional minimal model results with QBO-like oscillations. Those self-aggregation models have very limited representation of the stratosphere and a very limited time integration period to obtain QBO-like oscillations, whereas our three-dimensional model has coarse horizontal resolution of 5 km. In the future, the radiative-convective equilibrium model intercomparison project (RCEMIP; Wing *et al.* 2018) could include a series of experiments that investigate the QBO-like oscillation obtained as a result of ST dynamical coupling in the tropics through organizations of moist convective systems.

Acknowledgments

We thank Chih-Pei Chang, the chief editor of this volume, for giving us this opportunity to present and write this summary report of our recent research results. This work was supported by JSPS KAKENHI Grants JP24224011 and JP17H01159, and JSPS Core-to-Core Program B (Asia-Africa Science Platforms) for FY2015–2017 "International research collaborations and networking on extreme weather in changing climate in the Maritime Continent".

References

Albers, J. R., G. N. Kiladis, T. Birner, and J. Dias, 2016: Tropical upper-tropospheric potential vorticity intrusions during sudden stratospheric warmings. *J. Atmos. Sci.*, **73**, 2361–2384.

Anstey, J., S. Yoden, M. Geller, S. Osprey, K. Hamilton, and N. Butchart, 2018: Report on the Joint SPARC Dynamics and Observations Workshop: SATIO-TCS, FISAPS and QBOi, Kyoto, Japan. *SPARC Newsletter*, **50**, 19–25.

Bal, S., S. Schimanke, T. Spangehl, and U. Cubasch, 2017: Variable influence on the equatorial troposphere associated with SSW using ERA-Interim. *J. Earth Syst. Sci.*, **126**, 19, doi: 10.1007/s12040-017-0802-6.

Baldwin, M. P. and T. J. Dunkerton, 2001: Stratospheric harbingers of anomalous weather regimes. *Science*, **294**, 581–584.

Baldwin and co-auters, 2018: 100 years of progress in understanding the stratosphere and mesosphere. *Meteorol. Monographs*, **59**, 27.1–27.62. doi: https://doi.org/10.1175/AMSMONOGRAPHS-D-19-0003.1.

Bretherton, C. S., P. N. Blossey, and M. Khairoutdinov, 2005: An energy-balance analysis of deep convective self-aggregation above uniform SST. *J. Atmos. Sci.*, **62**, 4273–4292.

Bui, H.-H., E. Nishimoto, and S. Yoden, 2017: Downward influence of QBO-like oscillation on moist convection in a two-dimensional minimal model framework. *J. Atmos. Sci.*, **74**, 3635–3655.

Bui, H.-H., S. Yoden, and E. Nishimoto, 2019: QBO-like oscillations in a three-dimensional minimal model framework of the stratosphere–troposphere coupled system. *SOLA*, **15**, 62–67, doi:10.2151/sola.2019-013.

Collimore, C. C., D. W. Martin, M. H. Hitchman, A. Huesmannet, and D. E. Waliser, 2003: On the relationship between the QBO and tropical deep convection. *J. Clim.*, **16**, 2552–2568.

Eguchi, N. and K. Kodera, 2010: Impacts of stratospheric sudden warming event on tropical clouds and moisture fields in the TTL: A case study. *SOLA*, **6**, doi: 10.2151/sola.2010-035.

Emanuel, K., S. Solomon, D. Folini, S. Davis, and C. C. Agnazzo, 2013: Influence of tropical tropopause layer cooling on Atlantic hurricane activity. *J. Clim.*, **26**, 2288–2301.

Geller, M., P. Haynes, S.-W. Son, and S. Yoden, 2017: Report on the International Workshop on Stratosphere–Troposphere Dynamical Coupling in the Tropics 22–24 October 2015, Kyoto, Japan. *SPARC Newsletter*, **48**, 17–21.

Held, I. M., R. S. Hemler, and V. Ramaswamy, 1993: Radiative-convective equilibrium with explicit two-dimensional moist convection. *J. Atmos. Sci.*, **50**, 3909–3927.

Hitchcock, P., P. H. Haynes, M. H. Hitchman, T.-Y. Koh, T. Sakazaki, and S. Yoden, 2019: Report on the SATIO-TCS side meeting and related discussions at the 6th SPARC General Assembly 2018. *SPARC Newsletter*, **52**, 24–26.

Hood, L. L., 2017: QBO/solar modulation of the boreal winter Madden Julian oscillation: A prediction for the coming solar minimum. *Geophys. Res. Lett.*, **44**, 3849–3857.

Kodera, K., B. M. Funatsu, C. Claud, and N. Eguchi, 2015: The role of convective overshooting clouds in tropical stratosphere-troposphere dynamical coupling. *Atmos. Chem. Phys.*, **15**, 6767–6774.

Lee, J. C. K. and N. P. Klingaman, 2018: The effect of the quasi-biennial oscillation on the Madden–Julian oscillation in the Met Office Unified Model Global Ocean Mixed Layer configuration. *Atmos Sci Lett.*, **19**, e816. https://doi.org/10.1002/asl.816.

Liess, S. and M. A. Geller, 2012: On the relationship between QBO and distribution of tropical deep convection. *J. Geophys. Res.*, **117**, D03108, doi:10.1029/2011JD016317.

Liu, C. and M. W. Moncrieff, 2001: Cumulus ensembles in shear: Implications for parameterization. *J. Atmos. Sci.*, **58**, 2832–2842.

Marshall, A. G., H. H. Hendon, S.-W. Son, and Y. Lim, 2017: Impact of the quasi-biennial oscillation on predictability of the Madden–Julian oscillation. *Clim. Dynam.*, **49**, 1365–1377.

Muller, C. J. and I. M. Held, 2012: Detailed investigation of the self-aggregation of convection in cloud-resolving simulations. *J. Atmos. Sci.*, **69**, 2551–2565.

Nie, J. and A. H. Sobel, 2015: Responses of tropical deep convection to the QBO: Cloud-resolving simulations. *J. Atmos. Sci.*, **72**, 3625–3638.

Nishimoto, E. and S. Yoden, 2017: Influence of the stratospheric quasi-biennial oscillation on the Madden–Julian oscillation during austral summer. *J. Atmos. Sci.*, **74**, 1105–1125.

Nishimoto, E., S. Yoden, and H.-H. Bui, 2016: Vertical momentum transports associated with moist convection and gravity waves in a minimal model of QBO-like oscillation. *J. Atmos. Sci.*, **73**, 2935–2957.

Osprey, S., M. Geller, and S. Yoden, 2018: The stratosphere and its role in tropical teleconnections. *Eos*, **99**, https://doi.org/10.1029/2018EO097387.

Shaw, T. A. and T. P. Lane, 2013: Toward an understanding of vertical momentum transports in cloud-system-resolving model simulations of multiscale tropical convection. *J. Atmos. Sci.*, **70**, 3231–3247.

Shepherd, T. G., 2002: Issues in stratosphere–troposphere coupling. *J. Met. Soc. Jpn.*, **80-4B**, 769–792.

Wing, A. A. and K. A. Emanuel, 2014: Physical mechanisms controlling self-aggregation of convection in idealized numerical modeling simulations. *J. Adv. Model Earth Syst.*, **6**, 59–74.

Wing, A. A., K. Emanuel, C. E. Holloway, and C. Muller, 2017: Convective self-aggregation in numerical simulations: A review. *Surv. Geophys.*, **38**, 1173–1197.

Wing A. A., K. A. Reed, M. Satoh, B. Stevens, S. Bony, and T. Ohno, 2018: Radiative-convective equilibrium model intercomparison project. Geosci. *Model Dev.*, **11**, 793–813.

Yoden, S., H.-H. Bui, and E. Nishimoto, 2014: A minimal model of QBO-like oscillation in a stratosphere–troposphere coupled system under a radiative-moist convective quasi-equilibrium state. *SOLA*, **10**, 112–116.

Yoden, S., M. Taguchi, and Y. Naito, 2002: Numerical studies on time variations of the troposphere–stratosphere coupled system. *J. Met. Soc. Jpn.*, **80-4B**, 811–830.

Yoo, C. and S.-W. Son, 2016: Modulation of the boreal wintertime Madden–Julian oscillation by the stratospheric quasibiennial oscillation. *Geophys. Res. Lett.*, **43**(3), 1392–1398.

Zhang, L., K. B. Karnauskas, J. B. Weiss, and L. M. Polvani, 2018: Observational evidence of the downstream impact on tropical rainfall from stratospheric Kelvin waves. *Clim. Dynam.*, **50**, 3775–3782.

Chapter 18

Extreme Precipitation from Tropical Cyclones

Michael M. Bell
Department of Atmospheric Science, Colorado State University, Fort Collins, Colorado, USA
mmbell@colostate.edu

Tropical cyclones often produce large amount of rainfall over a number of days, causing large disasters that can be more damaging than other types of extreme rainfall that last over a number of hours. This paper reviews the multiple processes that are involved in the production of tropical cyclone rainfall, including multiple scale processes ranging from large to microphysical, and interactions with terrain. A field experiment to improve scientific understanding and forecast is described in the conclusion section.

1. Introduction

Tropical cyclones (TCs) are some of the world's most prodigious rainmakers. The 24-h world record for precipitation was set during Tropical Storm Denise when 1,825 mm of rain fell over Reunion Island in 1966, and a staggering 5,678 mm of rain fell over Reunion Island during a 10-day period from Cyclone Hyacinthe in 1980 (Rogers *et al.* 2009). A 4-day accumulation of up to 2,965 mm fell over Taiwan island during Typhoon Morakot in 2009, making it one of the worst flooding disasters in the past 50 years and one of the most studied TCs (Fig. 1, Wu 2013). More recently, significant flooding disasters in Japan in 2015 associated with Tropical Storm Etau produced 668-mm peak totals over a 2-day period, and in the US in 2017 Hurricane Harvey produced 1,538 mm over an 8-day period. Significant property damage and loss of life has resulted from extreme TC precipitation, with over 25% of direct TC fatalities in the US caused by freshwater flooding (Rappaport 2014). While some TC rainfall can be beneficial (Rogers *et al.* 2009), extreme events that produce too much rain remain an important societal problem, highlighting the continued need for basic and applied research for accurate quantitative precipitation forecasts (QPFs). This paper reviews recent studies in the area of extreme precipitation from TCs, highlighting progress and future challenges.

2. Ingredients for Extreme Rainfall

At a basic level, extreme rainfall results from high-intensity rain rates, long-duration events, or a combination of these two factors. This can be succinctly stated as "the heaviest precipitation occurs where the rainfall rate is the highest for the longest time" (attributed to C.F. Chappell), and can be described mathematically as $R = I \times D$, where R is rainfall accumulation, I is rainfall intensity, and D is duration. Doswell *et al.* (1996) developed an ingredient-based framework that provides a physical basis for extreme rainfall, but does not depend on the specifics of the synoptic or mesoscale conditions. In this framework, I and D can be considered as two axes of a rainfall accumulation phase space, with the presence of various ingredients

The Multiscale Global Monsoon System
Edited by C. P. Chang et al.

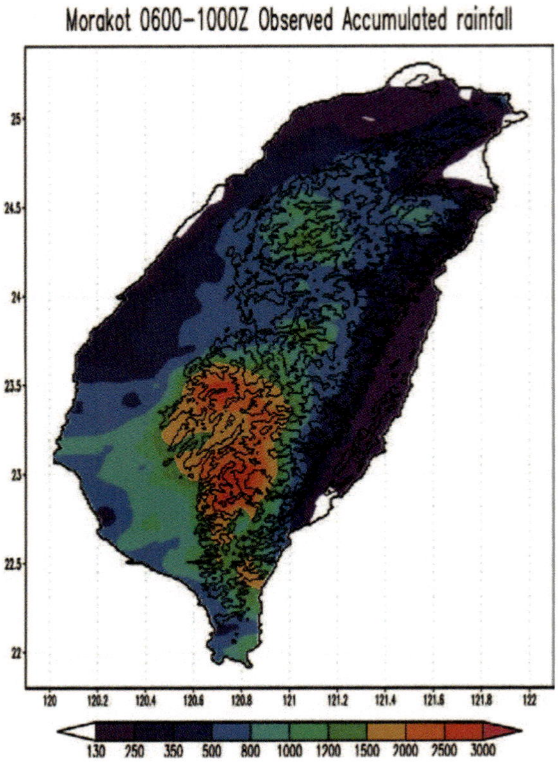

Fig. 1. The terrain of Taiwan (solid contour line) and the accumulated rainfall (mm; shaded) of Typhoon Morakot observed during the four days from 0000 UTC 6 August to 0000 UTC 10 August 2009. The solid contour line marks the height of 1,000 m of the topography. The x and y axes are the longitude (°E) and latitude (°N), respectively (*source*: T.-H. Yen and colleagues, adapted from Wu 2013).

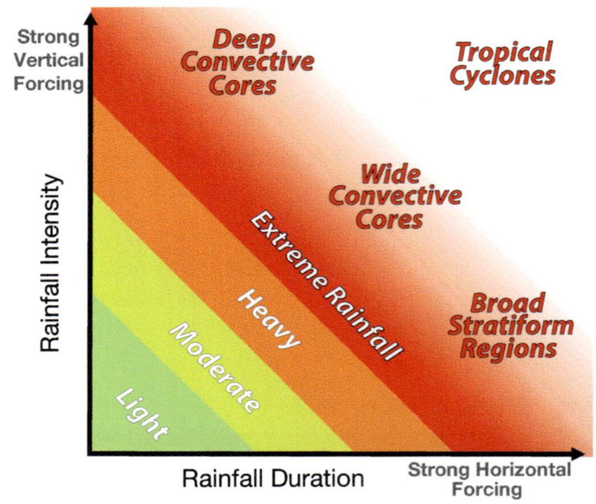

Fig. 2. Rainfall accumulation phase space with axes of intensity and duration. Global weather phenomena capable of producing extreme rainfall denoted in the upper right of the phase space.

producing different types of precipitation events (Fig. 2). The vertical axis of Fig. 2, representing I, is proportional to Ewq, where E is precipitation efficiency, w is vertical velocity, and q is water vapor mixing ratio. The precipitation efficiency E is the ratio of rainwater output to water vapor input, which is determined by the relative humidity and microphysical processes. The horizontal axis of Fig. 2, representing D, is proportional to L/C, where L is a system length scale and C is the system speed in the direction of the length scale. The atmospheric *ingredients* are then the components that contribute to I or D and combine nonlinearly to produce a spectrum of extreme precipitation events, and the *processes* are the multiscale interactions that focus them and modulate their magnitude. In the upper-left corner of Fig. 2 is the limiting case of high-intensity events in which D is not a significant factor in the accumulation. These events can be characterized by localized intense rain rates over periods of a few hours that can cause flash flooding (e.g., Caracena *et al.* 1979). In the lower-right corner of Fig. 2 is the limiting case of long-duration events in which I is not a significant factor in the accumulation. These events can be characterized by moderate rain rates over periods of many hours, days, or even weeks that can lead to river flooding (e.g., Houze *et al.* 2011, 2017a). Primary ingredients in this limiting case are larger, slow-moving weather systems with strong horizontal forcing that continually replenishes the water vapor removed by precipitation.

The most complex and dangerous part of this rainfall phase space is in the middle and upper-right corners of Fig. 2, where high-intensity events can occur over a long duration. Training and back-building convective cells from organized deep convection (Schumacher and Johnson

2005; Rasmussen *et al.* 2014), strongly forced orographic precipitation (Lin *et al.* 2001), and tropical cyclones (Rogers *et al.* 2009; Cheung *et al.* 2018) cannot only produce some of the highest rain accumulations but are also the most difficult to understand and predict. These more complex scenarios can be understood as the result of a combination of the ingredients found in the limiting cases, thereby allowing insight into the processes that produce extreme rainfall across different spatial and temporal scales. The conceptual framework can be applied universally in order to improve our basic understanding of extreme rainfall and its predictability across multiple regions of the world and does not rely on fixed thresholds or weather patterns that are specific to the regional environment in which the weather occurs.

Four precipitation systems that meet a global definition of "extreme" are denoted in the extreme part of the rainfall phase space in Fig. 2: TCs, deep convective cores (DCCs), wide convective cores (WCCs), and broad stratiform regions (BSRs). The latter three systems are defined following Houze *et al.* (2007, 2015, hereafter H15) through an extended analysis of radar echoes from the Tropical Rainfall Measurement Mission Precipitation Radar (TRMM PR). The criteria for DCC, WCC, and BSR thresholds have been refined and developed over many years of research on TRMM PR data (Rasmussen and Houze 2011; H15; Rasmussen *et al.* 2016), and have also been applied to ground-based radar data to describe the evolution of convection over the course of multi-day rain events (Zuluaga and Houze 2013). DCCs are defined as contiguous three-dimensional convective echo objects that exceed a moderate threshold intensity of 30 dBZ above 8 km, and are associated with young and vigorous convection. Stronger thresholds set a more rigorous definition that limits DCCs to the tallest continental convection on Earth (H15) and are not necessarily those that produce the most rainfall (Hamada *et al.* 2015). Moderate DCCs are representative of systems in the upper-left corner of the rainfall phase space that produce high-intensity rainfall. A WCC is a spatially organized convection that has grown upscale and is defined by a contiguous 30-dBZ echo extending over 800 km^2 in a horizontal area. WCCs are representative of events near the middle of the rainfall phase space, containing a mix of convective and stratiform elements associated with maturing MCSs. Extreme rainfall events from WCCs include training and back-building convective cells that produce repeated heavy rain over a long duration (Rasmussen *et al.* 2014). BSRs are defined as contiguous stratiform echo objects exceeding $30,000\,\text{km}^2$ in the horizontal dimension. They are representative of events in the lower-right corner of the phase space that are large, mature systems with widespread moderate rain over a long duration, and are often associated with decaying organized convection or slow-moving frontal systems (H15). TCs are representative of events in the upper-right corner of the phase space and can produce some of the most extreme rainfall totals in the world.

The East Asian monsoon provides the necessary ingredients to produce numerous TCs, DCCs, WCCs, and BSRs in the western North Pacific region. In mid-May the quasi-stationary Meiyu front becomes established over the northern South China Sea and is associated with the generation of many MCSs (Chen *et al.* 2000). The "front" is largely a shear and moisture boundary with a weak temperature gradient near Taiwan in late May and early June, and acts as a focusing mechanism for heavy rainfall events composed of DCCs, WCCs, and BSRs (Ding and Chan 2005; Ninomiya and Shibagaki 2007; Sampe and Xie 2010). The Meiyu front becomes more baroclinic as it moves north reaching Korea in July, where it is referred to as the Chagma front. Extreme rainfall in the subtropics is caused by TCs and diurnally forced DCCs and WCCs.

3. Synoptic and Mesoscale Processes

Large-scale environmental wind and moisture variabilities play a primary role in determining the location of TC precipitation. To the first order, extreme precipitation is driven by the TC track and motion, as these determine the local circulation and moisture availability through interactions with the larger-scale flow and terrain. Small variations in TC location and speed can have a large influence on the resulting precipitation distribution (Xie and Zhang 2012; Elsberry *et al.* 2013). The direct impact of the inner core of a landfalling TC can produce heavy rainfall, but indirect interactions with the large-scale flow patterns can often produce even more rainfall (Gao *et al.* 2009). In east Asia, the southwest monsoon is an important contributor to the environmental influences on TC rainfall by providing the ingredients needed for extreme precipitation and affecting the steering flow. TCs can produce extreme precipitation well away from the center by enhancing the strong moisture transport provided by the southwest monsoonal flow (Ge *et al.* 2010; Wu 2013). The multiscale interaction is two way, with convective and mesoscale features in TCs influencing the larger-scale environment, including development of frontal boundaries and associated precipitation during extratropical transition (Atallah *et al.* 2007).

Vertical wind shear is a primary contributor to the organization of tropical convection and has one of the largest influences on the TC precipitation distribution (Chen *et al.* 2006). In weak vertical wind shear, eyewall and rainband precipitations are more symmetric around the center, but they become increasingly concentrated downshear and left of shear as the shear increases (Hence and Houze 2012). In the case of Morakot (2009), moderate to strong northeasterly vertical wind shear was an important factor in producing a persistent azimuthal wavenumber-1 asymmetry in the precipitation that was modulated by higher-order asymmetries associated with vortex Rossby waves (Hall *et al.* 2013). Horizontal shearing flows such as a deformation field can also affect the precipitation, resulting in potential vorticity redistribution and low-level isentropic lift (Deng *et al.* 2017). Other complex interactions with mesoscale jets and vertical shear can lead to conditions favorable for intense, upright convection (Tang *et al.* 2014, 2018) or convective redevelopment and backbuilding (Schumacher and Johnson 2005; Wang *et al.* 2015) that can lead to heavy rainfall accumulations in ways similar to non-TC mesoscale convective systems.

For high-intensity rainfall, strong vertical moisture flux and high precipitation efficiency are required. In a moisture-rich environment there is frequently sufficient convective available potential energy (CAPE) for buoyant, saturated ascent on the cloud scale, but the conditions for strong organized synoptic or mesoscale ascent sufficient to produce extreme rainfall are met much less frequently. A key result from the previous field campaigns of RAINEX (Houze *et al.* 2009) and SoWMEX/TiMREX (Lai *et al.* 2011; Davis and Lee 2012; Xu *et al.* 2012) is that only a weak cold pool or density gradient is required to lift very moist, highly unstable air to the level of free convection (Crook and Moncrieff 1988). Heavy rainfall along a low-level boundary produced by evaporation from prior precipitation, sea breeze convergence, or SST gradients (Tu *et al.* 2014; Toy and Johnson 2014) is often associated with veering winds indicating warm air advection and isentropic ascent (Schumacher and Johnson 2005; Callaghan and Power 2016). In the tropics and subtropics a thermal wind relationship can still apply due to quasi-gradient rather than quasi-geostrophic balance (Tory 2014), suggesting that the mesoscale ascent in these scenarios can be understood in terms of interactions between vertical wind shear (VWS) and potential vorticity (PV) (Hoskins *et al.* 2003), but with added forcing from conditional instability.

In a mature TC, VWS tilts a PV tower downshear left due to differential advection and rotation, inducing a vertical perturbation pressure gradient, asymmetric secondary circulation, and distortion of the isentropes through adiabatic cooling and heating (Foerster 2016). Air traveling cyclonically around the TC in both the eyewall and the primary rainband is lifted isentropically in the downshear-right quadrant, and parcels with some small conditional instability can initiate deep convection that matures and transitions to stratiform downshear left (Reasor *et al.* 2013; Hence and Houze 2012; Didlake and Houze 2009, 2013; DeHart *et al.* 2014; Foerster *et al.* 2014). A similar lifting mechanism can occur with a mesoscale convective vortex (MCV) as VWS interacts with cold air below a balanced mid-level PV anomaly (Raymond and Jiang 1990; Trier *et al.* 2000; Lai *et al.* 2011). Horizontal shearing and deformation can modulate the convection and rainfall through filamentation and redistribution of the PV (Rozoff *et al.* 2006; Deng *et al.* 2017). Low-level VWS interactions with cold pools can also modulate convective intensity in MCSs (Rotunno *et al.* 1988), and this mechanism may play an important role in determining the intensity of TC rainbands (Tang *et al.* 2014, 2018).

In contrast to the ingredients that determine the rainfall intensity axis of the phase space, the rainfall duration axis is determined instead by the size and speed of the precipitating system. Many long-duration flooding events are associated with large-scale integrated water vapor transport (IVT) through "atmospheric rivers" that can transport copious amounts of water vapor across long distances, sometimes for multiple days (Lavers and Villarini 2013; Mahoney *et al.* 2016). In East Asia, the summer monsoon provides this transport through southwest flow that brings deep tropical moisture to the subtropics (Tu *et al.* 2014, 2017).

TCs are a prime example of how both shear and moisture transports across a variety of spatial and temporal scales can produce extreme rainfall (Wu 2013; Elsberry *et al.* 2013; Cheung *et al.* 2018). Mesoscale interactions between the PV field, shear, IVT, and terrain can produce extreme precipitation well away from the center (Gao *et al.* 2009; Nguyen and Chen 2011; Wu 2013). Hurricane Harvey (2017) created a substantial flooding disaster in the US due to being nearly stationary for 8 days, and the peak accumulation of 1,538 mm occurred to the NE of the hurricane center associated with redeveloping rainbands forced by southwesterly VWS and warm, moist IVT from the Gulf of Mexico. Typhoon Morakot (2009) was one of the worst flooding disasters in Taiwan (Xie and Zhang 2012; Wu 2013) with a 4-day accumulation of up to 2,965 mm forced by northeasterly shear that enhanced the southwest monsoon flow interaction with Taiwan's CMR. Much of the flooding in Japan associated with Severe Tropical Storm Etau (2015) occurred well after the TC made landfall due in part to interactions with the distant circulation of Typhoon Kilo. A similar complexity of interactions can occur in MCSs or diurnally forced convection over the terrain that makes skilled mesoscale QPF from extreme events one of the most challenging and important forecast problems.

4. Microphysical Processes

The microphysics of precipitation also play an important role in extreme events. In the eyewall, warm rain collision and coalescence processes dominate, but ice processes such as melting graupel are comparable to the warm rain processes outside the eyewall (Yang *et al.* 2011). Recent modeling studies suggest that the precipitation efficiency (E) of TCs is from 45% to 80% over the open ocean, but has an even greater range near steep terrain, from below 30% on the leeward side to above 95% on the windward side (Huang *et al.* 2014; Xu *et al.* 2017). Research observations are required to better understand and constrain the wide range of observed and simulated E that both directly affect the rainfall intensity and indirectly affect PV generation, cold pool development, and storm morphology. Surface fluxes, horizontal transport, and variability of water vapor are important, with positive impacts shown in QPFs when assimilating radar velocity and reflectivity (Bao *et al.* 2017). However, significant sensitivity and uncertainties remain in the parameterization of microphysical processes in numerical weather prediction models (Hendricks *et al.* 2016; Brown *et al.* 2016, 2017), including sensitivities to spatial resolution and nonlinear feedbacks with the model dynamics.

It is well established that tropical rainfall can be broadly categorized into convective and stratiform components based on their dominant microphysical processes, which yield characteristic radar echoes and vertical divergence profiles (Houze 1997). Active convective precipitation is dominated by warm rain processes that operate on short timescales, such as collection and accretion, while decaying stratiform precipitation is dominated by slower ice processes such as vapor deposition and aggregation. However, this categorization does not adequately distinguish extreme rainfall from ordinary rainfall in a TC since both convective and stratiform components can contribute to either scenario. Some water budgets from models suggest that ice processes are of comparable relevance to warm rain processes in TC rainbands (Yang *et al.* 2011), but there are large variations with different model configurations (Hendricks *et al.* 2016). Polarimetric radar observations suggest that riming may be more important in outer rainbands compared to inner rainbands (Wu *et al.* 2018). Significant variability in drop size distributions can often be found within a single precipitation event, however, suggesting that different combinations of warm and cold cloud processes may alternately dominate (Zagrodnik *et al.* 2018).

5. Terrain Interactions

TC interactions with complex terrain can produce some of the most extreme precipitation events around the world. As noted in the introduction, several of the largest TC rainfall events have occurred over islands with steep terrain such as La Reunion, Taiwan, and the Hawaiian Islands (e.g., 1,321 mm by Hurricane Lane in 2018). A TC can provide the necessary ingredients for heavy orographic rainfall (Lin *et al.* 2001), greatly enhancing the potential for an extreme event. The direct lifting effect by the terrain is only one of many orographic effects on a TC (Houze 2012). While lifting by the terrain can directly enhance vertical motion, interactions with the terrain can be much more complex depending on the strength and direction of the wind flow, moisture distribution, diurnal radiation, and surface conditions (Houze 2012). Terrain can enhance or suppress vertical velocity, available water vapor, precipitation efficiency, and system speed (Lin *et al.* 2001; Smith *et al.* 2012; Xie and Zhang 2012; Huang *et al.* 2014; Nugent *et al.* 2014). The rainfall intensity can be strongly modulated by affecting microphysical processes (Rowe *et al.* 2011, 2012; DeHart and Houze 2017; Houze *et al.* 2017). Complex terrain can also affect the spatial distribution and duration of rainfall (Chou *et al.* 2011; Lin *et al.* 2016; Nugent and Rios-Berrios 2018). The microphysics can be strongly modulated by the flow over the terrain, resulting in changes to the dominant processes in upslope or downslope flow (DeHart and Houze 2017), as well as changes in the precipitation efficiency (Huang *et al.* 2014). In addition to modulating the precipitation intensity, topography can also affect the spatial distribution and duration by changing the TC track (Lin *et al.* 2016) and size of the eyewall (Chou *et al.* 2011). These changes in track and structure can also affect the TC intensity, leading to complex feedbacks in the system and nonlinear changes in the precipitation.

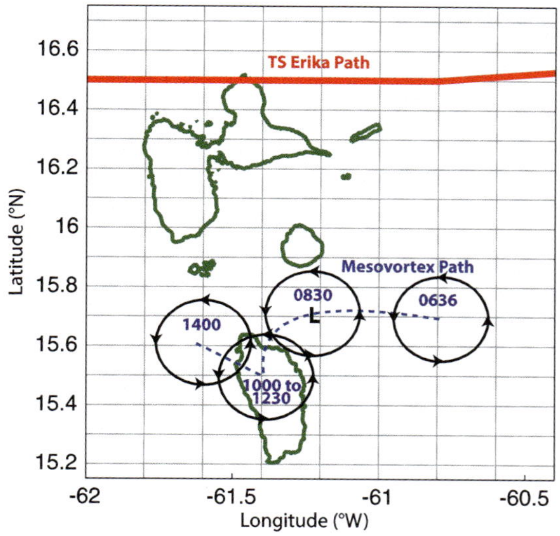

Fig. 3. TS Erika (2015) and the path of an embedded mesovortex that produced heavy rainfall over Dominica (adapted from Nugent and Rios-Berrios 2018).

Interactions with complex terrain are also very difficult to understand and forecast. An example of these complex interactions in Tropical Storm Erika (2015) from Nugent and Rios-Berrios (2018) is shown in Fig. 3. TS Erika moved through the Caribbean under the influence of northwesterly VWS, causing the precipitation to be concentrated to the south of the main TC center as it approached the island of Dominica. A mesovortex in the outer rainbands well away from the storm center interacted with the steep topography of the island and became locked to the terrain for several hours, enhancing both the rainfall intensity and duration, which resulted in record-breaking rainfall, flooding, landslides, and loss of life. The mesovortex propagated at a slower speed and different direction as compared to the TC center, suggesting that topographic effects played a dominant role in its location. The strong effects of the terrain on TCs make attribution of individual processes or factors difficult, and changes in TC track and motion may dominate long-term statistics of increasing TC rainfall in some regions due to the terrain interactions (Chang *et al.* 2013, Chu *et al.* 2013). Further research on complex terrain interactions is needed to resolve these uncertainties.

6. Summary and Future Directions

Recent studies and review articles (Chen *et al.* 2010; Wu 2013; Elsberry *et al.* 2013; Cheung *et al.* 2018) all emphasize the multiscale nature of TC precipitation, and how processes ranging from the synoptic to microscale can affect the total rainfall accumulation. A schematic that highlights several of the multiscale interactions described in the current review is shown in Fig. 4. At the synoptic scale, interactions with the monsoon can enhance long-range moisture transport and moist southwesterly flow and produce heavy rainfall far away from the TC center (Chen *et al.* 2010). Environmental vertical wind shear due to the large-scale flow is climatologically favored from the north in the western Pacific due to the monsoon (Wang and Wu 2016; Finocchio and Majumdar 2017), which

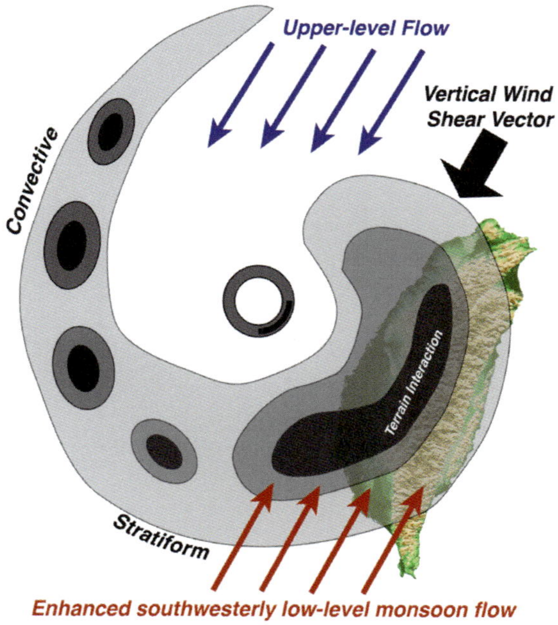

Fig. 4. Schematic of TC scenario producing extreme rainfall in the East Asian monsoon.

also strongly affects the precipitation distribution and vertical motion (Chen *et al.* 2006; Hence and Houze 2012). At the mesoscale, the TC track and vortex structure play a dominant role in determining the location and intensity of precipitation (Elsberry *et al.* 2013). At the microscale, highly efficient warm rain collection processes can rapidly convert condensed cloud water to high-intensity rainfall, with ice processes in stratiform precipitation also contributing to extreme rainfall totals over longer durations. Interactions with the terrain can modulate all of the above processes, changing the TC track, precipitation duration and intensity, and the microphysical processes (Wu 2013; Huang *et al.* 2014). While the schematic in Fig. 4 is idealized, it is representative of the conditions that led to record-breaking rainfall totals in Typhoon Morakot (2009).

The complexity of the multiscale interactions that can produce extreme rainfall makes QPFs from TCs very difficult. Given the overall higher rainfall accumulations of extreme events, high-resolution mesoscale models can obtain good QPF skill for lower rainfall thresholds (Wang 2015), but it remains challenging to deterministically predict the upper end of low-probability, high-impact events (Sukovich *et al.* 2014). Recent studies have begun to explore the limits of predictability in the context of TC track and intensity (e.g., Emanuel and Zhang 2016, 2017; Judt *et al.* 2016; Munsell *et al.* 2017), extreme rainfall from MCSs (Nielsen and Schumacher 2016), and the scales that contribute to forecast error growth (Zhang *et al.* 2007; Durran and Weyn 2016). Ensembles show promise for improved probabilistic forecasts of TC extreme precipitation (Zhang *et al.* 2010), including those that include multiple moist physics representations (Fang and Kuo 2013), but continued research and microphysical observations are necessary to reduce uncertainties in model physics parameterizations. Technological upgrades in polarimetric radar capability have led to recent studies that show great promise in helping to validate and improve parameterizations of warm rain processes (Brown *et al.* 2016, 2017; Feng and Bell 2018) and ice processes (Kalina *et al.* 2017; Didlake and Kumjian 2017) in NWP models.

To address these important scientific and forecast challenges, a joint international field experiment to study extreme rainfall will be conducted in spring/summer 2021 in tropical east Asia. With support from the US National Science Foundation, the Prediction of Rainfall Extremes Campaign In the Pacific (PRECIP) aims to improve our understanding of the multiscale dynamic, thermodynamic, and microphysical processes that produce extreme precipitation through new measurements with research polarimetric radars, disdrometers, and radiosondes. The primary objective of PRECIP is to simplify the complexity of multiscale interactions by identifying key ingredients and processes in the limiting cases of high-intensity and long-duration events in a moisture-rich environment. A comprehensive modeling effort will be undertaken before, during, and after the campaign to inform observational strategies as well as to identify model deficiencies to improve prediction of extreme rainfall. The campaign will be conducted with international partner projects during the period leading up to and including the Tokyo Olympics as part of an international effort to address the global problem of extreme weather and improve prediction of extreme rainfall from tropical cyclones.

References

Atallah, E., L. F. Bosart, and A. R. Aiyyer, 2007: Precipitation distribution associated with landfalling tropical cyclones over the eastern United States. *Mon. Wea. Rev.*, **135**, 2185–2206.

Bao, X., D. Wu, X. Lei, L. Ma, D. Wang, K. Zhao, and B. J. D. Jou, 2017: Improving the extreme rainfall forecast of Typhoon Morakot (2009) by assimilating radar data from Taiwan Island and mainland China. *J. Meteor. Res.*, **31**, 747–766.

Brown, B. R., M. M. Bell, and A. J. Frambach, 2016: Validation of simulated hurricane drop size distributions using polarimetric radar. *Geophys. Res. Lett.*, **43**, 910–917, doi:10.1002/2015GL067278.

Brown, B. R., M. M. Bell, and G. Thompson, 2017: Improvements to the snow melting process in a partially double moment microphysics parameterization. *J. Adv. Model. Sys.*, **9**, 1150–1166.

Callaghan, J. and S. B. Power, 2016: A vertical wind structure that leads to extreme rainfall and major flooding in southeast Australia. *J. South Hemisphere Earth Syst. Sci.*, **66**, 380–401.

Caracena, F., R. A. Maddox, L. R. Hoxit, and C. F. Chappell, 1979: Mesoanalysis of the Big Thompson storm. *Mon. Wea. Rev.*, **107**, 1–17.

Chang, C., Y. Yang, and H. Kuo, 2013: Large increasing trend of tropical cyclone rainfall in Taiwan and the roles of terrain. *J. Clim.*, **26**, 4138–4147.

Chang, W.-Y., W.-C. Lee, and Y.-C. Liou, 2015: The kinematic and microphysical characteristics and associated precipitation efficiency of subtropical convection during SoWMEX/TiMREX. *Mon. Wea. Rev.*, **143**, 317–340.

Chen, C.-S., Y.-L. Chen, C.-L. Liu, P.-L. Lin, and W.-C. Chen, 2007: Statistics of heavy rainfall occurrences in Taiwan. *Wea. Forecast.*, **22**, 981–1002.

Chen, S. J., W. Wang, K.-H. Lau, Q.-H. Zhang, and Y.-S. Chung, 2000: Mesoscale convective systems along the Meiyu front in a numerical model. *Meteorol. Atmos. Phys.*, **75**, 149–160.

Chen, S. S., J. A. Knaff, and F. D. Marks, 2006: Effects of vertical wind shear and storm motion on tropical cyclone rainfall asymmetries deduced from TRMM. *Mon. Wea. Rev.*, **134**, 3190–3208.

Chen, T.-C. and C.-C. Wu, 2016: The remote effect of Typhoon Megi (2010) on the heavy rainfall over northeastern Taiwan. *Mon. Wea. Rev.*, **144**, 3109–3131.

Cheung, K. Z., Yu, R. L. Elsberry, M. M. Bell, H. Jiang, T. C. Lee, K.-C. Lu, Y. Oikawa, L. Qi, R. F. Rogers, and K. Tsuboki, 2018: Recent advances in research and forecasting of tropical cyclone rainfall. *Trop. Cycl. Res. Rev.*, **7**(2), 106–127.

Chou, K. H., C. C. Wu, and Y. Wang, 2011: Eyewall evolution of typhoons crossing the Philippines and Taiwan: An observational study. *Terr. Atmos. Ocean. Sci.*, **22**, 535–548.

Chu, P.-S., D.-J. Chen, and P.-L. Lin, 2014: Trends in precipitation extremes during the typhoon season in Taiwan over the last 60 years. *Atmos. Sci. Lett.*, **15**, 37–43.

Crook, N. A. and M. W. Moncrieff, 1988: The effect of large-scale convergence on the generation and maintenance of deep moist convection. *J. Atmos. Sci.*, **45**, 3606–3624.

Davis, C. A. and W.-C. Lee, 2012: Mesoscale analysis of heavy rainfall episodes from SoWMEX/TiMREX. *J. Atmos. Sci.*, **69**, 521–537.

DeHart, J. C. and R. A. Houze, 2017: Orographic modification of precipitation processes in Hurricane Karl (2010). *Mon. Wea. Rev.*, **145**, 4171–4186.

Deng, D., N. E. Davidson, L. Hu, K. J. Tory, M. C. N. Hankinson, and S. Gao, 2017: Potential vorticity perspective of vortex structure changes of tropical cyclone Bilis (2006) during a heavy rain event following landfall. *Mon. Wea. Rev.*, **145**, 1875–1895.

Didlake, A. C. and R. A. Houze, 2009: Convective-scale downdrafts in the principal rainband of Hurricane Katrina (2005). *Mon. Wea. Rev.*, **137**, 3269–3293.

Didlake, A. C. and R. A. Houze, 2013: Dynamics of the stratiform sector of a tropical cyclone rainband. *J. Atmos. Sci.*, **70**, 1891–1911.

Didlake, A. C. and M. R. Kumjian, 2017: Examining polarimetric radar observations of bulk microphysical structures and their relation to vortex kinematics in Hurricane Arthur (2014). *Mon. Wea. Rev.*, **145**, 4521– 4541.

Ding, Y. and J. C. L. Chan, 2005: The East Asian summer monsoon: An overview. *Meteor. Atmos. Phys.*, **89**, 117–142.

Doswell, C. A., H. E. Brooks, and R. A. Maddox, 1996: Flash flood forecasting: An ingredients-based methodology. *Wea. Forecast.*, **11**, 560–581.

Durran, D. R. and J. A. Weyn, 2016: Thunderstorms do not get butterflies. *Bull. Amer. Meteor. Soc.*, **97**, 237– 243.

Elsberry, R. L., L. Chen, J. Davidson, R. Rogers, Y. Wang, and L. Wu, 2013: Advances in understanding and forecasting rapidly changing phenomena in tropical cyclones. *Trop. Cycl. Res. Rev.*, **2**, 13–24.

Emanuel, K. and F. Zhang, 2016: On the predictability and error sources of tropical cyclone intensity forecasts. *J. Atmos. Oceanic Technol.*, **73**, 3739–3747.

Emanuel, K. and F. Zhang, 2017: The role of inner-core moisture in tropical cyclone predictability and practical forecast skill. *J. Atmos. Sci.*, **74**, 2315–2324.

Fang, X. and Y.-H. Kuo, 2013: Improving ensemble-based quantitative precipitation forecasts for topography-enhanced typhoon heavy rainfall over Taiwan with a modified probability-matching technique. *Mon. Wea. Rev.*, **141**, 3908–3932.

Feng, Y.-C. and M. M. Bell, 2018: "Microphysical characteristics of an asymmetric eyewall in major Hurricane Harvey (2017)". *Geophys. Res. Lett.*, doi: 10.1029/2018gl080770.

Foerster, A. M. 2016: Radar-derived thermodynamic structure of a major hurricane in vertical wind shear. Ph.D. thesis, University of Hawaii, 101 pp.

Foerster, A. M., M. M. Bell, P. A. Harr, and S. C. Jones, 2014: Observations of the eyewall structure of Typhoon Sinlaku (2008) during the transformation stage of extratropical transition. *Mon. Wea. Rev.*, **142**, 3372–3392.

Galarneau, T. J., L. F. Bosart, and R. S. Schumacher, 2010: Predecessor rain events ahead of tropical cyclones. *Mon. Wea. Rev.*, **138**, 3272–3297.

Gao, S., Z. Meng, F. Zhang, and L. F. Bosart, 2009: Observational analysis of heavy rainfall mechanisms associated with severe tropical storm Bilis (2006) after its landfall. *Mon. Wea. Rev.*, **137**, 1881–1897.

Ge, X., T. Li, S. Zhang, and M. S. Peng, 2010: What causes the extremely heavy rainfall in Taiwan during Typhoon Morakot (2009)? *Atmos. Sci. Lett.*, **11**, 46–50.

Hall, J. D., M. Xue, L. Ran, and L. M. Leslie, 2013: High-resolution modeling of Typhoon Morakot (2009): Vortex Rossby waves and their role in extreme precipitation over Taiwan. *J. Atmos. Sci.*, **70**, 163–186.

Hamada, A., Y. N. Takayabu, C. Liu, and E. J. Zipser, 2015: Weak linkage between the heaviest rainfall and tallest storms. *Nature Comm.*, **6**, 6213 EP, doi:10.1038/ncomms7213.

Hence, D. A. and R. A. Houze, 2012: Vertical structure of tropical cyclone rainbands. *J. Atmos. Sci.*, **69**, 2644–2661.

Hendricks, E. A., Y. Jin, J. R. Moskaitis, J. D. Doyle, M. S. Peng, C.-C. Wu, and H.-C. Kuo, 2016: Numerical simulations of Typhoon Morakot (2009) using a multiply nested tropical cyclone prediction model. *Wea. Forecast.*, **31**, 627–645.

Hoskins, B., M. Pedder, and D. W. Jones, 2003: The omega equation and potential vorticity. *Quart. J. Roy. Meteor. Soc.*, **129**, 3277–3303.

Houze, R. A. 1997: Stratiform precipitation in regions of convection: A meteorological paradox? *Bull. Amer. Meteor. Soc.*, **78**, 2179–2196.

Houze, R. A. 2012: Orographic effects on precipitating clouds. *Rev. Geophys.*, **50**, RG1001, doi:doi:10.1029/ 2011RG000365.

Houze, R. A., W.-C. Lee, and M. M. Bell, 2009: Convective contribution to the genesis of Hurricane Ophelia (2005). *Mon. Wea. Rev.*, **137**, 2778–2800.

Houze, R. A., L. A. McMurdie, K. L. Rasmussen, A. Kumar, and M. M. Chaplin, 2017: Multiscale aspects of the storm producing the June 2013 flooding in Uttarakhand, India. *Mon. Wea. Rev.*, **145**, 4447–4466.

Houze, R. A., K. L. Rasmussen, S. Medina, S. R. Brodzik, and U. Romatschke, 2011: Anomalous atmospheric events leading to the summer 2010 floods in Pakistan. *Bull. Amer. Meteor. Soc.*, **92**, 291–298.

Houze, R. A., K. L. Rasmussen, M. D. Zuluaga, and S. R. Brodzik., 2015: The variable nature of convection in the tropics and subtropics: A legacy of 16 years of the Tropical Rainfall Measuring Mission (TRMM) satellite. *Rev. Geophys.*, **53**, doi:10.1002/2015RG000 488.

Houze, R. A., D. C. Wilton, and B. F. Smull, 2007: Monsoon convection in the Himalayan region as seen by the TRMM Precipitation Radar. *Quart. J. Roy. Meteor. Soc.*, **133**, 1389–1411.

Huang, H.-L., M.-J. Yang, and C.-H. Sui, 2014: Water budget and precipitation efficiency of Typhoon Morakot (2009). *J. Atmos. Sci.*, **71**, 112–129.

Judt, F., S. S. Chen, and J. Berner, 2016: Predictability of tropical cyclone intensity: Scale-dependent forecast error growth in high-resolution stochastic kinetic-energy backscatter ensembles. *Quart. J. Roy. Meteor. Soc.*, **142**, 43–57, doi:10.1002/qj.2626.

Kalina, E. A. *et al.*, 2017: The ice water paths of small and large ice species in Hurricanes Arthur (2014) and Irene (2011). *J. Appl. Meteor. Climatol.*, **56**, 1383–1404.

Lai, H.-W., C. A. Davis, and B. J.-D. Jou, 2011: A subtropical oceanic mesoscale convective vortex observed during SoWMEX/TiMREX. *Mon. Wea. Rev.*, **139**, 2367–2385.

Lavers, D. A. and G. Villarini, 2013: Atmospheric rivers and flooding over the central United States. *J. Clim.*, **26**, 7829–7836.

LeMone, M. A., E. J. Zipser, and S. B. Trier, 1998: The role of environmental shear and thermodynamic conditions in determining the structure and evolution of mesoscale convective systems during TOGA COARE. *J. Atmos. Sci.*, **55**, 3493–3518.

Lin, Y.-L., S.-H. Chen, and L. Liu, 2016: Orographic influence on basic flow and cyclone circulation and their impacts on track deflection of an idealized tropical cyclone. *J. Atmos. Sci.*, **73**, 3951–3974.

Lin, Y.-L., S. Chiao, T.-A. Wang, M. L. Kaplan, and R. P. Weglarz, 2001: Some common ingredients for heavy orographic rainfall. *Wea. Forecast.*, **16**, 633–660.

Mahoney, K. *et al.*, 2016: Understanding the role of atmospheric rivers in heavy precipitation in the southeast United States. *Mon. Wea. Rev.*, **144**, 1617–1632.

Munsell, E. B., F. Zhang, J. A. Sippel, S. A. Braun, and Y. Weng, 2017: Dynamics and predictability of the intensification of Hurricane Edouard (2014). *J. Atmos. Sci.*, **74**, 573–595.

Nguyen, H. V. and Y.-L. Chen, 2011: High-resolution initialization and simulations of Typhoon Morakot (2009). *Mon. Wea. Rev.*, **139**, 1463–1491.

Nielsen, E. R. and R. S. Schumacher, 2016: Using convection-allowing ensembles to understand the predictability of an extreme rainfall event. *Mon. Wea. Rev.*, **144**, 3651–3676.

Ninomiya, K. and Y. Shibagaki, 2007: Multi-scale features of the Meiyu-Baiu front and associated precipitation systems. *J. Meteor. Soc. Jpn.*, **85B**, 103–122.

Nugent, A. D. and R. Rios-Berrios, 2018: Factors leading to extreme precipitation on Dominica from tropical storm Erika (2015). *Mon. Wea. Rev.*, **146**, 525–541.

Nugent, A. D., R. B. Smith, and J. R. Minder, 2014: Wind speed control of tropical orographic convection. *J. Atmos. Sci.*, **71**, 2695–2712.

Rappaport, E. N., 2014: Fatalities in the United States from Atlantic tropical cyclones: New data and interpretation. *Bull. Amer. Meteor. Soc.*, **95**, 341–346.

Rasmussen, K. L., M. M. Chaplin, M. D. Zuluaga, and R. A. Houze, 2016: Contribution of extreme convective storms to rainfall in South America. *J. Hydrometeorol.*, **17**, 353–367.

Rasmussen, K. L. and R. A. Houze, 2011: Orogenic convection in South America as seen by the TRMM satellite. *Mon. Wea. Rev.*, **139**, 2399–2420.

Rasmussen, K. L., M. D. Zuluaga, and R. A. Houze, 2014: Severe convection and lightning in subtropical South America. *Geophys. Res. Lett.*, **41**, 7359–7366.

Raymond, D. J. and H. Jiang, 1990: A theory for long-lived mesoscale convective systems. *J. Atmos. Sci.*, **47**, 3067–3077.

Reasor, P. D., R. Rogers, and S. Lorsolo, 2013: Environmental flow impacts on tropical cyclone structure diagnosed from airborne Doppler radar composites. *Mon. Wea. Rev.*, **141**, 2949–2969.

Rogers, R., J. F. D. Marks, and T. Marchok, 2009: Tropical cyclone rainfall. In *Encyclopedia of Hydrological Sciences*, Ed., M. Anderson, John Wiley & sons, ltd., Chichester, UK.

Rotunno, R., J. B. Klemp, and M. L. Weisman, 1988: A theory for strong, long-lived squall lines. *J. Atmos. Sci.*, **45**, 463–485.

Rowe, A. K., S. A. Rutledge, and T. J. Lang, 2011: Investigation of microphysical processes occurring in isolated convection during NAME. *Mon. Wea. Rev.*, **139**, 424–443.

Rowe, A. K., S. A. Rutledge, and T. J. Lang, 2012: Investigation of microphysical processes occurring in organized convection during NAME. *Mon. Wea. Rev.*, **140**, 2168–2187.

Rozoff, C. M., W. H. Schubert, B. D. McNoldy, and J. P. Kossin, 2006: Rapid filamentation zones in intense tropical cyclones. *J. Atmos. Sci.*, **63**, 325–340.

Sampe, T. and S.-P. Xie, 2010: Large-scale dynamics of the Meiyu-Baiu rainband: Environmental forcing by the westerly jet. *J. Climate*, **23**, 113–134.

Schumacher, R. S. and R. H. Johnson, 2005: Organization and environmental properties of extreme-rain-producing mesoscale convective systems. *Mon. Wea. Rev.*, **133**, 961–976.

Smith, R. B. *et al.*, 2012: Orographic precipitation in the tropics: The Dominica Experiment. *Bull. Amer. Meteor. Soc.*, **93**, 1567–1579.

Sukovich, E. M., F. M. Ralph, F. E. Barthold, D. W. Reynolds, and D. R. Novak, 2014: Extreme quantitative precipitation forecast performance at the Weather Prediction Center from 2001 to 2011. *Wea. Forecasting*, **29**, 894–911.

Tang, X., W.-C. Lee, and M. Bell, 2014: A squall-line-like principal rainband in Typhoon Hagupit (2008) observed by airborne Doppler radar. *J. Atmos. Sci.*, **71**, 2733–2746.

Tang, X., W.-C. Lee, and M. Bell, 2018: Sub-rainband structure and dynamic characteristics in the principal rainband of Typhoon Hagupit (2008). *Mon. Wea. Rev.*, in press.

Tao, D. and F. Zhang, 2015: Effects of vertical wind shear on the predictability of tropical cyclones: Practical versus intrinsic limit. *J. Adv. Model. Earth Syst.*, **7**, 1534–1553.

Tory, K. J. 2014: The turning winds with height thermal advection rainfall diagnostic: Why does it work in the tropics? *Aus. Meteor. and Ocean. J.*, **64**, 231–238.

Toy, M. D. and R. H. Johnson, 2014: The influence of an SST front on a heavy rainfall event over coastal Taiwan during TiMREX. *J. Atmos Sci.*, **71**, 3223–3249.

Trier, S. B., C. A. Davis, and J. D. Tuttle, 2000: Long-lived mesoconvective vortices and their environment. Part I: Observations from the central United States during the 1998 warm season. *Mon. Wea. Rev.*, **128**, 3376–3395.

Tu, C.-C., Y.-L. Chen, C.-S. Chen, P.-L. Lin, and P.-H. Lin, 2014: A comparison of two heavy rainfall events during the Terrain-Influenced Monsoon Rainfall Experiment (TiMREX) 2008. *Mon. Wea. Rev.*, **142**, 2436–2463.

Tu, C.-C., Y.-L. Chen, S.-Y. Chen, Y.-H. Kuo, and P.-L. Lin, 2017: Impacts of including rain-evaporative cooling in the initial conditions on the prediction of a coastal heavy rainfall event during TiMREX. *Mon. Wea. Rev.*, **145**, 253–277.

Wang, C.-C. 2015: The more rain, the better the model performs — the dependency of quantitative precipitation forecast skill on rainfall amount for typhoons in Taiwan. *Mon. Wea. Rev.*, **143**, 1723–1748.

Wang, C. C., H. C. Kuo, R. H. Johnson, C. Y. Lee, S. Y. Huang, and Y. H. Chen, 2015: A numerical study of convection in rainbands of Typhoon Morakot (2009) with extreme rainfall: Roles of pressure perturbations with low-level wind maxima. *Atmos. Chem. Phys.*, **15**, 11097–11115.

Wu, C.-C. 2013: Typhoon Morakot: Key findings from the journal TAO for improving prediction of extreme rains at landfall. *Bull. Amer. Meteor. Soc.*, **94**, 155–160.

Wu, D., K. Zhao, M. R. Kumjian, X. Chen, H. Huang, M. Wang, A. C. Didlake, Y. Duan, and F. Zhang, 2018: Kinematics and Microphysics of Convection in the Outer Rainband of Typhoon Nida (2016) Revealed by Polarimetric Radar. *Mon. Wea. Rev.*, **146**, 2147–2159.

Xie, B. and F. Zhang, 2012: Impacts of typhoon track and island topography on the heavy rainfalls in Taiwan associated with Morakot (2009). *Mon. Wea. Rev.*, **140**, 3379–3394.

Xu, H., G. Zhai, and X. Li, 2017: Precipitation efficiency and water budget of Typhoon Fitow (2013): A particle trajectory study. *J. Hydrometeorol.*, **18**, 2331–2354.

Xu, W., E. J. Zipser, Y.-L. Chen, C. Liu, Y.-C. Liou, W.-C. Lee, and B. J.-D. Jou, 2012: An orography-associated extreme rainfall event during TiMREX: Initiation, storm evolution, and maintenance. *Mon. Wea. Rev.*, **140**, 2555–2574.

Yang, M.-J., S. A. Braun, and D.-S. Chen, 2011: Water budget of Typhoon Nari (2001). *Mon. Wea. Rev.*, **139**, 3809–3828.

Zagrodnik, J. P., L. A. McMurdie, and R. A. Houze, 2018: Stratiform precipitation processes in cyclones passing over a coastal mountain range. *J. Atmos. Sci.*, **76**, 1521–1546.

Zhang, F., N. Bei, R. Rotunno, C. Snyder, and C. C. Epifanio, 2007: Mesoscale predictability of moist baroclinic waves: Convection-permitting experiments and multistage error growth dynamics. *J. Atmos. Sci.*, **64**, 3579–3594.

Zhang, F., Y. Weng, Y.-H. Kuo, J. S. Whitaker, and B. Xie, 2010: Predicting Typhoon Morakot's catastrophic rainfall with a convection-permitting mesoscale ensemble system. *Wea. Forecast.*, **25**, 1816–1825.

Part III
Intraseasonal Oscillation and Forecast

Chapter 19

A Review of Contemporary MJO Theories

Da Yang*,†,**, Ángel F. Adames‡, Boualem Khouider§, Bin Wang¶ and Chidong Zhang||

*University of California, Davis, USA

†Lawrence Berkeley National Laboratory, USA

‡University of Michigan, Ann Abor, USA

§University of Victoria, Victoria, British Columbia, Canada

¶University of Hawaii, Manoa, USA

||NOAA/Pacific Marine Environmental Laboratory, Seattle, Washington, USA

**dyang@alumni.caltech.edu

The Madden–Julian oscillation (MJO) is a planetary-scale subseasonal rainfall pattern that initiates in the Indian Ocean and slowly propagates eastward at about 5 m s^{-1}. The reasons for its slow eastward propagation speed and large spatial scale have remained elusive. Our lack of understanding of the MJO ranks among the most important problems in tropical meteorology. Several theories have been proposed to explain the MJO dynamical and physical features. Here, four contemporary MJO theories are reviewed and compared. The four theories are the gravity wave theory, the trio-interaction theory, the moisture wave theory, and the skeleton theory. We describe the essential ingredients and assumptions of the four theories and discuss their success and limitations.

1. Introduction

The Madden–Julian oscillation (MJO) consists of a planetary-scale envelope of enhanced and suppressed rainfall activity and associated circulation. This rainfall envelope initiates over the Indian Ocean and slowly propagates eastward at about 5 m s^{-1}. The MJO has been referred to as the Storm King in the tropics (Hand 2015) due to its strong influence on both local and global weather as well as climate variabilities, including frequency of wildfires, flooding, tropical cyclones, and the strength of El Niño events (Zhang 2013). However, current weather models cannot accurately forecast the MJO and many climate models cannot reproduce its basic statistical behavior (Hung *et al.* 2013). Meanwhile, there is no universally accepted MJO theory.

In this chapter, we review and compare four contemporary theories of the MJO. Our focus will be on the processes that lead to the MJO's slow eastward propagation and select its planetary scale. The four models include the gravity wave model (Yang and Ingersoll 2013, 2014), the trio-interaction model (Wang *et al.* 2016), the moisture wave model (Adames and Kim 2016; Sobel and Maloney 2013), and the skeleton model (Majda and Stechmann 2009). This chapter includes four schematic diagrams to illustrate the key mechanisms proposed by these theories. We aim to provide some basic physical intuition of MJO dynamics.

We first overview potential key ingredients for the MJO (Section 2) and then discuss individual theories (recipes) that are constructed using different combinations of the ingredients (Sections 3–6).

The Multiscale Global Monsoon System
Edited by C. P. Chang et al.

2. Key Ingredients

The tropical atmosphere obeys the conservation of momentum, mass, energy, and tracers (e.g., water). A complete set of conservation laws and accurate physics should be able to reproduce the MJO as well as other phenomena. However, for understanding the essential physics, we would like to identify a minimum set of ingredients that are key to the MJO's formation, slow eastward propagation, and spatial-scale selection. The key ingredients may include equatorial waves, moist convection, cloud–radiative feedbacks, and boundary layer dynamics. These ingredients are identified based on theoretical considerations, observational, and modeling evidence.

Ingredient 1: *Equatorial waves.* Observations of intraseasonal variations in wind, temperature, and geopotential reveal that the MJO could be a wave-like propagating signal with rich, multiscale dynamical features (Hendon and Salby 1994; Zhang 2005). These observations suggest the importance of wave dynamics. Linear equatorial waves are among the most fundamental form of atmospheric flows in the tropics and should be considered as a part of our minimum ingredients. An analytical theory was developed to explain basic features of equatorial waves with the gravity and/or beta effect as their restoring forces (Matsuno 1966). This theory uses the conservation of momentum, mass, and energy to obtain linear free (uncoupled to convection) wave solutions.

In spite of its success in explaining many observed equatorial phenomena, none of these equatorial wave solutions match observed spectral characteristics of the MJO (Wheeler and Kiladis 1999). Thus, processes such as diabatic heating or conservation of moisture may be needed to explain the MJO.

Ingredient 2: *Moist convection.* The MJO is observed to be associated with intense, organized thunderstorms, indicating a fundamental role of moist convection in its dynamics (Madden and Julian 1971). Representing moist convection in MJO theories is, however, challenging. This challenge is largely because the spatial scale of thunderstorms is much smaller than the typical scale of the MJO flow. Thus, moist convection has to be parameterized based on an empirical physical understanding.

The four MJO models discussed here can be divided into two categories according to their convection parameterization: the gravity wave model (Section 3) and the large-scale models. The gravity wave model considers that the MJO is a large-scale envelope of small-scale short-lived gravity waves (a multiscale view). As there is no scale separation between individual convective storms and gravity waves, convection was treated as a triggered process: convective instability can be accumulated and then released when a certain threshold is reached (Ingredient 2.1) (Emanuel 1994; Yang and Ingersoll 2013). All of the other models consider that the MJO is a large-scale mode, so they can represent convection by emphasizing its effect on large-scale circulations (Ingredient 2.2). This is often referred to as a quasi-equilibrium (QE) convection scheme: convective instability is consumed once it is generated (Emanuel *et al.* 1994), and convective heating can at any moment be taken to be an instantaneous function of the thermodynamic state (Sobel and Maloney 2012). The details on how this instability is consumed vary between theories. Convection parameterization includes a *prognostic moisture equation* in the QE schemes presented in this paper. In particular, this moisture equation is the only prognostic equation in the moisture wave model.

Ingredient 3: *Radiative feedbacks.* Radiation might be important in maintaining the amplitude of the MJO. This is most evident in numerical simulations where the amplitude of the MJO is significantly reduced when longwave radiative cooling is homogenized horizontally (Andersen and Kuang 2012). Both the propagation speed and horizontal scale of the MJO, however, remained the same (Arnold and Randall 2015). This process is included in the trio-interaction model and (Section 4) the moisture wave model (Section 5).

Ingredient 4: *Boundary layer processes.* This includes both dynamics (Ingredient 4.1 Wang and Rui 1990) and thermal dynamics, such as the wind-induced surface heat exchange (Ingredient 4.2 WISHE) (Emanuel 1987; Neelin and Held 1987). It has been documented that there is boundary layer moisture convergence leading (to the east) the MJO convection center (Rui and Wang 1990). It is a major source of water vapor for MJO convection and may play a key role in its eastward propagation (Wang and Lee 2017). Boundary layer dynamics is explicit in the trio-interaction model (Section 4) and is parameterized in the moisture wave model (Section 5). The WISHE process has been included in the moisture wave model (Section 5).

3. The Gravity Wave Model

Yang and Ingersoll (2013, 2014) proposed that the MJO could be a large-scale envelope of small-scale, westward (W-) and eastward (E-) inertia–gravity (IG) waves that resonate with convective storms (I1 + I2.1). Based on this hypothesis, Yang and Ingersoll (2013, 2014) successfully simulated MJO-like variability using one-dimensional and two-dimensional shallow water models with triggered convection (the eastward moving large-scale envelope in Fig 1a; the intense spectral peak in Fig. 1b). In their models, convection is parameterized as a mass source to the upper troposphere and is triggered when the layer thickness of the upper troposphere is lower than a threshold (Fig. 1c and d). This can be interpreted as weak static stability favoring convection.

The zonal asymmetry between EIG and WIG leads to the eastward propagation of the MJO. As an envelope of EIG and WIG, the MJO would be a standing wave if EIG and WIG propagate at the same speed (the dashed line in Fig. 2a and b). However, EIG propagates faster than WIG due to the beta effect, and this asymmetry can lead to the eastward propagation of the MJO (the solid line in Fig. 2a and 2c):

$$c = (c_E - c_W)/2, \tag{1}$$

where c is the propagation speed of the MJO, and c_E and c_W are the propagation speeds of EIG and WIG, respectively. This scaling relation is tested against 60+ shallow water simulations, and the theoretically derived propagation speed fits simulation results (Fig. 2d).

The mean free path (MFP) of gravity waves determines the horizontal scale of the MJO in the Yang-Ingersoll model (Fig. 2e). The MFP is defined as the distance that the newly triggered gravity waves can travel before being interfered by the next convective storm. This distance increases with the gravity wave speed and decreases with the number density of convective storms (Fig. 2e–g). In the 1D model, a scaling relation is given by

$$k \sim 1/\lambda \sim \sqrt{S_C/c}, \tag{2}$$

where k is the MJO wavenumber, λ is the MFP, and S_C is the number density of convection. This scaling relation was obtained by dimensional analysis and verified against both one-dimensional and two-dimensional shallow water models (Fig. 2h).

The MJO is a form of convective self-aggregation in the gravity wave model, as its existence and development do not rely on any asymmetries of the background fields and boundary conditions (Bretherton *et al.* 2005; Emanuel *et al.* 2014; Held *et al.* 1993; Yang 2018b, 2018a). The aggregation mechanism is via resonance between convection and gravity waves. One would think that convection, which is a small-scale, short-lived event, has difficulties in generating a large-scale, long-lasting response (MJO). This would be true if the model is forced at short timescales by convection, especially

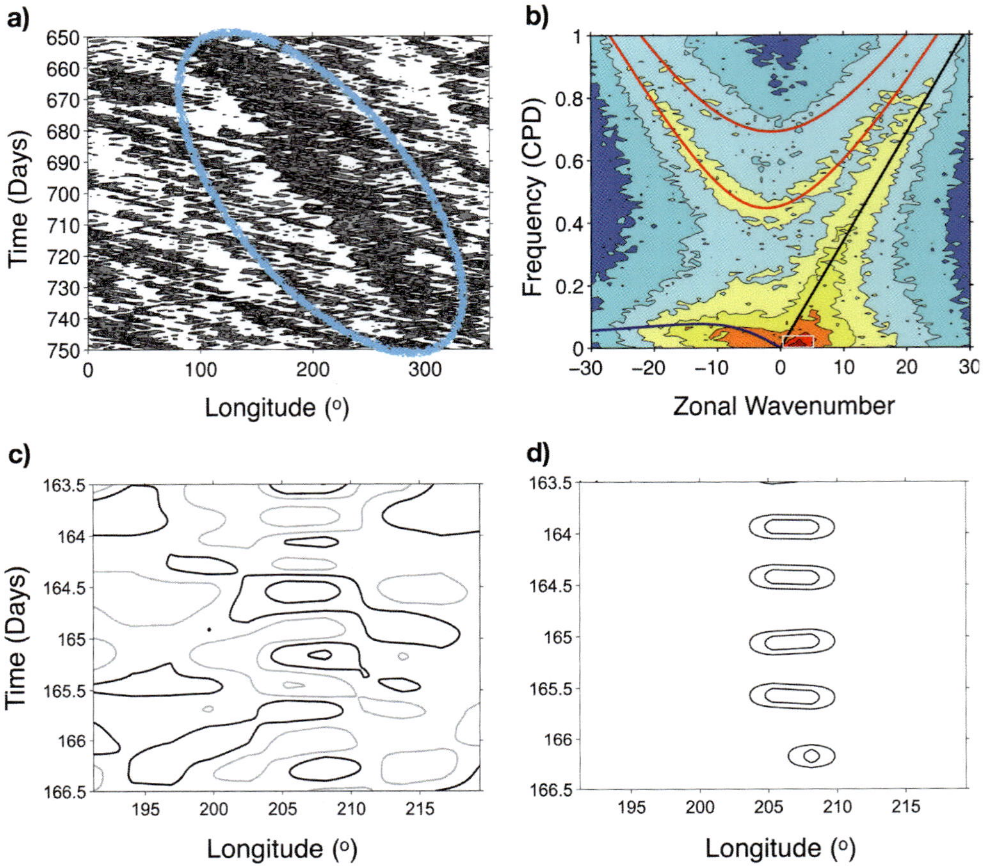

Fig. 1. Simulation results of the Yang–Ingersoll model (2D). (a) Time-longitude plot of zonal wind. Black is positive and white is negative. The blue envelope labels one MJO event. (b) 2D power spectrum of zonal wind. Red is high and blue is low. Solid lines are dispersion curves for equatorial waves. The white box labels the MJO spectral power. (c) Zoomed in time–longitude plot of geopotential. Black is positive and gray is negative. (d) Zoomed in time–longitude plot of convection. This figure is adopted from Yang and Ingersoll (2013).

since the advection terms are negligible (linear dynamics). This model is, however, self-excited as opposed to being externally forced. Convection excites quasi-standing IG waves, which triggers the next-cycle convection in the vicinity of recent convection events. At the statistically steady state, there is a large-scale, slowly varying envelope of convection events. Because convection is positive only, this large-scale aggregate appears as a new signal in the spectral domain. We refer to this effect as rectification of the direct-current (DC) convection. A similar rectification effect was introduced in Yang and Ingersoll (2011) to explore if the MJO is a wave packet of mixed Rossby–gravty (MRG) waves and was also used in Adames and Kim (2016) to explain the spatial-scale selection of the MJO.

This model predicts faster, bigger, and stronger MJOs in warmer climates. This is because gravity waves become faster with warming, leading to faster MJOs (Fig. 2a–c); faster gravity waves and stronger (fewer) storms lead to longer MFP, leading to bigger MJOs (Fig. 2e–g); stronger storms lead to stronger MJOs. These predictions are consistent with

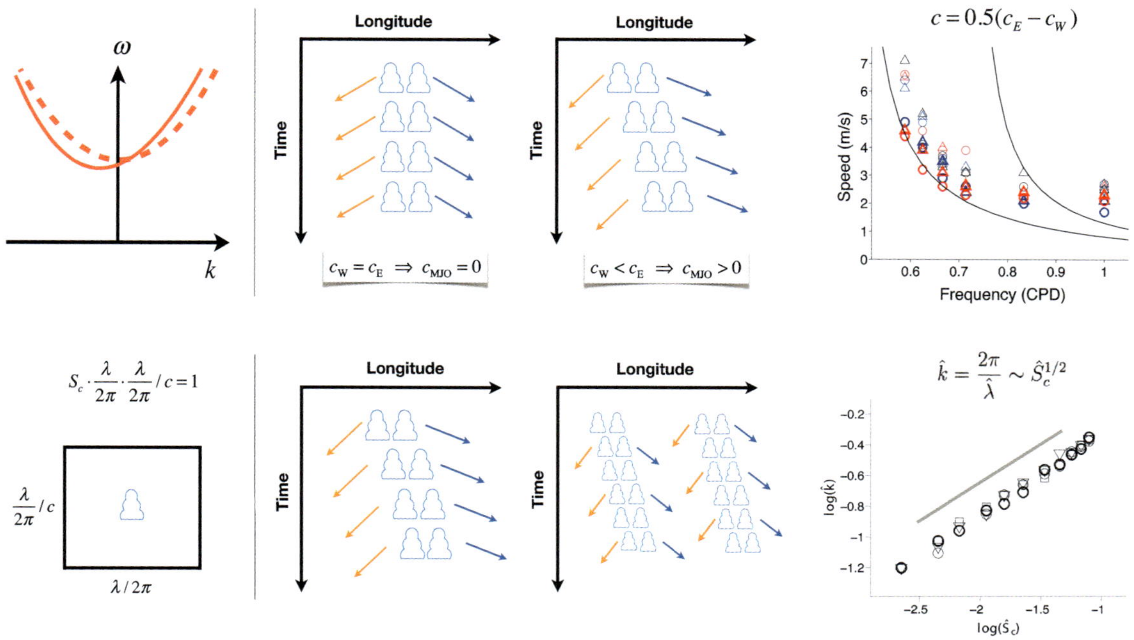

Fig. 2. Physical mechanisms of the Yang-Ingersoll model. (a) Dispersion relation of gravity waves. The dashed line represents the dispersion relation symmetric about east and west. The solid line represents the dispersion relation with an eastward tilt, as in Earth's tropical atmosphere. (b) Standing waves. (c) Eastward-propagating waves. (d) Propagation scaling. All markers are simulation results, and the solid lines are based on Eq. (3.1). (e) The time–space area occupied by one convection event. The zonal scale in this plot is the mean free path (MFP) of convective storms. (f) Wavenumber one case: strong storm and fast waves. (g) Wavenumber two case: weak storm and slow waves. (h) Wavenumber scaling in a 10-based log–log plot. Hatted variables are nondimensionalized. All markers are simulation results. The solid line shows the slope of 0.5 as predicted by Eq. (2). (d) is adopted from Yang and Ingersoll (2013), and (h) is adopted from Yang and Ingersoll (2014).

SPCAM simulation results in a wide range of climates (Pritchard and Yang 2016).

This model emphasized the importance of triggered convection in simulating the MJO, suggesting that the success of SPCAM in simulating the MJO lies in its explicit convection representation without assuming that convection is in energy quasi-equilibrium with the large-scale environment. This may point to a new way to parameterize convection in climate models.

Significant gravity wave signals have been observed in the high-frequency branch of the tropical wave spectrum, and its strength covaries with the MJO spectral power (Kikuchi 2014). This supports the gravity wave model. However, in the physical space, there is limited evidence suggesting that gravity waves are collocated with the MJO envelope. One possible explanation is that the previous studies did not use high-frequency data that can fully resolve gravity waves when examining their relationship with the MJO. This deserves more work in the future.

4. The Frictionally Coupled Rossby–Kelvin Wave and Trio-interaction Model

Wang and Rui (1990) proposed that the MJO is a large-scale, unstable coupled Rossby–Kelvin (CRK) wave packet that has the maximum growth rate at the planetary scale. The ingredients of this model include free troposphere wave dynamics (I1), quasi-equilibrium convection with a diagnostic moisture equation (I2.2), and boundary layer dynamics (I4). The CRK theory integrated three convection–dynamics feedback mechanisms that were proposed in the late 1980s: the wave–CISK (Convective Instability of the Second Kind) (Lau and Peng 1987), the evaporation–wind feedback (Emanuel 1987; Neelin *et al.* 1987; Wang 1988a), and the BL convergence feedback (Wang 1988b). Therefore, the CRK theory is essentially a "dynamic mode" theory. The major instability comes from the boundary layer moisture convergence (BLMC), which is located to the east of the major precipitation center. MJO disturbance grows through generation of eddy available potential energy (EAPE). The diabatic forcing collocated with temperature anomalies can amplify the MJO, while the diabatic forcing leading the temperature anomalies can propagate the MJO (Fig. 3b). The generation of EAPE is more efficient for longwaves in this model, selecting the planetary scale of the MJO. The BLMC also couples Rossby and Kelvin waves together with convective heating, yielding the CRK wave structure (Fig. 3c). The drawback of the CRK model is the omission of convection–moisture feedback process.

The recently proposed trio-interaction theory (Wang *et al.* 2016) is a combination of the CRK dynamic mode and the moisture mode theories (Section 5). The CRK theory overlooked moisture feedback, while the moisture mode neglected dynamic wave and BL feedback. The trio-interaction mode is essentially a "dynamic moisture mode". The ingredients of the trio-interaction model include free troposphere wave dynamics (I1), QE convection with a prognostic moisture equation (I2.2), the cloud–radiative heating (I3), and boundary layer dynamics (I4) (Fig. 3a).

The essence of trio-interaction concept is illustrated in Fig. 3a. The essential MJO physics is rooted in the trio interaction among diabatic heating, moisture, and dynamics (equatorial waves and boundary layer flows). The observed MJO structure suggests that the MJO involves equatorial Kelvin and Rossby waves and BL dynamics, as well as their interactions with diabatic heating. Diabatic heating includes

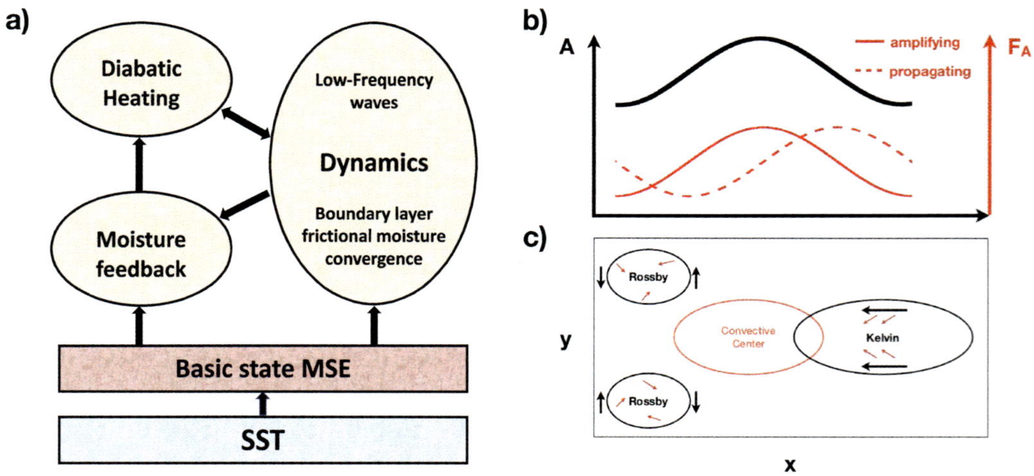

Fig. 3. Schematic diagram illustrating the trio-interaction model and the frictionally coupled Rossby–Kelvin wave (CRK) model. Panel (a) describes the essential ingredients and is adopted from Wang *et al.* (2016). Panel (b) illustrates the basic idea of the CRK theory: generation of eddy available energy is key to the MJO. Diabatic forcing (red line) collocated with temperature anomalies (black line) can amplify the MJO; diabatic forcing leading the temperature anomalies can propagate the MJO. Panel (c) illustrates the spatial structure of the coupled Rossby–Kelvin wave mode in both the free troposphere (black arrows) and in the boundary layer (orange arrows).

parameterized convective heating and a cloud–radiation feedback, which involves convective interaction with moisture (moisture feedback) and cloud–radiation feedback. The moisture variation is determined by the moisture convergence induced by the wave and BL dynamics. These constitute a trio-interaction system. The energy released from convective heating comes from the basic state moist static energy that is largely controlled by the surface air specific humidity or the basic state SST (Wang 1988b). The trio-interaction theory for MJO dynamics extends the Matsuno (1966)–Gill (1980) model by including interactive convective heating, moisture conservation, and BL dynamics.

Properties of the trio-interaction mode were elucidated by searching wave-like normal mode solutions for linear heating (Liu and Wang 2017) and solving an initial value problem for positive-only nonlinear heating (Wang *et al.* 2016) in a model with the simplified Betts–Miller cumulus scheme. The trio interaction produces a planetary-scale unstable MJO mode that not only propagates eastward slowly with a period of 30–60 days but also exhibits a coupled Kelvin–Rossby wave structure with its major convection led by BL low pressure. The theory explain why the MJO has a preferred planetary zonal scale, what controls the eastward propagation speed and prevailing periodicity, why the MJO possesses a coupled Kelvin–Rossby wave structure, why the MJO precipitation is led by the BL low pressure, and what controls amplification and decay of the MJO.

The trio-interaction theory uniquely reveals (i) the dependence of relative strengths of Rossby and Kelvin waves on the forms of cumulus schemes (Fig. 4a and b) and (ii) the dependence of the MJO propagation on its structural asymmetry with a stronger Kelvin wave component favoring faster eastward propagation (Fig. 4c). The theory also uniquely elaborates the roles of four major feedback and interaction processes. The convection–moisture feedback (Bretherton

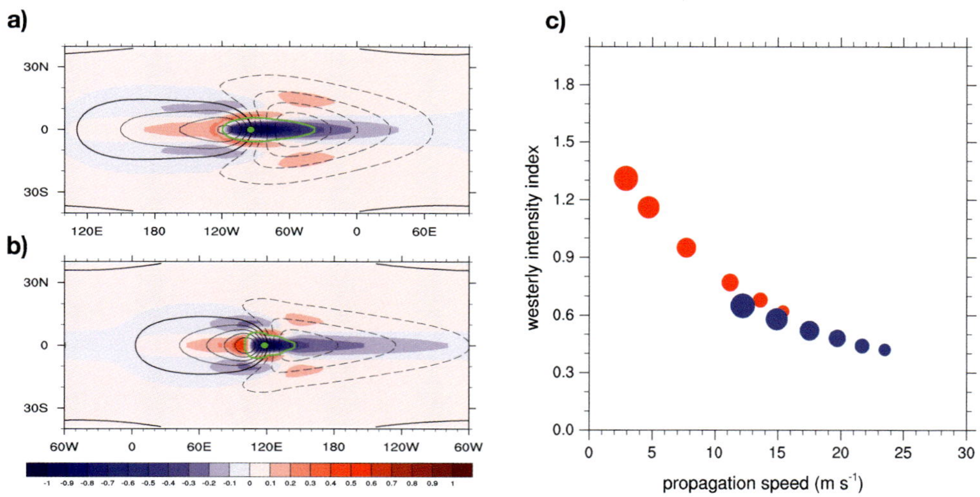

Fig. 4. Spatial structure and propagation speed of the MJO. (a) Simulation results with the Kuo Scheme (Kuo 1974). (b) Simulation results with the Betts–Miller Scheme (Betts and Miller 1986). All fields are normalized by their respective maxima (*absolute values*) at each panel. Colors represent boundary layer divergence (red is positive; blue is negative). The *green lines outline* the region where the normalized precipitation rate is larger than 0.1. The *thin solid* (*dashed*) *contours* indicate positive (negative) low-level zonal wind speed with a contour interval of 0.2. The *thick black solid line* denotes the zero contours. The *green dot* in each panel represents the location of maximum precipitation. (c) Propagation speed and westerly intensity index in both simulations. The Westerly intensity index is defined as the ratio of the maximum MJO westerly versus the maximum MJO easterly speed averaged between 5°S and 5°N. The sizes of the cycles are proportional to the SST (or basic state moist static energy at the surface), which varies from 27.0°C to 29.5°C, with a 0.5°C interval (adopted from Wang and Chen 2016).

et al. 2004; Tompkins 2001) mainly acts to reduce the propagation speed and growth rate of the short waves, producing a more realistic horizontal structure (Wang *et al.* 2016). The convection–equatorial wave feedback plays an important role in slowing down eastward propagation and increasing growth rate for planetary waves. The cloud–radiative feedback can generate instability by increasing moisture convergence, but it cannot produce a preferred planetary-scale unstable mode unless assuming the strength of radiative feedbacks is wavenumber dependent (Chen and Wang 2018). The convection–BL convergence feedback plays critical roles in generation of planetary-scale unstable modes and creation of the coupled Kelvin–Rossby wave structure of the MJO.

The sensitivity of MJO propagation and the instability to achieve cumulus parameterization in the theory explains why a variety of MJO behaviors have been produced in global models and why change of cumulus parameterization can effectively alter MJO simulations. The trio-interaction theory has led to a new dynamics-oriented diagnostic metrics for MJO simulations (Wang *et al.* 2018). Based on the trio-interaction theory, numerical modeling of MJO may be improved if the interaction between the BL convergence and shallow convective heating can be represented correctly.

5. The Moisture Wave Model

The moisture wave model (Adames and Kim 2016) considers that the MJO is a large-scale unstable mode that arises from a conservation law of atmospheric moisture. The model ingredients include I1, I2.2, I3, and I4. The instability mainly comes from longwave cloud–radiative feedbacks; the eastward propagation results mainly from horizontal and vertical advection of moisture. The rectification of cloud–radiative feedbacks produces the planetary scale of the MJO.

The concept behind the moisture wave model arises from observations revealing that moist thermodynamics are an important regulator for the occurrence and strength of tropical convection (Emanuel *et al.* 1994; Raymond 1995). Based on this observation, it was hypothesized that free tropospheric relative humidity plays an important role in the organization of tropical convection (Raymond 2000). Satellite observations revealed that tropical precipitation is an exponential function of column relative humidity, the ratio between column water vapor and its saturation value (Bretherton *et al.* 2004). This coherence is largest at the intraseasonal timescales (Yasunaga and Mapes 2012). It was later shown by Adames (2017) that precipitation anomalies, in the simplest case, equal moisture anomalies divided by its relaxation timescale. Thus, the evolution of intraseasonal moisture anomalies, to the first order, can be used to diagnose the evolution of MJO-related precipitation anomalies.

Numerous observational and numerical studies have suggested that the following moist processes are central to the propagation (1 and 2) and maintenance (3) of the MJO.

(1) Horizontal and vertical moisture advection is predominantly responsible for the propagation of the moisture anomalies (Andersen and Kuang 2012; Benedict and Randall 2007; Kiranmayi and Maloney 2011). In particular, advection of mean moisture by the MJO flow has been shown to be important (Kim *et al.* 2014; Pritchard and Yang 2016). Frictional moisture convergence has been suggested to be the main process that drives propagation from vertical moisture advection (Hsu and Li 2012). How these processes induce eastward propagation is shown schematically in Fig. 5.
(2) Horizontal moisture advection by modulation of high-frequency activity and WISHE has also been suggested to be important, but these processes are more highly dependent

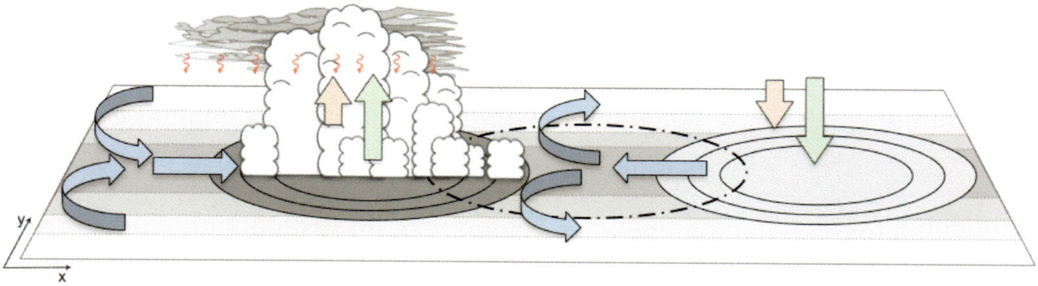

Fig. 5. Schematic showing the structure and propagation mechanism of the MJO as interpreted by the moisture mode framework. Regions of enhanced and suppressed column moisture (dark and light gray-shaded ovals, respectively) are collocated with enhanced and suppressed convection (green arrows), respectively. Suppression of outgoing longwave radiation (thin wiggling arrows) acts to warm the troposphere. This warming is balanced by enhanced vertical motion, which advects moisture upward, maintaining the region of enhanced precipitation (orange arrows). The inverse occurs in the region of suppressed convection. The anomalous heating and associated patterns of divergence lead to planetary wave responses (blue arrows). These modulate the distribution of moisture through horizontal and vertical moisture advection. For example, horizontal moisture advection induces moistening by advecting moisture down the mean meridional moisture gradient. The sum of all these moist processes results in a moisture tendency (dash-dot line) and as such the precipitation anomalies propagate eastward.

on MJO phase (Adames 2017; Andersen and Kuang 2012; Maloney 2009).

(3) Interactions between longwave radiation and clouds are key to the maintenance of the intraseasonal precipitation anomalies. High cirrus clouds resulting from deep convection reduce the outgoing longwave radiation. This causes net warming, which is balanced by upward motion and moisture import into the free troposphere. This radiatively driven ascent increases the vertical flux water vapor into the troposphere, largely offsetting the loss of moisture from condensation (see schematic in Fig. 5).

Adames and Kim (2016) incorporated these processes into a simple dynamical model in which moisture is the only prognostic variable. Having moisture as the only prognostic variable may not be realistic, but it allows us to more clearly understand the dynamics of the moisture wave. The processes in points 1 and 2 are either included explicitly or parameterized in terms of the MJO wind field as an approximation. Point 3 is parameterized from a linear regression of precipitation and outgoing longwave radiation. The result is a wave solution whose propagation features resemble those of a Rossby wave, but the propagation mechanism is due to processes 1 and 2, rather than meridional advection of planetary vorticity by the anomalous winds. The resulting dispersion is as follows:

$$\mathrm{Re}\,(\omega) = \frac{A_{\mathrm{KR}}\tilde{p}}{\tau_c k}, \tag{3}$$

where A_{KR} is the total moistening induced by the anomalous flow, which arises from horizontal and vertical moisture advection and modulation of surface latent heat fluxes. τ_c is the convective moisture adjustment timescale, a measure of the sensitivity of precipitation to changes in moisture (Betts 1986), and ρ is a frictional dissipation weight. A qualitatively similar dispersion relation was found by Fuchs and Raymond (2017), although the mechanism for scale selection is different in their model.

A salient feature of the models was their mechanism for planetary-scale selection. Feedbacks between clouds and radiation were found to be scale selective. In a moist atmosphere, upper tropospheric clouds expand far away from a region of precipitation (blue arrow in Fig. 6). This region reduces the outgoing longwave radiation, effectively warming the troposphere.

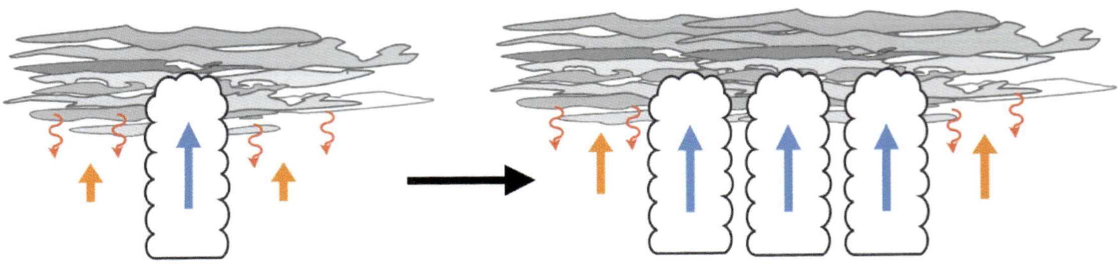

Fig. 6. Schematic describing the mechanism in which the interactions between convection and radiation may lead to planetary-scale selection. Blue arrows correspond to ascent occurring with association with deep convection while orange arrows correspond to ascent in association with radiative heating. Note that the radiative heating covers a greater zonal extent than the regions of convective heating. This non-local nature of radiatively driven ascent can cause growth of large-scale rainfall at the largest scales.

Upward motion (orange arrow) is obtained in the periphery of the precipitating region (blue arrow). This upward motion advects moisture upward, moistening the troposphere. The net result is that convection becomes more favorable over a region that is wider than the region of anomalous precipitation, thus favoring growth at a larger scale.

While the moisture wave model can explain several of the MJO's observed features above, there are several caveats. Although cloud–radiative feedbacks are suggested to be the main destabilization mechanism, model simulations in which these interactions are turned off still simulate an MJO, albeit weakened (Arnold and Randall 2015; Ma and Kuang 2016). Furthermore, although Adames and Kim (2016) showed that the MJO may be characterized by a westward group velocity, Chen and Wang (2018) showed that this group velocity may be a result of the filtering processes. Their study suggests that the MJO's group velocity is close to zero. A more comprehensive study of the MJO's propagation can be obtained from idealized modeling studies, as discussed in Section 7.

6. The Skeleton Model

Majda and Stechmann (2009) proposed that the MJO could be a neutral large-scale wave envelope, where the energy comes from smaller scales. In their model, the small scales, however, are parameterized, and only the large-scale envelope was explicit. The key ingredients include planetary-scale wave dynamics (I1) and QE convection (I2.2 with both moisture and precipitation equations). It is worth noting that although their convection parameterization represents the collective effect of individual storms on large-scale circulations (QE, I2.2), their parameterization significantly differs from the QE schemes in the moisture mode and CRK models as explained below.

The Skeleton model of Majda and Stechmann is based on the idea that the MJO is essentially a planetary-scale envelope of synoptic and mesoscale convective systems (Fig. 7). The main assumptions of the skeleton model include the

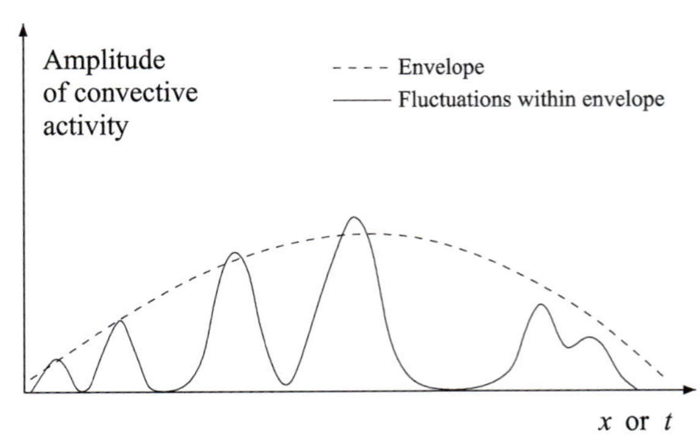

Fig. 7. MJO modulated wave activity as an envelope of synoptic and mesoscale cloud clusters (adapted from Majda and Stechmann 2016).

following:

(1) The planetary-scale structure of the MJO consists of super-positioned Gill-type Rossby and Kelvin waves (Gill 1980) under the equatorial longwave approximation.
(2) Synoptic and mesoscale convective systems (wave activity) are organized within a planetary-scale envelope of the MJO, as illustrated in Fig. 6.
(3) The wave activity tendency is proportional to the product of the wave activity and low-level moisture.

In this fashion the low-level moisture anomalies are set to oscillate freely against the wave activity on the intraseasonal scale, in the manner of a harmonic oscillator, together driving the Kelvin–Rossby wave dynamics. The skeleton model equations (not reproduced here for simplicity) result in a nearly dispersionless neutrally stable intraseasonal MJO mode in addition to a slower and westward moving Rossby mode.

The MJO dispersion relation takes the following form:

$$\omega \simeq \sqrt{\Gamma},$$

where $Q \leq 1$ is the background moisture gradient, Γ is a dimensional parameter that sets the strength of the moisture and wave activity interaction, and R is the background radiative cooling rate.

In essence the skeleton model suggests the following:

(1) The quadrupole vortex structure is fundamental to the MJO.
(2) The slow eastward propagation of the MJO comes from positive low-level moisture anomalies east of the convection center (*a priori*).
(3) Both nonlinearity and stochastic noise can give rise to MJO irregularity.
(4) The group velocity of the MJO is zero.
(5) The MJO is neutrally stable, which cannot self-initiate and self-sustain without upscale fluxes emanating from naturally unstable synoptic and mesoscale convective systems.

Figure 8 shows the results in a linear Skeleton model, as well as in a stochastic version. There are two oscillating solutions in the skeleton model: one propagating westward at

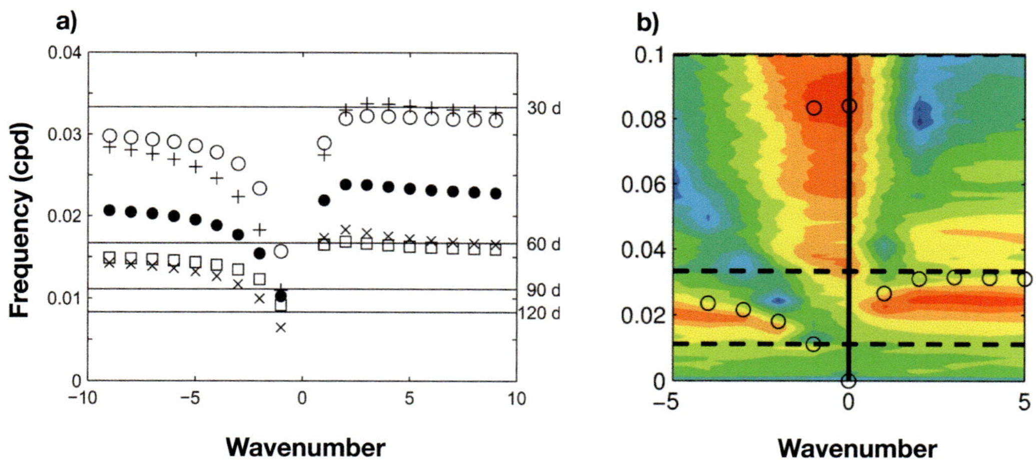

Fig. 8. Results of the Skeleton model. (a) The linear theory. Filled circles denote results with the standard parameter values. Other markers denote results with one change made to the standard parameter values: $Q = 0.8$ (open circles), $Q = 0.95$ (crosses), $\Gamma = 0.5$ (squares), and $\Gamma = 2$ (pluses). Horizontal lines in A denote oscillation periods of 30, 60, 90, and 120 days. (b) The stochastic simulation. Color shading represents spectral power density of water vapor, and circles mark the dispersion curves. (a) is adopted from Majda and Stechmann (2009) and (b) is adopted from Thual *et al.* (2014).

timescales longer than the MJO's, and the other propagating eastward with intraseasonal timescales using their parameter values. Because they are neutral modes, there is no apparent scale selection mechanism — amplitudes of different wavenumbers are comparable. Synoptic and mesoscale wave activity, which is the main source of growth, accumulates at the planetary scale and oscillates against the low-level moisture as demonstrated by numerical simulations using the stochastic version of this model (Thual *et al.* 2014). Furthermore, by adding other dynamical processes (or muscles to the skeleton) to this model (Majda and Stechmann 2011), such as nonlinear dynamics, stochasticity, and refined vertical structure, additional features such as the intermittency and vertical tilt can be captured by this model (Thual and Majda 2015). The addition of stochastic forcing in particular can capture the MJO intermittency, amount of primary and subsequent MJO events, and termination (Thual *et al.* 2014).

Chen and Wang (2017) found that large-scale precipitation and moisture anomalies of the MJO are in phase in satellite and reanalysis datasets as opposed to being in quadrature in the skeleton model. This result challenges the fundamental assumption in their convection parameterization. However, Chen and Wang (2018) found that the MJO's dispersion relation may be characterized by a group velocity of $\sim 0\,\mathrm{m\,s^{-1}}$, which seems to support the dispersion relation predicted by the skeleton model (Fig. 8a).

7. Summary

We summarize the key findings and ingredients of the four theories in Table 1. The four theories do not converge on all key aspects, but they agree that the interaction between equatorial waves and moist convection (I1 and I2) is essential to the MJO. The first three theories also agree that the differential rotation

Table 1. Summary of the four MJO models.

	Ingredients	Summary
Gravity wave theory	I1 + I2.1	1. The MJO is a large-scale envelope of small-scale, high-frequency EIG and WIG waves; 2. The MJO propagates eastward because EIG propagates faster than WIG; 3. The zonal scale of the MJO is set by the mean free path of gravity waves.
Trio-interaction theory	I1 + I2.2 + I3 + I4.1	1. The MJO is a large-scale, unstable, coupled Rossby–Kelvin wave via boundary layer moisture convergence; 2. The MJO propagates eastward because the moisture convergence leads convective center; 3. The zonal scale of the MJO is selected by optimal APE production (linear model), or by the travel distance of free Rossby and Kelvin waves (nonlinear model).
Moisture mode theory	I1 + I2.2 + I3 + I4	1. The MJO is a large-scale, unstable wave that owes its existence to moisture evolution; 2. The MJO propagates eastward because horizontal advection of moisture leads existing moisture anomalies; 3. The zonal scale is primarily set by a scale-dependent radiative feedback.
Skeleton theory	I1 + I2.2	1. The MJO is a large-scale envelope of synoptic-scale waves; 2. The MJO propagates eastward because low-level moisture anomalies lead convective center; 3. The zonal-scale selection can arise when the stochastic process and a warm pool are present.

of Earth's atmosphere (beta effect) is responsible for the eastward propagation of the MJO, as it is responsible for the small asymmetry between EIG and WIG (the gravity wave model) and the asymmetry between Rossby and Kelvin waves (the CRK and moisture wave models). Fast free troposphere wave dynamics is proposed to "select" the spatial scale of the MJO in the gravity wave and CRK models. Radiative feedbacks are, however, essential for the scale selection only in the moisture wave model.

Systematically testing these theories using the same set of model simulations and/or observational data products could identify their limitations and prove or disprove their predictions. We suggest designing a set of mechanism denial experiments to test the four MJO theories, in which key ingredients can be switched off one by one. If the basic MJO features are preserved in these experiments, the "key ingredients" are not essential. We also suggest developing a set of diagnostic tools that examine the role of key elements in individual theories. This will not only help to test theories but could also be used to evaluate climate models.

Acknowledgments

D.Y. was supported by Laboratory Directed Research and Development (LDRD) funding from Berkeley Lab, provided by the Director, Office of Science, of the U.S. Department of Energy under contract DE-AC02-05CH11231. C.Z. was supported by PMEL contribution # 4774. ÁFA was supported by the University of Michigan's startup package and by the National Science Foundation's grant AGS-1841559.

References

Adames, Á. F. 2017: Precipitation budget of the Madden–Julian oscillation. *J. Atmos. Sci.*, **74**(6), 1799–1817, https://doi.org/10.1175/JAS-D-16-0242.1

Adames, Á. F. and D. Kim, 2016: The MJO as a dispersive, convectively coupled moisture wave: Theory and observations. *J. Atmos. Sci.*, **73**(3), 913–941, https://doi.org/10.1175/JAS-D-15-0170.1

Andersen, J. A. and Z. Kuang, 2012: Moist static energy budget of MJO-like disturbances in the atmosphere of a zonally symmetric aquaplanet. *J. Clim.*, **25**(8), 2782–2804, https://doi.org/10.1175/JCLI-D-11-00168.1

Arnold, N. P. and D. A. Randall, 2015: Global-scale convective aggregation: Implications for the Madden-Julian oscillation. *J. Adv. Model. Earth Syst.*, **7**(4), 1499–1518, https://doi.org/10.1002/2015MS000498

Benedict, J. J. and D. A. Randall, 2007: Observed characteristics of the MJO relative to maximum rainfall. *J. Atmos. Sci.*, **64**(7), 2332–2354, https://doi.org/10.1175/JAS3968.1

Betts, A. K. and M. J. Miller, 1986: A new convective adjustment scheme. Part II: Single column tests using GATE WAVE, BOMEX, ATEX and ARCTIC air-mass data-sets. *Quar. J. Roy. Meteorol. Soc.*, **112**(473), 693–709, https://doi.org/10.1002/qj.49711247308

Bretherton, C. S., M. E. Peters, and L. E. Back, 2004: Relationships between water vapor path and precipitation over the tropical oceans. *J. Clim.*, **17**(7), 1517–1528, https://doi.org/10.1175/1520-0442(2004)017<1517:RBWVPA>2.0.CO;2

Bretherton, C. S., P. N. Blossey, and M. Khairoutdinov, 2005: An energy-balance analysis of deep convective self-aggregation above uniform SST. *J. Atmos. Sci.*, **62**(12), 4273–4292, https://doi.org/10.1175/JAS3614.1

Chen, G. and B. Wang, 2017: Reexamination of the wave activity envelope convective scheme in theoretical modeling of MJO. *J. Clim.*, **30**(3), 1127–1138, https://doi.org/10.1175/JCLI-D-16-0325.1

Chen, G. and B. Wang, 2018: Does the MJO have a westward group velocity? *J. Clim.*, **31**(6), 2435–2443, https://doi.org/10.1175/JCLI-D-17-0446.1

Emanuel, K., A. A. Wing, and E. M. Vincent, 2014: Radiative-convective instability. *J. Adv. Model. Earth Sys.*, **6**(1), 75–90, https://doi.org/10.1002/2013MS000270

Emanuel, K. A. 1987: An air–sea interaction model of intraseasonal oscillations in the tropics. *J. Atmos. Sci.*, **44**(16), 2324–2340, https://doi.org/10.1175/1520-0469(1987)044<2324:AASIMO>2.0.CO;2

Emanuel, K. A. 1994: *Atmospheric Convection*, Oxford University Press. Retrieved from https://global.oup.com/academic/product/atmospheric-convection-9780195066302?q=convectionemanuel&lang=en&cc=us

Emanuel, K. A., J. David Neelin, and C. S. Bretherton, 1994: On large-scale circulations in convecting atmospheres. *Quart. J. Roy. Meteorol. Soc.*, **120**(519), 1111–1143, https://doi.org/10.1002/qj.49712051902

Fuchs, Ž. and D. J. Raymond, 2017: A simple model of intraseasonal oscillations. *J. Adv. Model. Earth Syst.*, **9**(2), 1195–1211, https://doi.org/10.1002/2017MS000963

Gill, A. E. (1980). Some simple solutions for heat-induced tropical circulation. *Quart. J. R. Met. Soc.*, **106**, 447–462.

Hand, E. 2015: The storm king. *Sci.(New York, N.Y.)*, **350**(6256), 22–5, https://doi.org/10.1126/science.350.6256.22

Held, I. M., R. S. Hemler, and V. Ramaswamy, 1993: Radiative-convective equilibrium with explicit two-dimensional moist convection. *J. Atmos. Sci.*, **50**(23), 3909–3927, https://doi.org/10.1175/1520-0469(1993)050<3909:RCEWET>2.0.CO;2

Hendon, H. H. and M. L. Salby, 1994: The life cycle of the Madden–Julian oscillation. *J. Atmos. Sci.*, https://doi.org/10.1175/1520-0469(1994)051<2225:TLCOTM>2.0.CO;2

Hsu, P. and T. Li, 2012: Role of the boundary layer moisture asymmetry in causing the eastward propagation of the Madden–Julian oscillation. *J. Clim.*, **25**(14), 4914–4931, https://doi.org/10.1175/JCLI-D-11-00310.1

Hung, M. P., J. L. Lin, W. Wang, D. Kim, T. Shinoda, and S. J. Weaver, 2013: MJO and convectively coupled equatorial waves simulated by CMIP5 climate models. *J. Clim.*, **26**(17), 6185–6214, https://doi.org/10.1175/JCLI-D-12-00541.1

Kikuchi, K. 2014: An introduction to combined Fourier-wavelet transform and its application to convectively coupled equatorial waves. *Clim. Dynam.*, **43**(5–6), 1339–1356, https://doi.org/10.1007/s00382-013-1949-8

Kim, D., J.-S. Kug, and A. H. Sobel, 2014: Propagating versus nonpropagating Madden–Julian oscillation events. *J. Clim.*, **27**(1), 111–125, https://doi.org/10.1175/JCLI-D-13-00084.1

Kiranmayi, L. and E. D. Maloney, 2011: Intraseasonal moist static energy budget in reanalysis data. *J. Geophys. Res.*, **116**(D21), D21117, https://doi.org/10.1029/2011JD016031

Kuo, H. L. 1974: Further studies of the parameterization of the influence of cumulus convection on large-scale flow. *J. Atmos. Sci.*, **31**, 1232–1240, https://doi.org/10.1175/1520-0469(1974)031⟨1232:fsotpo⟩2.0.co;2

Liu, F. and B. Wang, 2017: Roles of the moisture and wave feedbacks in shaping the Madden–Julian oscillation. *J. Clim.*, **30**, 10275–10291, https://doi.org/10.1175/JCLI-D-17-0003.1

Ma, D. and Z. Kuang, 2016: A mechanism-denial study on the Madden–Julian oscillation with reduced interference from mean state changes. *Geophys. Res. Lett.*, **43**(6), 2989–2997, https://doi.org/10.1002/2016GL067702

Madden, R. A. and P. R. Julian, 1971: Detection of a 40–50 day oscillation in the zonal wind in the tropical Pacific. *J. Atmos. Sci.*, **28**(5), 702–708, https://doi.org/10.1175/1520-0469(1971)028<0702:DOADOI>2.0.CO;2

Majda, A. J. and S. N. Stechmann, 2009: A Simple dynamical model with features of convective momentum transport. *J. Atmos. Sci.*, **66**(2), 373–392, https://doi.org/10.1175/2008JAS2805.1

Majda, A. J. and S. N. Stechmann, 2011: Nonlinear dynamics and regional variations in the MJO skeleton. *J. Atmos. Sci.*, **68**(12), 3053–3071, https://doi.org/10.1175/JAS-D-11-053.1

Maloney, E. D. 2009: The moist static energy budget of a composite tropical intraseasonal oscillation in a climate model. *J. Clim.*, **22**(3), 711–729, https://doi.org/10.1175/2008JCLI2542.1

Matsuno, T. 1966: Quasi-geostrophic motions in the equatorial area. *J. Meteorolo. Soc. Jpn.*, **44**(February), 25–43, https://doi.org/10.1002/qj.49710644905

Neelin, J. D. and I. M. Held, 1987: Modeling tropical convergence based on the moist static energy budget. *Month. Weath. Rev.*, https://doi.org/10.1175/1520-0493(1987)115<0003:MTCBOT>2.0.CO;2

Pritchard, M. S. and D. Yang, 2016: Response of the superparameterized Madden–Julian oscillation to extreme climate and basic state variation challenges a moisture mode view. *J. Clim.*, JCLI-D-15-0790.1, https://doi.org/10.1175/JCLI-D-15-0790.1

Raymond, D. J. 1995: Regulation of moist convection over the west Pacific warm pool. *J. Atmos. Sci.*, https://doi.org/10.1175/1520-0469(1995)052⟨3945:ROMCOT⟩2.0.CO;2

Raymond, D. J. 2000: The hadley circulation as a radiative–convective instability. *J. Atmos. Sci.*, **57**(9), 1286–1297, https://doi.org/10.1175/1520-0469(2000)057⟨1286:THCAAR⟩2.0.CO;2

Sobel, A. and E. Maloney, 2012: An idealized semi-empirical framework for modeling the Madden–Julian oscillation. *J. Atmos. Sci.*, **69**(5), 1691–1705, https://doi.org/10.1175/JAS-D-11-0118.1

Sobel, A. and E. Maloney, 2013: Moisture modes and the eastward propagation of the MJO. *J. Atmos. Sci.*, 120920104307002, https://doi.org/10.1175/JAS-D-12-0189.1

Thual, S. and A. J. Majda, 2015: A suite of skeleton models for the MJO with refined vertical structure. *Math. Clim. Weath. Forecast.*, **1**(1), https://doi.org/10.1515/mcwf-2015-0004

Thual, S., A. J. Majda, and S. N. Stechmann, 2014: A stochastic skeleton model for the MJO. *J Atmos. Sci.*, **71**(2), 697–715, https://doi.org/10.1175/JAS-D-13-0186.1

Tompkins, A. M. 2001: Organization of tropical convection in low vertical wind shears: The role of water vapor. *J. Atmos. Sci.*, **58**(6), 529–545, https://doi.org/10.1175/1520-0469(2001)058<0529:OOTCIL>2.0.CO;2

Wang, B. and Rui, H. 1990: Dynamics of the Coupled Moist Kelvin–Rossby Wave on an Equatorial β-Plane. *J. Atmos. Sci.*, **47**(4), 397–413, https://doi.org/10.1175/1520-0469(1990)047<0397:DOTCMK>2.0.CO;2

Wang, B., F. Liu, and G. Chen, 2016: A trio-interaction theory for Madden–Julian oscillation. *Geosci. Lett.*, **3**(1), 34, https://doi.org/10.1186/s40562-016-0066-z

Wang, B., S. S. Lee, D. E. Waliser, C. Zhang, A. Sobel, E. Maloney, *et al.* 2018: Dynamics-oriented diagnostics for the Madden–Julian oscillation. *J. Clim.*, https://doi.org/10.1175/JCLI-D-17-0332.1

Wheeler, M. and Kiladis, G. N. 1999: Convectively coupled equatorial waves: Analysis of clouds and temperature in the wavenumber–frequency domain. *J. Atmos. Sci.*, **56**(3), 374–399, https://doi.org/10.1175/1520-0469(1999)056<0374:CCEWAO>2.0.CO;2

Yang, D. 2018a: Boundary-layer diabatic processes, the virtual effect, and convective self-aggregation. *J. Adv. Model. Earth Sys.*, https://doi.org/10.1029/2017MS001261

Yang, D. 2018b: Boundary layer height and buoyancy determine the horizontal scale of convective self-aggregation. *J. Atmos. Sci.*, **75**(2), 469–478, https://doi.org/10.1175/JAS-D-17-0150.1

Yang, D. and A. P. Ingersoll, 2011: Testing the hypothesis that the MJO is a mixed Rossby–Gravity wave packet. *J. Atmos. Sci.*, **68**(2), 226–239, https://doi.org/10.1175/2010JAS3563.1

Yang, D. and A. P. Ingersoll, 2013: Triggered convection, gravity waves, and the MJO: A Shallow-Water model. *J. Atmos. Sci.*, **70**(8), 2476–2486, Retrieved from http://dx.doi.org/10.1175/JAS-D-12-0255.1

Yang, D. and A. P. Ingersoll, 2014: A theory of the MJO horizontal scale. *Geophys. Res. Lett.*, **41**(3), 1059–1064, https://doi.org/10.1002/2013GL058542

Yasunaga, K. and B. Mapes, 2012: Differences between more divergent and more rotational types of convectively coupled equatorial waves. Part I: Space–time spectral analyses. *J. Atmos. Sci.*, **69**(1), 3–16, https://doi.org/10.1175/JAS-D-11-033.1

Zhang, C. 2005: Madden–Julian oscillation. *Rev. Geophys.*, **43**(2), RG2003, https://doi.org/10.1029/2004RG000158

Zhang, C. 2013: Madden-julian oscillation: Bridging weather and climate. *Bulle. Am. Meteorolo. Soc.*, **94**(12), 1849–1870, https://doi.org/10.1175/BAMS-D-12-00026.1

Chapter 20

Air-Sea Interactions in the Madden–Julian Oscillation

Charlotte A. DeMott*,‡ and Nicholas P. Klingaman†

**Department of Atmospheric Science, Colorado State University, Fort Collins, USA*
†National Centre for Atmospheric Science-Climate, University of Reading, UK
‡demott@atmos.colostate.edu

The Madden–Julian oscillation (MJO) controls intraseasonal variations in tropical atmospheric circulation and rainfall. Recent research has pointed to air-sea interactions as one mechanism, among several, that can maintain the MJO and enhance its eastward propagation across the equatorial Warm Pool. MJO atmospheric forcing affects the ocean mixed-layer structure, through variations in surface heat, moisture, and momentum fluxes between active (cloudy, windy) and break (clear, calm) periods, which produce sea surface temperature (SST) anomalies in quadrature with MJO convection. Variations in atmospheric forcing act on many scales. A strong diurnal cycle in the MJO suppressed phase can produce shallow mixed layers, warm SST anomalies, and encourage atmospheric shallow convection leading to the next MJO active phase. MJO atmospheric forcing can also interact with ocean dynamics to excite oceanic equatorial waves, which may interact with subsequent MJO events through SST anomalies. MJO-driven intraseasonal SST variability can feed back to the atmosphere through modulations in surface fluxes, which in turn maintain MJO-associated moist static energy anomalies, encouraging (discouraging) equatorial (off-equatorial) convection. Most coupled models show improved MJO fidelity relative to their atmosphere-only counterparts. However, the value of modeling results is often compromised by large systematic errors in the basic state and in the MJO itself, as well as the lack of sensitivity studies to pinpoint the mechanisms by which the atmospheric MJO responds to SST variability.

1. Introduction

The Madden–Julian oscillation (MJO; Madden and Julian, 1972) is a large-scale (O(10,000 km)), intraseasonal (30–70 day), transient tropical convective disturbance. The MJO is a multiscale phenomenon that manifests as quasi-cyclic rainy and dry periods within the equatorial Indo-Pacific. Its "active" phase is characterized by a large-scale (~5,000 km) "envelope" of deep convection that typically forms in the western Indian Ocean and propagates east. MJO convection weakens in the Maritime Continent (MC) and re-intensifies over the open West Pacific. The heaviest MJO-related precipitation often dissipates at the eastern Warm Pool edge, while the anomalous circulation accelerates east across the equatorial Western Hemisphere, allowing the MJO to affect weather across the Tropics.

Significant progress in the past decade has advanced our understanding of processes that contribute to MJO maintenance and propagation. MJO convection is destabilized and maintained by cloud radiative feedbacks (e.g., Arnold and Randall 2015), while the slow eastward propagation is regulated by anomalous moisture advection (Maloney 2009; Andersen and Kuang 2012). These findings are consistent with the "moisture mode theory" of the MJO, in which the disturbance is regulated by moisture variations.

The Multiscale Global Monsoon System
Edited by C. P. Chang et al.

The interactions among convection, equatorial waves, and boundary layer processes (e.g., Hsu and Li 2012) are entwined with the joint interactions of clouds and column moisture (Thayer-Calder and Randall 2009; Kim *et al.* 2009). Heat and moisture fluxes from the ocean surface play a secondary, and non-trivial role in column moistening.

Characterizing the role of ocean feedbacks to the MJO remains challenging. The MJO convective signal maximizes over the warm SSTs of the Indo-Pacific Warm Pool, and weakens over the Maritime Continent, where large islands reduce oceanic latent heat fluxes (e.g., Sobel *et al.* 2010). Support for the ocean's role in the MJO primarily comes from modeling studies, where including air-sea feedbacks in coupled models generally improves MJO fidelity over atmosphere-only configurations (DeMott *et al.* 2015 and references therein). However, understanding the role and relative importance of coupled air-sea feedbacks in models is complicated by the effects of mean state SST biases on tropical moisture distributions, which affect MJO propagation via moisture advection (Gonzalez and Jiang 2017; Lim *et al.* 2018; Kim 2017). Novel model configurations and experiment designs hold promise for progress.

2. Intraseasonal Atmosphere-ocean Coupled Processes

Air-sea feedbacks throughout the MJO life cycle follow a sequence of atmospheric forcing of the ocean, oceanic adjustment, and ocean feedback to the atmosphere (DeMott *et al.* 2015). This forcing–feedback loop spans daily to interannual timescales. During the MJO suppressed phase, reduced surface winds and enhanced solar heating stabilize the upper ocean (Anderson *et al.* 1996; Cronin and McPhaden 1997), reducing vertical mixing and accumulating energy in the top few meters, producing warm SST anomalies.

The SST response to surface heating is regulated by the ocean mixed layer depth. Neglecting vertical mixing, horizontal advection, and penetration of solar heating below the mixed layer, the SST tendency is approximated by $\partial T/\partial t = Q/(\rho C_p H)$, where T is the ocean mixed layer temperature, Q is the net surface energy flux into the ocean, ρ is water density, C_p is the specific heat of water, and H is the mixed layer depth. For a given surface heating (Q), SST will increase more for a shallow mixed layer than a deep mixed layer. Surface currents and mechanical mixing can limit the SST response to surface heating, but these processes, on average, play a secondary role in the upper-ocean heat budget over most of the Warm Pool (Halkides *et al.* 2015), except within about a 2.5° latitude about the Equator, where momentum forcing drives meridionally narrow surface currents.

Warm Pool climatological ocean mixed layer depths during boreal winter range from roughly 25–30 m, but deviate by ±10 m over the composite MJO life cycle (DeMott *et al.* 2015, and references therein). Mixed-layer depth perturbations can be substantially larger for individual events and exhibit significant spatial variations (e.g., Alappattu *et al.* 2017).

During calm conditions, extreme stabilization of the upper ocean can lead to diurnal shoaling (i.e., thinning) and warming of the ocean mixed layer (Weller and Anderson 1996; Soloviev and Lukas 1997; Matthews *et al.* 2014). These diurnal warm layers (DWLs) increase the daily average SST (Bellenger and Duvel 2009) and rectify into the intraseasonal SST cycle (Bernie *et al.* 2005). With the arrival of deep MJO convection, enhanced winds and decreased solar heating drive a rapid transfer of energy from the upper ocean to the atmosphere reflected by intense surface cooling (Zhang and McPhadden 2000).

The stability of the upper ocean can strongly modulate the ocean dynamic response to MJO convectively driven momentum surface flux forcing. An upper ocean stabilized by intense heating or freshening by rainfall can inhibit wind-driven vertical mixing, concentrating momentum fluxes in the upper ocean (Cronin and McPhaden 2002), resulting in strong surface currents, especially near the Equator. These equatorial oceanic jets can alter surface stresses, fluxes, and ocean warming for up to several weeks following the wind event, and influence the magnitude of oceanic feedbacks to subsequent MJO events (e.g., Moum *et al.* 2013, 2016).

Upper-ocean stability regulates the depth of wind-driven ocean mixing and the excitation of oceanic Kelvin (Kessler *et al.* 1995) and equatorial Rossby waves (e.g., Schott *et al.* 2009). The former interact with the Indonesian Throughflow (Sprintall *et al.* 2014; Drushka *et al.* 2010) and El Niño–Southern Oscillation cycles (Kessler 2005), while the latter can suppress the Seychelles Thermocline Ridge (e.g., Shinoda *et al.* 2013) and influence intraseasonal SST variability.

Intraseasonal SST anomalies shaped by the upper-ocean response to MJO forcing regulate surface latent and sensible heat fluxes (Fig. 1), which contribute to column moistening and drying through a variety of processes. To the east of the MJO convection, warm SSTs are theorized to augment surface pressure reductions and boundary layer convergence within the Kelvin wave part of the MJO (Wang and Xie 1998) and contribute to the MJO's eastward propagation. Diurnal SST cycles associated with DWLs can force a diurnal cycle of shallow convection that supplements lower tropospheric moistening by the large-scale flow (Seo *et al.* 2014; Ruppert and Johnson 2015). SST gradients on a variety of spatial scales induce boundary layer convergence that supports nascent convection (Li and Carbone 2012; Hsu and Li 2012).

During the MJO active phase, the transfer of energy from the upper ocean to the atmosphere cools the ocean surface. This cooling reduces surface latent and sensible heat fluxes by ~5 W m^{-2} for a 0.25-K surface temperature cooling (Webster *et al.* 1996). The reduced surface temperatures limit column moistening by

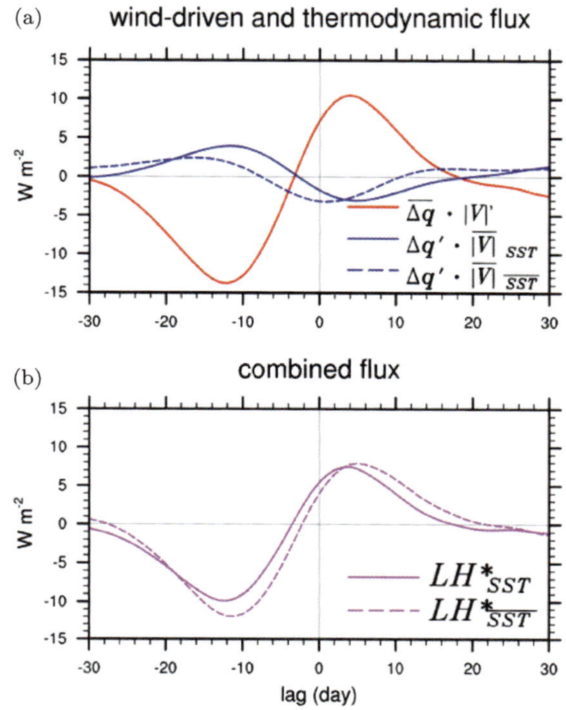

Fig. 1. Regression of (a) wind-driven (red) and thermodynamic (blue) perturbations to the surface latent flux onto 20–100-day filtered rainfall in the Indian Ocean. Thermodynamic perturbations calculated using the observed and the 61-day smoothed SST are plotted with solid and dashed lines, respectively. (b) The sum of wind-driven and thermodynamic latent heat flux perturbations. Intraseasonal SST perturbations reduce the overall amplitude of the flux (b), but shift the flux toward negative lags. This phase shift induced by SST perturbations enhances column moistening by surface fluxes during convective developing and active phases. Data from ECMWF-Interim reanalysis (Dee *et al.* 2011) for November through April, 1986–2013 and averaged 10°S–10°N and 60°E–70°E.

surface fluxes and, together with dry advection in the free troposphere, contribute to column drying west of the MJO convection. Thus, the warm SST feedbacks east of the MJO convection described above and the cold SST feedbacks to column moistening west of the MJO convection both contribute to the MJOs eastward propagation.

With the arrival of MJO convection, the transfer of heat from upper ocean to atmosphere can intensify both convection and the circulation response to its heating (e.g., Maloney and Sobel 2004). Off the Equator, ocean cooling by evaporation and wind mixing mutes surface fluxes, and may moderate growth of the Rossby wave part of the MJO (DeMott *et al.* 2016). Equatorial surface currents forced by MJO westerly winds sustain prolonged upper-ocean mixing that can limit antecedent SST increases and subsequent convection for the next MJO event (Moum *et al.* 2016).

Oceanic equatorial wave modes excited by the MJO provide other ocean feedback mechanisms to the atmosphere. In the Indian Ocean, surface westerly wind events following the passage of MJO convection can force oceanic Kelvin waves which, upon reaching Sumatra, reflect as downwelling oceanic equatorial Rossby waves. The warm SSTs that accompany downwelling oceanic Rossby waves have been linked to the onset of MJO convection in the western Indian Ocean (Webber *et al.* 2012; Rydbeck and Jensen 2017). In the western Pacific Ocean, MJO westerly wind events expand the eastern edge of the Warm Pool eastward via advection of warm waters (Drushka *et al.* 2015) and the excitation of oceanic Kelvin waves. Both these processes can contribute to the onset (McPhaden and Yu 1999) or maintenance (Kessler and Kleeman 2000) of El Niño events. The ensuing eastward expansion of the Warm Pool helps sustain MJO convection and propagation farther east than during ENSO neutral conditions (Pohl and Matthews, 2007).

The ocean-atmosphere feedbacks described above span a range of temporal scales involving both direct and indirect feedbacks to the MJO. We consider the direct ocean feedbacks to be those associated with SST modulation of surface fluxes (shown in Fig. 1), and their contributions to the overall MJO moist static energy budget (MSE; e.g., Maloney 2009), while indirect ocean feedbacks are those that affect MJO circulation anomalies or the background state condition.

DeMott *et al.* (2016) quantified the direct ocean feedbacks to the MJO by computing Reynolds-derived surface fluxes calculated from the observed and the 61-day smoothed SSTs:

$$LH \approx \rho L_v C_c[(\overline{\Delta q}|V|')' + (\Delta q'\overline{|V|})'],$$

where LH is the latent heat flux, ρ is air density, L_v is the latent heat of evaporation, C_e is the transfer coefficient, $|V|$ is the near-surface wind speed, and Δq is the difference between saturation specific humidity at T = SST and near-surface specific humidity. Overbars and primes indicate background and anomaly quantities, respectively. Fluxes computed with the "full" and "smoothed" SST are regressed onto the column-integrated MSE and its tendency. The difference in the regression coefficients for the "full SST" and "smoothed SST" flux terms (Fig. 2) represents the contribution of intraseasonal SST perturbations to MJO maintenance and propagation. The effect of SST perturbations on MJO maintenance (Fig. 2a) is to preferentially maintain MSE anomalies on the Equator in the western Indian and Pacific Oceans and suppress it off the Equator, especially in the southern Indian Ocean and east of the Philippines. SST perturbations also help

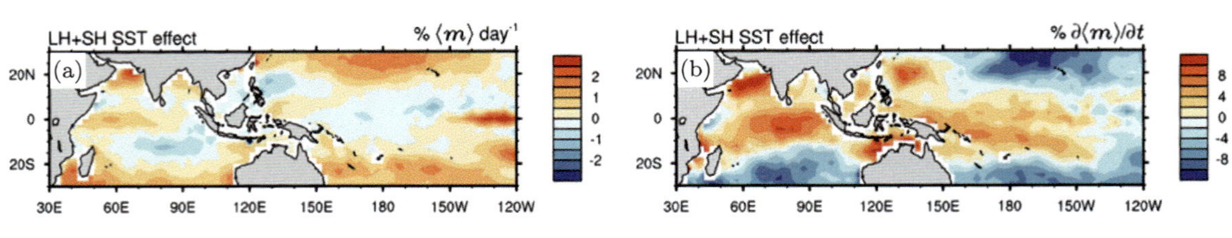

Fig. 2. The SST effect on (a) MSE (% MSE per day) and (b) MSE tendency (% dMSE/dt). Data from ECMWF-Interim reanalysis (Dee *et al.* 2011) for November through April, 1986–2013, and averaged 10°S–10°N and 60°E–70°E.

maintain MSE anomalies in the shallow waters northwest of Australia, which provide an important "passageway" for MJO convection during austral summer (Kim *et al.* 2017). The SST effect on MJO propagation is the difference in full SST and smoothed SST surface flux regression coefficients onto the MSE tendency (Fig. 2b). The SST modulation of surface fluxes accounts for as much as 10% of the total MSE tendency, consistent with the general improvement in MJO propagation characteristics when ocean feedbacks are included in the simulated MJO. Indirect ocean feedbacks on the MJO are more difficult to assess and likely involve complex interactions of surface fluxes, convection, wave dynamics, and the background state across multiple time and space scales (e.g., Wang and Chen 2017; Ramirez *et al.* 2017).

3. Ocean Feedbacks in Coupled Modeling Systems

A poor representation of the MJO has remained a key systematic error in general circulation models (GCMs) for decades (e.g., Lin *et al.* 2006; Kim *et al.* 2009; Hung *et al.* 2013; Ahn *et al.* 2017). Despite increases in resolution and the complexity of physical parameterizations, GCMs struggle to represent the organization, intensity, propagation, and period of the MJO, as well as associated boreal summer intraseasonal variability (Sabeerali *et al.* 2013). Because the MJO modulates sub-seasonal weather extremes globally, GCMs that fail to capture the MJO will also suffer from biases in their representations of high-impact weather (e.g., tropical cyclones, monsoon active–break cycles, and atmospheric rivers).

Understanding and improving the representation of the MJO in GCMs, including the sensitivity of that representation to changes in GCM configuration or physics, is critical to advancing predictions and projections of extreme weather around the globe. The representation of the MJO in models is sensitive to many factors, including horizontal (e.g., Crueger *et al.* 2013) and vertical resolution (e.g., Inness *et al.* 2001), convective momentum transport (e.g., Zhou *et al.* 2012), the vertical structure of convective heating (e.g., Seo and Wang 2010), the sensitivity of convection to environmental humidity (e.g., Hirons *et al.* 2013), and air-sea interactions (e.g., Klingaman and Woolnough 2014). It is difficult to compare the relative effects of these factors on the MJO, as they are often tested in isolation and their effects are highly model dependent. Two recent studies concluded that air-sea interactions were likely of secondary importance to convective physics (Crueger *et al.* 2013; Klingaman and Woolnough 2014).

Many air-sea coupled models (CGCMs) produce an improved MJO relative to their atmosphere-only (AGCM) counterparts, both in the amplitude of the convective perturbation and in the coherence and speed of its eastward propagation (e.g., DeMott *et al.* 2014; Zhu *et al.* 2017). CGCMs have greater prediction skill for the MJO in medium-range and sub-seasonal forecasts, compared to atmosphere-only models (e.g., Woolnough *et al.* 2007; Kim *et al.* 2008; Shelly *et al.* 2014).

In particular, the representation of the MJO in CGCMs is sensitive to the strength of the intraseasonal SST response to MJO convection. The diurnal ocean response to MJO atmospheric forcing rectifies strongly into the intraseasonal scale (Bernie *et al.* 2005; Shinoda 2005). Capturing the intraseasonal SST response to the MJO requires CGCMs with at most a 1-m upper-ocean vertical resolution and a 3-h air-sea coupling frequency, both of which improve MJO amplitude and propagation (e.g., Bernie *et al.* 2008; Seo *et al.* 2014; Tseng *et al.* 2015). Many contemporary CGCMs employ a 10-m vertical resolution and couple by the daily means of SST and surface fluxes. These models underestimate the role of diurnal and intraseasonal ocean feedbacks to the MJO, particularly in regions such as the western equatorial Indian Ocean and the Maritime Continent, where the MJO interacts strongly with the diurnal cycles of SST and atmospheric convection (Peatman *et al.* 2014). The strong effects of DWLs and intraseasonal variations in ocean stratification argue against the use of CGCMs with slab oceans for the MJO, which use a constant mixing depth that can also be employed to unphysically "tune" the strength of the SST response to atmospheric forcing anomalies (e.g., Maloney and Sobel 2004; Benedict and Randall 2011).

The role of air-sea feedbacks to the MJO in CGCMs is also sensitive to the simulated mean state SST. Many CGCMs produce cold SST biases in the tropics, particularly in the equatorial Pacific and Indian Oceans, which are associated with a preference for off-equatorial convection and easterly wind errors (e.g., Sperber *et al.* 2004). Off-equatorial convection favors Rossby-wave modes over MJO-like Kelvin wave modes, while easterly wind errors degrade the phase relationship between MJO winds, surface fluxes, and SST and inhibit moisture convergence into the MJO active phase. CGCM SST biases are also associated with errors in the background moisture distribution, such as enhanced moisture in the western Pacific, which produce errors in moisture transport by MJO-associated circulations and ultimately degrade the simulated MJO (e.g., Kim *et al.* 2017).

Correcting these SST biases, for example, through flux correction, can improve MJO simulation (e.g., Inness *et al.* 2003). Klingaman and Woolnough (2014) demonstrated that CGCM systematic errors influence the perceived effects of air-sea coupling, by employing the Met Office Unified Model (MetUM) coupled to the multi-column configuration of the K profile parameterization (KPP) one-dimensional ocean model. The key advantage of this framework is the ability to constrain the ocean mean state, through corrections to ocean temperature and salinity, to either the observations or the mean state from the MetUM CGCM. The effect of coupling on the MJO under the CGCM biased mean SST was considerably weaker than the effect of coupling under a near-observed mean SST. This suggests that systematic biases in the CGCM mean state, and hence in the representation of the MJO, may lead to biased estimates of the role of air-sea feedbacks.

Further, many studies have quantified the effects of air-sea coupling on the MJO by driving an AGCM with CGCM SSTs (e.g., DeMott *et al.* 2014), either using a mean annual cycle or a time-smoothed time series with interannual variability. While this technique ensures the CGCM and AGCM have a consistent mean state, it is able only to evaluate the influence of air-sea feedbacks on the MJO under the CGCM biased mean state. We recommend performing future experiments in a CGCM in which the ocean mean state can be controlled, for example, through the flux corrections described above.

Several studies have suggested a role for air-sea feedbacks in the MJO or related tropical intraseasonal variability through AGCM simulations alone, by comparing simulations forced by high-frequency (e.g., daily) and low-frequency (e.g., monthly) SSTs, in climate simulations or initialized predictions (e.g., Liess *et al.* 2004; Klingaman *et al.* 2008; de Boisseson *et al.* 2012). These results should be interpreted with caution, for not only do AGCMs neglect air-sea feedbacks but many AGCMs forced by high-frequency SSTs also develop incorrect, coincident, positive SST–surface flux–rainfall correlations, due to the tendency of convective parameterizations to maximize precipitation over warm SSTs (Pegion and Kirtman 2008). Thus, we strongly recommend against using AGCM experiments to infer the importance of air-sea interactions.

A final note of caution in the interpretation of CGCMs concerns the role of interannual variability. The Super-Parameterized Community Climate System Model (SPCCSM3) produces a robust, eastward-propagating MJO that strongly resembles observations (Fig. 3a,b). Air-sea interactions are hypothesized to play a role in the MJO in SPCCSM3, as the

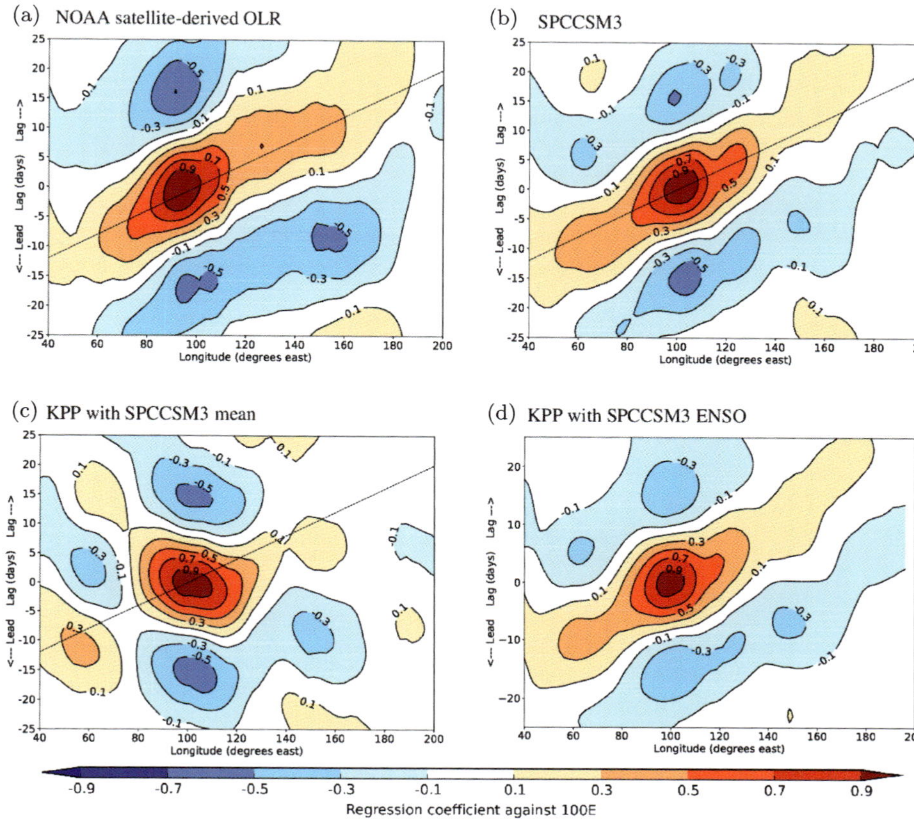

Fig. 3. Lead-lag regressions of 15S–15N averaged outgoing longwave radiation (OLR) against a base point at 100E, for November–April data from (a) NOAA satellite-derived OLR; (b) a 20-year simulation of SPCCSM3; (c) a 50-year simulation of SPCAM3-KPP, with KPP constrained to the SPCCSM3 ocean mean state; (d) a 60-year simulation of SPCAM3-KPP, with KPP constrained to a 3-year repeating cycle of ENSO variability derived from SPCCSM3. Data were bandpass filtered to 20–100 days. The dotted line shows an eastward propagation speed of $5\,\mathrm{m\,s^{-1}}$.

model generates an improved MJO relative to its atmosphere-only counterpart, the SPCAM3 (DeMott *et al.* 2014; Jiang *et al.* 2015).

However, our recent analysis shows that the improved MJO in SPCCSM is the product of an overly strong simulated teleconnection between the MJO and the El Niño–Southern Oscillation (ENSO), in which the MJO is very strong in El Niño years, but weak in La Niña and neutral in ENSO years (not shown). To demonstrate this, we coupled SPCAM3 to KPP to perform similar sensitivity experiments to the ocean mean state as in Klingaman and Woolnough (2014). When SPCAM3-KPP is constrained, through flux corrections, to reproduce the SPCCSM3 ocean mean state, SPCAM3-KPP fails to produce an eastward-propagating MJO (Fig. 3c). This simulation does not include ENSO variability, as KPP lacks ocean dynamics and the flux corrections constrain the ocean to the climatological SPCCSM3 ocean state. In a further SPCAM3-KPP experiment, the ocean is constrained to reproduce a repeating 3-year composite ENSO cycle derived from the SPCCM3 simulations. The mean state remains similar to the original SPCAM3-KPP simulation, by construction; only ENSO variability is added. In this simulation, SPCAM3-KPP produces a robust MJO (Fig. 3d) similar to SPCCSM3 (Fig. 3b). This suggests that ENSO feedbacks improve the MJO in SPCCSM3, not air-sea interactions within the MJO itself, and that isolating the role of air-sea interactions in the MJO in CGCMs requires removing the effects of interannual variability.

4. Summary and Outlook

Biases in CGCM representations of the MJO and its interactions with the underlying ocean impede our understanding of atmosphere–ocean feedbacks in the real-world MJO, necessitating modeling frameworks beyond standard CGCMs with dynamical oceans and diagnostic tools beyond well-heeled wavenumber–frequency spectra. GCM configurations with an atmosphere coupled to a one-dimensional mixed-layer ocean, with fine vertical resolution to capture DWLs and the ability to control the mean ocean state to reduce SST biases, show promise for bespoke sensitivity experiments to unravel atmosphere–ocean coupled processes in the MJO (e.g., Klingaman and Woolnough, 2014; Tseng *et al.* 2015). Diagnostic techniques that assess the contributions of intraseasonal SST variability to the MJO through the moist static energy (MSE) budget will help to reveal the mechanisms underlying the improved MJO in CGCMs (DeMott *et al.* 2016), including whether CGCMs overestimate the role of air-sea feedbacks, as compensation for weak or missing processes elsewhere in the model (e.g., in convective physics). Applications of these diagnostics to reanalysis data suggests that intraseasonal SST variability (1) maintains equatorial convection and offsets $\sim$10% of the MSE exported by convection during the MJO active phase, and (2) directly accounts for up to 10% of intraseasonal MSE tendency throughout the Warm Pool. Combining these diagnostics with novel CGCM experiments to isolate individual air-sea coupled processes will provide insight into the mechanisms through which the MJO atmospheric circulations force and respond to upper-ocean variability.

Improved understanding of ocean feedbacks to the MJO may be advanced through observations and modeling studies focused on the co-evolution of heat, moisture, salinity, and momentum in atmospheric and ocean mixed layers, which co-regulate the flux.

Acknowledgments

CAD was supported by the National Science Foundation grant AGS-1445191. NPK was supported by an Independent Research Fellowship from the UK Natural Environment Research Council (NE/L010976/1).

References

Ahn, M.-S., D. Kim, K. R. Sperber, I.-S. Kang, E. Maloney, D. Waliser, and H. Hendon, 2017: MJO simulation in CMIP5 climate models: MJO skill metrics and process-oriented diagnosis. *Clim. Dynam.*, doi:10.1007/s00382-017-3558-4.

Alappattu, D. P., Q. Wang, J. Kalogiros, N. Guy, and D. P. Jorgensen, 2017: Variability of upper ocean thermohaline structure during a MJO event from DYNAMO aircraft observations. *J. Geophys. Res. Oceans*, **122**, 1122–1140.

Andersen, J. A. and Z. Kuang, 2012: Moist static energy budget of MJO-like disturbance in the atmosphere of a zonally symmetric aquaplanet. *J. Clim.*, **25**(8), 2782–2804.

Anderson, S. P., R. A. Weller, and R. B. Lukas, 1996: Surface buoyancy forcing and the mixed layer of the western pacific warm pool: Observations and 1D model results. *J. Clim.*, **9**, 3056–3085.

Arnold, N. P. and D. A. Randall, 2015: Global-scale convective aggregation: Implications for the Madden–Julian oscillation. *J. Adv. Model. Earth Syst.*, **7**, 1499–1518.

Bellenger, H. and J.-P. Duvel, 2009: An analysis of tropical ocean diurnal warm layers. *J. Climate*, **22**, 3629–3646.

Benedict, J. J. and D. A. Randall, 2011: Impact of idealized air-sea coupling on Madden-Julian oscillation structure in the superparameterized CAM. *J. Atmos. Sci.*, **68**, 1990–2008.

Bernie, D. J., S. J. Woolnough, and J. M. Slingo, 2005: Modeling diurnal and intraseasonal variability of the ocean mixed layer. *J. Clim.*, **18**, 1190–1202.

Bernie, D. J., E. Guilyardi, G. Madec, J. M. Slingo, S. J. Woolnough, and J. Cole, 2008: Impact of resolving the diurnal cycle in an ocean-atmosphere GCM. Part 2: A diurnally coupled CGCM. *Clim. Dynam.*, **31**, 909–925.

Cronin, M. F. and M. J. McPhaden, 1997: The upper ocean heat balance in the western equatorial Pacific warm pool during September–December 1992. *J. Geophys. Res*, **102**, 8533–8553.

Cronin, M. F. and M. J. McPhaden, 2002: Barrier layer formation during westerly wind bursts, *J. Geophys. Res.*, **107**, 8020.

Crueger, T., B. Stevens and R. Brokopf, 2013: The Madden–Julian oscillation in ECHAM6 and the introduction of an objective MJO metric, *J. Clim.*, **26**, 3241–3257.

de Boisséson, E., M. A. Balmaseda, F. Vitart, and K. Mogensen 2012: Impact of the sea surface temperature forcing on hindcasts of Madden–Julian oscillation events using the ECMWF model. *Ocean Sci.*, **8**, 1071–1084.

Dee, D. P. and co-authors, 2011: The ERA-Interim reanalysis: Configuration and performance of the data assimilation system. *Q. J. R. Meteorol. Soc.*, **137**, 553–597.

DeMott, C. A., C. Stan, D. A. Randall, and M. D. Branson, 2014: Intraseasonal variability in Coupled GCMs: The roles of ocean feedbacks and model physics. *J. Clim.*, **27**(13), 4970–4995.

DeMott, C. A., J. J. Benedict, N. P. Klingaman, S. J. Woolnough, and D. A. Randall, 2016: Diagnosing ocean feedbacks to the MJO: SST-modulated surface fluxes and the moist static energy budget. *J. Geophys. Res. Atmos.*, **121**, 8350–8373.

DeMott, C. A., N. P. Klingaman, and S. J. Woolnough, 2015: Atmosphere-ocean coupled processes in the Madden–Julian oscillation. *Rev. Geophys.*, **53**, 1099–1154, 2014RG000478.

Drushka, K., H. Bellenger, E. Guilyardi, M. Lengaigne, J. Vialard, and G. Madec, 2015: Processes driving intraseasonal displacements of the eastern edge of the Warm Pool: The contribution of westerly wind events. *Clim. Dynam.*, **44**, 735–755.

Drushka, K., J. Sprintall, and S. T. Gille, 2010: Vertical structure of Kelvin Waves in the Indonesian Throughflow exit passages, *J. Phys. Oceanogr.*, **40**, 1965–1987.

Gonzalez, A. O. and X. Jiang, 2017: Winter mean lower tropospheric moisture over the maritime continent as a climate model diagnostic metric for the propagation of the maddenjulian oscillation. *Geophys. Res. Lett.*, **44**(5), 2588–2596, 2016GL072430.

Halkides, D. J., D. E. Waliser, T. Lee, D. Menemenlis, and B. Guan, 2015: Quantifying the processes controlling intraseasonal mixed-layer temperature variability in the tropical Indian Ocean. *J. Geophys. Res. Oceans*, **120**, 692–715.

Hirons, L. C., P. Inness, F. Vitart, and P. Bechtold, 2013: Understanding advances in the simulation of intraseasonal variability in the ECMWF model, Part II: The application of process-based diagnostics. *Q. J. R. Meteorol. Soc.*, **139**, 1417–1426.

Hsu, P.-C. and T. Li, 2012: Role of the boundary layer moisture asymmetry in causing the eastward propagation of the Madden–Julian oscillation. *J. Clim.*, **25**, 4914–4931.

Hung, M.-P., J.-L. Lin, W. Wang, D. Kim, T. Shinoda, and S. J. Weaver, 2013: MJO and convectively coupled equatorial waves simulated by CMIP5 climate models. *J. Clim.*, **26**(17), 6185–6214.

Inness, P. M., J. M. Slingo, E. Guilyardi, and J. Cole, 2003: Simulation of the Madden-Julian oscillation in a coupled general circulation model. Part II: The role of the basic state. *J. Clim.*, **17**, 365–382.

Kessler, W. S., M. J. McPhaden, and K. M. Weickmann, 1995: Forcing of intraseasonal Kelvin waves in the equatorial Pacific. *J. Geophys. Res. Oceans*, **100**, 10613–10631.

Kessler, W. S., 2005: The oceans, in *Intraseasonal Variability in the Atmosphere-Ocean Climate System*, Springer Praxis, Springer, Berlin, pp. 175–222.

Kim, D. *et al.*, 2009: Application of MJO simulation diagnostics to climate models. *J. Climate*, **22**, 6413–6436.

Kim, D., H. Kim, and M.-L. Lee, 2017: Why does the MJO detour the Maritime Continent during austral summer? *Geophys. Res. Lett.*, **44**, 2579–2587.

Kim, H.-M., 2017: The impact of the mean moisture bias on the key physics of MJO propagation in the ecmwf reforecast. *J. Geophys. Res.* **122**(15), 7772–7784.

Kim, H.-M., C. D. Hoyos, P. J. Webster, and I.-S. Kang, 2008: Sensitivity of MJO simulation and predictability to sea surface temperature variability. *J. Clim.*, **21**, 5304–5317.

Klingaman, N. P. and S. J. Woolnough, 2014b: The role of air-sea coupling in the simulation of the Madden–Julian oscillation in the Hadley Centre model. *Q. J. R. Meteorol. Soc.*, **140**, 2272–2286.

Klingaman, N. P., P. M. Inness, H. Weller, and J. M. Slingo, 2008: The importance of high-frequency sea-surface temperature variability to the intraseasonal oscillation of Indian monsoon rainfall. *J. Clim.*, **21**, 6119–6140.

Li, Y. and R. E. Carbone, 2012: Excitation of rainfall over the tropical Western Pacific. *J. Atmos. Sci.*, **69**, 2983–2994.

Liess, S., L. Bengtsson, and K. Arpe, 2004: The intraseasonal oscillation in ECHAM4. Part I: Coupled to a comprehensive ocean model. *Clim. Dynam.*, **22**, 671–688.

Lim, Y., S. W. Son, and D. Kim, 2018x: MJO prediction skill of the subseasonal-to-seasonal prediction models. *J. Adv. Model. Earth Sys.*

Lin, J. L. *et al.*, 2006: Tropical intraseasonal variability in 14 IPCC AR4 climate models. Part I: Convective signals. *J. Clim.*, **19**, 2665–2690.

Madden, R. A. and P. R. Julian, 1972: Description of global-scale circulation cells in the tropics with a 40–50 day period. *J. Atmos. Sci.*, **29**, 1109–1123.

Maloney, E. D., 2009: The moist statice energy budget of a composite tropical intraseasonal oscillation in a climate model. *J. Clim.*, **22**, 711–729.

Matthews, A. J., D. B. Baranowski, K. J. Heywood, P. J. Flatau, and S. Schmidtko, 2014: The surface diurnal warm layer in the Indian Ocean during CINDY/DYNAMO. *J. Clim.*, **27**, 9101–9122,

Maloney, E. D. and A. H. Sobel, 2004: Surface fluxes and ocean coupling in the tropical intraseasonal oscillation. *J. Clim.*, **17**, 4368–4386.

Maloney, E. D., 2009: The moist static energy budget of a composite tropical intraseasonal oscillation in a climate model. *J. Clim.*, **22**, 711–729.

Moum, J. N. *et al.*, 2013: Air-sea interactions from westerly wind bursts during the November 2011 MJO in the Indian Ocean, *Bull. Am. Meteorol. Soc.*, **95**, 1185–1199.

Moum, J. N., K. Pujiana, R.-C. Lien, and W. D. Smyth, 2016: Ocean feedback to pulses of the Madden–Julian Oscillation in the equatorial Indian Ocean. *Nat. Comm.*, **7**, 13203 (2016).

Peatman, S. C., A. J. Matthews, and D. P. Stevens, 2014: Propagation of the Madden–Julian oscillation through the Maritime Continent and scale interaction with the diurnal cycle of precipitation, *Q. J. R. Meteorol. Soc.*, **140**, 814–825.

Pegion, K. and B. P. Kirtman, 2008: The impact of air-sea interactions on the simulation of tropical intraseasonal variability, *J. Clim.*, **21**, 6616–6635.

Pohl, B. and A. J. Matthews, 2007: Observed changes in the lifetime and amplitude of the Madden–Julian oscillation associated with interannual ENSO sea surface temperature anomalies. *J. Clim.*, **20**, 2659–2674.

Ramirez, E., P. L. Da Silva Dias, and C. F. M. Raupp, 2017: Multiscale atmosphere-ocean interactions and the low-frequency variability in the Equatorial region. *J. Atmos. Sci.*, **74**, 2503–2523.

Ruppert, J. H. and R. H. Johnson, 2015: Diurnally modulated cumulus moistening in the pre-onset stage of the Madden–Julian oscillation during DYNAMO. *J. Atmos. Sci.*, **72**, 1622–1647.

Rydbeck, A. V. and T. G. Jensen, 2017: Oceanic impetus for convective onset of the Madden–Julian oscillation in the Western Indian Ocean. *J. Clim.*, **30**, 4299–4316.

Sabeerali, C. T., A. R. Dandi, A. Dhakate, K. Salunke, S. Mahapatra and S. A. Rao, 2013: Simulation of boreal summer intraseasonal oscillations in the latest CMIP5 coupled GCMs. *J. Geophys. Res.*, **118**, 4401–4420.

Schott, F. A., S.-P. Xie, and J. P. McCreary, 2009: Indian ocean circulation and climate variability. *J. Geophys. Res.*, **47**, RG1002.

Seo, H., A. C. Subramanian, A. J. Miller, and N. R. Cavanaugh, 2014: Coupled impacts of the diurnal cycle of sea surface temperature on the Madden–Julian oscillation. *J. Clim.*, **27**, 8422–8443.

Seo, K.-H. and W. Wang 2010: The Madden–Julian oscillation simulated in the NCEP Climate Forecast System model: The importance of stratiform heating. *J. Clim.*, **23**, 4770–4793.

Shelly, A., P. Xavier, D. Copsey, T. Johns, J. M. Rodriguez, S. Milton, and N. P. Klingaman, 2014: Coupled versus uncoupled hindcast simulations of the Madden–Julian oscillation in the Year of Tropical Convection. *Geophys. Res. Lett.*, **41**, 5670–5677.

Shinoda, T., T. G. Jenson, M. Flatau, S. Chen, W. Han, and C. Wang, 2013: Large-scale oceanic variability associated with the Madden–Julian oscillation during the CINDY/DYNAMO field campaign from satellite observations. *Remote Sens.*, **5**, 2072–2092.

Shinoda, T., 2005: Impact of diurnal cycle of solar radiation on intraseasonal SST variability in the western equatorial Pacific. *J. Clim.*, **18**, 2628–2636.

Sobel, A. H., E. D. Maloney, G. Bellon, and D. M. Frierson, 2010: Surface fluxes and tropical intraseasonal variability: A reassessment. *J. Adv. Model. Earth Sys.*, **2**, 2.

Soloviev, A. and R. Lukas, 1997: Sharp frontal interfaces in the near-surface layer of the ocean in the western equatorial Pacific warm pool. *J. Phys. Oceanogr.*, **27**, 999–1017.

Sperber, K. R., 2004: Madden–Julian variability in NCAR CAM2.0 and CCSM2.0. *Clim. Dynam.*, **23**, 259–278.

Sprintall, J., A. L. Gordon, A. Koch-Larrouy, T. Lee, J. T. Potemra, K. Pujiana, and S. E. Wijffels, 2014: The Indonesian seas and their role in the coupled ocean-climate system. *Nat. Geosci.*, **7**, 487–492.

Thayer-Calder, K. and D. A. Randall, 2009: The role of convective moistening in the Madden–Julian oscillation. *J. Atmos. Sci.*, **66**, 3297–3312.

Tseng, W.-L., B.-J. Tsuang, N. S. Keenlyside, H.-H. Hsu, and C.-Y. Tu, 2015: Resolving the upper-ocean warm layer improves the simulation of the Madden–Julian oscillation. *Clim. Dynam.*, **44**, 1487–1503.

Wang, B. and X. Xie, 1998: Coupled modes of the warm pool climate system. Part I: The role of air-sea interaction in maintaining Madden–Julian oscillations. *J. Clim.*, **8**, 2116–2135.

Wang, B. and G. Chen, 2017: A general theoretical framework for understanding essential dynamics of Madden–Julian oscillation. *Clim. Dynam.*, **49**, 2309–2328

Webber, B. G. M., A. J. Matthews, K. J. Heywood, and D. P. Steves, 2012: Ocean Rossby waves as a triggering mechanism for primary Madden–Julian events. *Q. J. R. Meteorol. Soc.*, **138**, 514–527.

Webster, P. J., C. A. Clayson, and J. A. Curry, 1996: Clouds, radiation, and the diurnal cycle of sea surface temperature in the tropical western Pacific. *J. Clim.*, **9**, 1712–1730.

Weller, R. A. and S. P. Anderson, 1996: Surface meteorology and air-sea fluxes in the western equatorial Pacific warm pool during the TOGA coupled ocean-atmosphere response experiment. *J. Clim.*, **9**, 1959–1990.

Woolnough, S. J., F. Vitart, and M. A. Balmaseda, 2007: The role of the ocean in the Madden–Julian oscillation: Implications for MJO prediction. *Q. J. R. Meteorol. Soc.*, **133**, 117–128.

Zhang, C. and M. J. McPhaden, 2000: Intraseasonal surface cooling in the equatorial western Pacific. *J. Clim.*, **13**, 2261–2276.

Zhang, C., M. Dong, S. Gualdi, H. H. Hendon, E. D. Maloney, A. Marshall, K. R. Sperber, and W. Wang 2006: Simulations of the Madden–Julian oscillation in four pairs of coupled and uncoupled global models. *Clim. Dynam.*, **27**, 573–592.

Zhu, J., W. Wang, and A. Kumar, 2017: Simulations of MJO propagation across the Maritime Continent: Impact of SST feedback. *J. Clim.*, **30**, 1689–1704.

Chapter 21

Review: MJO Propagation over the Maritime Continent

Daehyun Kim*,§, Eric D. Maloney†,¶ and Chidong Zhang‡,||

**Department of Atmospheric Sciences, University of Washington, Seattle, WA 98195, USA*
§daehyun@uw.edu
†Department of Atmospheric Science, Colorado State University, Fort Collins, CO 80521, USA
¶emaloney@atmos.colostate.edu
‡Pacific Marine Environmental Laboratory/NOAA, Seattle, WA 98115, USA
||chidong.zhang@noaa.gov

This review focuses on the propagation of the Madden–Julian oscillation (MJO) over the Indo-Pacific Maritime Continent (MC). The MJO tends to detour around the MC into the summer hemisphere. Moreover, the MJO's propagation comes to an end most frequently over the MC, which is often referred to as the "MC barrier effect". This peculiar behavior of the MJO in the MC is poorly represented in weather and climate models, partly due to our incomplete understanding of the interaction between the MJO and the MC. The MJO interacts with monsoon circulations, synoptic variability, and the diurnal cycle over the MC. Existing hypotheses for the MJO detour and the MC barrier effect involve the direct disruption of MJO convection by the presence of the MC islands and topography, processes associated with free tropospheric moistening, the diurnal cycle, and the basic state. The exact roles of these processes in the propagation of the MJO through the MC region need to be better understood.

1. Introduction

The Madden–Julian oscillation (MJO, Madden and Julian 1971, 1972) is the dominant mode of tropical intraseasonal (20–100 days) variability. It influences a wide range of weather, climate, and Earth system phenomena, including high-impact events (Zhang 2013). Despite its unmatched significance in the Earth system on the intraseasonal timescale, our understanding of the MJO remains incomplete.

The Earth provides the MJO with two interesting zonal asymmetries at the bottom boundary of the atmosphere. The first is the distribution of sea surface temperature (SST), which is considerably higher over the Indo-Pacific warm pool than over the other ocean basins. The effect of this zonal asymmetry in SST is well understood — MJO circulation anomalies are tightly coupled to convective anomalies over the warm pool region while they decouple from convection over cooler waters. The phase speed of the MJO over the warm pool is also much slower than that over the other ocean basins.

The second zonal asymmetry in the surface boundary condition comes from the Indo-Pacific Maritime Continent (MC), the largest archipelago on Earth, which is located right in the middle of the warm pool. While it has long been observed that the MJO exhibits distinct behavior over the MC (see Section 3), the mechanism through which the MC affects

The Multiscale Global Monsoon System
Edited by C. P. Chang et al.

the MJO's propagation is not well understood because propagation of the MJO over the MC had received relatively less attention than that over open oceans. Recently, a growing interest has developed in understanding the interaction of the MJO with the MC.

This chapter summarizes our current knowledge or lack thereof on the issue of MJO propagation over the MC. In Section 2, we survey the literature on multiscale interactions of the MJO with the monsoons, synoptic variability, and diurnal cycle over the MC. The peculiar behavior of the MJO over the MC is described in Section 3, and existing hypotheses for this behavior are reviewed in Sections 4 and 5. Section 6 discusses the remaining challenges in understanding the effect of the MC on the MJO's propagation, which is followed by conclusions in Section 7.

2. Interaction of the MJO with the Monsoons, Synoptic Variability, and Diurnal Cycle over the MC

The geographical characteristics of the MC make the area a unique place, where rich interactions among a wide variety of phenomena prevail. The complex land-sea distribution and topography embedded in a body of warm water — the Indo-Pacific warm pool — foster a persistent, strong diurnal cycle of convection and associated atmospheric circulation via diurnally varying differential heating between land and the ocean (e.g., Houze *et al.* 1981; Johnson and Priegnitz 1981). The Borneo vortex is a form of synoptic variability resulting from the interaction between topography and the circulation around Borneo. Located near the Equator in between the Asian and Australian continents, the circulation and convection in the MC are also influenced by the monsoon systems of the two continents as well as its own.

During its journey over the MC, the MJO inevitably interacts with these multiscale features: the diurnal cycle, synoptic variability, and monsoons. The modulation of the MC diurnal cycle occurs on multiple timescales (e.g., Yamanaka 2016). On the intraseasonal timescale, the MJO systematically influences the amplitude and phase of the MC diurnal cycle by modulating large-scale atmospheric conditions (Sui and Lau 1992; Ichikawa and Yasunari 2006; Tian *et al.* 2006; Rauniyar and Walsh 2010; Oh *et al.* 2012; Peatman *et al.* 2014, 2015; Birch *et al.* 2016; Vincent and Lane 2016; Sakeda *et al.* 2017). Observational and modeling studies have suggested that the MC land diurnal cycle strengthens (weakens) during the suppressed (enhanced) phase of the MJO, primarily due to increased (decreased) surface shortwave radiation (Peatman *et al.* 2014; Birch *et al.* 2016; Vincent and Lane 2016). The MC oceanic diurnal cycle responds to the MJO in the opposite manner, amplifying during the active phase of the MJO, essentially because the anomalously humid large-scale conditions provided by the MJO are favorable for the nighttime development of oceanic convection. As a result of the different relationships of land and oceanic diurnal cycles with the MJO phase, the arrival of an enhanced diurnal cycle of convection and daily mean precipitation over the MC islands is about one phase earlier than that over the ocean, which is often referred to as a "vanguard" of precipitation (Peatman *et al.* 2014).

The MJO interacts with synoptic systems over the MC (Chang *et al.* 2005, 2016). During boreal winter, the MC area experiences occasional intrusions of cold air masses from the north. These cold air masses originate from inland areas of the Asian continent and comprise part of the Asian winter monsoon system. The frequency of these monsoon cold surges, especially relatively weak ones, is affected by the MJO through its low-level meridional wind anomalies over the South China Sea. The MJO also influences the frequency and structure of the Borneo vortex. The frequency of Borneo vortex events increases with the intensity of cold surges, a connection that is heavily modulated by the MJO.

The MJO is known to interact with the Australian monsoon (Hendon and Liebmann 1990; Wheeler and McBride 2012) and the MC monsoon (Robertson 2011; Chang *et al.* 2016; Yoden *et al.* 2016). The timing and duration of active and break periods of both monsoon systems are affected by the MJO, although not every single active or break event is associated with the MJO. In the MC region, the modulation of rainfall by the MJO is more pronounced over the oceans than over the islands (Sobel *et al.* 2010; Zhang and Ling 2017). The monsoon circulations in turn provide the MJO with a background state. Recent studies have shown that the background state, especially that of low-level moisture, plays a critical role in the MJO's propagation (Jiang 2017; Gonzalez and Jiang 2017).

3. The MJO's Peculiar Behavior in the Maritime Continent

Intraseasonal variability of convection in the MC has been examined using satellite outgoing longwave radiation (OLR) data since the early 1980s (Lau and Chan 1983; Weickmann 1983; Lau and Chan 1985; Weickmann *et al.* 1985; Knutson *et al.* 1986; Wang and Rui 1990). The early studies found that intraseasonal variability of convection is weaker over the MC than the surrounding oceans (Weickmann *et al.* 1985) and that the eastward moving MJO often "detours" around the MC toward the summer hemisphere (Lau and Chan 1985; Knutson *et al.* 1986; Wang and Rui 1990, Fig. 1). Later studies have confirmed these findings using longer OLR datasets and satellite estimates of rainfall (Salby and Hendon 1994; Wang and Li 1994; Zhang and Dong 2004; Zhang and Ling

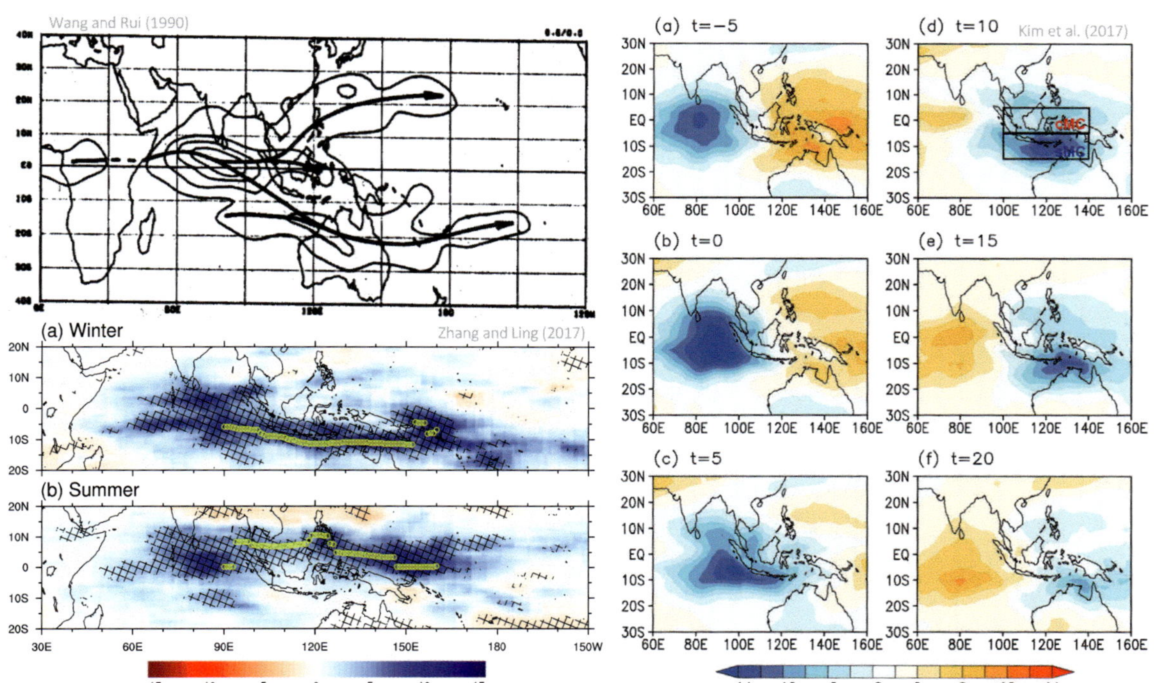

Fig. 1. (left) Preferred propagation paths of intraseasonal anomalies in (upper) OLR (Wang and Rui 1990) and (lower) precipitation (Zhang and Ling 2017). (right) Boreal winter (December–February) intraseasonal OLR anomalies regressed onto area-averaged intraseasonal OLR anomalies over the Indian Ocean (Kim *et al.* 2017a).

2017; Kim *et al.* 2017, Fig. 1). The right panel in Fig. 1 shows that during its passage over the MC, MJO anomalous convection during boreal winter is much stronger in the southern MC, especially over the ocean, than in the equatorial region, highlighting the detour of the MJO.

In addition to the weakening and detouring, MJO propagation is sometimes disrupted over the MC (Hirata *et al.* 2013; Ling *et al.* 2013; Kim *et al.* 2014a; Feng *et al.* 2015; Kerns and Chen 2016; Zhang and Ling 2017). Observations have shown that the chances for the termination of MJO propagation are higher over the MC than over other parts of the Indo-Pacific warm pool (Klingaman and Woolnough 2014; Peatman *et al.* 2015; Zhang and Ling 2017). For example, Zhang and Ling (2017) found that about half of the MJO events that form over the Indian Ocean terminate over the MC, though the rate is higher (about 77%) when only relatively strong events are considered (Feng *et al.* 2015). This MC barrier effect on the observed MJO suggests that the mechanisms detrimental to MJO propagation are specific to the MC. Some forecast models tend to overestimate the probability of MJO disruption over the MC (Vitart and Molteni 2010; Wang *et al.* 2014; Kim *et al.* 2014b). As an example, Vitart and Molteni (2010) showed that the percentage of MJO events progressing through consecutive phases when convection is over the MC (i.e., transitions from phase 4 to 5 and from 5 to 6) is underestimated in an earlier version of the ECMWF forecast model, suggesting that the detrimental effects are exaggerated in the models. Wang *et al.* (2014) and Kim *et al.* (2014b) found that MJO prediction skill is notably lower when forecasting MJO propagation across the MC in NCEP CFSv2, while later studies have found that such skill dependency is absent in other forecast models (Neena *et al.* 2014; Xiang *et al.* 2015; Lim *et al.* 2018).

Previous studies on the MC's effects on MJO propagation are reviewed in the following sections.

4. The MJO Detour

In this section, we group existing hypotheses for the MJO detour around the MC into two categories based on their core ideas. The first group of hypotheses — interference hypotheses — emphasizes the direct interference that the MJO would experience over the MC due to the presence of land masses, orography, and diurnal convection. The second group considers the regional differences in moistening processes as the key mechanism. We note that the two groups of ideas are not necessarily exclusive of each other.

4.1. *Interference hypotheses*

Hsu and Lee (2005) and Wu and Hsu (2009) showed that the *steep topography* in some MC islands distorts the low-level flow associated with the MJO over the MC. Wu and Hsu (2009) suggested that the north–south-oriented steep orography over Sumatra island blocks the propagation of the MJO circulation especially in the lower troposphere, preventing the development of subsequent convective anomalies in the central MC area. The observational results by Wu and Hsu (2009) are consistent with the modeling results of Inness and Slingo (2006), who showed that the low-level Kelvin wave component of the modeled MJO propagates more smoothly across the MC in their aqua-planet simulations when MC topography is removed.

Sobel *et al.* (2008, 2010) suggested that the increase of *surface latent heat flux* associated with the MJO under enhanced convection is limited in the MC islands due to substantially reduced heat capacity relative to ocean regions, the effect of which would weaken the MJO. MJO variability is reduced in mechanism denial experiments in which the interaction of large-scale circulation and surface latent heat flux is turned off (e.g., Maloney and Sobel 2004; Maloney *et al.* 2010; Kim *et al.* 2011), supporting the notion that the suppression of latent heat flux over MC islands might explain the weakening of the MJO there.

It has been hypothesized that a competition for instability exists between the *diurnal cycle of convection* and the MJO. That is, the strong diurnal cycle over the MC competes against the MJO for a finite amount of energy available for convection (Wang and Li 1994; Zhang and Hendon 1997; Neale and Slingo 2003; Oh *et al.* 2012; Hagos *et al.* 2016). While the exact quantity for which the diurnal cycle is competing with the MJO has remained elusive, modeling studies have shown that disabling or weakening the diurnal cycle over the MC islands enhances MJO propagation (Fig. 2, from Hagos *et al.* 2016), supporting the competition idea. Hagos *et al.* (2016) also emphasized the role of radiative feedbacks.

Majda and Yang (2016) emphasized the role of the upscale transport of momentum and heat by the diurnal cycle of MC convection. They showed that the rectified effects of the upscale momentum transfer strengthen lower (upper) level mean westerlies (easterlies) in the MJO circulation, while the upscale heat transfer weakens MJO temperature anomalies.

4.2. *Differential moistening hypothesis — Moisture mode thinking*

A growing school of thought considers the MJO to be a "moisture mode" (see Chapter 26).

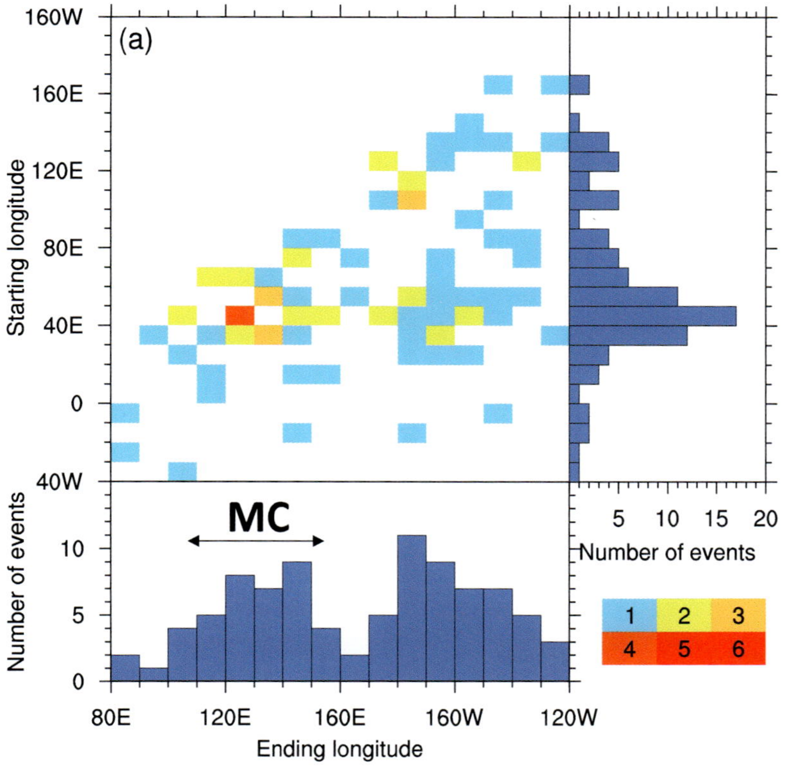

Fig. 2. Longitude–time diagram of equatorial precipitation from (a) TRMM observations and regional model simulations (b) with and (c) without the diurnal cycle over the MC_2 (from Hagos *et al.* 2016).

Under the moisture mode framework, maintenance and propagation of the MJO are explained by processes regulating the propagation and maintenance of moisture anomalies, with horizontal temperature gradients assumed to be small.

In an attempt to explain the MJO detour in the moisture mode framework, Kim *et al.* (2017) hypothesized that MJO convection detours the MC since moisture buildup is suppressed in the equatorial MC region, while moisture buildup occurs in the Timor and Arafura Seas (southern MC), thus favoring convection there. To test this "differential moistening hypothesis", Kim *et al.* (2017) analyzed the column-integrated moist static energy (MSE) budget of the MJO and showed that the growth of moisture anomalies in the central MC area is indeed much weaker than that in the southern MC area, primarily due to the difference in horizontal moisture advection (Fig. 3). Furthermore, they found that anomalous zonal moisture advection is larger in the southern MC area because the background zonal moisture gradient is much steeper there. During boreal winter, the Australian monsoon is active and it sets up the large-scale east–west gradient of the mean moisture in the southern MC region. Regarding meridional moisture advection, it was found that the main difference between the central and southern MC regions is not the background moisture gradient, but rather the pattern of MJO meridional wind anomalies. While the southern MC area features almost uniform northerly MJO wind anomalies during the period of moisture anomaly growth, bringing moist air from near 5°N to the south, the meridional wind anomalies in the central MC region show a wavy pattern with both northerly and southerly anomalies, weakening the MJO-scale moisture advection. It was hypothesized that the distortion of meridional wind anomalies in the central MC region is caused by the interaction of MJO flow with synoptic variability — particularly monsoon cold surges and the Borneo vortex.

5. MC Barrier Effects

As mentioned above, the MJO's propagation from the Indian Ocean to the west Pacific is often disrupted in the MC. Why do some MJO events stall over the MC? What distinguishes the MJO events that propagate across the MC from the non-propagating or blocked events? While the detour toward the summer hemi-

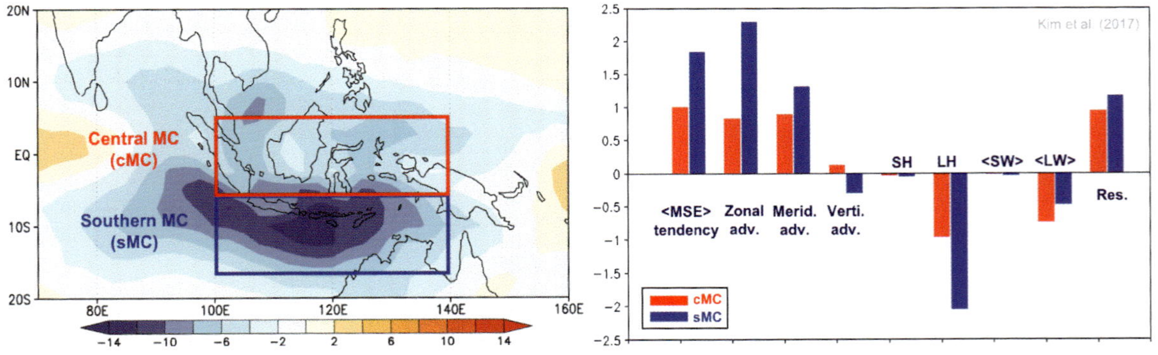

Fig. 3. (left) Intraseasonal OLR anomalies (W m^{-2}) when MJO is over the MC area. The central and southern MC area are indicated by red and blue box, respectively. (right) Column-integrated moist static energy budget normalized by the tendency in the central MC area during the MC moistening period. Red and blue bars represent that of central and southern MC areas respectively (figures are adopted from Kim *et al.* 2017).

sphere is a characteristic of many "propagating" or "MC-crossing" MJO events, it does not necessarily explain why some MJO events come to an end over the MC.

The MC barrier effect has been a subject of several recent observational and modeling studies (Hirata *et al.* 2013; Ling *et al.* 2013; Kim *et al.* 2014a, 2016; Feng *et al.* 2015; Kerns and Chen 2016; Zhang and Ling 2017). The key mechanisms proposed in these studies can be grouped into four categories, with two of them being closely related to the interference and differential moistening hypotheses for MJO detouring.

5.1. *Land vs. oceanic convection*

Zhang and Ling (2017) found that propagating MJO events differ from blocked events by the ratio of oceanic precipitation to land precipitation (Fig. 4). That is, MJO propagation is more likely to be stalled in the MC area when convection is strong over the islands. It was suggested that the development of oceanic convection is key to the propagation of the MJO across the MC and that MJO convection over the ocean is occasionally disrupted by anomalously strong land convection, similar to the interference hypotheses for MJO detouring that highlight the competition between land (diurnal cycle) and oceanic convection.

5.2. *Moisture mode thinking*

Kim *et al.* (2014a) analyzed the column-integrated MSE budget of propagating and non-propagating MJO events to examine their associated moistening/drying processes. Results show that meridional moisture advection, in particular the component from MJO perturbation winds acting upon the background meridional moisture gradient, is the primary factor that distinguishes propagating events from those that cease over the MC (Fig. 5). They also found that the meridional wind anomalies that are key to MJO propagation across the MC take the form of an equatorial Rossby wave response to negative heating anomalies (i.e., suppressed convection) over the west Pacific (Fig. 5).

Based on their findings, Kim *et al.* (2014a) suggested that the suppressed conditions over the west Pacific at the time of the MJO's onset over the Indian Ocean determine whether or not the identified MJO events propagate through the MC, emphasizing the essential role of the intrinsic dipole heating pattern of the MJO. Kim *et al.* (2017b) showed that the MJO prediction skill of the ECMWF model is substantially higher when the dipole heating pattern is more pronounced in the initial conditions. Feng *et al.* (2015) suggested that while suppressed conditions over the west Pacific might not be a sufficient condition for an MJO event to

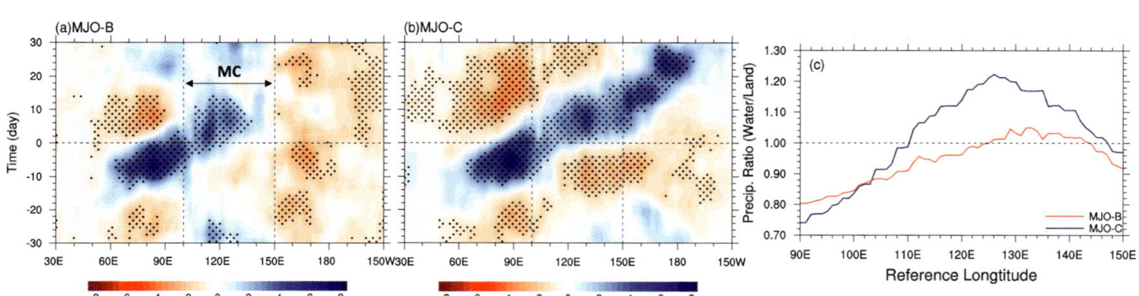

Fig. 4. Composite intraseasonal rain rate anomalies (mm day^{-1}) for MJO events that (a) stall over and (b) propagate across the MC region. (c) Longitudinal distribution of oceanic vs. land rain rate ratio (figures are adopted from Zhang and Ling 2017).

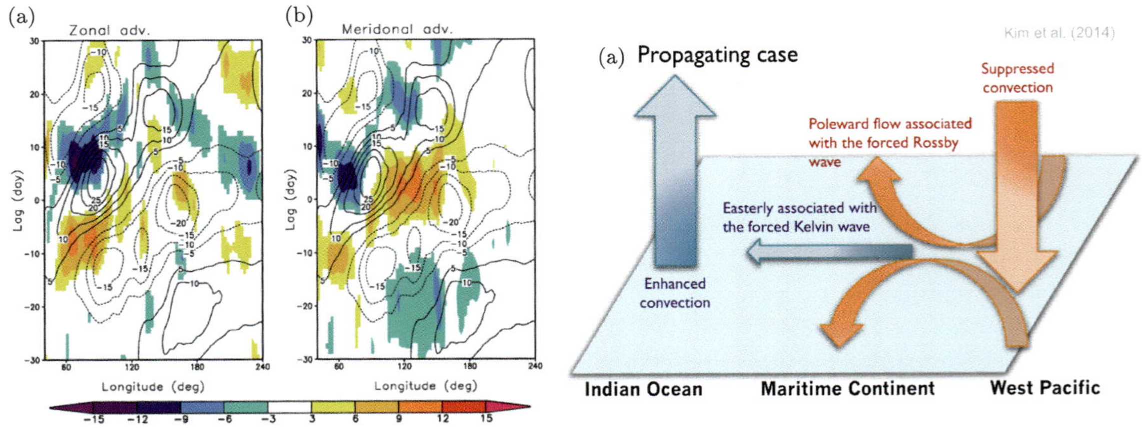

Fig. 5. (left) Composite longitude lag diagram of intraseasonal OLR ($\mathrm{W\,m^{-2}}$, contour) and (a) zonal and (b) meridional MSE advection anomalies ($\times 3 \times 10^5\ \mathrm{J\,m^{-2}}$). (right) Schematic diagram highlighting key components of MJO events that propagate across the MC area (from Kim *et al.* 2014a).

propagate across the MC, eastward propagation is much more pronounced when dry anomalies are present over the west Pacific.

5.3. *Role of air-sea coupling*

The possible effects of SST anomalies and air-sea coupling on MJO propagation across the MC have been discussed. Hirata *et al.* (2013) showed that pronounced eastward propagation of the MJO across the MC is associated with warmer SST anomalies over the west Pacific. They suggested that warmer ocean conditions favor eastward propagation of the MJO by promoting the development of convection over the west Pacific.

Zhang and Ling (2017) found that SST in the Timor Sea is higher in MJO events that propagate through the MC than in those that stall over the MC. This SST difference cannot be explained by the surface energy budget, suggesting potential roles of oceanic processes.

5.4. *Transient dry precursor*

Feng *et al.* (2015) examined the time evolution of moisture anomalies in individual non-propagating MJO cases. They found that, in a subset of the non-propagating events, moistening over the eastern MC and west Pacific area to the east of the Indian Ocean and western MC MJO convection is suppressed by intrusion of Rossby wave-type dry signals that originate in the eastern Pacific and propagate to the west. This finding was supported by DeMott *et al.* (2018), who named them "transient dry precursors". Whether the transient dry precursors occur coincidently or whether they are part of MJO circulation is currently an open question.

6. Remaining Challenges — Unveiling the Exact and Relative Roles of the MC Diurnal Cycle and the Mean State Moisture Gradient

While the studies surveyed in this chapter have made notable progress in understanding the MJO detours around the MC, MC barrier effects, and their implications for MJO propagation, many questions still remain unanswered. In this section, we discuss two of the challenging issues that need to be addressed to advance our understanding of the interaction between the MJO and MC.

First, how the MC diurnal cycle interferes with MJO propagation needs to be understood. While it has been hypothesized that MC land convection in the form of a strong, persistent diurnal cycle competes for a finite amount of moisture with MJO convection, the nature of the competition and the mechanism by which the diurnal cycle of convection interferes with the MJO remain elusive. Also, the extent to which the upscale momentum and heat transfer from the MC diurnal cycle affect the MJO circulation needs to be examined further using observations and high-resolution model simulations. Questions that need to be answered in this regard include the following: What controls the MC diurnal cycle? How does MC land convection affect the development of convection over the adjacent sea? What are the roles of local circulations, moisture, and temperature in the competition between land and oceanic convection over the MC?

Second, the prevailing diurnal cycle over the MC, as a persistent heat source, could also affect the basic state. The fact that most rainfall over the MC islands is associated with the diurnal cycle (e.g., Yamanaka 2016) suggests that the diurnal cycle contributes significantly to the seasonal mean diabatic heating. Bergemann *et al.* (2015) also showed that up to 40–60% of mean rainfall in certain MC regions could be caused by coastal effects including offshore propagation of strong diurnal convective systems. In other words, without the diurnal cycle, the distribution of mean precipitation over the MC would be very different. Previous studies that conducted a set of GCM experiments with and without the MC islands found notable changes in mean precipitation over the MC region when the MC islands were removed (Neale and Slingo 2003; Zhu *et al.* 2010; Tseng *et al.* 2017), supporting the notion that a tight link exists between the MC diurnal cycle and the mean state.

Recent observational and modeling studies of MJO propagation suggest that to correctly represent the propagation of the MJO over the MC, a model is required to maintain an accurate basic state, especially the large-scale moisture gradient. Gonzalez and Jiang (2017) showed that among different models, those with a more realistic meridional moisture gradient in the MC have a better representation of MJO propagation. As mentioned above, Kim *et al.* (2017) also suggested that the zonal moisture gradient in the Timor Sea is important for propagation of detouring MJO events through that region, and hence models should also get this zonal gradient correct to simulate realistic propagation. This result highlights that an accurate representation of the seasonal mean background state is a prerequisite for an accurate representation of MJO propagation, consistent with other studies arguing that the mean state is what determines the propagation characteristics of the MJO (Zhang and Dong 2004; Zhu *et al.* 2010; Li 2014; Adames *et al.* 2016; Jiang *et al.* 2018).

The possible link between the MC diurnal cycle and the mean state together with the robust relationship between the mean state and MJO simulation fidelity in models suggests that a GCM bias in the diurnal cycle over the MC would likely cause mean state biases, which would in turn affect the propagation of the MJO.

In order to better understand the role that the MC diurnal cycle plays in shaping the mean state, the following questions need to be answered: What determines the basic state (e.g., moisture distribution) and its seasonal cycle? To what extent does the MC diurnal cycle affect the background humidity distribution? What are the relative roles of the diurnal cycle and the monsoons in determining the mean state?

7. Summary and Conclusions

This chapter reviews research efforts stimulated by the observed detour of the MJO around the MC islands toward the summer hemisphere and MC barrier effects on MJO propagation. The different behavior of the MJO over the MC region relative to the open ocean can also be a test bed for theories for the MJO and our understanding of tropical atmosphere.

Our review outlines several hypotheses for the MJO detour around the MC and the MC barrier effect. Some of these hypotheses may reinforce each other, while others compete, reflecting our limited understanding of MJO propagation over the MC. Interference with MJO propagation due to complex land-sea contrasts, steep mountains, and the strong diurnal cycle of land convection over the MC has been suggested. Under the moisture mode framework, which is not necessarily exclusive of interference ideas, regional differences in moistening processes and MSE budget differences between propagating and non-propagating MJO events have been analyzed, highlighting horizontal moisture advection as a key process.

This review indicates the need for further research on the key processes underlying the hypotheses discussed in this chapter. In particular, a better understanding of the competition between land and oceanic convection in the MC and the effect of the MC diurnal cycle on the basic state will help improve our ability to model MJO propagation through this region. Efforts are warranted to further understand the complex multiscale interaction between monsoons, MJO, synoptic variability, and diurnal cycle over the MC, by using observations from the recent and ongoing field campaigns — Years of the Maritime Continent (YMC) and Propagation of Intra-Seasonal Tropical Oscillations (PISTON).

Acknowledgments

We appreciate C. P. Chang's leadership role in organizing IWM-VI. D. Kim was supported by NASA grant 80NSSC17K0227 and NOAA grant NA18OAR4310300. E. Maloney acknowledges support from the Office of Naval Research grant N00014-16-1-3087 and NASA CYGNSS project NNX17AH77G. PMEL contribution number 4787.

References

Adames, Á. F., J. M. Wallace, and J. M. Monteiro, 2016: Seasonality of the structure and propagation characteristics of the MJO. *J. Atmos. Sci.*, **73**, 3511–3526.

Bergemann, M., C. Jakob, and T. P. Lane, 2015: Global detection and analysis of coastline-associated rainfall using an objective pattern recognition technique. *J. Clim.*, **28**, 7225–7236.

Birch, C. E., S. Webster, S. C. Peatman, D. J. Parker, A. J. Matthews, Y. Li, and M. E. E. Hassim, 2016: Scale interactions between the MJO and the Western Maritime Continent. *J. Clim.*, **29**, 2471–2492.

Chang, C.-P., P. A. Harr, and H.-J. Chen, 2005: Synoptic disturbances over the equatorial South China Sea and Western Maritime Continent during boreal winter. *Mon. Weather Rev.*, **133**, 489–503.

Chang, C.-P., M.-M. Lu, and H. Lim, 2016: Monsoon convection in the Maritime Continent: Interaction of large-scale motion and complex terrain. *Meteorol. Monogr.*, **56**, 6.1–6.29.

Feng, J., T. Li, and W. Zhu, 2015: Propagating and nonpropagating MJO events over Maritime Continent. *J. Clim.*, **28**, 8430–8449.

Gonzalez, A. O. and X. Jiang, 2017: Winter mean lower tropospheric moisture over the Maritime Continent as a climate model diagnostic metric for the propagation of the Madden–Julian oscillation. *Geophys. Res. Lett.*, **44**, 2588–2596.

Hagos, S. M., C. Zhang, Z. Feng, C. D. Burleyson, C. De Mott, B. Kerns, J. J. Benedict, and M. N. Martini, 2016: The impact of the diurnal cycle on the propagation of Madden–Julian Oscillation convection across the Maritime Continent. *J. Adv. Model. Earth Syst.*, n/a-n/a.

Hendon, H. H., and B. Liebmann, 1990: The Intraseasonal (30–50 day) Oscillation of the Australian Summer Monsoon. *J. Atmos. Sci.*, **47**, 2909–2924.

Hirata, F. E., P. J. Webster, and V. E. Toma, 2013: Distinct manifestations of austral summer tropical intraseasonal oscillations. *Geophys. Res. Lett.*, **40**, 3337–3341.

Houze Jr, R. A., S. G. Geotis, F. D. Marks Jr., and A. K. West, 1981: Winter Monsoon Convection in the Vicinity of North Borneo. Part I: Structure and time variation of the clouds and precipitation. *Mon. Wea. Rev.*, **109**, 1595–1614.

Hsu, H.-H. and M.-Y. Lee, 2005: Topographic effects on the eastward propagation and initiation of the Madden–Julian oscillation. *J. Clim.*, **18**, 795–809.

Ichikawa, H. and T. Yasunari, 2006: Time–space characteristics of diurnal rainfall over Borneo and surrounding oceans as observed by TRMM-PR. *J. Clim.*, **19**, 1238–1260.

Inness, P. M. and J. M. Slingo, 2006: The interaction of the Madden–Julian oscillation with the Maritime Continent in a GCM. *Q. J. R. Meteorol. Soc.*, **132**, 1645–1667.

Jiang, X., 2017: Key processes for the eastward propagation of the Madden–Julian oscillation based on multimodel simulations. *J. Geophys. Res. Atmos.*, n/a-n/a.

Jiang, X., Á. F. Adames, M. Zhao, D. Waliser, and E. Maloney, 2018: A unified moisture mode framework for seasonality of the Madden–Julian oscillation. *J. Clim.*, **31**, 4215–4224.

Johnson, R. H. and D. L. Priegnitz, 1981: Winter Monsoon Convection in the Vicinity of North Borneo. Part II: Effects on large-scale fields. *Mon. Wea. Rev.*, **109**, 1615–1628.

Kerns, B. W. and S. S. Chen, 2016: Large-scale precipitation tracking and the MJO over the Maritime Continent and Indo-Pacific warm pool. *J. Geophys. Res. Atmos.*, **121**, 8755–8776.

Kim, D., A. H. Sobel, and I.-S. Kang, 2011: A mechanism denial study on the Madden–Julian oscillation. *J. Adv. Model. Earth Syst.*, **3**, M12007.

Kim, D., J.-S. Kug, and A. H. Sobel, 2014a: Propagating versus nonpropagating Madden–Julian oscillation events. *J. Clim.*, **27**, 111–125.

Kim, D., H. Kim, and M.-I. Lee, 2017: Why does the MJO detour the Maritime Continent during austral summer? *Geophys. Res. Lett.*, **44**, 2579–2587.

Kim, H.-M., P. J. Webster, V. E. Toma, and Kim, 2014b: Predictability and prediction skill of the MJO in two operational forecasting systems. *J. Clim.*, **27**, 5364–5378.

Klingaman, N. P., and S. J. Woolnough, 2014: The role of air-sea coupling in the simulation of the Madden–Julian oscillation in the Hadley Centre model. *Q. J. R. Meteorol. Soc.*, **140**, 2272–2286.

Knutson, T. R., K. M. Weickmann, and J. E. Kutzbach, 1986: Global-scale intraseasonal oscillations of outgoing longwave radiation and 250 mb zonal wind during Northern Hemisphere Summer. *Mon. Weather Rev.*, **114**, 605–623.

Lau, K.-M., and P. H. Chan, 1983: Short-term climate variability and atmospheric teleconnections from satellite-observed outgoing longwave radiation. Part I: simultaneous relationships. *J. Atmos. Sci.*, **40**, 2735–2750.

Lau, K.-M., and P. H. Chan, 1985: Aspects of the 40–50 Day Oscillation during the Northern Winter as inferred from outgoing longwave radiation. *Mon. Weather Rev.*, **113**, 1889–1909.

Lee, J.-E., and Coauthors, 2012: Reduction of tropical land region precipitation variability via transpiration. *Geophys. Res. Lett.*, **39**, n/a–n/a.

Li, T., 2014: Recent advance in understanding the dynamics of the Madden–Julian oscillation, *J. Meteor. Res.*, **28**, 1–33.

Lim, Y., S.-W. Son, and D. Kim, 2018: MJO Prediction skill of the subseasonal-to-seasonal prediction models. *J. Clim.*, **31**, 4075–4094.

Ling, J., C. Zhang, and P. Bechtold, 2013: Large-scale distinctions between MJO and non-MJO convective initiation over the Tropical Indian Ocean. *J. Atmos. Sci.*, **70**, 2696–2712.

Madden, R. A., and P. R. Julian, 1972: Description of global-scale circulation cells in the Tropics with a 40–50 day period. *J. Atmos. Sci.*, **29**, 1109–1123.

Majda, A. J., and Q. Yang, 2016: A multiscale model for the intraseasonal impact of the diurnal cycle over the Maritime Continent on the Madden–Julian oscillation. *J. Atmos. Sci.*, **73**, 579–604.

Maloney, E. D., and A. H. Sobel, 2004: Surface fluxes and ocean coupling in the Tropical Intraseasonal Oscillation. *J. Clim.*, **17**, 4368–4386.

Maloney, E. D., A. H. Sobel, and W. M. Hannah, 2010: Intraseasonal variability in an aquaplanet general circulation model. *J. Adv. Model. Earth Syst.*, **2**, n/a-n/a.

Neale, R., and J. Slingo, 2003: The Maritime Continent and its role in the global climate: A GCM study. *J. Clim.*, **16**, 834–848.

Neena, J. M., J. Y. Lee, D. Waliser, B. Wang, and X. Jiang, 2014: Predictability of the Madden–Julian oscillation in the intraseasonal variability hindcast experiment (ISVHE). *J. Clim.*, **27**, 4531–4543.

Oh, J.-H., K.-Y. Kim, and G.-H. Lim, 2012: Impact of MJO on the diurnal cycle of rainfall over the western Maritime Continent in the austral summer. *Clim. Dyn.*, **38**, 1167–1180.

Peatman, S. C., A. J. Matthews, and D. P. Stevens, 2014: Propagation of the Madden–Julian oscillation through the Maritime Continent and scale interaction with the diurnal cycle of precipitation. *Q. J. R. Meteorol. Soc.*, **140**, 814–825.

Peatman, S. C., A. J. Matthews, and D. P. Stevens, 2015: Propagation of the Madden–Julian oscillation and scale interaction with the diurnal cycle in a high-resolution GCM. *Clim. Dynam.*, **45**, 2901–2918.

Rauniyar, S. P., and K. J. E. Walsh, 2011: Scale interaction of the diurnal cycle of rainfall over the Maritime Continent and Australia: influence of the MJO. *J. Clim.*, **24**, 325–348.

Robertson, A. W., 2011: *The Global Monsoon System*, Eds. C.-P. Chang, Y. Ding, N.-C. Lau, R. H. Johnson, B. Wang, and T. Yasunari, World Scientific Publishing Co. Pte. Ltd.

Sakaeda, N., G. Kiladis, and J. Dias, 2017: The diurnal cycle of tropical cloudiness and rainfall associated with the Madden–Julian oscillation. *J. Clim.*, **30**, 3999–4020.

Salby, M. L., and H. H. Hendon, 1994: Intraseasonal behavior of clouds, temperature, and motion in the tropics. *J. Atmos. Sci.*, **51**, 2207–2224.

Sobel, A. H., E. D. Maloney, G. Bellon, and D. M. Frierson, 2008: The role of surface heat fluxes in tropical intraseasonal oscillations. *Nat. Geosci*, **1**, 653–657.

Sobel, A. H., E. D. Maloney, G. Bellon, and D. M. Frierson, 2010: Surface fluxes and tropical intraseasonal variability: A reassessment. *J. Adv. Model. Earth Syst.*, **2**, 2.

Sui, C.-H., and K.-M. Lau, 1992: Multiscale phenomena in the tropical atmosphere over the Western Pacific. *Mon. Weather Rev.*, **120**, 407–430.

Tian, B. J., D. E. Waliser, and E. J. Fetzer, 2006: Modulation of the diurnal cycle of tropical deep convective clouds by the MJO. *Geophys. Res. Lett.*, **33**.

Tseng, W.-L., H.-H. Hsu, N. Keenlyside, C.-W. June Chang, B.-J. Tsuang, C.-Y. Tu, and L.-C. Jiang, 2017: Effects of surface orography and land–sea contrast on the Madden–Julian oscillation in the Maritime Continent: A numerical study using ECHAM5-SIT. *J. Clim.*, **30**, 9725–9741.

Vincent, C. L. and T. P. Lane, 2016: Evolution of the diurnal precipitation cycle with the passage of a Madden–Julian oscillation event through the Maritime Continent. *Mon. Weather Rev.*, **144**, 1983–2005.

Vitart, F. and F. Molteni, 2010: Simulation of the Madden–Julian oscillation and its teleconnections in the ECMWF forecast system. *Q. J. R. Meteorol. Soc.*, **136**, 842–855.

Wang, B. and H. Rui, 1990: Synoptic climatology of transient tropical intraseasonal convection anomalies: 1975–1985. *Meteorol. Atmos. Phys.*, **44**, 43–61.

Wang, B. and T. Li, 1994: Convective interaction with boundary-layer dynamics in the development of a tropical intraseasonal system. *J. Atmos. Sci.*, **51**, 1386–1400.

Wang, W., M.-P. Hung, S. J. Weaver, A. Kumar, and X. Fu, 2014: MJO prediction in the NCEP climate forecast system version 2. *Clim. Dynam.*, **42**, 2509–2520.

Weickmann, K. M., 1983: Intraseasonal circulation and outgoing longwave radiation modes during Northern Hemisphere Winter. *Mon. Weather Rev.*, **111**, 1838–1858.

Weickmann, K. M., G. R. Lussky, and J. E. Kutzbach, 1985: Intraseasonal (30–60 Day) Fluctuations of outgoing longwave radiation and 250 mb streamfunction during Northern Winter. *Mon. Weather Rev.*, **113**, 941–961.

Wheeler, M. C. and J. L. McBride, 2012: Australasian monsoon. *Intraseasonal Variability in the Atmosphere-Ocean Climate System*, Springer Berlin Heidelberg, Berlin, Heidelberg, pp. 147–197.

Wu, C.-H. and H.-H. Hsu, 2009: Topographic influence on the MJO in the Maritime Continent. *J. Clim.*, **22**, 5433–5448.

Xiang, B., M. Zhao, X. Jiang, S.-J. Lin, T. Li, X. Fu, and G. Vecchi, 2015: The 3–4-Week MJO Prediction skill in a GFDL coupled model. *J. Clim.*, **28**, 5351–5364.

Yamanaka, M. D., 2016: Physical climatology of Indonesian maritime continent: An outline to comprehend observational studies. *Atmos. Res.*, **178–179**, 231–259.

Zhang, C. and J. Ling, 2017: Barrier effect of the Indo-Pacific Maritime Continent on the MJO: Perspectives from tracking MJO precipitation. *J. Clim.*, **30**, 3439–3459.

Zhang, C. and H. H. Hendon, 1997: Propagating and standing components of the intraseasonal oscillation in tropical convection. *J. Atmos. Sci.*, **54**, 741–752.

Zhang, C. and M. Dong, 2004: Seasonality in the Madden–Julian oscillation. *J. Clim.*, **17**, 3169–3180.

Zhu, W., T. Li, X. Fu, and J.-J. Luo, 2010: Influence of the Maritime Continent on the Boreal Summer Intraseasonal Oscillation. *J. Meteorol. Soc. Japan. Ser. II*, **88**, 395–407.

Chapter 22

The Moisture Mode Framework of the Madden–Julian Oscillation

Ángel F. Adames*,¶, Daehyun Kim†,‖, Eric D. Maloney‡,** and Adam H. Sobel§,††

*Department of Climate and Space Science and Engineering, University of Michigan, Ann Arbor, MI.
†Department of Atmospheric Sciences, University of Washington, Seattle, WA
‡Department of Atmospheric Sciences, Colorado State University, Fort Collins, CO
§Departments of Applied Physics and Applied Mathematics and Earth and Environmental Sciences, Lamont-Doherty Earth Observatory, Columbia University, New York, NY, USA
¶afadames@umich.edu
‖daehyun@uw.edu
**emaloney@atmos.colostate.edu
††ahs129@columbia.edu

Moist thermodynamics play a central role in regulating tropical deep convection. In particular, tropospheric humidity has been found to be highly spatially and temporally correlated with precipitation, particularly at the intraseasonal (20–100 day) timescale and the planetary scale (zonal wavenumbers 1–5). This and other evidence support the hypothesis that tropospheric moisture variations are central to the dynamics of the Madden–Julian Oscillation (MJO), the most prominent mode of tropical intraseasonal variability. Numerous studies have developed this hypothesis into a view of the MJO in which it is considered to be a "moisture mode". The intraseasonal moist static energy budget in combination with the weak temperature gradient approximation has been used to understand the processes that maintain and propagate the intraseasonal precipitation anomalies. These analyses indicate that horizontal and vertical moisture advection play essential roles in the propagation of the MJO while longwave radiative heating is a key process in the maintenance of the MJO.

1. Introduction

The Madden–Julian oscillation (MJO), named after Madden and Julian's (1971, 1972) seminal work, is a prominent mode of intraseasonal variability in the tropics. It was initially described as a large-scale envelope of enhanced convection several thousand kilometers across, coupled with a zonal overturning circulation in the equatorial plane that propagates eastward at $\sim 5\,\mathrm{m\,s^{-5}}$ in a period of 30–90 days. Subsequent work showed that the MJO's wind structure exhibits a horizontal structure that resembles a coupled Kelvin and Rossby wave response to a quasi-steady equatorial heat source (Matsuno 1966; Gill 1980; Kiladis *et al.* 2005). None of Matsuno's (1966) free propagating waves exhibit the robust Kelvin–Rossby wave couplet structure that has been observed in the MJO, suggesting that the MJO is driven by processes not included in his dry shallow water model. Subsequent studies indicated that one of the missing processes is the interaction between large-scale motions and heating associated with deep convection (Lau and Lau 1986; Hayashi and Golder 1986; Lau *et al.* 1988; Rui and Wang 1990, among others).

Many early studies attempted to capture the salient features of the MJO in simple theoretical models (e.g., Chang 1977; Lau and Peng

The Multiscale Global Monsoon System
Edited by C. P. Chang et al.

1987; Neelin *et al.* 1987; Emanuel 1987; Wang and Rui 1990). A number of reviews summarize subsets of this work (e.g., Wang *et al.* 2005; Sobel *et al.* 2010). While each study claims some degree of success, the essential mechanisms portrayed by the various models differ quite widely, indicating a lack of agreement within the field. Some early theories shared the common theme that deep convection was a direct response to low-level mass and moisture convergence in the large-scale circulation. Some others used "quasi-equilibrium" thinking in which near constancy of convective available potential energy or some similar quantity was a key closure assumption (e.g., Emanuel *et al.* 1994) and feedbacks between surface turbulent fluxes and convection were critical.

Subsequent studies, however, revealed that free-tropospheric relative humidity was important to the organization of deep convection (e.g., Brown and Zhang 1997; Sherwood 1999; Raymond 2000; Derbyshire *et al.* 2004; Bretherton *et al.* 2004; Sobel *et al.* 2004). Studies with comprehensive models showed that MJO simulation was substantially improved if deep convection parameterizations were made more sensitive to free-tropospheric relative humidity (Kim *et al.* 2012, 2014a; Jiang *et al.* 2015, see a review by Kim and Maloney 2017). Other studies concluded that radiative feedbacks, perhaps together with surface flux feedbacks, played a role in the MJO, supported by evidence from theory (Raymond 2001), model simulations (e.g., Andersen and Kuang 2011; Kim *et al.* 2011), and the observation that the MJO weakens over land (Sobel *et al.* 2008, 2010). These insights together have led to a view of the MJO that has become broadly known as *moisture mode theory*.

Moisture mode theory posits that the MJO is distinct from other convectively coupled equatorial waves (CCEWs). While dry versions of the CCEWs can be obtained from Matsuno's shallow equations, moisture modes require a prognostic equation for moisture for their existence. The processes that lead to the evolution of moisture are those that lead to the evolution of the moisture mode.

In this chapter, we summarize the moisture mode framework and some of the insights we have obtained from applying it to the MJO. Section 2 describes the coupling of relative humidity and precipitation. Section 3 describes the initial moisture mode theories and the weak temperature gradient approximation. Section 4 describes how the moisture and moist static energy (MSE) budgets have been used to understand the MJO. Section 5 discusses how mechanism denial experiments have supported the moisture mode view. Section 6 discusses more recent theoretical developments. This chapter is summarized in Section 7, and some outstanding questions are outlined.

2. Coupling of Precipitation and Column Relative Humidity

The deep tropics (equatorward of 10°N/S) are characterized by weak spatial and temporal variability in free-tropospheric temperature and large variability in free-tropospheric relative humidity (Charney 1963, Sobel *et al.* 2001). Using a two-layer model, Neelin and Held (1987) related the climatological mean low-level convergence to the net energetic input into the atmosphere (surface fluxes plus radiative heating) and a quantity they called the gross moist stability (GMS) which relates the vertically integrated mass transport to the MSE transport (e.g., Raymond *et al.* 2009). Neelin and Held (1987) defined the GMS as the difference between lower-tropospheric latent energy ($L_v q$) and upper-tropospheric dry static energy (s). Because temperature varies little in the tropics, the inverse relationship between rainfall and the GMS implies that rainfall is largest in regions where the water vapor content is high.

While the details of the argument of Neelin and Held (1987) have not been supported

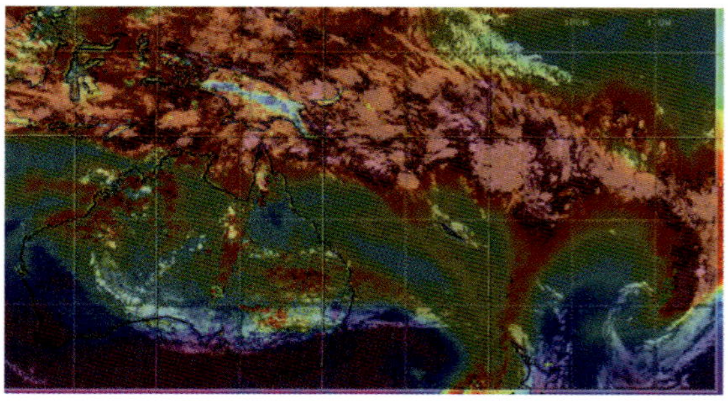

Fig. 1. Total column water vapor overlaid by infrared satellite imagery from HIWAMARI-8 for an active MJO event occurring in January 2018. Red shading indicates areas where column water vapor exceeds 60 mm. Imagery was obtained from http://tropic.ssec.wisc.edu/.

by subsequent studies — variations in vertical structure they neglected in the GMS are important, as are variations in energy input — this study both drew attention to the free- tropospheric humidity and established a theoretical framework that would be subsequently expanded and widely used. A subsequent study by Held *et al.* (1993) noted that regions of precipitation are characterized by enhanced column-integrated water vapor, similar to the satellite image of an MJO event shown in Fig. 1.

Based on these studies, Raymond (2000) hypothesized that precipitation P is highly sensitive to column water vapor $\langle q \rangle$, where angle brackets denote column integration from 100 to 1000 hPa. Satellite observations confirmed this hypothesis (Bretherton *et al.* 2004, Rushley *et al.* 2018), revealing that precipitation is approximately an exponential function of column relative humidity

$$P = P_0 \exp(aRH), \tag{1}$$

where $RH = \langle q \rangle / \langle q_s \rangle$, where q_s is the saturation specific humidity. A histogram of precipitation frequency as a function of precipitation intensity and RH is shown in Fig. 2. Because variations in temperature in the tropics are generally small, fluctuations in RH are largely due to changes in column water vapor $\langle q \rangle$.

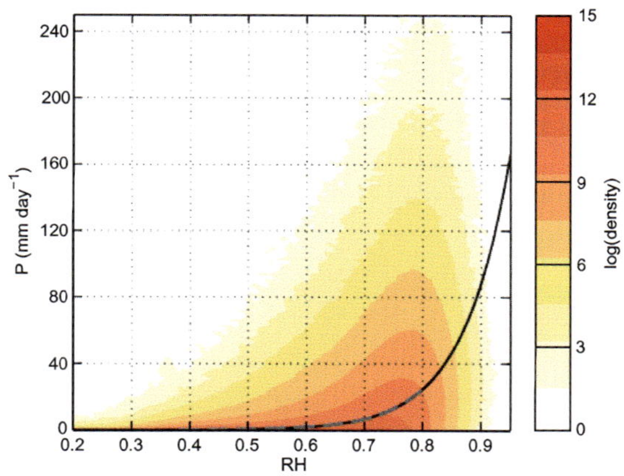

Fig. 2. Joint probability density distribution of TRMM 3b42 P (Huffman *et al.* 2007) and ERA-Interim RH (Dee *et al.* 2011) for the period 1998–2013. The shading is the logarithm of the number of observations of a bin of precipitation and RH. The nonlinear fit (solid line), following Eq. (1), is shown as a solid line. Bins are every 0.01 RH and 2 mm day^{-1}.

The relationship between RH and P is not homogenously coherent across all spatial and temporal scales. A space–time spectral analysis by Yasunaga and Mapes (2012) revealed that RH and P are most coherent at the intraseasonal timescale, as shown in Fig. 3. A satellite-based analysis by Myers and Waliser (2003) revealed that water vapor and precipitation are indeed highly correlated in space in the MJO.

To further elucidate how RH and P are related at the intraseasonal timescale, we linearize Eq. (1) with respect to a climatological-mean precipitation $\bar{P}$. This linearization yields the following equation for P':

$$P' = a\bar{P}RH'. \tag{2}$$

The scaling by $\bar{P}$ implies that perturbations in RH produce larger precipitation anomalies in the climatologically rainy areas of the tropics.

Furthermore, the scaling by a also implies that the amplitude of P' is related to the steepness of the pickup curve in Fig. 2, with a steeper pickup associated with larger intraseasonal precipitation anomalies. Steeper curves indicate that deep convection is highly suppressed at

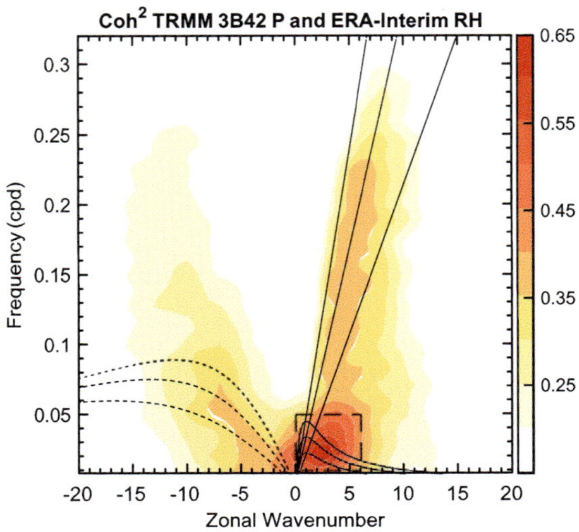

Fig. 3. Coherence squared (shading) between TRMM 3B42 precipitation and ERA-Interim column-integrated water vapor. The dispersion curves correspond to the Kelvin and equatorial Rossby waves (dashed lines) for equivalent depths of 12, 25, 50, and 90 m. The solid lines correspond to a moisture mode dispersion curve as identified in Adames and Kim (2016). The dashed box encloses timescales of 20–100 days and eastward-propagating wavenumbers 1–6, where MJO activity is strongest.

low RH, while it is strongly enhanced at high RH.

3. Initial Moisture Mode Theories and the Weak Temperature Gradient Approximation

The relationship between water vapor and precipitation discussed in the previous section motivated Raymond (2001) to propose a theoretical framework for the MJO in which the intraseasonal precipitation anomalies are a result of enhanced column water vapor. He found that such a mode would be unstable due to cloud–radiative interactions and enhanced by wind-induced surface heat exchange (WISHE) (Emanuel 1987; Neelin *et al.* 1987). Raymond (2001) found that this *moisture mode* has a significant rotational flow component, akin to balanced modes such as Rossby waves but in contrast to the highly divergent circulations found in equatorial Kelvin waves, to which the MJO was thought be more closely related in many earlier studies.

A nearly concurrent study by Sobel *et al.* (2001) derived a balanced dynamical framework on a hypothetical near-equatorial f-plane. Their model emphasized the fact that tropical temperature fluctuations are small, which results in a thermodynamic balance between vertical advection of dry static energy s and apparent heating,

$$\omega \frac{\partial s}{\partial p} \simeq Q \tag{3}$$

which is referred to as the *weak temperature gradient approximation* (Charney 1963; Held and Hoskins 1985; Sobel and Bretherton 2000; Sobel *et al.* 2001). In the tropics, Q is approximately sum of convective Q_c and radiative heating Q_r. Sobel *et al.* (2001) showed that in this framework, balanced moisture waves arise, which bear similarity to those in the model proposed by Raymond (2001). These waves are balanced and propagate due to horizontal advection of mean moisture by the anomalous winds. Because temperature fluctuations are small, it follows that excursions in RH in Eqs. (1) and (2) are predominantly due to changes in moisture.

4. Moisture and MSE Budgets of the MJO

The studies by Raymond (2001) and Sobel *et al.* (2001) provided several hypotheses that could be tested in observations, reanalysis, and modeling studies. For example, if the MJO is a moisture mode, then the processes that lead to the evolution of the moisture field are central to its dynamics. An example of how moisture can impact MJO evolution is shown schematically in Fig. 4. Fluctuations in RH shape the large-scale distribution of convection (top left), which leads to a gravity wave response that acts to adjust the thermodynamic profile toward WTG balance (bottom left). Convergence/divergence in

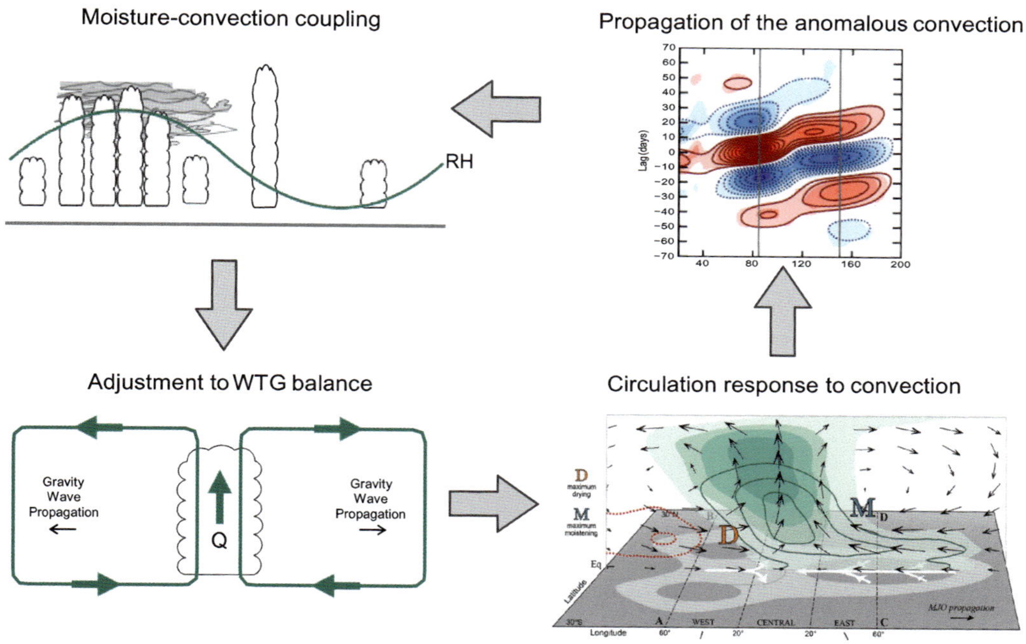

Fig. 4. Schematic summarizing the propagation evolution of the MJO as suggested by the moisture mode framework and with application of the WTG approximation. The lower-right diagram is adapted from Adames and Wallace (2015).

convection drives a large-scale circulation (bottom right), which then causes changes in the distribution of *RH*, resulting in the propagation of the convection (top right).

While one can view these processes through the moisture budget alone, tight coupling of the moisture and temperature equations has led many authors to prefer working with variables that are conserved under condensation, namely, MSE or moist entropy, to understand the MJO. The column-integrated MSE budget is written as follows:

$$\frac{\partial \langle m' \rangle}{\partial t} = -\langle \mathbf{v} \cdot \nabla m \rangle' - \langle \omega \partial_p m \rangle' + \langle Q_r \rangle' + SH' + L_v E', \tag{4}$$

where the RHS terms are

- Horizontal MSE advection,
- Vertical MSE advection,
- Anomalous total column radiative heating, the sum of longwave and shortwave heating $\langle Q_r \rangle' = \langle Q_{SW} \rangle' + \langle Q_{LW} \rangle'$,
- Surface sensible heat flux, and
- Surface latent heat flux, where L_v is the latent energy of vaporization.

When the WTG approximation is applied, perturbations in column-integrated MSE are approximately equivalent to those in the latent energy in moisture, $\langle m \rangle' \simeq L_v \langle q \rangle'$, although the prognostic equations for the two quantities are different. For example, the MSE equation contains a radiative heating term, whereas the moisture budget per se does not (Wolding and Maloney 2015). This difference is due to how these processes are expressed in these budgets. While radiative heating is explicit in the MSE budget, it also has an impact in vertical moisture advection due to the vertical motion that such heating produces (Eq. (3)). Thus, the net impact radiation has in the moisture and MSE budgets is the same, but it is distributed in the tendency terms differently. As we will see below, a more succinct way to describe the role of radiation in moisture modes can be obtained through a "WTG-based" moisture budget.

Numerous studies have analyzed the moisture and MSE budgets of the MJO in the context of observations (Sobel *et al.* 2014; Yokoi and Sobel 2015), reanalysis (Benedict and Randall 2007; Kiranmayi and Maloney 2011), and GCMs (Maloney 2009; Andersen and Kuang 2012). These studies agree on the following results:

(a) Horizontal and vertical MSE advections are the main processes responsible for the propagation of the MJO (Haertel *et al.* 2008, Maloney *et al.* 2010; Andersen and Kuang 2012; Kim *et al.* 2014; Pritchard and Bretherton 2014; Adames and Wallace 2015; Jiang 2017). Longwave radiative heating and surface latent heat fluxes act to retard this eastward propagation (Crueger and Stevens 2015; Ma and Kuang 2016; DeMott *et al.* 2016).

(b) Longwave radiative heating and, to a lesser extent, surface latent heat fluxes are largely responsible for the maintenance of the intraseasonal precipitation anomalies (Maloney and Sobel 2004; Andersen and Kuang 2012; Kim *et al.* 2014b, 2015; Crueger and Stevens 2015; Ma and Kuang 2016).

Hovmöller diagrams of the MSE budget terms in ERA-Interim are shown in Fig. 5. It is clear that horizontal MSE advection leads the MSE anomalies, thus contributing to its propagation. Longwave radiative heating is largely in phase with the MSE anomalies and is the leading MSE maintenance process.

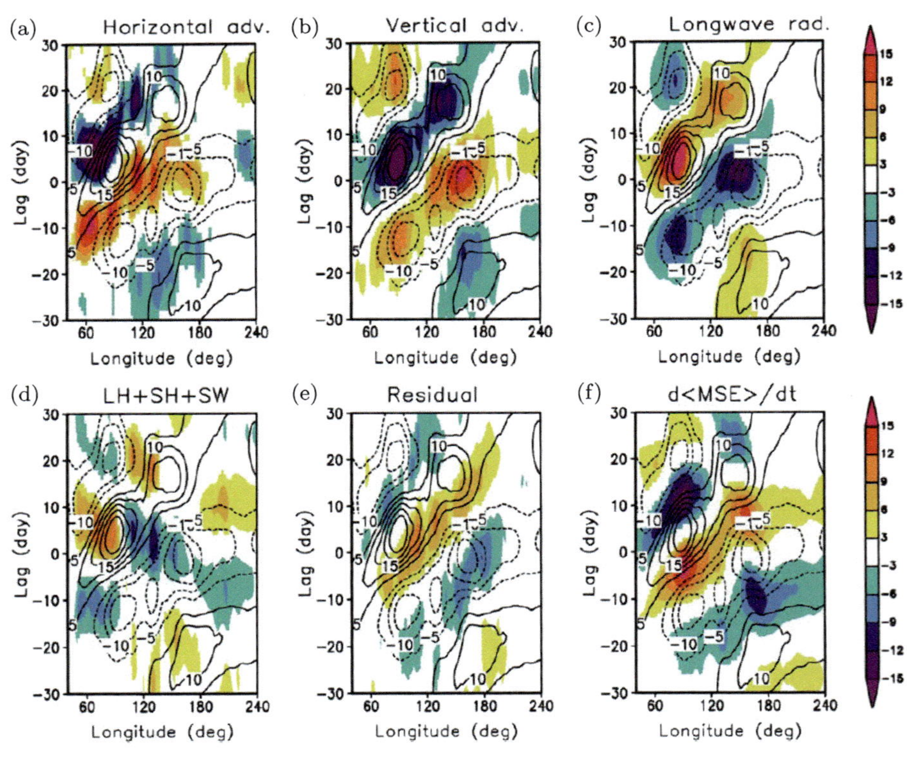

Fig. 5. Lag composite Hovmöller diagram of anomalous (a) horizontal and (b) vertical advection terms of the column-integrated MSE budget, (c) column longwave radiative heating, (d) sum of surface latent and sensible heat flux and column shortwave radiative heating, (e) residual term, and (f) tendency term for propagating MJO events. Shaded fields are in units of W m^{-2}. Contoured field is the column MSE anomaly (adapted from Kim *et al.* 2014).

A well-documented residual exists in the ERA-Interim MSE budget (Fig. 5e), which results from either missing moistening processes or from numerical errors in the calculation of the advection terms.

While the column-integrated budgets have provided useful insights into the MJO, they conceal some aspects of the mechanisms by which processes such as radiative heating moisten the atmosphere. These processes were elucidated by Chikira (2014), who proposed a vertically resolved version of the MSE budget that makes use of the WTG approximation. By substituting ω in the thermodynamic energy equation into the moisture budget he obtained the following "WTG-based" moisture budget:

$$\frac{\partial q'}{\partial t} = -(\mathbf{v}\cdot\nabla q)' - (1-\alpha)Q_c' L_v^{-1} + \alpha Q_c' L_v^{-1} + \partial_p(\overline{q''\omega''})' \tag{5}$$

where the terms on the right-hand side are intraseasonal anomalies of the following processes:

- horizontal moisture advection,
- convective column process: the moistening by vertical moisture advection by convection and loss of moisture through microphysical processes, where $\alpha = L_v\left(\frac{\partial q}{\partial p}\right)\left(\frac{\partial s}{\partial p}\right)^{-1}$ is the ratio between the vertical latent energy gradient and the vertical dry static energy gradient,
- vertical moisture advection by radiative heating, and
- vertical moisture flux convergence by unresolved eddies.

Figure 6 shows longitude–height cross sections of the MJO as represented by the Super-parameterized Community Earth System Model

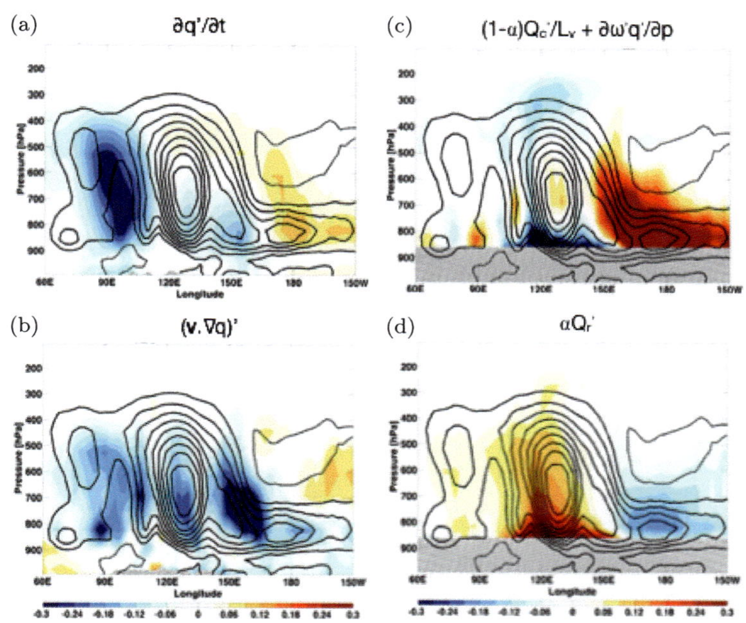

Fig. 6. Longitude–height sections of meridionally averaged (5°N–5°S) composite intraseasonal specific humidity (contour) and moisture budget terms (shading). The panels show (a) the Eulerian moisture tendency, (b) horizontal moisture advection, (c) vertical moisture advection by convection, subgrid-scale vertical eddy moisture flux convergence and loss of moisture through microphysical processes, and (d) vertical moisture advection by radiative heating. Contour interval is 0.1 g kg^{-1} (adapted from Wolding *et al.* 2016).

(SP-CESM) (Wolding *et al.* 2016). In the MJO phase shown in Fig. 6, vertical moisture advection from convective processes Q_c' is the largest contributor to moistening to the east of the anomalously moist region, while horizontal moisture advection contributes to drying to the west. Radiative heating is in phase with moisture itself, indicating that it contributes to the maintenance of the moisture anomalies. The WTG-based moisture budget reveals that anomalous longwave radiative heating in regions of deep convection induces anomalous ascent in the lower free troposphere, which moistens the layer through vertical moisture advection. This moistening offsets the net drying effect of deep convection (Fig. 6c). Wolding and Maloney (2016) showed that these mechanisms are robust throughout the simulated MJO cycle although horizontal moisture advection plays a larger role than what is shown in Fig. 6 (see their Fig. 10).

Studies of the WTG moisture budget using reanalysis (Wolding and Maloney 2015, Adames 2017) and cloud-resolving models (Janiga and Zhang 2016) have obtained results similar to those shown in Figs. 5 and 6.

5. Insights from GCM Sensitivity Studies

The moisture mode idea incorporates knowledge from GCM studies on the effects of specific GCM parameters, especially those in the convection scheme, into MJO simulation. By making precipitation strongly (in some cases exclusively) dependent on column relative humidity, moisture mode theory builds in the assumption that moisture–convection coupling is a fundamental building block in the MJO. While an accurate simulation of the MJO has been a challenging task for GCMs (e.g., Slingo *et al.* 1996; Ahn *et al.* 2017), it has repeatedly been reported since the late 1980s that GCM MJOs can be enhanced by inhibiting parameterized deep convection in the presence of a dry free troposphere (see the review by Kim and Maloney 2017). It was later found that inhibiting parameterized deep convection tightens the coupling between moisture and precipitation in model simulations (e.g., Tokioka *et al.* 1988; Wang and Schlesinger 1999; Maloney and Hartmann 2001; Benedict and Randall 2009; Hannah and Maloney 2011; Kim *et al.* 2011; Del Genio *et al.* 2012; Hannah and Maloney 2014), and recent multi-model studies showed that models with a tighter moisture–convection coupling tended to better simulate the MJO (Kim *et al.* 2014a; Jiang *et al.* 2015; Ahn *et al.* 2017), consistent with the moisture mode theory.

The models with good MJO simulations have also been used to test hypotheses proposed by the moisture mode theory about the key processes in the MJO. Mechanism denial experiments have revealed that simulations where moisture or MSE is nudged to their Zonal mean values eliminate MJO-like activity (see Fig. 7a and b) (Andersen and Kuang 2012, Shi *et al.* 2018). Experiments where interactive longwave cloud–radiative interactions are turned off result in weakened MJO-like activity (Fig. 7c) (Kim *et al.* 2011; Landu and Maloney 2011; Andersen and Kuang 2012; Arnold and Randall 2015; Crueger and Stevens 2015; Ma and Kuang 2016; Wang *et al.* 2016).

However, not all modeling studies support a moisture mode view. A study by Pritchard and Yang (2016) challenged this framework by showing that intraseasonal variability in a suite of model simulations that covered a wide range of climates exhibited behavior that was more consistent with Kelvin waves. While some studies indicate that the moisture mode view can be used to understand the MJO's behavior in different climates (Wolding *et al.* 2017; Adames *et al.* 2017; Bui and Maloney 2018; Rushley *et al.* 2019), a study attempting to bridge these results to those of Pritchard and Yang (2016) has not been performed.

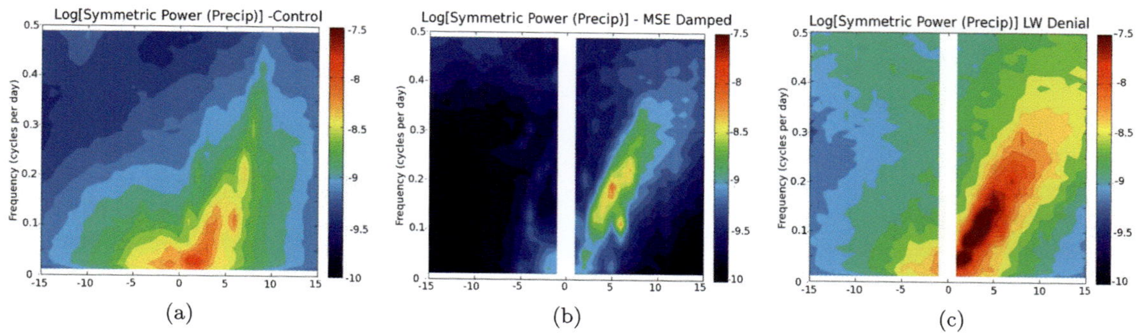

Fig. 7. Base-10 logarithm of the equatorially symmetric spectral power of precipitation for (a) a control run of the SPCAM model (b) a run where MSE damped into its time and zonal mean value and (c) a simulation where cloud LW radiative heating is turned off. The power spectrum is averaged over the region 15°S to 15°N (adapted from Andersen and Kuang 2012).

6. Recent Theoretical Developments

Since the initial theoretical foundations of the moisture mode were developed (Raymond 2001; Sobel *et al.* 2001), observation and modeling studies have motivated further elaboration of the theoretical arguments of the moisture mode (Fuchs and Raymond 2005; 2017; Raymond and Fuchs 2007; Sugiyama 2009; Sobel and Maloney 2012, 2013; Sukhatme 2014; Adames and Kim 2016).

One of the most recent theoretical treatments of the moisture mode is the model of Adames and Kim (2016), who expanded on the original models of Sobel and Maloney (2012, 2013). In their model, moisture is the only prognostic variable and the wind and geopotential fields are diagnosed as the Matsuno–Gill response to a quasi-steady equatorial heat source. While this assumption may not always be valid during the MJO cycle, it allows us to understand the propagation of the moisture anomalies within a simplified framework. They linearized the processes in Eq. (5) associated with horizontal moisture advection, surface latent heat fluxes, and frictional moisture convergence, and expressed them in terms of the anomalous lower-tropospheric flow. The moistening from all these processes is summarized in terms of a moisture advection parameter A_{KR}.

The real component of the dispersion relation of Adames and Kim (2016) takes the following form

$$\omega_r \simeq \frac{\tilde{p} A_{KR}}{\tau_c k}, \tag{6}$$

where τ_c is the convective moisture adjustment timescale, the rate in which precipitation relaxes moisture back to its climatological value, k is the zonal wavenumber, and $\tilde{p}$ is a dissipation coefficient. An example of this type of dispersion curve is shown in Fig. 3. It is reminiscent of the dispersion of a Rossby wave, but its direction of propagation is eastward. This similarity is due to these processes being analogous. In Rossby waves, the horizontal winds advect the background potential vorticity, which induces westward propagation of the vorticity anomalies. In the moisture mode, advection of background moisture by the anomalous winds induces eastward propagation of the moisture and precipitation anomalies.

Growth in their model is due to the impact that cloud–radiative feedbacks have on the moisture anomalies. As mentioned in the previous sections, radiative heating can cause the total vertical moisture advection to equal or exceed the amount of moisture lost through condensation. The difference between total vertical moisture advection and precipitation can be understood in terms of an effective gross moist stability (EGMS), $EGMS \simeq (\langle \omega \partial_p q \rangle - P')/P'$. When the total net import of moisture exceeds the loss through precipitation, the EGMS is negative, and the moisture anomalies grow through *moisture mode instability* (Raymond and Fuchs 2009). In the model of Adames and Kim (2016), cloud–radiative feedbacks also select the planetary scale of the MJO.

Another relevant model is the one developed by Fuchs and Raymond (2017) (FR17). In their model WISHE is responsible for the propagation of the moisture anomalies, and the wind and temperature anomalies are also prognostic. They obtain a dispersion relation that is similar to that of Adames and Kim (2016) with a notable difference. A is the only mechanism responsible for propagation in the FR17 model, and also induces the growth of the moisture anomalies, even when the EGMS is positive. While radiation plays a role in maintaining the moisture anomalies, it is of secondary importance in their model. A can also explain the MJO's planetary scale in their model. A debatable aspect of their model is that, like that of Emanuel (1987), it requires a background state of surface easterlies so that WISHE can amplify an eastward-propagating mode.

Other notable models in which moisture is included as a prognostic variable include the MJO skeleton model (Majda and Stechmann 2009) and the trio interaction model (Wang *et al.* 2016). Both models are further discussed in Chapter YY.

7. Summary and Outstanding Questions

This chapter summarizes the basic aspects of the moisture mode framework and its application to the MJO. This includes the relationship between column water vapor and precipitation, the use of moisture and MSE budgets to understand MJO evolution, and how GCM studies and theory have refined our understanding of the MJO.

The moisture mode framework is not a theory that was simply proposed and tested thereafter. Rather, results from observational studies and GCM simulations have stimulated, supported, and challenged the moisture mode view. At the same time, the theoretical foundations of the moisture mode have been modified and refined, providing new insights with which to interpret GCM results. This dynamic interaction between observations, model, and theory has provided insights into the physical processes that drive the MJO.

Numerous outstanding questions remain. There is no universally agreed upon process for MJO scale selection, even within the moisture mode framework. From a moisture mode view, the MJO may be related to the process of convective self-aggregation since longwave radiative heating is central to both (Arnold and Randall 2015). However, this relationship has not been fully elucidated upon.

MJO initiation remains poorly understood. It is likely that different processes that lead to MJO initiation come from a multitude of sources, including circumnavigating waves and extratropical sources (Powell and Houze 2015; Zhao *et al.* 2013; Sakaeda and Roundy 2016). It is likely that MJO initiation involves changes in tropospheric CAPE, convective inhibition, and circulation (Matthews 2008; Straub 2013; Powell 2016). The WTG approximation is likely inapplicable at this time in the MJO cycle.

While the theory described above idealizes the MJO as a "pure" moisture mode, dynamically distinct from the Kelvin wave, it is possible that the observed MJO is a hybrid of some sort. Roundy (2012) presented evidence that many observed disturbances fall on a spectrum between a strictly defined MJO and Kelvin waves. To the extent that a pure moisture mode exists, it does so most clearly over the Indian ocean, where the MJO's phase speed is slowest and convective activity is strong. As the MJO proceeds eastward into the Pacific, it speeds up, dries out, and comes to resemble a Kelvin wave more closely (e.g., Sobel and Kim 2012). Extant theory has yet to describe the hybrid disturbances suggested by these observations.

It is also not clear why MJO activity is intermittent, with some years exhibiting enhanced activity and others being quiescent. Influences of ENSO (Tam and Lau, 2005) and the QBO (Son *et al.* 2017; Nishimoto and Yoden 2017; Zhang and Zhang 2018) may play a role in this intermittency. The mechanisms in which the QBO affects the MJO are likely complex, and the moisture mode framework may not be able to explain them.

Several studies have shown a relationship between the MJO and coupled ocean-atmosphere processes (Hendon 2000; Sobel and Gildor 2003; Maloney and Sobel 2004; Sobel *et al.* 2010, among others). More recent studies have found links between the MJO and regional ocean processes, the diurnal cycle of the upper ocean, and interactions between the ocean and the mean state (see reviews by DeMott *et al.* 2015; Hendon 2005; Chapter ZZ). Similarly, interactions between the MJO and the landmasses of the Maritime Continent (MC), the so-called "MC-barrier effect", remain a topic of intense research (Kim *et al.* 2016, 2017; Zhang *et al.* 2017, see also Chapter XX). The moisture mode framework may be useful in explaining some of the new developments in these topics.

Acknowledgments

ÁFA was supported by the National Oceanic and Atmospheric Administration (NOAA) grant NA15OAR4310099 and the University of Michigan's startup package. DK was supported by the National Aeronautics and Space Administration grant 80NSSC17K0227 and by the Korea Meteorological Administration Research and Development Program under Grant KMI2018-03110. AHS acknowledges support from National Science Foundation grant AGS-1543932. EDM was supported by NASA CYGNSS project NNX17AH77G and ONR PISTON project N00014-16-1-3087.

References

Adames, Á. F. and J. M. Wallace, 2015: Three-dimensional structure and evolution of the moisture field in the MJO. *J. Atmos. Sci.*, **72**, 3733–3754, doi:10.1175/JAS-D-15-0003.1.

Adames, Á. F. and D. Kim, 2016: The MJO as a dispersive, convectively coupled moisture wave: Theory and observations. *J. Atmos. Sci.*, **73**, 913–941, doi:10.1175/JAS-D-15-0170.1.

Adames, Á. F., 2017: Precipitation budget of the Madden–Julian oscillation. *J. Atmos. Sci.*, **74**(6), 1799–1817.

Adames, Á. F., D. Kim, A. H. Sobel, A. Del Genio, and J. Wu, 2017: Characterization of moist processes associated with changes in the propagation of the MJO with increasing CO2, *J. Adv. Model. Earth Syst.*, **9**, 2946–2967, https://doi.org/10.1002/2017MS001040.

Ahn, M. S. *et al.*, 2017: MJO simulation in CMIP5 climate models: MJO skill metrics and process-oriented diagnosis. *Clim. Dynam.*, **49**, 4023, https://doi.org/10.1007/s00382-017-3558-4.

Andersen, J. A. and Z. Kuang, 2012: Moist static energy budget of MJO-like disturbances in the atmosphere of a zonally symmetric aquaplanet. *J. Clim.*, **25**, 2782–2804, doi:10.1175/ JCLI-D-11-00168.1.

Arnold, N. P. and D. A. Randall, 2015: Global-scale convective aggregation: Implications for the Madden–Julian oscillation. *J. Adv. Model. Earth Syst.*, **7**, 1499–1518, https://doi.org/10.1002/2015MS000498.

Benedict, J. J. and D. A. Randall, 2007: Observed characteristics of the MJO relative to maximum rainfall. *J. Atmos. Sci.*, **64**, 2332–2354, doi:10.1175/JAS3968.1.

Benedict, J. J. and D. A. Randall, 2009: Structure of the Madden–Julian oscillation in the superparameterized CAM. *J. Atmos. Sci.*, **66**, 3277–3296.

Bretherton, C. S., M. E. Peters, and L. E. Back, 2004: Relationships between water vapor path and precipitation over the tropical oceans. *J. Clim.*, **17**, 1517–1528, doi:10.1175/1520-0442(2004)017,1517:RBWVPA.2.0.CO;2.

Brown, R. G. and C. Zhang, 1997: Variability of midtropospheric moisture and its effect on cloud-top height distribution during TOGA COARE. *J. Atmos. Sci.*, **54**, 2760–2774.

Charney, J. G., 1963: A note on large-scale motions in the tropics. *J. Atmos. Sci.*, **20**, 607–609, doi:10.1175/1520-0469(1963)020,0607:ANOLSM.2.0.CO;2.

Chikira, M., 2014: Eastward-propagating intraseasonal oscillation represented by Chikira–Sugiyama cumulus parameterization. Part II: Understanding moisture variation under weak temperature gradient balance. *J. Atmos. Sci.*, **71**, 615–639, doi:10.1175/JAS-D-13-038.1.

Crueger, T. and B. Stevens, 2015: The effect of atmospheric radiative heating by clouds on the Madden–Julian oscillation. *J. Adv. Model. Earth Syst.*, **7**, 854–864, doi:10.1002/2015MS000434.

Dee, D. P. and Coauthors, 2011: The ERA-Interim reanalysis: Configuration and performance of the data assimilation system. *Quart. J. Roy. Meteor. Soc.*, **137**, 553–597, doi:10.1002/qj.828.

Del Genio, A. D., Y.-H. Chen, D. Kim, and M.-S. Yao, 2012: The MJO transition from shallow to deep convection in CloudSat/CALIPSO data and GISS GCM simulations. *J. Clim.*, **25**, 3755–3770, doi:10.1175/JCLI-D-11-00384.1.

DeMott, C. A., N. P. Klingaman, and S. J. Woolnough, 2015: Atmosphere-ocean coupled processes in the Madden–Julian oscillation. *Rev. Geophys.*, **53**(4), 1099–1154, 2015. ISSN 1944-9208. doi: 10.1002/2014RG000478.

DeMott, C. A., J. J. Benedict, N. P. Klingaman, S. J. Woolnough, and D. A. Randall, 2016: Diagnosing ocean feedbacks to the MJO: SST-modulated surface fluxes and the moist static energy budget, *J. Geophys. Res. Atmos.*, **121**, 8350–8373, doi:10.1002/2016JD025098.

Emanuel, K. A. 1987: An air-sea interaction model of intraseasonal oscillations in the tropics. *J. Atmos. Sci.*, **44**, 2324–2340.

Emanuel, K. A., J. D. Neelin, and C. S. Bretherton, 1994: On large-scale circulations in convecting atmospheres. *Quart. J. Roy. Meteorol. Soc.*, **120**, 1111–1143, doi:10.1002/qj.49712051902.

Fuchs, Ž. and D. J. Raymond, 2005: Large-scale modes in a rotating atmosphere with radiative-convective instability and WISHE. *J. Atmos. Sci.*, **62**, 4084–4094.

Fuchs, Ž. and Raymond, D. J., 2017: A simple model of intraseasonal oscillations. *J. Adv. Model. Earth Syst.*, **9**, 1195–1211, https://doi.org/10.1002/2017MS000963.

Gill, A. E. 1980: Some simple solutions for heat-induced tropical circulation. *Quart. J. Roy. Meteor. Soc.*, **106**, 447–462, doi:10.1002/qj.49710644905.

Hannah, W. M. and E. D. Maloney, 2011: The Role of Moisture–Convection Feedbacks in Simulating the Madden–Julian oscillation. *J. Clim.*, **24**, 2754–2770.

Hannah, W. M. and E. D. Maloney, 2014: The moist static energy budget in NCAR CAM5 hindcasts during DYNAMO. *J. Adv. Model. Earth Syst.*, **6**, 420–440, doi: 10.1002/2013MS000272.

Hayashi, Y. and D. G. Golder, 1986: Tropical intraseasonal oscillations appearing in a GFDL general circulation model and FGGE data. Part I: Phase propagation. *J. Atmos. Sci.*, **43**, 3058–3067.

Held, I. M., R. S. Hemler, and V. Ramaswamy, 1993: Radiative-convective equilibrium with explicit two-dimensional moist convection. *J. Atmos. Sci.*, **50**, 3909–3927, doi:10.1175/1520-0469(1993)050,3909:RCEWET.2.0.CO;2.

Held, I. M. and Hoskins, B. J., 1985. Large-scale eddies and the general circulation of the troposphere. In *Advances in Geophysics* Vol. 28, Elsevier, pp. 3–31.

Held, I. M., 2000: Impact of air-sea coupling on the Madden-Julian oscillation in a general circulation model. *J. Atmos. Sci.*, **57**, 3939–3952.

Hendon, H. H., 2005: Air-Sea Interaction, in *Intraseasonal Variability in the Atmosphere-Ocean Climate System*, Springer Praxis, Springer, Berlin, pp. 223–246.

Hu, Q. and D. A. Randall, 1994: Low-frequency oscillations in radiative–convective systems. *J. Atmos. Sci.*, **51**, 1089–1099.

Huffman, G. J. and Coauthors, 2007: The TRMM Multisatellite Precipitation Analysis (TMPA): Quasi-global, multiyear, combined-sensor precipitation estimates at fine scales. *J. Hydrometeor.*, **8**, 38–55, doi:10.1175/JHM560.1.

Janiga, M. A. and C. Zhang, 2016: MJO moisture budget during DYNAMO in a cloud-resolving model. *J. Atmos. Sci.*, **73**, 2257–2278, doi:10.1175/JAS-D-14-0379.1.

Jiang, X. *et al.*, 2015: Vertical structure and physical processes of the Madden–Julian oscillation: Exploring key model physics in climate simulations, *J. Geophys. Res. Atmos.*, **120**, 4718–4748, doi:10.1002/2014JD022375.

Jiang, X. 2017: Key processes for the eastward propagation of the Madden–Julian Oscillation based on multimodel simulations. *J. Geophys.*

Res. Atmos., **122**, 755–770, doi:10.1002/2016 JD025955.

Kiladis, G. N., K. H. Straub, and P. T. Haertel, 2005: Zonal and Vertical Structure of the Madden–Julian Oscillation. *J. Atmos. Sci.*, **62**, 2790–2809, https://doi.org/10.1175/JAS3520.1.

Kim, D., A. H. Sobel, A. D. Del Genio, Y. Chen, S. J. Camargo, M.-S. Yao, M. Kelley, and L. Nazarenko, 2012: Tropical subseasonal variability simulated in the NASA GISS general circulation model. *J. Clim.*, **25**, 4641–4659.

Kim, D. and Coauthors, 2014a: Process-oriented MJO simulation diagnostic: Moisture sensitivity of simulated convection. *J. Clim.*, **27**, 5379–5395, doi:10.1175/JCLI-D-13-00497.1.

Kim, D., J.-S. Kug, and A. H. Sobel, 2014b: Propagating versus nonpropagating Madden–Julian oscillation events. *J. Clim.*, **27**, 111–125, doi:10.1175/JCLI-D-13-00084.1.

Kim, D., M.-S. Ahn, I.-S. Kang, and A. D. Del Genio, 2015: Role of longwave cloud–radiation feedback in the simulation of the Madden–Julian oscillation. *J. Clim.*, **28**, 6979–6994, doi:10.1175/JCLI-D-14-00767.1.

Kim, D., H. Kim, and M.-I. Lee, 2017: Why does the MJO detour the Maritime Continent during austral summer? *Geophys. Res. Lett.*, **44**, 2579–2587, doi:10.1002/2017GL072643.

Kim, D., A. H. Sobel, and I.-S. Kang, 2011: A mechanism denial study on the Madden–Julian Oscillation. *J. Adv. Model. Earth Syst.*, **3**, M12007, doi:10.1029/2011MS000081.

Kim, D. and E. D. Maloney, 2017: Review: Simulation of the Madden–Julian Oscillation using general circulation models. *The Global Monsoon System*, Vol. 9 of *World Scientific Series on Asia-Pacific Weather and Climate*, World Scientific, pp. 119–130.

Kim, H., D. Kim, F. Vitart, V. E. Toma, J. Kug, and P. J. Webster, 2016: MJO Propagation across the Maritime Continent in the ECMWF Ensemble Prediction System. *J. Clim.*, **29**, 3973–3988, https://doi.org/10.1175/JCLI-D-15-0862.

Kiranmayi, L. and E. D. Maloney, 2011: Intraseasonal moist static energy budget in reanalysis data. *J. Geophys. Res.*, **116**, D21117, doi:10.1029/2011JD016031.

Klingaman, N. P. and S. J. Woolnough, 2014: The role of air-sea coupling in the simulation of the Madden–Julian Oscillation in the Hadley Centre Model. *Quart. J. Roy. Meteorol. Soc.*, **140**, 10.1002/qj.2295.

Landu, K. and E. D. Maloney, 2011: Understanding intraseasonal variability in an aquaplanet GCM. *J. Meteor. Soc. Jpn.*, **89**, 195–210, doi:10.2151/jmsj.2011-302.

Lau, N.-C. and K.-M. Lau, 1986: The structure and propagation of intraseasonal oscillations appearing in a GFDL general circulation model. *J. Atmos. Sci.*, **43**, 2023–2047.

Lau, K. M. and L. Peng, 1987: Origin of low-frequency intra-seasonal oscillations in the tropical atmosphere. Part I: Basic theory. *J. Atmos. Sci.*, **44**, 950–972.

Lau, N., I. M. Held, and J. D. Neelin, 1988: The Madden–Julian oscillation in an idealized general circulation model. *J. Atmos. Sci.*, **45**, 3810–3832, https://doi.org/10.1175/1520-0469(1988)045⟨3810:TMJOIA⟩2.0.CO;2.

Ma, D. and Z. Kuang, 2016: A mechanism-denial study on the Madden–Julian oscillation with reduced interference from mean state changes, *Geophys. Res. Lett.*, **43**, 2989–2997, doi:10.1002/2016GL067702.

Majda, A. J. and S. N. Stechmann, 2009: The skeleton of tropical intraseasonal oscillations. *Proc. Natl. Acad. Sci. USA*, **106**, 8417–8422, doi:10.1073/pnas.0903367106.

Maloney, E. D. and D. L. Hartmann, 2001: The Madden–Julian oscillation, barotropic dynamics, and North Pacific tropical cyclone formation. Part I: Observations. *J. Atmos. Sci.*, **58**, 2545–2558.

Maloney, E. D. and A. H. Sobel, 2004: Surface fluxes and ocean coupling in the tropical intraseasonal oscillation. *J. Clim.*, **17**, 4368–4386, https://doi.org/10.1175/JCLI-3212.1.

Maloney, E. D. 2009: The moist static energy budget of a composite tropical intraseasonal oscillation in a climate model. *J. Clim.*, **22**, 711–729, doi:10.1175/2008JCLI2542.1.

Maloney, E. D., A. H. Sobel, and W. M. Hannah, 2010: Intraseasonal variability in an aquaplanet general circulation model. *J. Adv. Model. Earth Syst.*, **2**, 5, doi: 10.3894/JAMES.2010.2.5.

Matthews, A. J. 2008: Primary and successive events in the Madden–Julian oscillation. *Quart. J. Roy. Meteorol. Soc.*, **134**, 439–453, doi:10.1002/qj.224.

Matsuno, T. 1966: Quasi-geostrophic motions in the equatorial area. *J. Meteor. Soc. Jpn.*, **44**, 25–43.

Myers, D. S. and D. E. Waliser, 2003: Three-dimensional water vapor and cloud variations associated with the Madden–Julian oscillation during Northern Hemisphere winter. *J. Clim.*,

16, 929–950, doi:10.1175/1520-0442(2003)016, 0929: TDWVAC.2 0.CO;2.

Neelin, J. D. and I. M. Held, 1987: Modeling tropical convergence based on the moist static energy budget. *Mon. Wea. Rev.*, **115**, 3–12, doi:10.1175/1520-0493(1987)115,0003:MTCBOT.20.CO;2.

Neelin, J. D., I. M. Held, and K. H. Cook, 1987: Evaporation-wind feedback and low-frequency variability in the tropical atmosphere. *J. Atmos. Sci.*, **44**, 2341–2348.

Nishimoto, E. and S. Yoden, 2017: Influence of the stratospheric quasi-biennial oscillation on the Madden–Julian oscillation during Austral Summer. *J. Atmos. Sci.*, **74**, 1105–1125, https://doi.org/10.1175/JAS-D-16-0205.1.

Pritchard, M. S. and C. S. Bretherton, 2014: Causal evidence that rotational moisture advection is critical to the superparameterized Madden–Julian oscillation. *J. Atmos. Sci.*, **71**, 800–815, doi:10.1175/JAS-D-13-0119.1.

Pritchard, M. S. and D. Yang, 2016: Response of the Superparameterized Madden–Julian oscillation to extreme climate and basic-state variation challenges a moisture mode view. *J. Clim.*, **29**, 4995–5008, https://doi.org/10.1175/JCLI-D-15-0790.1.

Powell, S. W. and R. A. Houze, 2015: Effect of dry large-scale vertical motions on initial MJO convective onset. *J. Geophys. Res.*, **120**, 4783–4805. doi: 10.1002/2014JD022961.

Powell, S. W. 2016: Updraft buoyancy within and moistening by cumulonimbi prior to MJO convective onset in a regional model. *J. Atmos. Sci.*, **73**(7), 2913–2934.

Raymond, D. J. 1995: Regulation of moist convection over the west Pacific warm pool. *J. Atmos. Sci.*, **52**, 3945–3959.

Raymond, D. J. 2000: Thermodynamic control of tropical rainfall. *Quart. J. Roy. Meteorol. Soc.*, **126**, 889–898.

Raymond, D. J. 2001: A new model of the Madden–Julian oscillation. *J. Atmos. Sci.*, **58**, 2807–2819, doi:10.1175/1520-0469(2001)058, 2807:ANMOTM.2.0.CO;2.

Raymond, D. J. and Ž. Fuchs, 2007: Convectively coupled gravity and moisture modes in a simple atmospheric model. *Tellus A*, **59**, 627–640. doi:10.1111/j.1600-0870.2007.00268.x.

Raymond, D. J. and Ž. Fuchs, 2009: Moisture modes and the Madden–Julian oscillation. *J. Clim.*, **22**, 3031–3046, doi:10.1175/2008JCLI2739.1.

Roundy, P. E. 2012: Observed structure of convectively coupled waves as a function of equivalent depth: Kelvin waves and the Madden–Julian oscillation. *J. Atmos. Sci.*, **69**, 2097–2106.

Rui, H. and B. Wang, 1990: Development characteristics and dynamic structure of tropical intraseasonal convection anomalies. *J. Atmos. Sci.*, **47**, 357–379.

Rushley, S. S., D. Kim, C. S. Bretherton, and M.-S. Ahn, 2018: Reexamining the nonlinear moisture-precipitation relationship over the tropical oceans. *Geophys. Res. Lett.*, **45**. https://doi.org/10.1002/2017GL076296.

Rushley, S., S. Kim, and Á. F. Adames, 2019: Changes in the MJO under greenhouse gas-induced warming in CMIP5 models. *J. Clim.*, Revised.

Sakaeda, N. and P. E. Roundy, 2016: The development of upper-tropospheric geopotential height anomaly in the western hemisphere during MJO convective initiations. *Quart. J. Roy. Meteorol. Soc.*, **142**(695), 942–956.

Sherwood, S. C. 1999: Convective precursors and predictability in the tropical western Pacific. *Mon. Wea. Rev.*, **127**, 2977–2991.

Shi, X., D. Kim, Á. F. Adames, and J. Sukhatme, 2018: WISHE-moisture mode in an aquaplanet simulation. *J. Adv. Model. Earth Sys.*, **10**, https://doi.org/10.1029/2018MS001441.

Slingo, J. M. and Coauthors, 1996: Intraseasonal oscillations in 15 atmosphericgeneral circulation models: Results from an AMIP diagnostic subproject. *Clim. Dynam.*, **12**, 325–357.

Sobel, A. H. and E. Maloney, 2012: An idealized semi-empirical frame-work for modeling the Madden–Julian oscillation. *J. Atmos. Sci.*, **69**, 1691–1705, doi:10.1175/JAS-D-11-0118.1.

Sobel, A. H. and E. Maloney, 2013: Moisture modes and the eastward propagation of the MJO. *J. Atmos. Sci.*, **70**, 187–192, doi:10.1175/ JAS-D-12-0189.1.

Sobel, A. H. and C. S. Bretherton, 2000: Modeling tropical precipitation in a single column. *J. Clim.*, **13**, 4378–4392, https://doi.org/10.1175/1520-0442(2000)013<4378:MTPIAS>2.0.CO;2.

Sobel, A. H., J. Nilsson, and L. M. Polvani, 2001: The weak temperature gradient approximation and balanced tropical moisture waves. *J. Atmos. Sci.*, **58**, 3650–3665, doi:10.1175/1520-0469(2001)058,3650:TWTGAA.2.0.CO;2.

Sobel, A. H. and D. Kim, 2012: The MJO-Kelvin wave transition. *Geophys. Res. Lett.*, **39**, L20808, doi:10.1029/2012GL053380.

Sobel, A. H., S. E. Yuter, C. S. Bretherton, and G. N. Kiladis, 2004: Large-scale meteorology and deep convection during TRMM KWAJEX. *Mon. Wea. Rev.*, **132**, 422–444.

Sobel, A. H., E. D. Maloney, G. Bellon, and D. M. W. Frierson, 2010: Surface fluxes and tropical intraseasonal variability: A reassessment. *J. Adv. Model. Earth Sys.*, **2**, doi:10.3894/JAMES.2010.2.2.

Sobel, A. H., S. Wang, and D. Kim, 2014: Moist static energy budget of the MJO during DYNAMO. *J. Atmos. Sci.*, **71**, 4276–4291, doi:10.1175/JAS-D-14-0052.1.

Son, S.-W., Y. Lim, C. Yoo, H. H. Hendon, and J. Kim, 2017: Stratospheric control of the Madden–Julian oscillation. *J. Clim.*, **30**(6), 1909–1922.

Straub, K. H., 2013: MJO Initiation in the real-time multivariate MJO Index. *J. Clim.*, **26**, 1130–1151, https://doi.org/10.1175/JCLI-D-12-00074.1.

Sugiyama, M. 2009: The moisture mode in the quasi-equilibrium tropical circulation model. Part I: Analysis based on the weak temperature gradient approximation. *J. Atmos. Sci.*, **66**, 1507–1523, doi:10.1175/2008JAS2690.1.

Sukhatme, J. 2014: Low-frequency modes in an equatorial shallow-water model with moisture gradients. *Quart. J. Roy. Meteor. Soc.*, **140**, 1838–1846, doi:10.1002/qj.2264.

Tam, C. and N. Lau, 2005: Modulation of the Madden–Julian oscillation by ENSO: Inferences from observations and GCM simulations. *J. Meteorol. Soc. Jpn. Ser. II*, **83**(5), 727–743.

Tokioka, T., K. Yamazaki, A. Kitoh, and T. Ose, 1988: The equatorial 30–60 day oscillation and the Arakawa–Schubert penetrative cumulus parameterization. *J. Meteor. Soc. Jpn.*, **66**, 883–901.

Wang, B. and H. Rui, 1990: Dynamics of coupled moist Kelvin-Rossby waves on an equatorial beta-plane. *J. Atmos. Sci.*, **47**, 397–413.

Wang, B. 2005: Theory. *Intraseasonal Variability in the Atmosphere-Ocean Climate System*, Praxis Publishing, pp. 307–360.

Wang, B., F. Liu, and G. Chen, 2016: A trio-interaction theory for Madden–Julian oscillation. *Geosci. Lett.*, **3**, 34, doi:10.1186/ s40562-016-0066-z.

Wang, S., A. H. Sobel, and J. Nie, 2016: Modeling the MJO in a cloud-resolving model with parameterized large-scale dynamics: Vertical structure, radiation, and horizontal advection of dry air. *J. Adv. Model. Earth Syst.*, **8**(1), 121–139, doi: 10.1002/2015MS000529.

Wang, W. and M. E. Schlesinger, 1999: The dependence of convective parameterization of the tropical intraseasonal oscillation simulated by the UIUC 11-layer atmospheric GCM. *J. Clim.*, **12**, 1423–1457.

Wolding, B. O. and E. D. Maloney, 2015: Objective diagnostics and the Madden–Julian oscillation. Part II: Application to moist static energy and moisture budgets. *J. Clim.*, **28**, 7786–7808, doi:10.1175/JCLI-D-14-00689.1.

Wolding, B. O., E. D. Maloney, and M. Branson, 2016: Vertically resolved weak temperature gradient analysis of the Madden–Julian Oscillation in SP-CESM. *J. Adv. Model. Earth Syst.*, **8**, 1586–1619, doi:10.1002/2016MS000724.

Wolding, B. O., E. D. Maloney, S. Henderson, and M. Branson, 2017: Climate change and the Madden–Julian Oscillation: A vertically resolved weak temperature gradient analysis. *J. Adv. Model. Earth Syst.*, **9**, 307–331. https://doi.org/10.1002/2016MS000843.

Yano, J. and M. H. Ambaum, 2017: Moist static energy: Definition, reference constants, a conservation law and effects on buoyancy. *Quart. J. Roy. Meteorol. Soc.*, **143**, 2727–2734. doi:10.1002/qj.3121.

Yasunaga, K. and B. Mapes, 2012: Differences between more divergent and more rotational types of convectively coupled equatorial waves. Part I: Space–time spectral analyses. *J. Atmos. Sci.*, **69**, 3–16.

Yokoi, S. and A. H. Sobel, 2015: Intraseasonal variability and seasonal march of the moist static energy budget over the eastern Maritime Continent during CINDY2011/DYNAMO. *J. Meteorol. Soc. Jpn. Ser. II*, **93**, 81–100.

Zhang, C. and J. Ling, 2017: Barrier effect of the Indo-Pacific Maritime Continent on the MJO: Perspectives from tracking MJO precipitation. *J. Clim.*, **30**, 3439–3459, https://doi.org/10.1175/JCLI-D-16-0614.1.

Zhang Chidong and Bosong Zhang, 2018: QBO-MJO Connection. *J. Geophys. Res.: Atmos.*, doi: http://dx.doi.org/10.1002/2017JD028171.

Zhu, H. and H. Hendon, 2015: Role of large scale moisture advection for simulation of the MJO with increased entrainment. *Quart. J. Roy. Meteor. Soc.*, **141**, 2127–2136, doi:10.1002/qj.2510.

Chapter 23

MJO Prediction: Current Status and Future Challenges

Hyemi Kim*,§, Frédéric Vitart† and Duane E. Waliser‡

*School of Marine and Atmospheric Sciences, Stony Brook University, NY, USA
†European Centre for Medium-Range Weather Forecasts, Reading, UK
‡Jet Propulsion Laboratory, California Institute of Technology, Pasadena, CA, USA
§hyemi.kim@stonybrook.edu

There has been a growing interest in forecasting the weather and climate within the sub-seasonal time range. The Madden–Julian oscillation (MJO), an organized envelope of tropical convection, is recognized as a primary source of sub-seasonal predictability. The MJO prediction skill in the dynamical forecast system has only recently exceeded the skill of empirical predictions. The improvement of MJO prediction in dynamical forecasting systems has been mainly due to more observations and computer resources, better data assimilation techniques, advances in theoretical understanding, and improved numerical models aided in part by multinational efforts through field campaigns and numerical model experiments. Nevertheless, estimates of the MJO predictability suggest that there is still considerable room for improvement. This paper synthesizes the progress that has been made in the past decade regarding our MJO prediction capabilities with dynamical prediction systems and our scientific understanding of its predictability, discusses the remaining challenges, and recommends new research avenues to improve the MJO prediction. This paper is a concise version of an extensive review on MJO prediction in Kim *et al.* (2018).

1. Introduction

The Madden–Julian oscillation (MJO) (Madden and Julian, 1971, 1972) is the dominant mode of variability in the tropics on timescales in excess of one week but less than a season. The MJO also has considerable impacts in the middle and high latitudes and is considered as a major source of global predictability on the sub-seasonal timescale (e.g., Waliser 2011). Therefore, skillful prediction of the MJO represents a key objective for global sub-seasonal prediction. Here, we review recent progress in MJO prediction research, current international collaborative activities for improving the sub-seasonal prediction, and issues and challenges in MJO prediction. This paper is a concise version of an extensive review on MJO prediction in Kim *et al.* (2018). Acronyms are given in Table 1.

2. International Field Campaigns and Model Experiments

There have been several internationally coordinated efforts on multi-model hindcast experiments with the aim of better understanding the sub-seasonal predictability and prediction (Zhang *et al.* 2013). The IntraSeasonal Variability Hindcast Experiment (ISVHE) launched in 2009 was the first attempt to coordinate multi-model long-term hindcast experiments focusing on sub-seasonal prediction. Eight ocean-atmosphere coupled models from six centers participated in this experiment (Neena *et al.* 2014). The sub-seasonal predictability and prediction skill from the ISVHE hindcasts have been documented in a series of papers (Zhang *et al.* 2013; Neena *et al.* 2014; Lee *et al.* 2015; Lee and Wang 2016).

The Multiscale Global Monsoon System
Edited by C. P. Chang et al.

Table 1. List of acronyms.

Acronyms	
20CRv2C	NOAA 20th Century Reanalysis v2C
BCC	Beijing Climate Center
BoM	Australian Bureau of Meteorology
CAM	Community Atmosphere Model
CESM	Community Earth System Model
CFS	Climate Forecast System
CMA	China Meteorological Administration
CPC	Climate Prediction Center
CYNDY	Cooperative Indian Ocean Experiment on Intraseasonal Variability in Year 2011
DYNAMO	Dynamics of the MJO
ECCC	Environment and Climate Change Canada
ECMWF	European Centre for Medium-Range Weather Forecasts
ERA Interim	ECMWF Interim reanalysis
FIM-iHYCOM	Flow-following Icosahedral Model coupled with an icosahedral-grid version of the Hybrid Coordinate Ocean Model
GASS	Global Energy and Water Exchanges Atmospheric System Study
GEFS	Global Ensemble Forecast System
GEOS-5	Goddard Earth Observing System Model Version 5
GFDL	Geophysical Fluid Dynamics Laboratory
GFS	Global Forecast System
GloSea5	Global Seasonal Forecasting System Version 5
IRI	International Research Institute for Climate and Society
MAPP	Modeling, Analysis, Predictions, and Projections
MJOTF	MJO Task Force
NASA	National Aeronautics and Space Administration
NCAR	National Center for Atmospheric Research
NCEP	National Centers for Environmental Prediction
NICAM	Nonhydrostatic ICosahedral Atmospheric Model
NOAA	National Oceanic and Atmospheric Administration
POAMA	Predictive Ocean-Atmosphere Model for Australia
SCOAR	Scripps Coupled Ocean-Atmosphere Regional
SNU	Seoul National University
SP-CAM	Super Parameterized Community Atmosphere Model
UH	University of Hawaii
UKMO	United Kingdom Met Office
WCRP	World Climate Research Programme
WWRP	World Weather Research Programme
YOTC	Year of Tropical Convection

The WWRP/WCRP Sub-seasonal to Seasonal Prediction Project (S2S Project) was launched in 2013. The first phase ended in November 2018 (Zhang *et al.* 2013; Vitart *et al.* 2017), and the World Meteorological Organization (WMO) has now approved the S2S Project for a second phase of 5 years. Near real-time forecasts and hindcasts from 11 operational centers have been delivered (Vitart *et al.* 2017). The majority of models demonstrated sub-seasonal skill improvement compared to their earlier versions (Vitart 2017).

Eleven models participated in the MJOTF/GASS multi-model experiment producing 20-day hindcasts of two strong MJO events during the Year of Tropical Convection (YOTC) period in boreal winter 2009–2010 (Klingaman *et al.* 2015a, b). From October 2011 to March 2012, an international field campaign CINDY/DYNAMO (DYNAMO) collected *in situ* observations in the tropical Indian Ocean (Yoneyama *et al.* 2013; Zhang *et al.* 2013). Three MJO events were captured by the radar, ship/mooring, and sounding observational network during the DYNAMO field campaign, and various aspects of the MJO have been documented. Several studies conducted hindcast experiments for the DYNAMO period to better understand the processes associated with the MJO and to assess the MJO prediction ability in global and regional models (e.g., Fu *et al.* 2013; Kerns and Chen 2014; Ling *et al.* 2014; Hannah *et al.* 2015; Xiang *et al.* 2015; Hagos *et al.* 2016; Janiga and Zhang 2016).

3. MJO Prediction Skill in Dynamical Forecast Systems

The most popular index used for MJO prediction studies is the Real-time Multivariate (RMM) MJO index developed by Wheeler and Hendon (2004). RMM1 and RMM2 are the first and second principal components of the combined empirical orthogonal functions (EOFs) of outgoing longwave radiation (OLR), with zonal wind at 200 hPa and 850 hPa averaged between 15°N and 15°S. They separate the MJO life cycle into eight different phases (phase 1–8) based on the location of convective and circulation field associated with MJO. The predicted RMMs are obtained by the projection of the predicted anomalies onto the observed combined EOF eigenvectors (Gottshalck *et al.* 2010). The majority of the studies on MJO simulations and forecasts use the RMM indices as a measure of the MJO. Unless otherwise stated explicitly, the RMM prediction skill refers to the time at which the bivariate correlation between the observed and predicted RMMs falls below 0.5.

The ensemble-mean RMM prediction skill scores (measured by signal-to-noise metrics) in the ISVHE hindcasts vary widely, between 1 and 4 weeks in the boreal winter, with most of the models having a skill of 2–3 weeks (Neena *et al.* 2014). During the past decade, some forecasting systems have shown a substantial improvement in MJO prediction that has been attributed to various factors (Rashid *et al.* 2011; Hudson *et al.* 2013; Neena *et al.* 2014; Viart 2014; Marshall *et al.* 2012, 2017; MacLachlan 2015). For example, continuous improvement of RMM skill in the ECMWF model was mainly due to the improved model physics (Vitart 2014). The recent hindcasts have RMM prediction skill scores varying widely between 10 and 38 days (Fig. 1), which exhibit an overall improvement over the ISVHE models.

Besides the long-term hindcast simulations, forecast experiments have been performed over individual MJO events during a specific period. For the two strong MJO events during the YOTC period (October 2009–January 2010), models have proven RMM skill (correlation 0.7) of 2 weeks, with several models maintaining skills beyond that (Klingaman *et al.* 2015a). During the DYNAMO period (September 2011–March 2012), the RMM skill (correlation 0.5) is about 13 days in GFS, 22 days in CFSv2, 27 days in BCC, 28 days in UH, and 29 days in the GFDL model (Fu *et al.* 2013; Xiang *et al.* 2015; Liu *et al.* 2017). With relatively short forecast

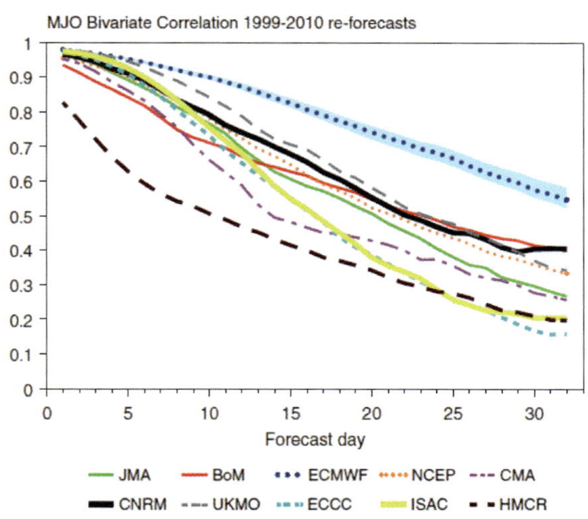

Fig. 1. RMM bivariate correlation between the model ensemble means and ERA-Interim for ten S2S models (Fig. 1 from Vitart 2017).

experiments during the DYNAMO period, the RMM skill (correlation 0.9) is 9 days in SPCAM, 7 days in NCAR CAM5 (Hannah and Maloney 2014; Hannah *et al.* 2015), and 5 days in NICAM (Nasuno 2013).

MJO amplitude and phase are both important in the assessment of MJO prediction skill. Most of the contemporary models exhibit faster decay of the ensemble-mean MJO forecast signal than the observation within a week (Vitart 2017). A amplitude approaches zero at long lead times as the phase of the individual ensemble members becomes random (Rashid *et al.* 2011, Marshall *et al.* 2017). Skill declines are attributed to MJO phase errors (Lim *et al.* 2018) rather than amplitude errors, meaning that the propagation (or stagnation) of the MJO is not accurately predicted (Rashid *et al.* 2011; Kim *et al.* 2014; Vitart 2014; Wang *et al.* 2014; Xiang *et al.* 2015; Wu *et al.* 2016; Liu *et al.* 2017). A consensus among studies is that models present slower propagation speeds than the observations (Vitart 2017). On the contrary, some models (GFDL and BCC) have shown rather faster propagation (Xiang *et al.* 2015; Liu *et al.* 2017).

4. Impact of Various Factors on MJO Prediction

4.1. *MJO amplitude and phase*

MJO prediction skill is sensitive to the initial MJO amplitude and phase. When the model is initialized with a strong MJO signal, the prediction skill tends to be higher than when initialized with weak or with no MJO signal (e.g., Rashid *et al.* 2011). Skill is higher in boreal winter when the MJO is generally more active (e.g., Rashid *et al.* 2011; Wang *et al.* 2014; Liu *et al.* 2017). The sensitivity of prediction skill to the initial MJO phases is highly model dependent (Lim *et al.* 2018). For example, ECMWF and GFDL hindcasts show relatively high skill when the model is initialized with an MJO signal in the Indian Ocean (RMM phase 2–3) (Xiang *et al.* 2015; Kim 2017), but in phase 3–4 in POAMA (Rashid *et al.* 2011), and phase 4–5 and phase 8–1 in CFSv2 (Wang *et al.* 2014).

4.2. *Low-frequency variability*

MJO predictability and prediction skill vary with changes in the low-frequency background and/or mean state (e.g., Waliser *et al.* 2001). With 30-year hindcasts from the POAMA, substantial extension of the MJO predictability and prediction skill is found during the easterly phase of the quasi-biennial oscillation (QBO) (Marshall *et al.* 2017). The extended prediction skill and predictability during the easterly QBO are not only due to more initially stronger MJO events (Yoo and Son 2016, Son *et al.* 2017) but also due to more coherent MJO events (Nishimoto and Yoden 2017). The interannual change of MJO prediction has received little attention, mainly due to the lack of hindcasts over a sufficiently long period. A better understanding of the large-scale basic state forcing on MJO predictability could potentially enhance the MJO prediction skill.

4.3. *Ensemble generation*

For a perfectly reliable system, the ensemble spread equals the root-mean-square (RMS) error between the ensemble mean and observations (e.g., Leutbecher and Palmer 2008). Ensemble predictions for MJO have shown to be underdispersive (or overconfident), indicating a lack of ensemble spread (Rashid *et al.* 2011; Kim *et al.* 2014; Neena *et al.* 2014; Xiang *et al.* 2015; Green *et al.* 2017; Liu *et al.* 2017; Lim *et al.* 2018). Figure 2 shows the relationship between the ensemble spread and the ensemble-mean RMS error for the MJO in ISVHE hindcasts (Neena *et al.* 2014). In all models, the spread is smaller than the RMS error, and models with relatively better dispersed ensembles exhibit higher ensemble-mean skill.

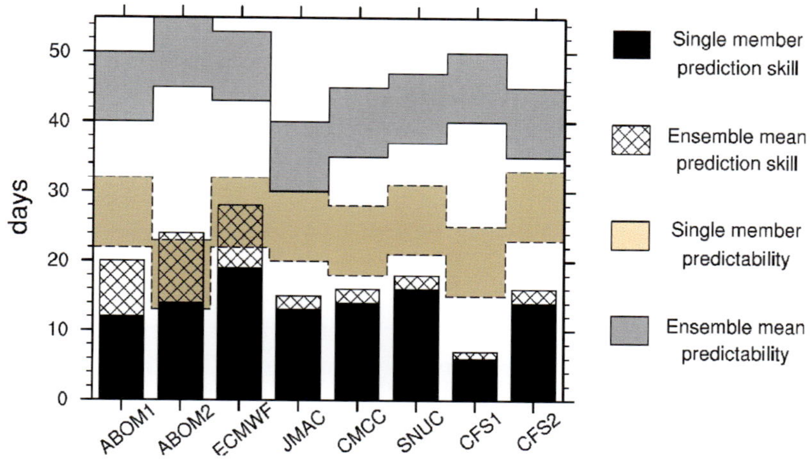

Fig. 2. The single-member prediction skill (black bar) and ensemble-mean prediction skill (hatched bar) estimates (days) for MJO for the eight models are shown along with their respective single-member (tan shaded area) and ensemble-mean (gray shaded area) estimates of MJO predictability (±5-day range) (Fig. 4 from Neena *et al.* 2014).

Several efforts are underway to improve the representation of uncertainties from initial conditions and the models. To generate initial perturbations suitable for MJO prediction, the breeding method (Liess *et al.* 2005; Chikamoto *et al.* 2007) and singular vectors (Molteni and Palmer 1993; Ham *et al.* 2012) have been explored/adopted. With 10-year hindcasts using NASA GEOS-5, Ham *et al.* (2012) showed 2–3-day extended RMM skill with an empirical singular vector approach rather than random perturbations. Hudson *et al.* (2013) showed improved MJO prediction skill in POAMA-2 using a coupled breeding approach, compared with the lagged ensemble approach in POAMA-1.5. Devising ensemble generation approaches tailored for the MJO might have a considerable impact on MJO prediction.

4.4. *Multi-model ensemble*

Recent studies have clearly shown that averaging multi-ensembles and multi-models can extend the MJO prediction skill (e.g., Fu *et al.* 2013; Kim *et al.* 2014; Neena *et al.* 2014; Ren *et al.* 2017). In all models, ensemble-mean skill exceeds single-ensemble member skill in MJO prediction (e.g., Neena *et al.* 2014; Vitart 2017). The size of the ensemble also impacts the ensemble-mean skill. The S2S models with relatively larger ensemble size show greater improvements of RMM skill in ensemble mean compared to their control forecasts (Vitart 2017).

With an equally weighted multi-model ensemble (MME) of CFSv2 and UH hindcasts, a significant extension of RMM skill (36 days) was found compared to their individual skills (22 and 28 days, respectively) (Fu *et al.* 2013). When two dynamical models with similar prediction skill and sufficient model diversity are combined, the MME shows improvements in RMM skill, although including a low-performance model in the MME can degrade the skill (Green *et al.* 2017). To extend the MJO skill, therefore, individual models need to be improved in tandem with developing an optimal strategy for MME.

4.5. *Ocean-atmosphere coupling*

The ocean-atmosphere interaction affects the MJO simulation and prediction. Many observational diagnoses have shown coherent variations in sea surface temperature, surface fluxes, and convection associated with the MJO (review in DeMott *et al.* 2015 and references therein). In hindcasts over short periods during major field campaigns, about 3–5 days of improved MJO skill was found in the coupled configuration compared to the uncoupled forecasts in NCEP CFSv2 and ECHAM (Fu *et al.* 2013) and Met Office Unified model (Shelly *et al.* 2014). With hindcasts over a longer period (>10 years), turning off the coupled feedback resulted in a loss of 8 days of RMM skill in FIM-iHYCOM (Green *et al.* 2017).

Although most studies show improvement in MJO prediction by including the ocean-atmosphere interaction, the role of coupling is case dependent. Among five MJO events that occurred during DYNAMO, only two observed MJO events were strongly coupled with the ocean while some of them were largely controlled by the atmospheric internal dynamics (Fu *et al.* 2015). The atmospheric response to the ocean also varies with the model physics (convective parameterization) (Wang *et al.* 2015). Various aspects of the model configurations, such as model physics and resolution, could affect the response of the MJO to ocean-atmosphere coupling (Seo and Wang 2010; Crueger *et al.* 2013).

4.6. *Model physics*

Due to the huge computational costs for a long record of extended range hindcasts, only a handful of studies have performed sensitivity tests of the MJO skill to model physics or resolution based on case studies. Using the operational GFS during the DYNAMO period, Wang *et al.* (2015) found the MJO forecasts to be sensitive to the cumulus convection scheme. In particular, the Simplified Arakawa–Schubert version 2 (SAS2) convection scheme (Han and Pan 2011), which has been used in the NCEP operational GFS since 2011, leads to a much weaker MJO amplitude and lower forecasting skill than the Relaxed Arakawa–Schubert (RAS, Moorthi and Suarez 1992, 1999) and SAS (used in the NCEP CFSv2) schemes. This is mainly due to the drier lower troposphere caused by a persistent weak shallow convective moistening and stronger drying associated with the deep convection in SAS2, which lead to less intense convective activity. The significant increase of the MJO prediction skill in the ECMWF model was mainly due to the improved model physics (Vitart 2014), although the improvement cannot be tied to a single change of the forecast system. With CAM5, which is the atmospheric component of the NCAR CESM1, sensitivity experiments were performed for two MJO events during the DYNAMO period (Hanna and Maloney 2014). They showed that the RMM skill increases from 12 days to 20 days via an enhancement to the entrainment rate for deep convection, which tends to strengthen the MJO amplitude (Bechtold *et al.* 2008; Klingaman and Woolnough 2014).

5. Challenges and Recommendation

This paper synthesizes the key advances in MJO prediction studies during the past decade. Theoretical models and field campaigns have led to a better understanding of the processes which are critical to simulating the MJO in numerical models. This has guided improvements in numerical weather prediction models, leading to significant advances in MJO prediction. Nevertheless, estimates of the predictability of the MJO suggest that many challenges remain to improve the dynamical forecasting systems and to fully realize the predictability of the MJO. Along with forecast system improvements highlighted in previous sections, some additional specific issues and recommendations are discussed below.

5.1. *Maritime continent prediction barrier*

The Maritime Continent is the largest archipelago on the planet with complex land-sea distribution and orography and multiscale ocean-atmospheric interactions. When the MJO propagates from the Indian Ocean to the western Pacific, it often weakens or completely breaks down when it reaches the Maritime Continent (e.g., Rui and Wang 1990; Hendon and Salby 1994), due to orography, strong diurnal convection, disrupted atmosphere–ocean feedbacks, and many other factors. Compared to nature, this Maritime Continent barrier effect is exaggerated in initialized forecasts (Vitart 2017). Simulating the MJO propagation across the Maritime Continent is one of the major hurdles to overcome to improve the MJO prediction skill. Because of the existence of this barrier in dynamical models, there has been a growing interest in understanding the critical processes involved in the MJO propagation, which has led to the establishment of the Years of the Maritime Continent (YMC) international project.

5.2. *Process-based hindcast evaluation*

Many GCMs still exhibit shortcomings in simulating realistic MJO characteristics including its amplitude, propagation, as well as horizontal and vertical structures (e.g., Jiang *et al.* 2015, 2018; Ahn *et al.* 2017). Up until now, however, evaluations of MJO fidelity in sub-seasonal forecasts over multiple years had focused mostly on performance-based (e.g., correlation, RMS error) rather than process-based metrics. During recent years, great efforts have been made to understand the critical processes for better MJO simulation by developing a so-called "process-based diagnosis" of GCMs (review in Jiang and Kim 2017). Process-based evaluation (e.g., moisture-convection, air-sea interaction processes) will shed light on which physical processes in the atmosphere–ocean system need to be better represented in numerical models to produce better MJO predictions.

6. Conclusion

Sub-seasonal forecasts are particularly important since many management decisions, including those related to water, food, and hazard considerations, fall into this range. Skillful predictions of anomalous weather events, such as extreme precipitation and heat wave, in the sub-seasonal timescale could provide policymakers, emergency managers, and stakeholders with advanced warnings to prepare mitigating actions. Since the MJO is regarded as a major source of sub-seasonal predictability, the continuous improvement of MJO prediction during the past few decades warrants an optimistic view on forewarning for MJO-related extreme and hazardous as well as fair-weather events. In particular, the recent collaborative efforts on sub-seasonal prediction (S2S Project, NOAA SubX, NOAA S2S Task Force) provide an unprecedented opportunity to facilitate addressing the challenges with process-level understanding for MJO and associated weather and climate events. A more extensive review on MJO predictability and prediction can be found in Kim *et al.* (2018).

Acknowledgments

Constructive and valuable comments from two anonymous reviewers are greatly appreciated. Kim was supported by NSF AGS-1652289, NOAA NA16OAR4310070, and KMA R&D Program KMI2018-03110. D. Waliser's contributions were carried out on behalf of the Jet Propulsion Laboratory, California Institute of Technology, under a contract with NASA, including support from the NASA Modeling, Analysis, and Prediction Program.

References

Ahn, M. S. *et al.*, 2017: MJO simulation in CMIP5 climate models: MJO skill metrics and process-oriented diagnosis. *Clim. Dynam.*, doi:10.1007/s00382-017-3558-4.

Bechtold, P., M. Köhler, T. Jung, F. Doblas-Reyes, M. Leutbecher, M. J. Rodwell, F. Vitart, and G. Balsamo, 2008: Advances in simulating atmospheric variability with the ECMWF model: From synoptic to decadal time-scales. *Quart. J. Roy. Meteor. Soc.*, **134**, 1337–1351, doi:10.1002/qj.289.

Chikamoto, Y., H. Mukougawa, T. Kubota, H. Sato, A. Ito, and S. Maeda, 2007: Evidence of growing bred vector associated with the tropical intraseasonal oscillation. *Geophys. Res. Lett.*, **34**, L04806, doi:10.1029/2006GL028450.

Crueger, T., B. Stevens, and R. Brokopf, 2013: The Madden–Julian Oscillation in ECHAM6 and the Introduction of an Objective MJO Metric. *J. Clim.*, **26**, 3241–3257, https://doi.org/10.1175/JCLI-D-12-00413.1.

DeMott, C. A., N. P. Klingaman, and S. J. Woolnough, 2015: Atmosphere–ocean coupled processes in the Madden–Julian oscillation. *Rev. Geophys.*, **53**, doi:10.1002/2014RG000478.

Fu, X., B. Wang, D. E. Waliser, and L. Tao, 2007: Impact of atmosphere–ocean coupling on the predictability of monsoon intraseasonal oscillations. *J. Atmos. Sci.*, **64**(1), 157–174.

Fu, X., B. Wang, J. Y. Lee, W. Q. Wang, and L. Gao, 2011: Sensitivity of dynamical intraseasonal prediction skills to different initial conditions. *Mon. Wea. Rev.*, **139**, 2572–2592.

Fu, X., J. Y. Lee, P. C. Hsu, H. Taniguchi, B. Wang, W. Wang, and S. Weaver, 2013: Multi-model MJO forecasting during DYNAMO/CINDY period. *Clim. Dynam.*, **41**, 1067–1081.

Fu, X., W. Wang, J.-Y. Lee, B. Wang, K. Kikuchi, J. Xu, J. Li, and S. Weaver, 2015: Distinctive roles of air-sea coupling on different MJO events: A new perspective revealed from the DYNAMO/CINDY field campaign. *Mon. Wea. Rev.*, **143**, 794–812.

Gottschalck, J. and coauthors, 2010: A framework for assessing operational Madden-Julian Oscillation forecasts: A CLIVAR MJO Working Group project. *Bull. Amer. Meteor. Soc.*, **91**, 1247–1258.

Green, B. W., S. Sun, R. Bleck, S. G. Benjamin, and G. A. Grell, 2017: Evaluation of MJO predictive skill in multiphysics and multimodel global ensembles. *Mon. Wea. Rev.*, **145**, 2555–2574, https://doi.org/10.1175/MWR-D-16- 0419.1

Hagos, S. M., C. Zhang, Z. Feng, C. D. Burleyson, C. De Mott, B. Kerns, J. J. Benedict, and M. N. Martini, 2016: The impact of the diurnal cycle on the propagation of Madden–Julian Oscillation convection across the Maritime Continent. *J. Adv. Model. Earth Syst.*, **8**, 1552–1564, doi:10.1002/2016MS000725.

Ham, Y., S. Schubert, and Y. Chang, 2012: Optimal initial perturbations for ensemble prediction of the Madden–Julian oscillation during Boreal Winter. *J. Clim.*, **25**, 4932–4945, https://doi.org/10.1175/JCLI-D-11-00344.1.

Hamill, T. M. and G. N. Kiladis, 2014: Skill of the MJO and Northern Hemisphere blocking in GEFS medium-range reforecasts. *Mon. Wea. Rev.*, **142**, 868–885, doi:10.1175/MWR-D-13-00199.1.

Han, J. and H. L. Pan, 2011: Revision of convection and vertical diffusion schemes in the NCEP Global Forecast System. *Wea. Forecast.*, **26**, 520–533, doi:10.1175/WAF-D-10-05038.1.

Hannah, W. M. and E. D. Maloney, 2014: The moist static energy budget in NCAR CAM5 hindcasts during DYNAMO. *J. Adv. Model. Earth Syst.*, **6**, 420–440.

Hannah, W. M., E. D. Maloney, and M. S. Pritchard, 2015: Consequences of systematic model drift in DYNAMO hindcasts with SP-CAM and CAM5. *J. Adv. Model. Earth Sys.*, **7**.

Hendon, H. H. and M. L. Salby, 1994: The life cycle of the Madden–Julian oscillation. *J. Atmos. Sci.*, **51**, 2225–2237.

Hudson, D., A. G. Marshall, Y. Yin, O. Alves, and H. H. Hendon, 2013: Improving intraseasonal prediction with a new ensemble generation strategy. *Mon. Weather Rev.*, **141**, 4429–4449.

Janiga, M. A. and C. Zhang, 2016: MJO Moisture Budget during DYNAMO in a cloud-resolving model. *J. Atmos. Sci.*, **73**, 2257–2278, https://doi.org/10.1175/JAS-D-14-0379.1.

Jiang X. 2017: Key processes for the eastward propagation of the Madden–Julian oscillation based on multi-model simulations. *J. Geophys. Res. Atmos.*, **122**, doi:10.1002/2016JD025955.

Jiang, X., D. E. and coauthors, 2015: Vertical structure and diabatic processes of the Madden-Julian oscillation: exploring key model physics in climate simulations. *J. Geophys. Res. Atmos.*, **120**, 4718–4748.

Jiang, X. and D. Kim, 2017: Progress and status of MJO simulation in climate models and process-oriented diagnostics, in *The Global Monsoon Systems, Sixth International Workshop on Monsoons (IWM-VI)*, November, 2017, Singapore, pp. 13–17. https://www.wmo.int/pages/prog/arep/wwrp/new/documents/IWM6AbstractsVolume.pdf

Jiang, X., A. F. Adames, M. Zhao, D. E. Waliser, and E. D. Maloney, 2018: A unified moisture moist framework for seasonality of MJO propagation. *J. Clim.*, https://doi.org/10.1175/JCLI-D-17-0671.1.

Kang, I. S., P. H. Jang, M. Almazroui, 2014: Examination of multi-perturbation methods for ensemble prediction of the MJO during boreal summer. *Clim. Dynam.*, **42**, 2627, https://doi.org/10.1007/s00382-013-1819-4.

Kerns, B. W. and S. S. Chen, 2014: Equatorial dry air intrusion and related synoptic variability in MJO initiation during DYNAMO. *Mon. Wea. Rev.*, **142**, 1326–1343, https://doi.org/10.1175/MWR-D-13-00159.1

Kim, D. and E. D. Maloney, 2017: Review: Simulation of the Madden–Julian oscillation using general circulation models. In *The Global Monsoon System*, 3rd Edition, Eds. C.-P. Chang *et al.*

Kim, H. M., P. J. Webster, V. E. Toma, and D. Kim, 2014: Predictability and prediction skill of the MJO in two operational forecasting systems. *J. Clim.*, **27**(14), 5364–5378.

Kim, H. M., D. Kim, F. Vitart, V. E. Toma, J. S. Kug, and P. J. Webster, 2016: MJO propagation across the Maritime Continent in the ECMWF ensemble prediction system. *J. Clim.*, **29**(11), 3973–3988.

Kim, H., F. Vitart, and D. E. Waliser, 2018: Prediction of the Madden–Julian oscillation: A review. *J. Clim.*, https://doi.org/10.1175/JCLI-D-18-0210.1.

Klingaman, N. P. and S. J. Woolnough, 2014: Using a case-study approach to improve the Madden-Julian oscillation in the Hadley Centre model. *Q. J. R. Meteorol. Soc.*, **140**, 2491–2505.

Klingaman, N. P. and coauthors, 2015a: Vertical structure and physical processes of the Madden–Julian Oscillation: Linking hindcast fidelity to simulated diabatic heating and moistening. *J. Geophys. Res. Atmos.*, 10.1002/2014JD022374.

Klingaman, N. P., X. Jiang, P. K. Xavier, J. Petch, D. Waliser, and S. J. Woolnough, 2015b: Vertical structure and physical processes of the Madden–Julian oscillation: Synthesis and summary. *J. Geophys. Res. Atmos.*, 10.1002/2015JD023196.

Lau, W. K. M., and D. E. Waliser, Eds., 2011: *Intraseasonal Variability of the Atmosphere–Ocean Climate System, 2nd Edition.* Springer, Heidelberg, Germany, pp. 613.

Lee, J. Y., X. Fu, and B. Wang, 2017: Predictability and prediction of the Madden-Julian oscillation: a review on progress and current status. In *The Global Monsoon System: Research and Forecast*, pp. 147–159.

Lee, S.-S., B. Wang, D. E. Waliser, J. M. Neena, and J.-Y. Lee, 2015: Predictability and prediction skill of the boreal summer intraseasonal oscillation in the Intraseasonal Variability Hindcast Experiment. *Clim. Dynam.*, **45**, 2123–2135, DOI: 10.1007/s00382-014-2461-5.

Lee, S.-S. and B. Wang, 2016: Regional boreal summer intraseasonal oscillation over Indian Ocean and Western Pacific: Comparison and predictability study. *Clim. Dynam.*, **46**, 2213–2229, doi: 10.1007/s00382-015-2698-7.

Leutbecher, M. and T. N. Palmer, 2008: Ensemble forecasting. *J. Comp. Phys.*, **227**, 3515–3539.

Liess, S., D. E. Waliser, and S. D. Schubert, 2005: Predictability studies of the intraseasonal oscillation with the ECHAM5 GCM. *J. Atmos. Sci.*, **62**(9), 3320–3336.

Lim, Y., S. Son, and D. Kim, 2018: MJO prediction skill of the subseasonal-to-seasonal prediction models. *J. Clim.*, https://doi.org/10.1175/JCLI-D-17-0545.1.

Ling J., P. Bauer, P. Bechtold, A. Beljaars, R. Forbes, F. Vitart, M. Ulate, and C. Zhang, 2014: Global versus local MJO forecast skill of the ECMWF model during DYNAMO. *Mon. Wea. Rev.*, **142**, 2228–2247.

Ling, J., C. Li, T. Li, X. Jia, B. Khouider, E. Maloney, F. Vitart, Z. Xiao, and C. Zhang, 2017: Challenges and Opportunities in MJO studies. *Bull. Amer. Meteor. Soc.*, **98**, ES53–ES56, https://doi.org/10.1175/BAMS-D-16-0283.1.

Liu, X. and coauthors, 2017: MJO prediction using the sub-seasonal to seasonal forecast model of Beijing Climate Center. *Clim. Dynam.*, **48**, 3283–3307.

MacLachlan, C. and coauthors, 2015: Global Seasonal forecast system version 5 (GloSea5): a high-resolution seasonal forecast system. *Q. J. R. Meteorol. Soc.*, **141**, 1072–1084, doi:10.1002/qj.2396.

Madden, R. A. and P. R. Julian, 1971: Detection of a 40–50 day oscillation in the zonal wind in the tropical Pacific. *J. Atmos. Sci.*, **28**, 702–708.

Madden, R. A. and P. R. Julian, 1972: Description of global-scale circulation cells in the tropics with a 40–50 day period. *J. Atmos. Sci.*, **29**, 1109–1123.

Marshall, A. G., H. H. Hendon, and D. Hudson, 2016: Visualizing and verifying probabilistic forecasts of the Madden-Julian oscillation. *Geophys. Res. Lett.*, **43**, 12,278–12,286, doi:10.1002/2016GL071423.

Marshall, A. G., H. H. Hendon, S.-W. Son, and Y. Lim, 2017: Impact of the quasi-biennial oscillation on predictability of the Madden–Julian oscillation. *Clim. Dynam.*, **49**, 1365–1377.

Miyakawa, T. and coauthors, 2014: Madden–Julian oscillation prediction skill of a new-generation global model. *Nat. Commun.*, **5**, 3769, doi:10.1038/ncomms4769.

Molteni, F. and T. N. Palmer, 1993: Predictability and finite time instability of the northern winter circulation. *Q. J. R. Meteorol. Soc.*, **119**, 269–298.

Moorthi, S. and M. J. Suarez, 1992: Relaxed Arakawa–Schubert: A parameterization of moist convection for general circulation models. *Mon. Wea. Rev.*, **120**, 978–1002, doi:10.1175/1520-0493(1992)120,0978:RASAPO.2.0.CO;2.

Moorthi, S. and M. J. Suarez, 1999: Documentation of version 2 of relaxed Arakawa–Schubert cumulus parameterization with convective downdrafts. A Office Note 99-01, pp. 44.

Nasuno, T. 2013: Forecast skill of Madden-Julian Oscillation events in a global nonhydrostatic model during the CINDY2011/DYNAMO observation period. *SOLA*, **9**, 69–73, doi:10.2151/sola.2013-016.

Neena, J. M., J. Y. Lee, D. Waliser, B. Wang, and X. Jiang, 2014: Predictability of the Madden–Julian Oscillation in the Intraseasonal Variability Hindcast Experiment (ISVHE). *J. Climate*, **27**, 4531–4543.

Nishimoto, E. and S. Yoden, 2017: Influence of the Stratospheric Quasi-Biennial Oscillation on the Madden–Julian Oscillation during Austral Summer. *J. Atmos. Sci.*, **74**(4), 1105–1125.

Rashid, H. A., H. H. Hendon, M. C. Wheeler, and O. Alves, 2011: Prediction of the Madden–Julian oscillation with the POAMA dynamical prediction system. *Clim. Dynam.*, **36**, 649–661.

Ren, H. L. *et al.*, 2017: Prediction of primary climate variability modes in Beijing climate center, *J. Meteor. Res.* **31**, 204–223.

Rui, H. and B. Wang, 1990: Development characteristics and dynamic structure of tropical intraseasonal convection anomalies. *J. Atmos. Sci.*, **47**, 357–337.

Seo, K. H., W. Wang, J. Gottschalck, Q. Zhang, J-K. Schemm, W. Higgins, and A. Kumar, 2009: Evaluation of MJO forecast skill from several statistical and dynamical forecast models. *J. Clim.*, **22**, 2372–2388.

Seo, K. H. and W. Wang, 2010: The Madden–Julian Oscillation Simulated in the NCEP Climate Forecast System Model: The Importance of Stratiform Heating. *J. Clim.*, **23**, 4770–4793, https://doi.org/10.1175/2010JCLI2983.1.

Shelly, A., P. Xavier, D. Copsey, T. Johns, J. M. Rodriguez, S. Milton, and N. P. Klingaman, 2014: Coupled versus uncoupled hindcast simulations of the Madden-Julian oscillation in the Year of Tropical Convection. *Geophys. Res. Lett.*, **41**, 5670–5677, doi:10.1002/2013GL059062.

Son, S.-W., Y. Lim, C. Yoo, H. H. Hendon, and J. Kim, 2017: Stratospheric control of the Madden–Julian oscillation, *J. Climate*, **30**(6), 1909–1922, doi:10.1175/JCLI-D-16-0620.1.

Vitart, F. 2014: Evolution of ECMWF sub-seasonal forecast skill. *Q. J. R. Meteorol. Soc.*, doi: 10.1002/qj.2256.

Vitart, F. 2017: Madden—Julian oscillation prediction and teleconnections in the S2S database. *Q. J. R. Meteorol. Soc.*, **143**, 2210–2220, doi:10.1002/qj.3079.

Vitart, F. and T. Jung, 2010: Impact of the northern Hemisphere extratropics on the skill in predicting the Madden–Julian oscillation. *Geophys. Res. Lett.*, **37**, L23805, doi:10.1029/2010GL045465.

Vitart, F. and F. Molteni, 2010: Simulation of the Madden–Julian Oscillation and its teleconnections in the ECMWF forecast system. *Q. J. R. Meteorol. Soc.*, **136**, 842–855.

Vitart, F. and coauthors, 2017: The sub-seasonal to seasonal prediction (S2S) project database. *Bull. Amer. Meteor. Soc.*, doi: 10.1175/BAMS-D-16-0017.1.

Waliser, D. E. 2006a: Intraseasonal Variability. In *The Asian Monsoon*, Chapter 5, Ed. Bin Wang, Springer, Heidelberg, Germany, p. 787.

Waliser, D. E. 2006b: Predictability of Tropical Intraseasaonal Variability. In *Predictability of Weather and Climate*, Chapter 11, Eds., T. Palmer and R. Hagedorn, Cambridge University Press, 718.

Waliser, D. E. 2011: Predictability and Forecasting. In *Intraseasonal Variability of the Atmosphere-Ocean Climate System*, Chapter 12, 2nd Edition, Eds., W. K. M. Lau and D. E. Waliser, Springer, Heidelberg, Germany, pp. 613.

Waliser, D., K. M. Lau, and J.-H. Kim, 1999: The influence of coupled sea surface temperatures on the Madden–Julian oscillation: A model perturbation experiment. *J. Atmos. Sci.*, **56**, 333–358.

Waliser, D., Z. Zhang, K. M. Lau, and J. H. Kim, 2001: Interannual sea surface temperature variability and the predictability of tropical intraseasonal variability. *J. Geophys. Res. Atmos.*, **58**, 2595–2614.

Waliser, D., K. M. Lau, W. Stern, and C. Jones, 2003: Potential predictability of the Madden-Julian oscillation. *Bull. Amer. Meteor. Soc.*, **84**, 33–50.

Waliser, D., K. Weickmann, R. Dole, S. Schubert, O. Alves, C. Jones, M. Newman, H. L. Pan, A. Roubicek, S. Saha, C. Smith, H. van den Dool, F. Vitart, M. Wheeler, and J. Whitaker, 2006: The experimental MJO prediction project. *Bull. Amer. Meteor. Soc.*, **87**(4), 425–431.

Waliser, D. E. and coauthors, 2012: The "Year" of Tropical Convection (May 2008 to April 2010): Climate Variability and Weather Highlights. *Bull. Amer. Meteor. Soc.*, Vol 93, No 8, DOI:10.1175/2011BAMS3095.1.

Wang, W., M. P. Hung, S. J. Weaver, A. Kumar, and X. Fu, 2014: MJO prediction in the NCEP Climate Forecast System version 2, *Clim. Dynam.*, 10.1007/s00382-013-1806-9.

Wang, W., A. Kumar, J.X. Fu, and M. Hung, 2015: What is the Role of the Sea Surface Temperature Uncertainty in the Prediction of Tropical Convection Associated with the MJO? *Mon. Wea. Rev.*, **143**, 3156–3175, https://doi.org/10.1175/MWR-D-14-00385.1.

Wheeler, M. C. and H. H. Hendon, 2004: An all-season real-time multivariate MJO index: Development of an index for monitoring and prediction. *Mon. Wea. Rev.*, **132**, 1917–1932.

Wheeler, M. C., H. J. Kim, J. Y. Lee, and J. C. Gottschalck, 2017a: Real-time forecasting of modes of tropical intraseasonal variability: The Madden–Julian and Boreal Summer Intraseasonal Oscillations. In *The Global Monsoon System*: 3rd Edition, pp. 131–138.

Wheeler, M. C., H. Zhu, A. H. Sobel, D. Hudson, and F. Vitart, 2017b: Seamless precipitation prediction skill comparison between two global models. *Q. J. R. Meteorol. Soc.*, **143**, 374–383, doi:10.1002/qj.2928.

Wu, J., H.-L. Ren, J. Q. Zuo, *et al.*, 2016: MJO prediction skill, predictability, and teleconnection impacts in the Beijing Climate Center Atmospheric General Circulation Model. *Dyn. Atmos. Oceans*, **75**, 78–90, doi: 10.1016/j.dynatmoce.2016.06.001.

Xiang, B., M. Zhao, X. Jiang, S. Lin, T. Li, X. Fu, and G. Vecchi, 2015: The 3–4-Week MJO Prediction Skill in a GFDL Coupled Model. *J. Clim.*, 28, 5351–5364.

Yoneyama, K., C. Zhang, and C. N. Long, 2013: Tracking Pulses of the Madden–Julian Oscillation. *Bull. Amer. Meteor. Soc.*, **94**, 1871–1891, https://doi.org/10.1175/BAMS-D-12-00157.1.

Yoo, C. and S. W. Son, 2016: Modulation of the boreal wintertime Madden-Julian oscillation by the stratospheric quasi-biennial oscillation. *Geophys. Res. Lett.*, **43(3)**, 1392–1398, doi:10.1002/2016GL067762.

Zhang, C. 2005: Madden–Julian oscillation, *Rev. Geophys.*, **43**, RG2003, doi:10.1029/2004RG000158, 2005.

Zhang, C. 2013: Madden–Julian oscillation: Bridging weather and climate. *Bull. Amer. Met. Soc.*, **94**, 1849.1870, doi:10.1175/BAMS-D-12-00026.1.

Zhang, C., J. Gottschalck, E. D. Maloney, M. Moncrieff, F. Vitart, D. E. Waliser, B. Wang, and M. C. Wheeler, 2013: Cracking the MJO nut. *Geophys. Res. Lett.*, **40**, 1223–1230.

Zhang, C. and J. Ling, 2017: Barrier Effect of the Indo-Pacific Maritime Continent on the MJO: Perspectives from Tracking MJO Precipitation. *J. Clim.*, **30**, 10.1175/JCLI-D-16-0614.1, 3439–3459.

Chapter 24

The Association between the Phase Speed of the Madden–Julian Oscillation and Atmospheric Circulation

Paul E. Roundy

Department of Atmospheric and Environmental Sciences
University at Albany, State University of New York, 1400 Washington Ave., Albany, NY 12222, USA
proundy@albany.edu

Convective and circulation features associated with the southern summer Madden–Julian oscillation (MJO) may move eastward due to a variety of factors. This work assesses the association between the phase speed of MJO convective and circulation anomalies and slowly evolving background wind patterns during December through February. A real-valued space–time wavelet transform is applied to create a time index of MJO signals at different phase speeds. Simple linear regression is then applied to diagnose the global circulation patterns linearly correlated with the base index. Results suggest that advection of intraseasonal wind by the background wind and advection of background wind by intraseasonal wind contribute substantially to the variability in the phase speed in global circulation patterns associated with the MJO. A second regression algorithm based on a common MJO index during periods of high or low background wind in select regions suggests substantial variation of phase speed of convective anomalies associated with variations in background wind.

1. Introduction

The Madden–Julian oscillation (MJO) couples moist deep convection across the tropics to the global atmospheric circulation (Zhang 2005). The MJO is associated with eastward- and poleward-moving alternating regions of suppressed and enhanced rainfall mainly over the tropical ocean warm pool regions (Madden and Julian 1994). These regions of anomalous rainfall interact with propagating and standing oscillations of geopotential height across the subtropics and the middle to higher latitudes (e.g., L'Heureux and Higgins 2008; Roundy *et al.* 2010; Frederiksen and Lin 2013; Garfinkel *et al.* 2014; Adames *et al.* 2014; and many others). The MJO helps organize breaking Rossby wave events across the globe (Moore *et al.* 2010; MacRitchie and Roundy 2015; Henderson *et al.* 2016), modulates tropical cyclone activity (e.g., Maloney and Hartmann 2001; Roundy and Frank 2004; Aiyyer and Molinari 2008; Camargo *et al.* 2009), and even helps organize severe weather events such as violent tornado outbreaks in the United States (e.g., Thompson and Roundy 2013).

During most MJO events, circulation signals in the upper troposphere emerge from active convection over the west Pacific basin and propagate rapidly eastward across the tropics of the Western Hemisphere, with some similarity at times to dry and convectively coupled Kelvin waves (Hendon and Salby 1994; Matthews 2000; Straub and Kiladis 2003; Sobel and Kim 2012; Roundy 2012b, 2014; Sakaeda and Roundy 2015). Sakaeda and Roundy (2015) suggest that beyond Kelvin waves, these signals include substantial contributions from advection terms including advection of and by background zonal wind along with eddy fluxes of momentum from synoptic and intraseasonal waves arriving from the extratropics and breaking on the

The Multiscale Global Monsoon System
Edited by C. P. Chang et al.

tropopause. The intraseasonal portion of the Rossby wave signal includes the response to previous MJO convection to the west. Kelvin waves have geopotential height anomalies in phase with zonal wind anomalies, while the sign of zonal flow on the equatorward sides of Rossby wave gyres is opposite the associated geopotential height anomalies. In many cases, the zonal phase relationship in the circumnavigating signal of the MJO between pressure and wind is directly opposite that of Kelvin waves, as Rossby waves dominate the flow. The west Pacific MJO radiates intraseasonal Rossby waves into the extratropical atmosphere, which then propagate eastward across most of the temperate Pacific basin and then eastward and equatorward over the East Pacific and Atlantic Oceans as well as across Africa and southwestern Asia (Roundy 2014; Seo and Son 2012). These wave patterns influence the organization of convection (Sakaeda and Roundy 2015, 2016b).

Therefore, not only does the MJO force the extratropical circulation but it responds to it as well (e.g., Ray and Zhang 2010; Ray and Li 2013; Zhao *et al.* 2013; Roundy 2014; Sakaeda and Roundy 2015). This work does not aim to demonstrate whether the MJO could exist without such two-way interactions. Instead, it focuses on how the observed interactions associate with different behaviors in the MJO.

Recent works arising from analysis of reanalysis products and data collected during the DYNAMO field campaign suggest that many MJO convective events are initiated by wave activity arriving over the Indian Ocean from the west, which includes a mix of Kelvin waves and extratropical Rossby waves breaking onto the equator of the Western Hemisphere and Africa (Roundy 2014; Powell and Houze 2015; Sakaeda and Roundy 2015, 2016a,b; Zhang *et al.* 2017). The Kelvin waves induce upward motion, which cools the deep troposphere. Cooling in the region near the top of the atmospheric boundary layer reduces convective inhibition, allowing deep convection to begin organizing (e.g., Powell and Houze 2015). Studies like Ray and Zhang (2010), Ray and Li (2013), and Zhao *et al.* (2013) have discounted the Kelvin wave as important, while others support its relevance (Powell and Houze 2015; Sakaeda and Roundy 2015; Haertel *et al.* 2015; and others). Some modeling studies blocked Kelvin waves from crossing the Atlantic basin, and the MJO signals still being organized were associated with extratropical wave intrusions (Ray and Zhang 2010; Ray and Li 2013; Zhao *et al.* 2013). However, these extratropical waves themselves would have induced Kelvin waves east of the blocked region and west of the Indian Ocean (Sakaeda and Roundy 2015), so Kelvin waves would have arrived there anyway.

Tapering anomalous flow across the subtropics to zero silenced MJO activity in some of these studies, though more recent similar model experiments have suggested that the reduction of MJO activity in the models resulted from cooling of the model tropics. Ma and Kuang (2016) showed that artificially maintaining the warmth of the tropics in spite of the disconnect from the extratropics maintains the MJO spectral peak. These results suggest that the MJO might not fundamentally depend on its connection to the extratropics, though it is difficult to imagine what an MJO circulation pattern lacking connection across the subtropical jet stream might look like. Much of the relevant interaction with the extratropical atmosphere may occur over the Western Hemisphere (e.g., Roundy 2014; Sakaeda and Roundy 2015), so that its impacts could have leaked into the model Indian basin of Ma and Kuang from the west. In any case, it is clear that the observed wave forcing does trigger MJO convective onset (or set its timing) and might influence other characteristics of the observed MJO, even if the MJO might exist without extratropical wave precursors.

Some authors have argued that some MJO events (labeled as primary events, Matthews 2008) may not be initiated by Kelvin waves arriving from the west, though diagnosis of an event as primary depends on the index used (e.g., Straub 2013), and preliminary analysis still suggests that most events diagnosed as primary do include wave precursors from the west (either Kelvin, extratropical, or both).

Although the *basic* large-scale mechanism of initiation of most observed MJO convective events may be a solved problem (if not yet broadly agreed upon), the field has yet to reach broad agreement on the dynamics of propagation of the MJO. Numerous mostly non-exclusive hypotheses suggest possible mechanisms, and many of them may explain aspects of MJO dynamics, but no single mechanism seems to explain all major aspects of the movement of the MJO (e.g., Matthews 2000; Seo and Kim 2003). A summary of previous works on MJO propagation mechanisms is available from Li and Hsu (2017). Although most MJO events share some set of distinguishing characteristics, every event includes important large-scale patterns that differ from the average event. Many mechanisms may be relevant in different combinations at different times in the life cycle of an event.

Different MJO events propagate at different speeds; the convection and circulation signals may have some freedom to migrate with respect to each other; and the phase speeds of the different elements vary across the lifetimes of events and as they cross different regions of the world.

This work is the beginning of a much broader study to test mechanisms whereby interaction with the subtropical jet stream or other aspects of the background circulation can modulate the eastward propagation speed of circulation and convective signals associated with the MJO. In that sense, it represents a work in progress, but one that has already produced some compelling results. Different intensities of the jet streams and other background winds would constitute different base states across which circulation signals associated with the MJO evolve, and these different states must impact the movement of the intraseasonal circulation anomalies. Observed associations might suggest that these altered signals could then feed back on the convection.

2. Wavelet and Momentum Budget Analysis

Wavenumber–frequency wavelet analysis (e.g., Roundy 2012a,b; Roundy 2017) is used to extract time indexes associated with the MJO or atmospheric waves at selected zonal wavenumbers and frequencies, allowing the user to roughly specify the phase speed of the target extracted signal. These time series are then used as predictors in regression models designed to diagnose associated signals in fields of data, which can be different for MJO events characterized by different phase speeds. This section describes how the wavelet time indexes are calculated. A time series of wavelet-filtered NOAA interpolated outgoing longwave radiation (OLR, Liebmann and Smith 1996) anomalies from the background mean and seasonal cycle (including four harmonics) is extracted by projecting data onto a wavelet function in the longitude–time domain. A range of 1974–2010 was used for the climatology. For a longitude–time array of OLR anomaly data $X(x,t)$ averaged from 7.5°S to 7.5°N, the wavelet index w at a given base longitude x_0 and time t_0 is given by (1).

$$W(x_0,t_0) = \sum_{x}^{n_x} \sum_{t=-100}^{100} \frac{1}{\gamma}\cos(2\pi(k(x-x_0) - F_c(t-t_0)))\exp\left(\frac{-(x-x_0)^2}{F_{Bx}} + \frac{-(t-t_0)^2}{F_{Bt}}\right) X(t-t_0, x-x_0), \quad (1)$$

where

$$\gamma = \sum_{x=0}^{n_x} \sum_{t=-100}^{100} \exp\left(\frac{-(x-x_0)^2}{F_{Bx}} + \frac{-(t-t_0)^2}{F_{Bt}}\right).$$

Here, k is made equivalent to zonal wavenumber 2, the dominant convective signal in the MJO over the warm pool. F_C is set to the frequency corresponding to a disturbance at wavenumber 2 evolving eastward at integer phase speeds from 1 to $8\,\mathrm{ms}^{-1}$. n_x is the number of grid points in the x direction, which includes three cycles around the world. x_0 is somewhere in the center global segment, so that the Gaussian tapering in the zonal direction (represented by the exp() function in (1)) can taper approximately to zero in spite of proximity of x_0 to the eastern or western edge of the single global data array. F_{Bx} and F_{Bt} are the Gaussian structure bandwidth parameters in longitude and time, set to 1,000 and 10,000 units, respectively. These parameters set the localization of the target signal, with larger values localizing the wavelet less closely about t_0 and x_0. Over-localized signals produce unacceptable biases away from extracting signals at the target wavenumber and frequency. Many other values were tested and results were not found highly sensitive to changes except at much smaller values. The result of (1) for each t_0 approximately traces a signal at the target phase speed crossing the base longitude in the 7.5°N to 7.5°S latitude band.

Regression of fields of data against indexes w selected for specific zonal scales and phase speeds allows for comparison of the associated structures in those fields, showing the differences at different phase speeds. Data were obtained from the NCEP/NCAR reanalysis for the same climatology as the OLR data. The regression algorithm follows Wheeler *et al.* (2000) except that it includes all data from December through February, with no test to include only periods of time with above-average variance. The base longitude is 80°E, near the maximum in amplitude in MJO convection over the Indian Ocean. After calculating the field of regression slope coefficients, the amplitude of the base point OLR anomaly is set at −2 standard deviations to reconstruct an estimate of a high-amplitude event. A test of the full wavelet regression algorithm on artificial data is given in Appendix A.

Figure 1a shows the upper-tropospheric circulation and convective pattern associated with $1\,\mathrm{ms}^{-1}$ convective signal at wavenumber 2 centered over the eastern Indian Ocean. Amplitudes are scaled to −2 standard deviations in the base index. Figure 1b shows the result at $7\,\mathrm{ms}^{-1}$. This analysis reveals only the part of the circulation and convection evolving at the same timescale as the index signal, so it excludes synoptic and low-frequency background circulations. In bulk, the faster signal is associated with more amplitude in the subseasonal signal downstream over the Pacific region across to the N. Atlantic basin. The slower signal is associated with greater amplitude in geopotential height anomalies over Europe and Asia. Note the reversal of the height pattern over the N. Atlantic, consistent with reversal of the North Atlantic Oscillation (NAO), consistent with Yadav and Straus (2017), though their results were plotted with some time lag (which matters less for the slowest events). Also, their composites would include any background state circulation signals that favor fast or slow events along with the part of the flow directly correlated with the MJO.

The results presented here show that the phase speed of the MJO is relevant to the structure of the expected response pattern. It is unclear whether this difference results from different forcing patterns in the vertical, a different extratropical base state, or different translation speeds of the convection relative to the circulation response to previous convection. The structure of the region of convection changes relatively little with phase speed but is strongest at $4\,\mathrm{ms}^{-1}$ (not shown). At still slower phase speeds, however, the intensity of the convection becomes smaller than or similar to that at $7\,\mathrm{ms}^{-1}$.

By reconstructing the advection terms approximately from daily mean 2.5° NCEP–NCAR reanalysis data (using a first-order centered finite differencing scheme) (and repeated for comparison using CFS reanalysis data from

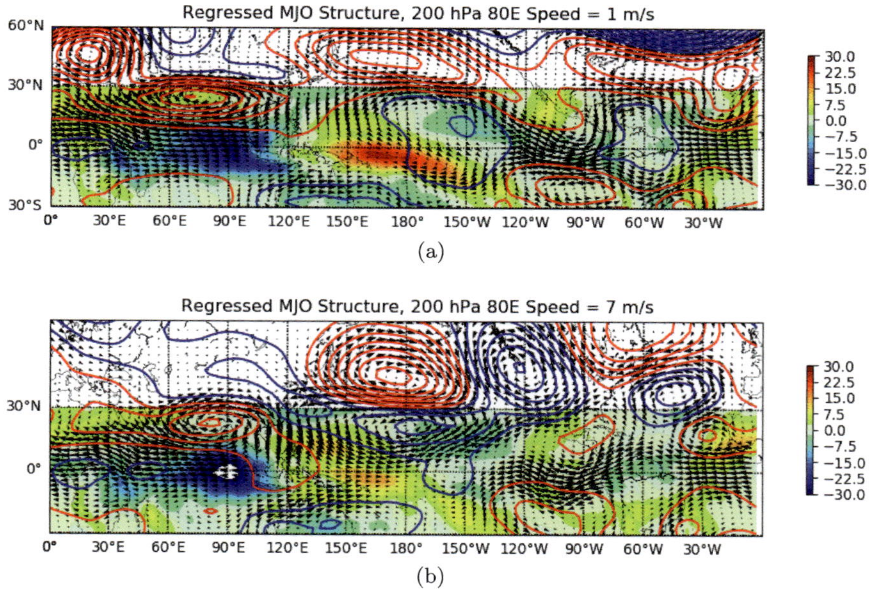

Fig. 1. Regression of 200-hPa geopotential height anomaly (contours, red is positive, with an interval of 20 m), 200-hPa winds (vectors), and OLR anomalies (Wm^{-2}, shaded) against wavelet-filtered OLR at 80°E and the equator at (a) $1\,ms^{-1}$ and (b) $7\,ms^{-1}$ and zonal wavenumber 2.

a shorter climatology, results not shown) and then regressing these against the wavelet-filtered indexes, the algorithm diagnoses the association between the advection terms and the phase speed of the OLR anomalies. Winds were broken into their synoptic (<20 days), intraseasonal (20–100 days), and background low-frequency (>100 days) components, and the advection of zonal wind was constructed including only the intraseasonal terms and the cross terms between intraseasonal and background wind (i.e., synoptic effects were excluded). Figure 2 shows 400-hPa wind anomalies, advection of zonal wind by the horizontal wind (contours, with westerly acceleration in red), and OLR anomalies regressed against four wavenumber–frequency wavelet indexes, i.e., nonlinear intraseasonal terms. In this case, phase velocities of 1, 3, 5, and $7\,ms^{-1}$, all at zonal wavenumber 2 eastward for negative OLR anomalies, are centered at 110°E. The red contours east of Asia represent westerly wind advection at the leading edge of the anomalous intraseasonal westerly wind associated with the extending northwest Pacific jet stream forced by MJO convection. Comparison against the actual rate of change of the regressed zonal wind shows that the horizontal advection is larger (not shown). Part of the difference is offset as geostrophic adjustment changes the mass distribution in response to the accelerated wind and the Coriolis force (not shown), because that Coriolis force is not balanced by a geopotential gradient force. The rate of advection north and east of the convection is substantially higher associated with faster eastward propagation in the OLR anomalies, consistent with the view that advection is a major contributing factor to the propagation of MJO circulation signals. However, Fig. 1 shows that fast signals also are associated with stronger subseasonal circulation anomaly amplitudes over the Pacific sector, which would also contribute to the rates of advection without affecting the phase speed (since size of advection terms scale with both the advecting wind and the size of the gradient of the advected wind).

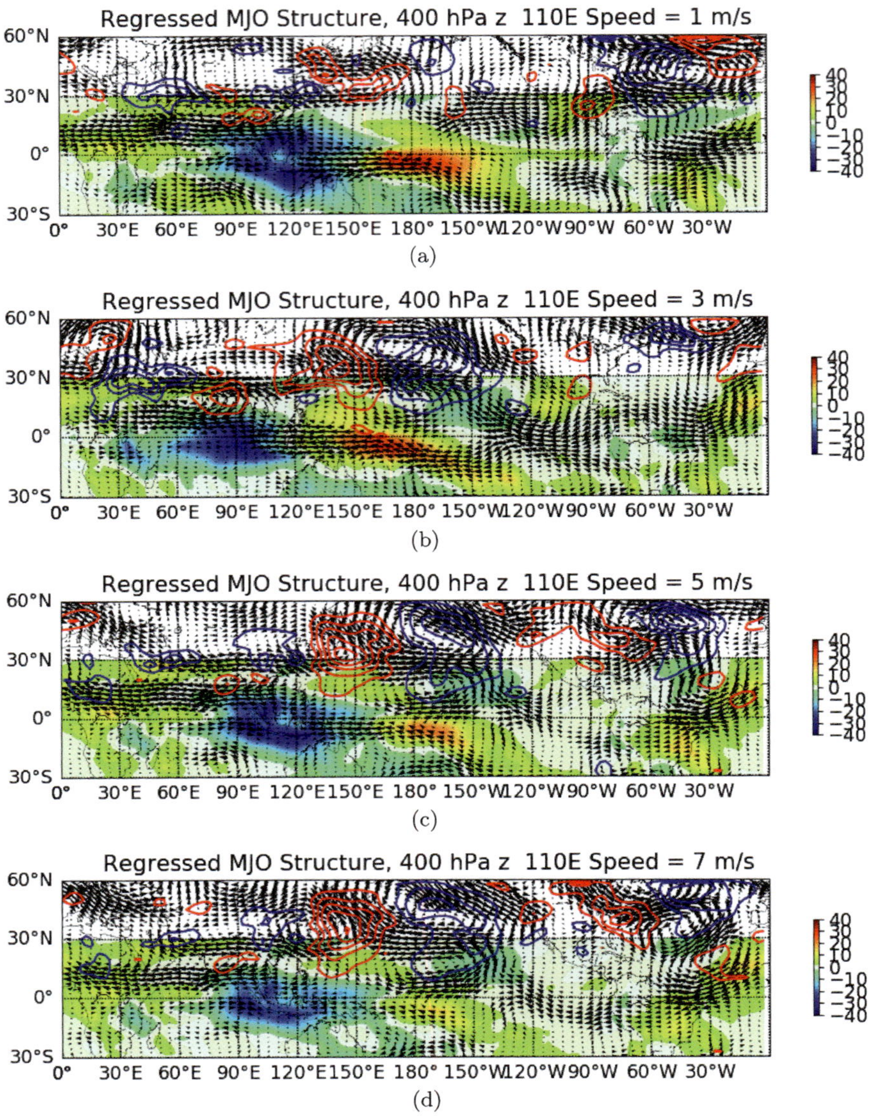

Fig. 2. Regressed OLR anomalies (shading, Wm^{-2} and 400-hpa wind anomalies (vectors). Contours represent advection of zonal wind anomaly by horizontal wind (positive in red, contour interval is $1.5 \times 10^{-5}\,\mathrm{ms}^{-2}$). The strongest wind vector is $7\,\mathrm{ms}^{-1}$.

Further momentum budget analysis (not shown), including expansion of each term into low-frequency, subseasonal, and transient components, as well as analysis of nonlinear terms, suggests that the leading term at most longitudes that contributes to these shifts in the intraseasonal circulation anomalies is advection by the background wind. However, when the intraseasonal zonal wind anomaly reaches an exit region in the background jet stream, advection of background wind by the intraseasonal wind becomes important. Then, as the intraseasonal westerly wind anomaly near 30°N continues eastward, the dominant factor in the observed propagation is advection of intraseasonal wind by the intraseasonal wind (not shown, this mechanism allows a jet filament to break off from the mean jet and continue to move eastward over the eastern North Pacific Ocean).

These results suggest that the phase speeds of MJO convection and circulation patterns may be modulated by advection by base state winds, especially those in the vicinity of the subtropical jet stream poleward of the regions of active or suppressed deep convection. Although previous authors have developed possible explanations of MJO evolution via interaction with the subtropical atmospheric circulation (e.g., Straus and Lindzen 2000), most recently proposed MJO propagation mechanisms rely on interaction between moisture and convection. This presently proposed dynamical mechanism of MJO propagation would not exclude mechanisms inherent to moisture, convection, or Kelvin wave dynamics in the tropics, as this new mechanism could occur simultaneously with other mechanisms. Further, the dynamical effects might influence the moisture recharge–discharge process by inducing horizontal or vertical motions and so mimic or motivate the internal effects.

3. Modified Regression Algorithm

A modified linear regression algorithm is used here to demonstrate the changes in the propagation characteristics of MJO convection and circulation that are associated with different upper-tropospheric background wind speeds in selected regions.

Linear regression is frequently applied to diagnose relationships between signals in a time index and time-varying patterns in a field of data. When the means and seasonal cycles of predictor and predictand variables have been subtracted so that the y intercept is zero, the technique provides a field of regression slope coefficients that can then be scaled to some amplitude in the predictor index. Yet, when the model is trained on data from different times of the year, different slope coefficients result. Recently, Roundy (2017) developed a new statistical algorithm for fitting such regression models in a way that allows the slope coefficients to vary continuously across the seasonal cycle.

In the standard simple linear regression approach, one calculates the slope coefficient by taking the ratio of the covariance between the predictor and the predictand to the variance in the predictor. This new algorithm fits a predictor matrix $\boldsymbol{X}$, consisting of a column of ones and other columns consisting of sine–cosine pairs representing the seasonal cycle and a few of its harmonics, to the array products of the predictor with itself and the predictor with the predictand. In these array products, the value at each time step going down a column is just the product between the data values occurring at the same time, i.e., $x_t x_t$ for the variance of the predictor x or $x_t y_t$ for the covariance between x and the predictand y. The shape of the matrix $\boldsymbol{X}$ is the number of days in the dataset by one plus twice the number of harmonics retained (including sine and cosine columns).

$$x * x = \boldsymbol{X} C_{\text{var}} + \delta_{xx}, \tag{2}$$

$$x * y = \boldsymbol{X} C_{\text{covar}} + \delta_{xy}, \tag{3}$$

C_{var} and C_{covar} are both vectors with length equal to the number of columns in $\boldsymbol{X}$. Solving for C_{var} and C_{covar} after setting the δs to zero, provides the linear fit of the seasonal cycle to the variance of x and the covariance between x and y. The slope coefficient favored on a given day of the year is just the estimate of the seasonally evolving covariance on that day divided by the estimate of the seasonally evolving predictor variance on the same day, given the row of $\boldsymbol{X}$ corresponding to that day of the year.

$$m(d) = \frac{\boldsymbol{X}(d) C_{\text{covar}}}{\boldsymbol{X}(d) C_{\text{var}}}, \tag{4}$$

for day of year d. If results for fitting the variance are not well behaved, gamma regression can be applied to fit the seasonal cycle to the squares of the predictor.

In this work, the same regression algorithm is modified to generate models in which the slope coefficients vary with quantities different from the seasonal cycle (following from the conclusions of Roundy 2017). In a regression model designed to diagnose global circulation patterns associated with the RMM index, the variance and covariance quantities can be modeled according to their linear relationship with some quantity that might modulate MJO propagation (i.e., replace the seasonal cycle and its harmonics in the matrix $\boldsymbol{X}$ with a single time series in a quantity on which MJO propagation characteristics might depend). The same proofs of efficacy demonstrated by Roundy (2017) in the regression algorithm for seasonally evolving slope coefficients would apply, as the technique is the same even if the quantities change to which the variance and covariance time series are fit.

For example, preliminary results had suggested that advection by the zonal wind in the vicinity of the subtropical jet stream influences the eastward phase speed of convection and geopotential height anomalies associated with the MJO over the Pacific Ocean. To illustrate, we construct a single time series vertical vector from the real-time multivariate MJO (RMM) index (Wheeler and Hendon 2004) that attains its temporal maxima in phase 6. This rotated index is made by taking 0.383*RMM1 + 0.92*RMM2. Assume that this index is the predictor $\boldsymbol{x}$, and the time series extracted from fields of outgoing longwave radiation (OLR, Liebmann and Smith 1996) anomalies and 200-hPa wind and geopotential height anomalies are the predictand vertical vectors $\boldsymbol{y}$. Construct the model based on data from the season December through February. To construct the model, replace the set of columns representing the seasonal cycle and its harmonics in the matrix $\boldsymbol{X}$ with a single time series, the 100-day low-pass filtered zonal wind averaged over the Pacific basin from 140°E to 180° and from 25°N to 40°N. This low pass filtered wind index retains its long-term mean and seasonal cycle. Retain the original column of ones in $\boldsymbol{X}$ to fit the intercept. Fit this new matrix $\boldsymbol{X}$ to the array product between $\boldsymbol{x}$ and $\boldsymbol{y}$. This approach fits the low-frequency wind index to the time series of the products between $\boldsymbol{x}$ and $\boldsymbol{y}$, so it constitutes an estimate of how the covariance between $\boldsymbol{x}$ and $\boldsymbol{y}$ varies with the low-frequency wind index. Also fit the matrix $\boldsymbol{X}$ to the time series of the squares of the values in $\boldsymbol{x}$. This fit finds the predictor variance as it depends on the low-frequency wind index. The slope coefficient relating the rotated RMM index $\boldsymbol{x}$ to the grid point time series $\boldsymbol{y}$ as modulated by some state in the low-frequency wind index is just the covariance evaluated at that low-frequency wind state divided by the predictor variance evaluated at the same state. The resulting model allows the user to estimate the relationship between the RMM index and the field variables as that relationship changes with background wind in the region of the west Pacific subtropical jet stream.

Figure 3a shows the pattern reconstructed for a lag of 5 days after RMM phase 6 with a background wind speed of 40 ms^{-1}, while Fig. 3b shows the same pattern settings but with a background wind speed of 20 ms^{-1} (these values are near the top and bottom of the range of intensities of the background mean subtropical jet stream in that region during December through February). The 5-day lag allows time for any difference in phase speed to become apparent in zonal shifts between the signals associated with the slow and fast jets. Results show intraseasonal anomalous features moving as the jet influences them, but the plotted results do not include the modified jet itself.

In the fast jet regression map (Fig. 3a), the whole anomalous circulation pattern over the North Pacific and North America is shifted eastward relative to the slow jet map. Examination of other time lags (not shown) reveals

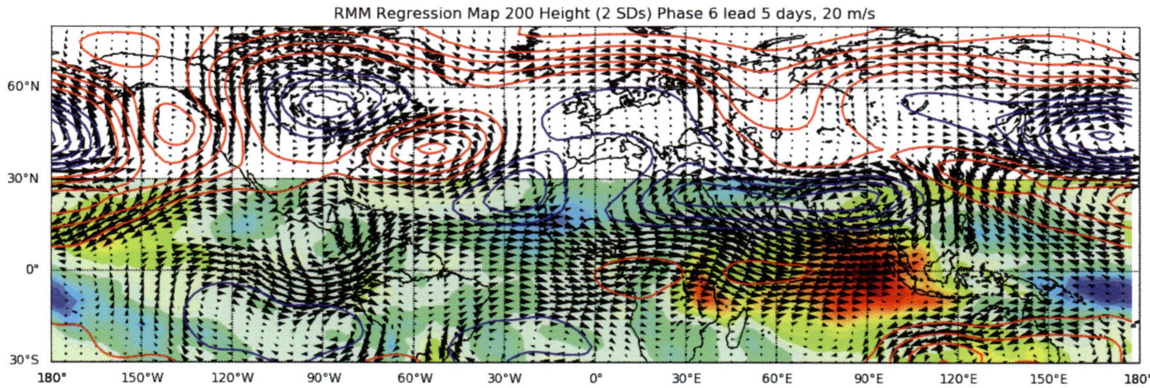

(a)

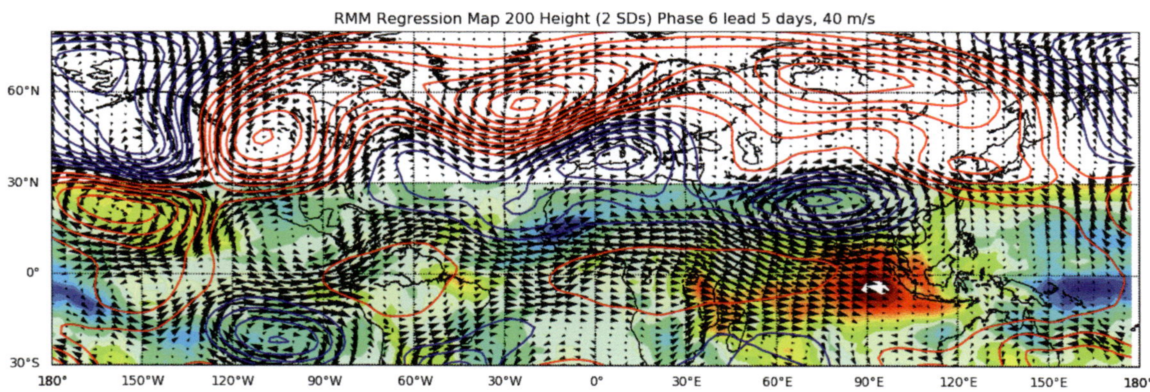

(b)

Fig. 3. Regressed 200-hpa geopotential height anomaly (contours, interval 20 m, red is positive), wind vectors, and OLR anomaly (shading, active convection suggested in blue), during December–February. (a) During periods with background northwest Pacific jet at 20 ms^{-1}. (b) With background Pacific jet at 40 ms^{-1}.

that the geopotential height patterns extend eastward more rapidly with the fast jet from the initial condition of a closer match at the time of phase 6 (though the two fields are not identical at lag = 0). The South Pacific OLR anomalies also extend eastward more rapidly in the fast jet map. The eastward shifting of the geopotential height anomaly pattern is consistent with the effects of advection by the background wind on propagation of the signal downstream (Seo and Son 2012; Seo and Lee 2017). It is possible that advection of these intraseasonal surges of zonal wind anomaly in this manner would influence the propagation of MJO convection because it would modulate the regional mass balance of the tropics. The basic hypothesis is as follows: Acceleration of upper-tropospheric westerly wind by advection would imply geostrophic adjustment in response to an unbalanced Coriolis force (because the circulation anomalies are much larger than the Rossby radius of deformation), leading to transfer of mass from north of the jet to south of the jet, suggesting upper-tropospheric mass convergence into the tropics east of the present region of convection. This mass convergence would reinforce subsidence there. Then, as the ridge anomaly forced by this mechanism and amplified by the convection advects eastward, the region of mass divergence that favors convection on the western and equatorward side of the ridge anomalies encourages eastward movement of the convection. Indeed, the Southern Hemisphere OLR anomalies in Fig. 3a suggest that the convection over the South Pacific extends farther eastward in the case with the enhanced low-frequency jet stream. Thus, this mechanism may be important for propagation of the observed MJO, even if there exist mechanisms internal to the interplay between moisture and convection that would also influence propagation.

As another illustration of the algorithm and its potential as a diagnostic tool, lagged regression models for OLR anomalies averaged from 5°N to 5°S are constructed as above, based on the RMM index rotated to target phase 3. As above in the Pacific jet example, the variance and covariance quantities in the linear regression are fit to a column of ones and a low-pass filtered upper-tropospheric zonal wind index at some location, to diagnose the slope coefficients between the RMM index and the OLR anomalies around the equatorial band, as modulated by wind at that location. Then, the longitude–time lag OLR diagrams are constructed based on zonal wind at both the 10th and the 90th percentiles of intensity in the background wind at the selected location. In Fig. 4a, when the low-frequency zonal wind is slow at the selected location, MJO convection tends to begin gradually over the eastern Indian Ocean and evolve slowly eastward from there (Fig. 4a, shading). When the low-frequency wind is strong westerly (contours), convection begins slowly over the western Indian Ocean and evolves rapidly eastward across the basin, slowing down near the Maritime Continent, but remaining faster than the signal when the low-frequency wind is weak westerly. Figure 4b shows a similar comparison given background wind at 152°E 10°N, which does not produce a statistically significant difference between the phase speeds of the negative OLR anomalies over the Indian Ocean. This result suggests that the phase speed of MJO convection may be modulated by wind speed in the subtropical jet region proximate to

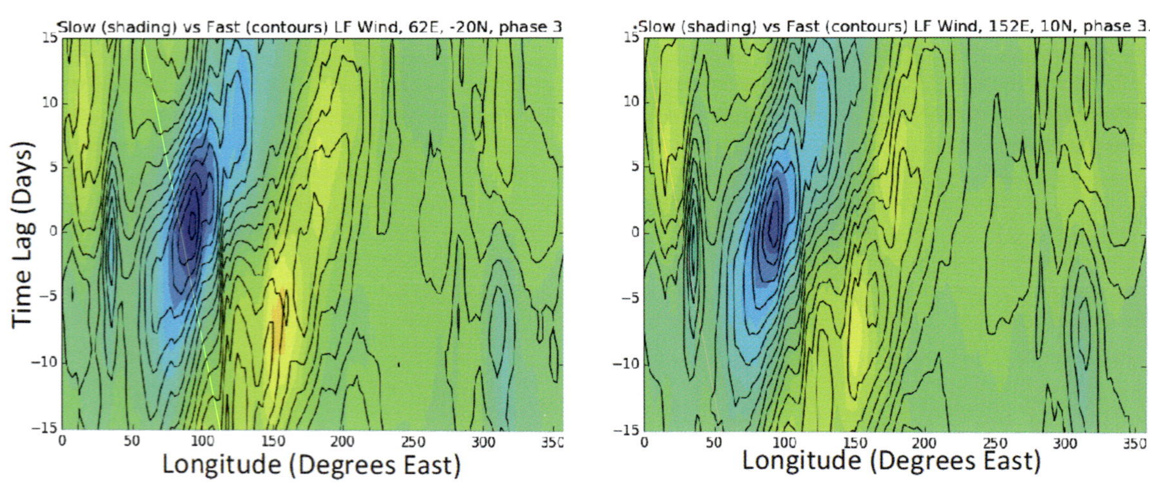

(a) (b)

Fig. 4. (a) shows regressed equatorial OLR anomalies centered at $t = 0$ at RMM phase 3 when background wind at 62.5°E 20°S is in the 10th percentile (shading) and the 90th percentile (contours, interval 5 Wm^{-2}). Active convection is suggested in blue (the scaling is arbitrary because at a given background wind state it constitutes a linearization of the subseasonal relationship about that state). (b) is similar but for wind at 152°E 10°N. It is included as a null case wherein the Indian basin OLR signals are not broadly statistically significantly different between slow and fast winds at that location.

the longitude range of the convective anomaly, whereas locations at longitudes far away do not necessarily produce similar direct effects (in the case of Fig. 4b, the location is not even in the subtropical jet stream, and is in a different basin).

4. Conclusions

Two regression algorithms, one based on wavelet filtering OLR anomalies for specific phase speeds to obtain time indexes at a base longitude relevant to signals crossing that location at the selected phase speed and the other based on a new type of linear regression algorithm, show that the phase speeds of circulation and convective signals associated with the MJO vary with the background flow in which the MJO is embedded. A more eastward background flow is associated with a more rapid eastward translation of the anomalous patterns associated with MJO events. Regression of selected reconstructed horizontal advection terms against the wavelet-filtered indexes for planetary-scale OLR anomalies at different phase speeds suggests that higher phase speeds of active convective regions over the Maritime Continent region are associated with higher rates of advection of intraseasonal wind by the background flow proximate to the subtropical jet stream. However, regressed circulation patterns associated with OLR anomalies of different speeds suggest that slower MJO signals are associated with more amplitude in the atmospheric circulation to the north and west of the convection, while faster MJO signals are associated with greater amplitude to the north and east of the convection. Thus, part of the difference in the advection rates would relate to larger amplitudes and gradients in the subseasonal circulation at higher phase speeds. Further analysis is planned to better separate the contributions of variability in background wind to MJO phase speed.

Acknowledgments

The NOAA ESRL provided OLR anomalies and NCEP/NCAR reanalysis data. Funding was provided by National Science Foundation Grants AGS1358214 and AGS1128779 to Paul Roundy.

Appendix A

Section 2 describes the wavelet regression algorithm used to extract the patterns in fields of data associated with OLR anomalies of a selected spatial scale crossing a base longitude at a selected phase speed. This appendix provides a demonstration of the algorithm on artificial datasets, intended to illustrate how it extracts signals at the target phase speed from other signals and can then reconstruct the relationship with a correlated signal in another variable at a phase shift. The idea is to create two databases, each containing two periods of activity in eastward-moving signals at two phase speeds (2 and $5\,\mathrm{ms}^{-1}$) at zonal wavenumber 2, plus some slightly smoothed Gaussian random noise. The $2\,\mathrm{ms}^{-1}$ signals are centered at the base longitude and days 220 and 420, and the $5\,\mathrm{ms}^{-1}$ signals are set at the base longitude and days 290 and 599. The corresponding dataset (symbolizing geopotential height anomaly) has the same wave signals shifted zonally by a quarter wavelength (sine instead of cosine), but with its own random noise. The periods of wave activity all overlap with each other in each dataset and with a base point (so that the signal can be detected there by the wavelet algorithm). Figure A.1 shows the data configuration.

Figure A.3 shows the diagnosed relationship between artificial OLR and height anomalies estimated via the regression algorithm. Figures A.2 and A.1 together show that the wavelet algorithm cleanly extracts the $5\,\mathrm{ms}^{-1}$ wave signal from the other with a minimum of noise (which appears as flecks in the contour lines). Figure A.3 shows a time lag regression

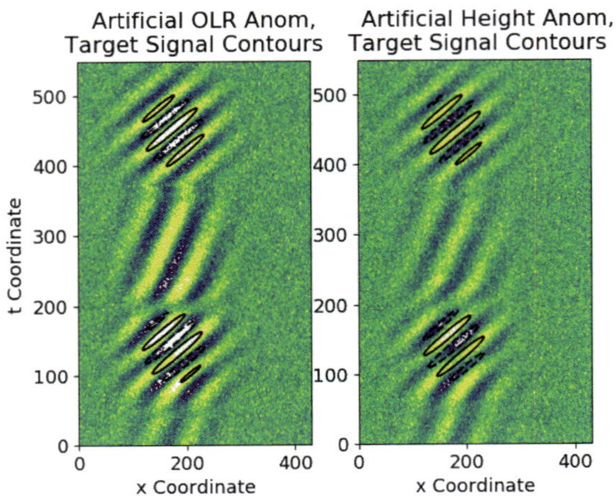

Fig. A.1. Artificial OLR (a) and height anomalies (b). Both are identical except wave signals in b are shifted by a quarter wavelength from a and both panels include different random noise. The contours highlight the target $5\,\mathrm{ms}^{-1}$ wave signals that the wavelet index algorithm is designed to extract (centered on x coordinate 175).

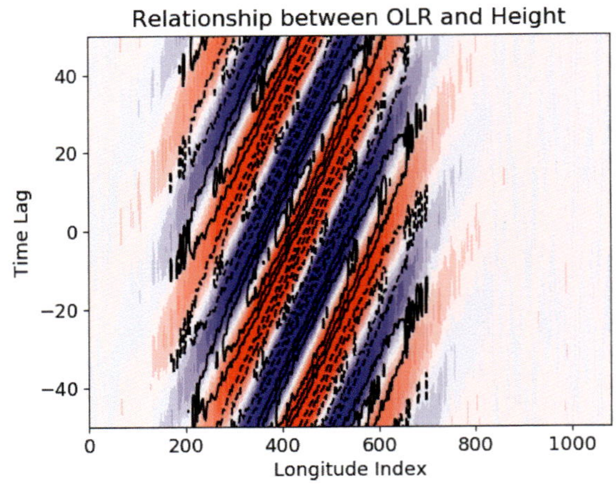

Fig. A.3. Lag regressed artificial OLR anomalies (shaded) and the corresponding artificial height anomalies (contours). The longitude index is in degrees, but repeats three times around the world to suit the algorithm. The slope of the regressed anomalies is consistent with the $5\,\mathrm{ms}^{-1}$ target phase speed as shown by the data.

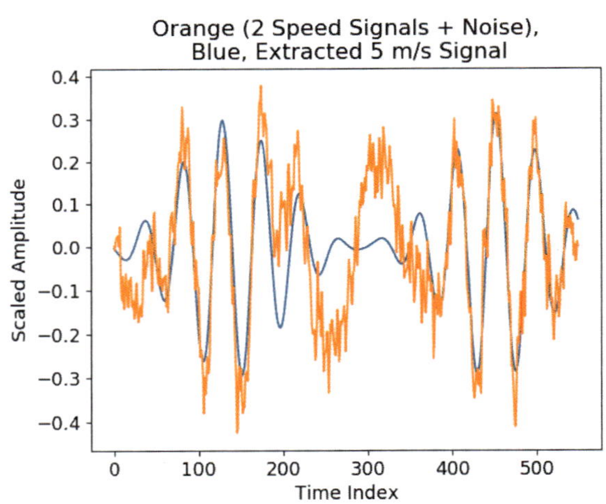

Fig. A.2. shows the time series data at the base longitude (orange) and the calculated wavelet index for wavenumber 2 signals moving eastward at $5\,\mathrm{ms}^{-1}$ (blue). Both time series are scaled up for convenience of plotting.

diagram, highlighting the regressed relationship between the artificial datasets, based on the wavelet "OLR" index at $5\,\mathrm{ms}^{-1}$. The regression algorithm is able to reconstruct the correlated pattern in the phase shifted "height anomaly" variable at the correct phase speed in spite of noise and the mixed signal at the lower speed. A much longer artificial dataset would reduce noise in the lag regressed signal (not shown).

References

Adames, A. F., J. Patoux, and R. C. Foster, 2014: The contribution of extratropical waves to the MJO wind field. *J. Atmos. Sci.*, **71**, 155–176, doi: http://dx.doi.org/10.1175/JAS-D-13-084.1

Aiyyer, A. and J. Molinari, 2008: MJO and Tropical cyclogenesis in the Gulf of Mexico and Eastern Pacific: Case study and idealized numerical modeling. *J. Atmos. Sci.*, **65**, 2691–2704.

Camargo, S. J., M. C. Wheeler, and A. H. Sobel, 2009: Diagnosis of the MJO modulation of tropical cyclogenesis using an empirical index. *J. Atmos. Sci.*, **66**, 3061–3074.

Frederiksen, J. S. and H. Lin, 2013: Tropical extratropical interactions of intraseasonal oscillations. *J. Atmos. Sci.*, **70**, 3180–3197.

Garfinkel, C. I., J. J. Benedict, and E. D. Maloney, 2014: Impact of the MJO on the boreal winter extratropical circulation. *Geophys. Res. Lett.*, **41**, 6055–6062, doi:10.1002/2014GL061094.

Haertel, P., K. Straub, and A. Budsock, 2015: Transforming circumnavigating Kelvin waves that initiate and dissipate the Madden–Julian Oscillation. *Q. J. R. Meteorol. Soc.*, **141**, 1586–1602, doi:10.1002/qj.2461

Henderson, S. A., E. D. Maloney, and E. A. Barnes, 2016: The influence of the Madden–Julian oscillation on Northern Hemisphere winter blocking. *J. Clim.*, **29**, 4597–4616.

Hendon, H. H. and M. L. Salby, 1994: The life cycle of the Madden–Julian oscillation. *J. Atmos. Sci.*, **51**, 2225–2237.

Li, T. and P.-c Hsu, 2017: Madden–Julian oscillation: Observations and mechanisms. In *Fundamentals of Tropical Climate Dynamics*, Springer Atmospheric Science Book Series.

Liebmann, B. and C. A. Smith, 1996: Description of a complete (interpolated) outgoing longwave radiation dataset. *Bull. Amer. Meteor. Soc.*, **77**, 1275–1277.

L'Heureux, M. L. and R. W. Higgins, 2008: Boreal winter links between the Madden–Julian oscillation and the Arctic Oscillation. *J. Clim.*, **21**, 3040–3050.

Ma, D. and Z. Kuang, 2016: A mechanism-denial study on the Madden–Julian oscillation with reduced interference from mean state changes. *Geophys. Res. Lett.*, **43**, 2989–2997, doi:10.1002/2016GL067702.

MacRitchie, K. and P. E. Roundy, 2016: The two-way relationship between the Madden–Julian oscillation and anticyclonic wave breaking. *Q.J.R. Meteorol. Soc.*, **142**, 2159–2167, doi:10.1002/qj.2809

Madden, R. A. and P. R. Julian, 1994: Observations of the 40–50-day tropical oscillation-A review. *Mon. Wea. Rev.*, **122**, 814–837.

Maloney, E. D. and D. L. Hartmann, 2001: The Madden–Julian oscillation, barotropic dynamics, and north pacific tropical cyclone formation. Part I: observations. *J. Atmos. Sci.*, **58**, 2545–2558, doi: http://dx.doi.org/10.1175/1520-0469(2001)058⟨2545:TMJOBD⟩2.0.CO;2

Matthews, A. J., 2000: Propagation mechanisms for the Madden–Julian oscillation. *Q. J. Roy. Meteorol. Soc.*, **126**, 2637–2651.

Matthews, A. J., 2008: Primary and successive events in the Madden–Julian oscillation. *Quart. J. Roy. Meteorol. Soc.*, **134**, 439–453.

Narapusetty B., T. DelSole, and M. Tippett, 2009: Optimal estimation of the climatological mean. *J. Clim.*, **22**, 4845–4859, https://doi.org/10.1175/2009JCLI2944.1.

Moore, R. W., O. Martius, and T. Spengler, 2010: The modulation of the subtropical and extratropical atmosphere in the Pacific Basin in response to the Madden–Julian oscillation. *Mon. Wea. Rev.*, **138**, 2761–2779.

Powell, S. W. and R. A. Houze Jr., 2015: Effect of dry large-scale vertical motions on initial MJO convective onset. *J. Geophys. Res. Atmos.*, **120**, 4783–4805, doi: 10.1002/2014JD022961.

Ray, P. and T. Li, 2013: Relative roles of circumnavigating waves and extratropics on the MJO and its relationship with the mean state. *J. Atmos. Sci.*, **70**, 876–893, doi: http://dx.doi.org/10.1175/JAS-D-12-0153.1

Ray, P. and C. Zhang, 2010: A case study on the mechanisms of extratropical influence on the Madden–Julian oscillation. *J. Atm. Sci.*, **67**, 515–528.

Roundy, P. E. 2017: Diagnosis of seasonally varying regression slope coefficients and application to the MJO. *Q.J.R. Meteorol. Soc.*, **143**, 1946–1952, doi:10.1002/qj.3054

Roundy, P. E. 2012a: The spectrum of convectively coupled Kelvin waves and the Madden–Julian oscillation in regions of low-level easterly and westerly background flow. *J. Atmos. Sci.*, **69**, 2107–2111.

Roundy, P. E., 2012b: Observed structure of convectively coupled waves as a function of equivalent depth: Kelvin waves and the Madden–Julian oscillation. *J. Atmos. Sci.*, **69**, 2097–2106.

Roundy, P. E., 2014: Some aspects of western hemisphere circulation and the Madden–Julian oscillation. *J. Atmos. Sci.*, **71**, 2027–2039, doi: http://dx.doi.org/10.1175/JAS-D-13-0210.1

Roundy, P. E., 2017: Diagnosis of seasonally varying regression slope coefficients and application to the MJO. *Q.J.R. Meteorol. Soc.*, **143**, 1946–1952, doi:10.1002/qj.3054

Roundy, P. E., 2018: A wave-number frequency wavelet analysis of convectively coupled equatorial waves and the MJO over the Indian Ocean. *Q.J.R. Meteorol. Soc.*, doi:10.1002/qj.3207

Roundy, P. E. and W. M. Frank, 2004: A climatology of waves in the equatorial region. *J. Atmos. Sci.*, **61**, 2105–2132, doi: http://dx.doi.org/10.1175/1520-0469(2004)061⟨2105:ACOWIT⟩2.0.CO;2

Roundy, P. E., K. MacRitchie, J. Asuma, and T. Melino, 2010: Modulation of the global atmospheric circulation by combined activity in the Madden–Julian oscillation and the El Niño/Southern Oscillation during Boreal winter. *J. Clim.*, **23**, 4045–4059, doi:10.1175/2010

Sakaeda, N. and P. E. Roundy, 2015: The development of upper-tropospheric wind over the western hemisphere in association with MJO convective initiation. *J. Atmos. Sci.*, **72**, 3138–3160, https://doi.org/10.1175/JAS-D-14-0293.1

Sakaeda, N. and P. E. Roundy, 2016a: The intraseasonal atmospheric angular momentum associated with MJO convective initiations. *Q.J.R. Meteorol. Soc.*, **142**: 1371–1384, doi:10.1002/qj.2740

Sakaeda, N. and P. E. Roundy, 2016b: The development of upper-tropospheric geopotential height anomaly in the Western Hemisphere during MJO convective initiations. *Q.J.R. Meteorol. Soc.*, **142**, 942–956, doi:10.1002/qj.2696

Seo, K.-H. and K.-Y. Kim, 2003: Propagation and initiation mechanisms of the Madden–Julian oscillation. *J. Geophys. Res.*, **108**, 4384, doi:10.1029/2002JD002876, D13.

Seo, K. and S. Son, 2012: The global atmospheric circulation response to tropical diabatic heating associated with the Madden–Julian oscillation during Northern winter. *J. Atmos. Sci.*, **69**, 79–96, https://doi.org/10.1175/2011JAS3686.1

Seo, K. and H. Lee, 2017: Mechanisms for a PNA-like teleconnection pattern in response to the MJO. *J. Atmos. Sci.*, **74**, 1767–1781, https://doi.org/10.1175/JAS-D-16-0343.1

Sobel, A. and Kim, 2012: The MJO-Kelvin wave transition. *Geophys. Res. Lett.*, **39**, L20808, doi:10.1029/2012GL053380.

Straub, K. H. 2013: MJO Initiation in the real-time multivariate MJO Index. *J. Clim.*, **26**, 1130–1151.

Straub, K. H. and G. N. Kiladis, 2003: Interactions between the Boreal summer intraseasonal oscillation and higher-frequency tropical wave activity. *Mon. Wea. Rev.*, **131**, 945–960.

Straus, D. M. and R. S. Lindzen, 2000: Planetary-scale baroclinic instability and the MJO. *J. Atmos. Sci.*, **57**, 3609–3626, https://doi.org/10.1175/1520-0469(2000)057⟨3609:PSBIAT⟩2.0.CO;2

Thompson, D. B. and P. E. Roundy, 2013: The relationship between the Madden–Julian oscillation and U.S. violent tornado outbreaks in the spring. *Mon. Wea. Rev.*, **141**, 2087–2095, doi: http://dx.doi.org/10.1175/MWR-D-12-00173.1

Wheeler, M. and H. H. Hendon, 2004: An all-season real-time multivariate MJO index: Development of an index for monitoring and prediction. *Mon. Wea. Rev.*, **123**, 1917–1932.

Yadav, P. and D.M. Straus, 2017: Circulation response to fast and slow MJO episodes. *Mon. Wea. Rev.*, **145**, 1577–1596, https://doi.org/10.1175/MWR-D-16-0352.1

Zhang, C. 2005: The Madden–Julian oscillation. *Rev. Geophys.*, **43**, RG2003, doi:10.1029/2004RG000158.

Zhang, F., S. Taraphdar, and S. Wang, 2017: The role of global circumnavigating mode in the MJO initiation and propagation. *J. Geophys. Res. Atmos.*, **122**, 5837–5856, doi:10.1002/2016JD025665.

Zhao, C., T. Li, and T. Zhou, 2013: Precursor signals and processes associated with MJO initiation over the tropical Indian ocean. *J. Clim.*, **26**, 291–307.

Chapter 25

Progress and Status of MJO Simulation in Climate Models and Process-Oriented Diagnostics

Xianan Jiang[*,§], Daehyun Kim[†,¶] and Eric D. Maloney[‡,||]

[*] *Joint Institute for Regional Earth System Science & Engineering, University of California, Los Angeles, CA 900035, USA*
[†] *Department of Atmospheric Sciences, University of Washington, Seattle, WA 98195, USA*
[‡] *Department of Atmospheric Science, Colorado State University, Fort Collins, CO 80521, USA*
[§] *xianan@ucla.edu*
[¶] *daehyun@uw.edu*
[||] *emaloney@atmos.colostate.edu*

Despite its tremendous influence on extreme weather worldwide, realistic simulations of the Madden–Julian oscillation (MJO) have been a grand challenge for global climate models. In the first part of this chapter, a brief overview of recent advances in modeling the MJO is provided by particularly highlighting improved MJO simulations achieved through implementations of stochastic cumulus approaches. In the second part, the most recent community efforts in the development of process-oriented diagnostics and metrics for MJO simulations are briefly reviewed. These diagnostics and metrics have been built upon the process understanding of key MJO physics, providing important guidance to expose critical model deficiencies in simulating the MJO. These processes include convective sensitivity to environment moisture, convection–circulation interactions, cloud–radiation feedbacks, and large-scale control (e.g., the lower-tropospheric mean moisture pattern).

1. Introduction

The Madden–Julian oscillation (MJO, Madden and Julian 1971, 1994) plays a crucial role in the earth's hydrological cycle (Lau and Waliser 2012) by exerting tremendous influence on global climate and weather extremes over both tropical and extra-tropical regions (Zhang 2013). Due to its quasi-periodic behavior on intraseasonal timescales, the MJO represents one of the primary predictability sources for short-term climate prediction (Vitart *et al.* 2012; NAS 2016) and provides a fundamental basis for "seamless prediction" (e.g., Hoskins 2013). However, the MJO remains poorly represented in most of our latest generation global climate models (GCMs; Hung *et al.* 2013; Jiang *et al.* 2015; Ahn *et al.* 2017). For the limited number of GCMs that are able to capture the bulk characteristics of the MJO, the reasons for their good MJO simulations are also not well understood (e.g., Klingaman *et al.* 2015a). The poor model capability in representing the MJO leaves us great disadvantages in projecting future activity of weather and climate systems that are significantly modulated by the MJO.

The great challenges in simulating the MJO pose an urgent need for improvement in understanding the fundamental physics of the MJO. Most recently, great efforts have been undertaken in the community on process-oriented diagnosis of GCMs, aimed at identifying critical

The Multiscale Global Monsoon System
Edited by C. P. Chang et al.

processes that discriminate good MJO models from poor ones. For example, development of the process-oriented MJO metrics is one subproject of the WGNE MJO Task Force[1] and the NOAA's Climate Program Office Model Diagnostics Task Force.[2] The results of these process-oriented diagnoses are expected to inform changes of model parametrizations that are desirable for improved MJO simulations (e.g., Kim *et al.* 2009; Waliser *et al.* 2009; Sperber and Kim 2012; Wheeler and Maloney 2013; Kim *et al.* 2014).

This chapter briefly reviews advances in modeling the MJO using GCMs, with a specific focus on recent modeling development that applies a stochastic convection parameterization approach, and progress in developing process-oriented diagnostics for MJO simulations.

2. Progress of MJO Simulation in Climate Models

Many previous modeling studies suggested that simulations of the MJO are sensitive to model representation of subgrid cumulus processes. For example, improved MJO simulations can be achieved by inhibiting deep convection in model cumulus schemes through enhanced cumulus entrainment rate or increased rain reevaporation (see a review by Kim and Maloney 2017). However, the improved MJO representation achieved by tuning these parameters in model cumulus schemes often occurs at the cost of degraded model mean state and other climate phenomena (e.g., Kim *et al.* 2011). To avoid this issue and to comprehensively improve climate simulations including the MJO, explicit cloud-resolving models (CRMs) have been implemented in global models by either adopting a super-parameterization approach (Khairoutdinov and Randall 2003; Randall *et al.* 2003), i.e., replacing the model cumulus parameterization with a vertical 2D CRM, or expanding high-resolution CRMs over the global domain (Miura *et al.* 2007; Satoh *et al.* 2008). While success of this CRM approach for improvement of MJO simulations has been widely reported, even these CRM models could have substantial mean state biases (e.g., Stan *et al.* 2010; Kim *et al.* 2011). Moreover, due to the required large computing resources, currently it is not practical to extensively employ CRMs for climate projection studies that need long-term integrations with multiple ensemble members, although the super-parameterized GCMs have been used to understand MJO characteristics under idealized or perturbed climate (e.g., Pritchard and Yang 2016; Wolding *et al.* 2017).

While largely we have observed limited progress in improving MJO simulations in GCMs since the last IWM-V in Macao, in 2013 (see review chapters for the IWM-V workshop by Kim and Maloney 2017; Klingaman *et al.* 2017), one recent noteworthy development is the implementation of the stochastic convective parameterization approach in GCMs (e.g., Deng *et al.* 2015; Deng *et al.* 2016; Wang *et al.* 2016; Goswami *et al.* 2017a, b; Peters *et al.* 2017), based on the earlier stochastic modeling concept by introducing subgrid cumulus variability into deterministic parameterization scheme-based coarse-resolution GCMs (e.g., Buizza *et al.* 1999; Lin and Neelin 2003). One of these stochastic convective schemes, the stochastic multi-cloud model (SMCM), which was first developed and tested based on idealized frameworks (e.g., Khouider *et al.* 2010; Frenkel *et al.* 2013) and aqua-planet global atmospheric models (Deng *et al.* 2015), has been recently implemented to several different GCMs with full complexity (Goswami *et al.* 2017b, a; Peters *et al.* 2017). The SMCM is constructed

[1]http://wgne.meteoinfo.ru/activities/on-going-activities/wgne-mjo-task-force/.
[2]http://cpo.noaa.gov/Meet-the-Divisions/Earth-System-Science-and-Modeling/MAPP/MAPP-Task-Forces/Model-Diagnostics-Task-Force.

on rectangular lattices within each GCM grid to mimic the observed life cycle of organized tropical convective systems (e.g., Johnson *et al.* 1999; Mapes *et al.* 2006). Each element of the lattice is characterized either by one particular cloud type (congestus, deep, or stratiform) or clear sky conditions. Transition probabilities from one cloud type to another as well as formation and decay rates of different cloud types are linked to large-scale conditions, such as mid-tropospheric moisture and vertical velocity, and convective available potential energy (CAPE), depicting interactions between clouds at each lattice and their large-scale environment. The fraction of each cloud type within each GCM grid can then be obtained by a Markov stochastic process on each lattice, and total heating and drying effects by cumulus processes can be derived by prescribed Q_1 and Q_2 profiles associated with each cloud type.

By replacing the cumulus parameterization scheme with the SMCM in the Climate Forecast System version 2 (CFSv2, i.e., CFS–SMCM), improved simulations of both convectively coupled equatorial waves (CCEWs) and the MJO are obtained (Goswami *et al.* 2017a,b). In particular, both the MJO's eastward propagation beyond the Maritime Continent during boreal winter and its northward propagation during boreal summer over the Asian monsoon region are much improved in the CFS–SMCM (Goswami *et al.* 2017b). Note that cloud–radiation feedbacks have not been taken into account in this version of CFS–SMCM, which suggests a potential for further improvement of the MJO and CCEWs when using this SMCM approach in GCMs.

In another recent implementation of the SMCM to the convection parameterization of the ECHAM6.3 (Peters *et al.* 2017), an SMCM-based closure for deep convection triggering on a large-scale environment, such as vertical velocity and relative humidity at 500-hPa, was used to replace the standard CAPE-based closure in the original ECHAM6.3 (Nordeng 1994). Despite this simple implementation of SMCM, the eastward propagation of the MJO is also significantly improved in ECHAM6-SMCM over the original ECHAM6.3.

Compared to the conventional ways of tuning parameters in the convection schemes, one advantage of this SMCM approach is that the dominant parameters affecting model MJO variability tend to be different from those controlling the model mean state (Goswami *et al.* 2017a, Peters *et al.* 2017). Therefore, unlike the known parameter tuning strategies that give an improved MJO at the expense of the mean state, good model mean state can be largely retained in the CFS-SMCM and ECHAM6-SMCM along with improved MJO simulations (Goswami *et al.* 2017b, a; Peters *et al.* 2017). However, a drawback of the SMCM implementation to GCMs is the complicated calibration process of the SMCM which involves many parameters in depicting transition probability among different cloud types. While the model MJO simulations are sensitive to these parameters, such as the stratiform cloud decay timescale, many of these parameters are subject to observational constraints. Additionally, plausible dependence of these parameters on the large-scale environment needs to be considered, particularly for climate projection studies.

Another form of a stochastic convection scheme was also recently implemented in the National Center for Atmospheric Research (NCAR) atmospheric community model CAM5 (Wang *et al.* 2016). This stochastic approach was based on the one proposed by Plant and Craig (2008) that used a Poisson distribution for the number of plumes in a convective ensemble conditional on the CAPE and an exponentially distributed cloud base mass flux based on the theory of equilibrium statistical mechanics. It was shown that the PDF of precipitation intensity and the MJO eastward propagation are significantly improved by adding stochastic nature to the deep convection scheme. While there is still great room for improvement in the stochastic convection schemes described above, these initial results suggest great potential for improvement of MJO representation with stochastic cumulus schemes, which could serve as a less computationally expensive alternative to the CRM approach in representing subgrid cumulus variability.

3. Process-oriented Metrics for the MJO

The strategy of the process-oriented diagnosis for the MJO is first to come up with a scalar measure for processes that are suggested to be important in simulating the MJO, i.e., process-oriented metrics, and then to examine the relationship between these process-oriented metrics and an MJO fidelity score in multi-model simulations. A robust relationship between a process-oriented metric and the MJO fidelity score suggests that the corresponding process is likely a key process operating in the group of models that contributes to their MJO simulation fidelity. These metrics that depict important processes for the MJO can help diagnose process-level model deficiencies and thereby help guide improvement in specific model parameterization schemes toward improving MJO simulations. Motivated by various observational, modeling, and theoretical studies on the MJO, several process-oriented metrics for the MJO have been recently proposed, including diagnostic metrics of the Relative Humidity Composite based on Precipitation (RHCP) (Kim *et al.* 2014; Ahn *et al.* 2017), the Normalized Gross Moist Stability (NGMS; Benedict *et al.* 2014; Jiang *et al.* 2015; Ahn *et al.* 2017), the mean lower-tropospheric moisture pattern (Gonzalez and Jiang 2017; Jiang 2017), the Greenhouse Enhancement Factor (GEF, Kim *et al.* 2015; Ahn *et al.* 2017), vertical moistening profiles as a function of rain rate (Klingaman *et al.* 2015a, b), and the convective moisture adjustment timescale (e.g., Jiang *et al.* 2016). In this section, we provide a brief review of these metrics.

3.1. *Relative humidity composite based on precipitation (RHCP) diagnostic*

Based on observational evidence of the strong coupling between convection and moisture (e.g., Bretherton *et al.* 2004; Peters and Neelin 2006; Holloway and Neelin 2009; Yasunaga and Mapes 2012), and also the great model sensitivity of MJO simulations to moisture–convection coupling (Zhu *et al.* 2009; Xavier *et al.* 2010; Hannah and Maloney 2011; Del Genio *et al.* 2012; Kim *et al.* 2012; Kim and Kang 2012), an RHCP metric was proposed for MJO model diagnoses (Kim *et al.* 2014). Specifically, the RHCP metric is formulated as the low-level (850–700 hPa) RH difference between the upper 10% precipitation percentile and the lower 20% precipitation percentile derived over the Indo-Pacific warm pool (60°E–180°, 15°S–15°N), measuring the amount of low-level RH increase required for a transition from weak to strong rain regimes, thus representing model convection sensitivity to environment moisture. A statistically robust relationship between the RHCP and model MJO performance has been illustrated in multi-model simulations (Kim *et al.* 2014; Jiang *et al.* 2015; Klingaman *et al.* 2015a; Ahn *et al.* 2017). For example, a correlation of about 0.56 was obtained between the RHCP metric and model MJO skill scores, denoted by the ratio of eastward versus westward power in planetary-scale, intraseasonal equatorial precipitation anomalies following Kim *et al.* (2009), based on 28 model simulations of Coupled Model Intercomparison Project Phase 5 (CMIP5) (Fig. 1a, Ahn *et al.* 2017).

3.2. *Normalized gross moist stability (NGMS)*

Under the moisture mode framework for the MJO (e.g., Raymond and Fuchs 2009; Sobel and Maloney 2012; Sobel and Maloney 2013; Adames and Kim 2016), recent analyses of the

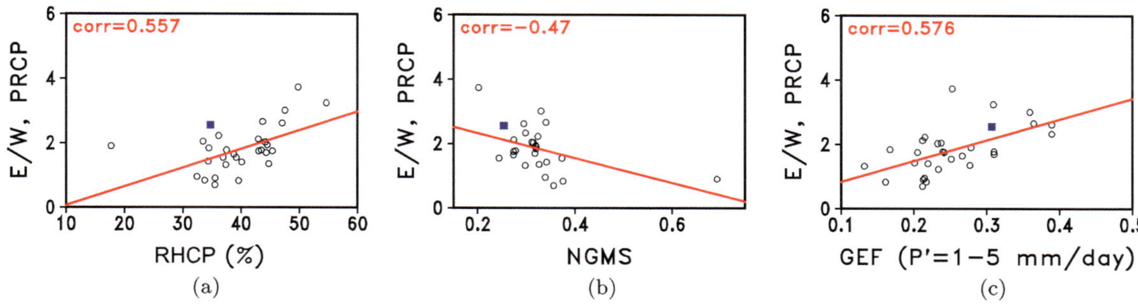

Fig. 1. Scatter plots of model skill score for the MJO eastward propagation, denoted by the eastward versus westward power spectrum ratio of equatorial precipitation, and (a) RHCP, (b) NGMS, and the GEF metric averaged over 1–5 mm day^{-1} rainfall regime in 28 IPCC CMIP5 model simulations. See text for details of these metrics. Adapted from Ahn *et al.* (2017).

moist static energy (MSE) budget have yielded critical insights into key processes for the maintenance and propagation of the MJO (readers are referred to the chapter on the MJO moisture mode theory by Adames *et al.* in this collection for details). The results of the MSE budget analysis suggest that (i) radiative heating and surface fluxes serve as two primary energy sources in sustaining the MJO, while both horizontal and vertical MSE advections export energy in regions of MJO convection, and (ii) horizontal and vertical MSE advections help the MJO to propagate eastward by increasing/decreasing MSE to the west/east of the enhanced convection. A diagnostic metric of the NGMS (Neelin and Held 1987; Raymond *et al.* 2009), which is defined as MSE export through vertical and/or horizontal MSE advection per unit convective activity and measures the efficiency of the large-scale circulation in discharging MSE out of the atmospheric column, can be used as a metric for MJO instability. It is hypothesized that the NGMS should be small or negative in order to sustain strong MJO convection (Raymond and Fuchs 2009; Hannah and Maloney 2011; Sobel and Maloney 2012). By applying the NGMS diagnosis for six GCM simulations, Benedict *et al.* (2014) illustrated a robust relationship between the vertical component of the time-mean NGMS over the Indo-Pacific warm pool and MJO simulation skill. Statistically significant anti-correlations between the winter mean NGMS over the Indo-Pacific warm pool and MJO skill based on multi-model simulations are also obtained in Jiang *et al.* (2015) and Ahn *et al.* (2017). As shown in Fig. 1b, a correlation of about −0.47 was found between the vertical component of NGMS and model MJO skill scores in 28 model simulations (Ahn *et al.* 2017). Moreover, a strong anti-correlation (−0.9) between summer mean vertical NGMS and the amplitude of the summer MJO over the eastern Pacific was also noted in eight simulations based on three GCMs (Maloney *et al.* 2014).

3.3. *Greenhouse enhancement factor (GEF)*

Motivated by previous studies that feedback between the longwave radiation and clouds plays a critical role in destabilizing the MJO (e.g., Raymond 2001; Lin *et al.* 2004; Bony and Emanuel 2005; Jiang *et al.* 2011; Andersen and Kuang 2012, Sobel and Maloney 2012, 2013; Chikira 2014; Wolding and Maloney 2015; Wolding *et al.* 2016), a GEF, defined by the negative ratio of anomalous outgoing longwave radiation to anomalous precipitation, is used to represent the strength of longwave radiation and cloud interactions associated with the MJO (Kim *et al.* 2015). It was shown that a GEF metric, derived by the weighted average of GEF over a 1–5 mm day^{-1} precipitation anomaly regime, exhibits a statistically significant correlation to model MJO simulation skill across multi-model simulations (Fig. 1c, Kim *et al.* 2015; Ahn *et al.* 2017).

3.4. *Lower-tropospheric mean moisture pattern*

This diagnostic is motivated by the finding that the advection of column MSE or equivalently the lower-tropospheric moisture, particularly its horizontal component, plays a critical role in driving the eastward propagation of the winter MJO (e.g., Maloney 2009; Maloney *et al.* 2010; Andersen and Kuang, 2012; Kim *et al.*, 2014; Sobel *et al.* 2014 Chikira 2014; Adames and Wallace 2015; Arnold *et al.* 2015; Jiang 2017). Under this process, the spatial distribution of the winter mean lower-tropospheric moisture over the equatorial Indo-Pacific region (Fig. 2a) holds a key for moistening (drying) to the east (west) of the MJO convection through advection by MJO anomalous winds, thus promoting the eastward propagation of the winter MJO. The critical role of the mean lower-tropospheric moisture pattern for the eastward propagation of the MJO has been further confirmed by recent modeling studies based on multi-model simulations from the MJO Task Force/GEWEX GASS MJO model comparison project (Gonzalez and Jiang 2017; Jiang 2017). The model skill in representing the low-level (900–650 hPa) mean moisture pattern over the Maritime Continent region (red rectangle in Fig. 2a) exhibits a very high correlation (about 0.8) with model MJO propagation skill in about 25 climate model simulations (Fig. 2b; Gonzalez and Jiang 2017). The critical role of the mean lower-tropospheric moisture pattern for MJO prediction skill is also illustrated based on model hindcasts (e.g., Kim 2017; Lim *et al.* 2017).

It is noteworthy that a recent multi-model analysis for the boreal summer MJO over the Indian Ocean also suggests that, similar to the eastward propagation of the winter MJO, the summer mean lower-tropospheric moisture pattern is also critical for the northward propagation of the boreal summer MJO due to moistening/drying to the north/south of the boreal summer MJO by horizontal advection of mean moisture by the MJO circulation (Jiang *et al.* 2018). These results, therefore, suggest that the distinct propagation characteristics of the MJO between the boreal winter and summer are largely regulated by seasonal migration in the low-level mean moisture pattern (Jiang *et al.* 2018).

3.5. *Other metrics*

Other process-oriented metrics have also been explored and linked to MJO skill in multi-model simulations. A net moistening metric, depicted by a vertical profile of the moisture tendency as a function of rain rate, was found to be highly correlated to model MJO simulation skill (Klingaman *et al.* 2015a,b). It was hypothesized that the net moistening in the mid-troposphere under the moderate rain regimes (2–9 mm day^{-1}) is crucial for high-quality MJO simulations.

A multi-model analysis by Jiang *et al.* (2016) suggested that the convective moisture adjustment timescale in a model, defined by the ratio of intraseasonal perturbations of precipitable water and surface precipitation (e.g., Bretherton *et al.* 2004; Sobel and Maloney 2013), can be an excellent metric for MJO amplitude in model simulations. The convective moisture adjustment timescale depicts how rapidly precipitation must increase to remove excess column water vapor, or alternately the efficiency of surface precipitation generation per unit column water vapor anomaly. Rushley *et al.* (2018) found that the convective moisture adjustment timescale is commonly overestimated in

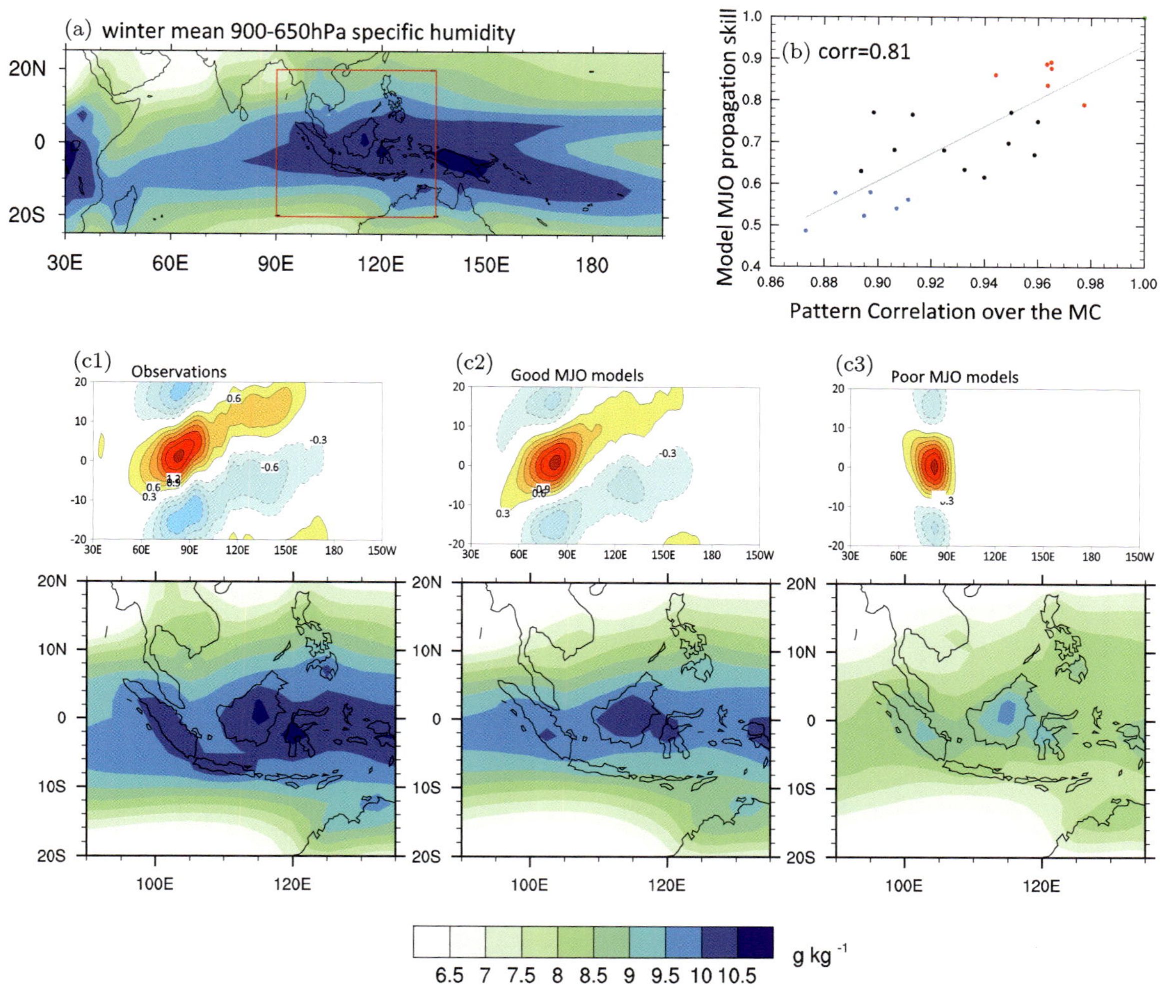

Fig. 2. (a) Winter mean 900–650 hPa specific humidity based on ERA-Interim; (b) scatter plot of model skill for eastward propagation of the MJO vs model skill of mean 900–650 hPa moisture over the Maritime Continent (MC; red rectangle in a) based on MJOTF/GASS model simulations. Red (blue) dots denote good (poor) MJO models. Lower panels: Time–longitude rainfall diagrams along the equator (10°S–10°N) and low-level (900–650 hPa) mean moisture over the MC in observations (c1), good (c2) vs poor (c3) MJO model composites. Hovmöller diagrams of rainfall anomalies associated with the MJO are derived based on lag-regression against an Indian Ocean box (75°–85°E; 5°S–5°N) for both observations and simulations (adapted from Jiang 2017 and Gonzalez and Jiang 2017).

GCMs. The key underlying physics regulating the convective moisture adjustment time scale needs to be further investigated (Jiang *et al.* 2016).

Additionally, analyses also suggested that the equatorial vertical-longitudinal distribution of diabatic heating, which depicts the multi-cloud structure of the MJO, could be used as a good metric for the MJO eastward propagation (e.g., Jiang *et al.* 2015; Wang *et al.* 2017). Meanwhile, the horizontal structure of boundary layer moisture convergence that provides moisture preconditioning in the lower troposphere to the east of the MJO convection center is also found to be highly correlated to MJO propagation across multi-model simulations (Wang and Lee 2017; Wang *et al.* 2018).

4. Summary and Discussion

Despite its tremendous influences on global weather and climate, the MJO remains poorly represented in the latest generation of GCMs. This grand challenge in modeling the MJO is largely due to model deficiencies in depicting cumulus processes and their interaction with the large-scale environment, as suggested by significant improvement of MJO simulations by replacing the cumulus parameterizations with explicit CRMs in these global GCMs, as well as other means of improving the representation of cumulus convection. While the CRM approach represents a future direction for model development, it is not practical in the near future for climate projection studies that need long-term integrations due to limitations in computing resources. Therefore, alternative affordable approaches to CRMs for improved depiction of subgrid cumulus processes in GCMs are worth investigating. In this chapter, an SMCM approach which provides a cheap way to represent interactions between multi-cloud structure within a GCM grid cell and large-scale conditions was briefly reviewed. Several recent implementations of an SMCM into GCMs suggest the great potential for the SMCM approach to improve simulations of the MJO and convectively coupled equatorial waves, although there is also room for refinement of the SMCM approach.

On the contrary, while improved MJO representation has been achieved in some GCMs including CRMs, an understanding of the essential model processes for realistic MJO simulations remains elusive. This gap has motivated recent efforts in the community on development of process-oriented metrics for the MJO as briefly introduced in this chapter. While statistically significant correlations between several process-oriented metrics and model skill for MJO propagation and/or amplitude are evident in multi-model simulations, uncertainties also exist in how these metrics are derived, as these correlations could be sensitive to changes in geographical domain, vertical levels, rain regimes, as well as observational and model datasets used for these metrics (e.g., Jiang *et al.* 2015; Kim *et al.* 2015; Ahn *et al.* 2017). Therefore, further investigations are needed to explore key model processes to discriminate good MJO models from poor ones, including further verification of these existing MJO metrics with independent model datasets. The upcoming CMIP6 will provide an excellent opportunity for continued efforts in process-oriented diagnoses of the MJO to obtain the knowledge necessary for improved MJO simulations in GCMs.

References

Adames, Á. F. and J. M. Wallace, 2015: Three-dimensional structure and evolution of the moisture field in the MJO. *J. Atmos. Sci.*, **72**, 3733–3754, doi:10.1175/JAS-D-15-0003.1.

Adames, Á. F. and D. Kim, 2016: The MJO as a Dispersive, convectively coupled moisture wave: Theory and observations. *J. Atmos. Sci.*, **73**, 913–941, doi:10.1175/JAS-D-15-0170.1.

Ahn, M.-S., D. Kim, K. R. Sperber, I.-S. Kang, E. Maloney, D. Waliser, and H. Hendon, 2017: MJO simulation in CMIP5 climate models: MJO skill metrics and process-oriented diagnosis. *Clim. Dynam.*, 1–23, doi:10.1007/s00382-017-3558-4.

Andersen, J. A. and Z. Kuang, 2012: Moist static energy budget of MJO-like disturbances in the atmosphere of a zonally symmetric aquaplanet. *J. Clim.*, **25**, 2782–2804, doi:10.1175/jcli-d-11-00168.1.

Arnold, N. P., M. Branson, Z. Kuang, D. A. Randall, and E. Tziperman, 2015: MJO intensification with warming in the superparameterized CESM. *J. Clim.*, **28**, 2706–2724, doi:10.1175/jcli-d-14-00494.1.

Benedict, J. J., E. D. Maloney, A. H. Sobel, and D. M. W. Frierson, 2014: Gross moist stability and MJO simulation skill in three full-physics GCMs. *J. Atmos. Sci.*, **71**, 3327–3349, doi:10.1175/JAS-D-13-0240.1.

Bony, S. and K. A. Emanuel, 2005: On the role of moist processes in tropical intraseasonal variability: Cloud–Radiation and moisture–convection feedbacks. *J. Atmos. Sci.*, **62**, 2770–2789, doi:10.1175/JAS3506.1.

Bretherton, C. S., M. E. Peters, and L. E. Back, 2004: Relationships between water vapor path and precipitation over the tropical oceans. *J. Clim.*, **17**, 1517–1528.

Buizza, R., M. Milleer, and T. N. Palmer, 1999: Stochastic representation of model uncertainties in the ECMWF ensemble prediction system. *Quart. J. Roy. Meteor. Soc.*, **125**, 2887–2908, doi:10.1002/qj.49712556006.

Chikira, M., 2014: Eastward-propagating intraseasonal oscillation represented by Chikira–Sugiyama cumulus parameterization. Part II: Understanding moisture variation under weak temperature gradient balance. *J. Atmos. Sci.*, **71**, 615–639, doi: 10.1175/JAS-D-13-038.1.

Del Genio, A. D., Y. Chen, D. Kim, and M.-S. Yao, 2012: The MJO transition from shallow to deep convection in CloudSat/CALIPSO data and GISS GCM simulations. *J. Clim.*, **25**, 3755–3770, doi:10.1175/JCLI-D-11-00384.1.

Deng, Q., B. Khouider, and A. J. Majda, 2015: The MJO in a Coarse-resolution GCM with a stochastic multicloud parameterization. *J. Atmos. Sci.*, **72**, 55–74, doi:10.1175/jas-d-14-0120.1.

Deng, Q., B. Khouider, A. J. Majda, and R. S. Ajayamohan, 2016: Effect of stratiform heating on the planetary-scale organization of tropical convection. *J. Atmos. Sci.*, **73**, 371–392, doi:10.1175/jas-d-15-0178.1.

Frenkel, Y., A. J. Majda, and B. Khouider, 2013: Stochastic and deterministic multicloud parameterizations for tropical convection. *Climate Dyn.*, **41**, 1527–1551, doi:10.1007/s00382-013-1678-z.

Gonzalez, A. O. and X. Jiang, 2017: Winter mean lower-tropospheric moisture over the Maritime Continent as a climate model diagnostic metric for the propagation of the Madden–Julian oscillation. *Geophys. Res. Lett.*, 10.1002/2016GL072430.

Goswami, B. B., B. Khouider, R. Phani, P. Mukhopadhyay, and A. J. Majda, 2017a: Implementation and calibration of a stochastic multicloud convective parameterization in the NCEP Climate Forecast System (CFSv2). *J. Adv. Model. Earth Sys.*, **9**, 1721–1739, doi:10.1002/2017MS001014.

Goswami, B. B., B. Khouider, R. Phani, P. Mukhopadhyay, and A. Majda, 2017b: Improving synoptic and intraseasonal variability in CFSv2 via stochastic representation of organized convection. *Geophys. Res. Lett.*, **44**, 1104–1113, doi:10.1002/2016GL071542.

Hannah, W. M. and E. D. Maloney, 2011: The role of moisture–convection feedbacks in simulating the Madden–Julian oscillation. *J. Clim.*, **24**, 2754–2770, doi:10.1175/2011jcli3803.1.

Holloway, C. E. and J. D. Neelin, 2009: Moisture vertical structure, column water vapor, and tropical deep convection. *J. Atmos. Sci.*, **66**, 1665–1683, doi:10.1175/2008jas2806.1.

Hoskins, B., 2013: The potential for skill across the range of the seamless weather-climate prediction problem: a stimulus for our science. *Quart. J. Roy. Meteor. Soc.*, **139**, 573–584, doi:10.1002/qj.1991.

Hung, M.-P., J.-L. Lin, W. Wang, D. Kim, T. Shinoda, and S. J. Weaver, 2013: MJO and convectively coupled equatorial waves simulated by CMIP5 climate models. *J. Clim.*, **26**, 6185–6214, doi:10.1175/JCLI-D-12-00541.1.

Jiang, X., 2017: Key processes for the eastward propagation of the Madden–Julian oscillation based on multimodel simulations. *J. Geophys. Res.: Atmos.*, doi:10.1002/2016JD025955.

Jiang, X., M. Zhao, E. D. Maloney, and D. E. Waliser, 2016: Convective moisture adjustment time scale as a key factor in regulating model amplitude of the Madden–Julian oscillation. *Geophys. Res. Lett.*, **43**, 10,412–10,419, doi:10.1002/2016GL070898.

Jiang, X., Á. F. Adames, M. Zhao, D. Waliser, and E. Maloney, 2018: A unified moisture moist framework for seasonality of MJO propagation. *J. Clim.*, in press.

Jiang, X., D. E. Waliser, W. S. Olson, W.-K. Tao, T. S. L'Ecuyer, K.-F. Li, Y. L. Yung, S. Shige, S. Lang, and Y. N. Takayabu, 2011: Vertical diabatic heating structure of the MJO: Intercomparison between recent reanalyses and TRMM estimates. *Mon. Weather Rev.*, **139**, 3208–3223, doi:10.1175/2011mwr3636.1.

Jiang, X., *et al.*, 2015: Vertical structure and physical processes of the Madden–Julian oscillation: Exploring key model physics in climate simulations. *J. Geophysical Research: Atmospheres*, **120**, 4718–4748, doi:10.1002/2014JD022375.

Johnson, R. H., T. M. Rickenbach, S. A. Rutledge, P. E. Ciesielski, and W. H. Schubert, 1999: Trimodal characteristics of tropical convection. *J. Clim.*, **12**, 2397–2418.

Khairoutdinov, M. F. and D. A. Randall, 2003: Cloud resolving modeling of the ARM summer 1997 IOP: Model formulation, results, uncertainties, and sensitivities. *J. Atmos. Sci.*, **60**, 607–625.

Khouider, B., J. Biello, and A. J. Majda, 2010: A stochastic multicloud model for tropical convection. *Commun. Math. Sci.*, **8**, 187–216.

Kim, D. and I.-S. Kang, 2012: A bulk mass flux convection scheme for climate model: description and moisture sensitivity. *Climate Dynam.*, **38**, 411–429, doi:10.1007/s00382-010-0972-2.

Kim, D. and E. Maloney, 2017: Simulation of the Madden–Julian oscillation using general circulation models. *The Global Monsoon System*, 3rd Edition, World Scientific, pp. 161–172.

Kim, D., M.-S. Ahn, I.-S. Kang, and A. D. D. Genio, 2015: Role of longwave cloud–radiation feedback in the simulation of the Madden–Julian oscillation. *J. Clim.*, **28**, doi:6979–6994, doi:10.1175/JCLI-D-14-00767.1.

Kim, D., A. H. Sobel, E. D. Maloney, D. M. W. Frierson, and I. S. Kang, 2011: A systematic relationship between intraseasonal variability and mean state bias in AGCM simulations. *J. Clim.*, **24**, 5506–5520, doi:10.1175/2011jcli4177.1.

Kim, D., A. H. Sobel, A. D. Del Genio, Y. Chen, S. J. Camargo, M.-S. Yao, M. Kelley, and L. Nazarenko, 2012: The tropical subseasonal variability simulated in the NASA GISS general circulation model. *J. Clim.*, **25**, 4641–4659, doi:10.1175/JCLI-D-11-00447.1.

Kim, D., P. Xavier, E. Maloney, M. Wheeler, D. Waliser, K. Sperber, H. Hendon, C. Zhang, R. Neale, Y.-T. Hwang, and H. Liu, 2014: Process-oriented MJO simulation diagnostic: Moisture sensitivity of simulated convection. *J. Clim.*, **27**, 5379–5395, doi:10.1175/JCLI-D-13-00497.1.

Kim, D., *et al.*, 2009: Application of MJO simulation diagnostics to climate models. *J. Clim.*, **22**, 6413–6436, doi:10.1175/2009jcli3063.1.

Kim, H.-M. 2017: The impact of the mean moisture bias on the key physics of MJO propagation in the ECMWF reforecast. *J. Geophys. Res.: Atmos.*, doi:10.1002/2017JD027005, 2017JD027005.

Klingaman, N. P., X. Jiang, P. K. Xavier, J. Petch, D. Waliser, and S. J. Woolnough, 2015a: Vertical structure and physical processes of the Madden–Julian oscillation: Synthesis and summary. *J. Geophys. Res.: Atmos.*, **120**, 4671–4689, doi:10.1002/2015JD023196.

Klingaman, N. P., *et al.*, 2015b: Vertical structure and physical processes of the Madden–Julian oscillation: Linking hindcast fidelity to simulated diabatic heating and moistening. *J. Geophys. Res.: Atmos.*, **120**, 4690–4717, doi:10.1002/2014JD022374.

Klingaman, N. P., X. Jiang, P. K. Xavier, J. Petch, D. Waliser, and S. J. Woolnough, 2017: Vertical structure and diabatic processes of the Madden–Julian oscillation. *The Global Monsoon System*, 3rd edition, World Scientific, pp. 161–172.

Lau, W. K.-M. and D. E. Waliser, 2012: *Intraseasonal Variability in the Atmosphere–Ocean Climate System*, 2nd edition, Springer, Heidelberg, Germany, 613p.

Lim, Y., S. W. Son, and D. Kim, 2017: MJO prediction skill of the subseasonal-to-seasonal prediction models. *J Clim.*, in press.

Lin, J., B. Mapes, M. Zhang, and M. Newman, 2004: Stratiform precipitation, vertical heating profiles, and the Madden–Julian oscillation. *J. Atmos. Sci.*, **61**, 296–309.

Lin, J. W.-B. and J. D. Neelin, 2003: Toward stochastic deep convective parameterization in general circulation models. *Geophys. Res. Lett.*, **30**, doi:10.1029/2002GL016203, n/a-n/a.

Madden, R. A. and P. R. Julian, 1971: Detection of a 40–50 day oscillation in zonal wind in tropical pacific. *J. Atmos. Sci.*, **28**, 702-&.

Madden, R. A. and P. R. Julian, 1994: Observations of the 40–50-day tropical oscillation: A review. *Mon. Weather Rev.*, **122**, 814–837.

Maloney, E. D., 2009: The moist static energy budget of a composite tropical intraseasonal oscillation in a climate model. *J. Clim.*, **22**, 711–729.

Maloney, E. D., A. H. Sobel, and W. M. Hannah, 2010: Intraseasonal variability in an aquaplanet general circulation model. *J. Adv. Model. Earth Syst.*, **2**, 10.3894/james.2010.2.5.

Maloney, E. D., X. Jiang, S.-P. Xie, and J. J. Benedict, 2014: Process-oriented diagnosis of east pacific warm pool intraseasonal variability. *J. Clim.*, **27**, 6305–6324, doi:10.1175/jcli-d-14-00053.1.

Mapes, B., S. Tulich, J. Lin, and P. Zuidema, 2006: The mesoscale convection life cycle: Building block or prototype for large-scale tropical waves? *Dynam. Atmos. Oceans*, **42**, 3–29, doi:10.1016/J.Dynatmoce.2006.03.003.

Miura, H., M. Satoh, T. Nasuno, A. T. Noda, and K. Oouchi, 2007: A Madden-Julian oscillation event realistically simulated by a global cloud-resolving model. *Science*, **318**, 1763–1765, doi:10.1126/Science.1148443.

NAS, 2016: Next generation earth system prediction: Strategies for subseasonal to seasonal forecasts. *National Research Council, National Academy of Sciences*, Washington DC, doi:ISBN-978-0-309-38880-1, 290 pages.

Neelin, J. D. and I. M. Held, 1987: Modeling tropical convergence based on the moist static energy budget. *Mon. Weather Rev.*, **115**, 3–12.

Nordeng, T. E., 1994: Extended versions of the convective parametrization scheme at ECMWF and their impact on the mean and transient activity of the model in the tropics. *ECMWF Tech. Memo.* 206, 41 pp.

Peters, K., T. Crueger, C. Jakob, and B. Möbis, 2017: Improved MJO-simulation in ECHAM6.3 by coupling a Stochastic Multicloud Model to the convection scheme. *J. Adv. Model. Earth Sys.*, **9**, 193–219, doi:10.1002/2016MS000809.

Peters, O. and J. D. Neelin, 2006: Critical phenomena in atmospheric precipitation. *Nat Phys*, **2**, 393–396.

Plant, R. S. and G. C. Craig, 2008: A stochastic parameterization for deep convection based on equilibrium statistics. *J. Atmos. Sci.*, **65**, 87–105, doi:10.1175/2007jas2263.1.

Pritchard, M. S. and D. Yang, 2016: Response of the superparameterized Madden–Julian oscillation to extreme climate and basic-state variation challenges a moisture mode view. *J. Clim.*, **29**, 4995–5008, doi:10.1175/jcli-d-15-0790.1.

Randall, D., M. Khairoutdinov, A. Arakawa, and W. Grabowski, 2003: Breaking the cloud parameterization deadlock. *Bull. Am. Meteorol. Soc.*, **84**, 1547–1564, doi:10.1175/BAMS-84-11-1547.

Raymond, D. J., 2001: A new model of the Madden-Julian oscillation. *J. Atmos. Sci.*, **58**, 2807–2819.

Raymond, D. J. and Ž. Fuchs, 2009: Moisture modes and the Madden–Julian oscillation. *J. Clim.*, **22**, 3031–3046, doi:10.1175/2008jcli2739.1.

Raymond, D. J., S. Sessions, A. Sobel, and Z. Fuchs, 2009: The mechanics of gross moist stability. *J. Adv. Model. Earth Sys.*, **1**, doi:10.3894/james.2009.1.9, 20 pp.

Rushley, S. S., D. Kim, C. S. Bretherton, and M. S. Ahn, 2018: Reexamining the nonlinear moisture-precipitation relationship over the tropical oceans. *Geophys. Res. Lett.*, **45**, 1133–1140, doi:10.1002/2017GL076296.

Satoh, M., T. Matsuno, H. Tomita, H. Miura, T. Nasuno, and S. Iga, 2008: Nonhydrostatic icosahedral atmospheric model (NICAM) for global cloud resolving simulations. *J. Comput. Phys.*, **227**, 3486–3514, doi:10.1016/Jjcp.2007.02.006.

Sobel, A. and E. Maloney, 2012: An idealized semi-empirical framework for modeling the Madden–Julian oscillation. *J. Atmos. Sci.*, **69**, 1691–1705, doi:10.1175/jas-d-11-0118.1.

Sobel, A. and E. Maloney, 2013: Moisture modes and the eastward propagation of the MJO. *J. Atmos. Sci.*, **70**, 187–192, doi:10.1175/Jas-D-12-0189.1.

Sobel, A., S. Wang, and D. Kim, 2014: Moist static energy budget of the MJO during DYNAMO. *J. Atmos. Sci.*, **71**, 4276–4291, doi:10.1175/JAS-D-14-0052.1.

Sperber, K. R. and D. Kim, 2012: Simplified metrics for the identification of the Madden–Julian oscillation in models. *Atmos. Sci. Lett.*, **13**, 187–193, doi:10.1002/asl.378.

Stan, C., M. Khairoutdinov, C. A. DeMott, V. Krishnamurthy, D. M. Straus, D. A. Randall, J. L. Kinter, and J. Shukla, 2010: An ocean–atmosphere climate simulation with an embedded cloud resolving model. *Geophys. Res. Lett.*, **37**, doi:10.1029/2009gl040822, -.

Vitart, F., A. Robertson, A. Kumar, H. Hendon, Y. Takaya, H. Lin, A. Arribas, J.-Y. Lee, D. Waliser, B. Kirtman, and H.-K. Kinm, 2012: Subseasonal to seasonal prediction: Research implementation plan. WWRP/THORPEX-WCRP Report.

Waliser, D. *et al.*, 2009: MJO simulation diagnostics. *J. Clim.*, **22**, 3006–3030, doi:10.1175/2008jcli2731.1.

Wang, B. and S.-S. Lee, 2017: MJO propagation shaped by zonal asymmetric structures: Results from 24 GCM simulations, **30**, 7933–7952, doi:10.1175/jcli-d-16-0873.1.

Wang, B., S.-S. Lee, D. E. Waliser, C. Zhang, A. Sobel, E. Maloney, T. Li, X. Jiang, and K.-J. Ha, 2018: Dynamics-oriented diagnostics for the Madden–Julian oscillation. *J. Clim.*, **31**, 3117–3135, doi:10.1175/jcli-d-17-0332.1.

Wang, L., T. Li, E. Maloney, and B. Wang, 2017: Fundamental causes of propagating and non-propagating MJOs in MJOTF/GASS models. *J. Clim.*, **30**, 3743–3769, doi:10.1175/jcli-d-16-0765.1.

Wang, Y., G. J. Zhang, and G. C. Craig, 2016: Stochastic convective parameterization improving the simulation of tropical precipitation variability in the NCAR CAM5. *Geophys. Res. Lett.*, **43**, 6612–6619, doi:10.1002/2016GL069818.

Wheeler, M. and E. Maloney, 2013: Madden–Julian oscillation (MJO) task force: A joint effort of the climate and weather communities. *CLIVAR Exchanges*. No. 61, **18**(1).

Wolding, B. O. and E. D. Maloney, 2015: Objective diagnostics and the Madden–Julian oscillation. Part II: Application to moist static energy and moisture budgets. *J. Clim.*, **28**, 7786–7808, doi:10.1175/jcli-d-14-00689.1.

Wolding, B. O., E. D. Maloney, and M. Branson, 2016: Vertically resolved weak temperature gradient analysis of the Madden–Julian oscillation in SP-CESM. *J. Adv. Model. Earth Sys.*, doi:10.1002/2016MS000724.

Wolding, B. O., E. D. Maloney, S. Henderson, and M. Branson, 2017: Climate change and the Madden-Julian oscillation: A vertically resolved weak temperature gradient analysis. *J. Adv. Model. Earth Sys.*, doi:10.1002/2016MS000843, n/a-n/a.

Xavier, P. K., J. P. Duvel, P. Braconnot, and F. J. Doblas-Reyes, 2010: An evaluation metric for intraseasonal variability and its application to CMIP3 twentieth-century simulations. *J. Clim.*, **23**, 3497–3508, doi:10.1175/2010jcli3260.1.

Yasunaga, K. and B. Mapes, 2012: Differences between more divergent and more rotational types of convectively coupled equatorial waves. Part I: Space–time spectral analyses. *J. Atmos. Sci.*, **69**, 3–16, doi:10.1175/jas-d-11-033.1.

Zhang, C., 2013: Madden–Julian oscillation: Bridging weather and climate. *Bull. Am. Meteorol. Soc.*, **94**, 1849–1870, doi:10.1175/bams-d-12-00026.1.

Zhu, H. Y., H. Hendon, and C. Jakob, 2009: Convection in a parameterized and superparameterized model and its role in the representation of the MJO. *J. Atmos. Sci.*, **66**, 2796–2811, doi:10.1175/2009jas3097.1.

Chapter 26

Monsoon Sub-Seasonal Prediction

Vincent Moron[*,†,**], Rodrigo Bombardi[†], Harry Hendon[§], Andrew Marshall[¶], Atul Kumar Sahai[||], and Rajib Chattopadhyay[||]

[*] *Aix Marseille Univ, CNRS, IRD, INRA, Coll France, CEREGE, Aix-en-Provence, France*

[†] *International Research Institute for Climate and Society, Columbia University, Palisades, USA*

[‡] *Department of Geography, Texas A&M University, College Station, USA*

[§] *Bureau of Meteorology, Melbourne, Australia*

[¶] *Bureau of Meteorology, Hobart, Australia*

[||] *Indian Institute for Tropical Meteorology, India Meteorological Department, Pune, India*

[**] *moron@cerege.fr*

This chapter summarizes current efforts to predict on the sub-seasonal timescale monsoon features such as the beginning/end dates, duration, accumulated precipitation, and active/break phases. We present analyses related to the spatial coherence of rainfall anomalies at sub-seasonal 7–90-day timescales, retrospective sub-seasonal forecasts of monsoon onset and demise dates across the tropics, and lastly two regional examples of retrospective sub-seasonal forecasts for India (June–September) and tropical Australia (November–March). The spatial coherence of rainfall anomaly variability may be viewed as an empirical estimate of the upper bound of the potential predictability (PP). Small spatially coherent regions suggest poor PP, due to the influence of unpredictable small-scale motions which are not canceled even a in weekly or a monthly average. In contrast, large spatially coherent regions of rainfall anomalies may reflect the impact of phenomena able to synchronize local-scale rainfall variations in space, including predictable atmospheric motions at sub-seasonal timescales. The spatially coherent regions are clearly larger across the oceans (especially over the Equatorial Pacific) than over the continents (especially over regions of complex topography). Spatially coherent regions are also usually larger during both onset or demise stages of the local monsoon and usually decrease during the peak monsoon period (especially across the continents), suggesting that the seasonal monsoonal precipitation may not always be the most predictable characteristic of the rainy season. The timing of the rainy season over monsoonal regions can be forecast using a hybrid approach that combines observation and retrospective forecasts from the sub-seasonal to seasonal database. Skillful forecasts are reached for the regional-scale onset and demise dates as much as one month in advance for some monsoonal regions including North and South America, East Asia, and Northern Australia. Over India, the success of skillful sub-seasonal predictions during the boreal summer monsoon, in June–September, is primarily related to the correct simulation and forecast of the northward-propagating intraseasonal oscillation. Current operational IITM and ECMWF prediction systems show promising skill, even with a lead time of 4 weeks, especially over Western and North-Western India. The skill also appears to be higher in June and September, around the regional-scale onset and demise of the boreal summer monsoon, than during its peak in July–August. Over tropical Australia, sub-seasonal variations during the November–March monsoon are primarily dominated by the impact of the planetary-scale Madden–Julian Oscillation (MJO). Retrospective forecasts using POAMA-2 show skill higher than 0.5 till week 5 for a

The Multiscale Global Monsoon System
Edited by C. P. Chang et al.

regional-scale wind index based on the zonal component of the 850-hPa wind anomaly. In contrast the forecast skill is lower for regional-scale rainfall. It suggests that useful sub-seasonal predictions may be issued when errors related to local depiction of the MJO in the monsoon rainfall have been reduced.

1. Introduction

Monsoons systems are primarily characterized by alternating dry and wet seasons, with the wet seasons peaking at, or just after, the summer solstice of the corresponding hemisphere. The location, shape, and size of the landmass as well as topography also play a role in the timing and strength of the monsoon. Due to warm surface temperature, and a large amount of precipitable water during the wet season, tropical rainfall is mostly convective. The basic unit of convection (i.e., cumulonimbus) is embedded into larger, and partly nested, systems, from thunderstorms to mesoscale convective clusters, tropical depressions, and ultimately an intertropical convergence zone (ITCZ). The annual cycle is strongly modulated by the distribution of energy in the earth system (Biasutti *et al.* 2018). The occurrence and intensity of these convective systems and associated rainfall are modulated, beyond the annual cycle, by a myriad of atmospheric waves, from 2–3-day Kelvin waves to the 20–60-day Madden–Julian oscillation (MJO) (Madden and Julian 1994; Wheeler and Kiladis 1999; Lubis and Jacob 2015). In turn, convection leads to a wide spectrum of waves, i.e., the Convectively coupled equatorial waves (Kiladis *et al.* 2009), which contribute to a large fraction of rainfall variations across the tropical zone. Over the continent, convective systems are also strongly controlled by the diurnal cycle (Ricciardulli and Sardeshmukh 2002). Tropical rainfall is thus produced by a complex interacting and multiscale system, leading to a rich and wide spectrum of space–time variations, for example, the most intense rainfall, which contribute the most to the seasonal amount, being usually related to the shortest and fastest scales.

Seasonal monsoon rainfall is, by definition, the most comprehensive variable since it sums all rainy events. In addition, its interannual variability emphasizes the influence of any slow systematic forcing such as the El Niño Southern oscillation (ENSO). In this context, it is assumed that the seasonal summation increases the spatial coherence of near-homogeneous anomalies at an interannual timescale. Moron *et al.* (2007) showed that daily mean intensity is strongly impacted by the wettest days of a season and is far less spatially coherent than the seasonal frequency of wet days on interannual timescales. The seasonal frequency of wet days could be further analytically decomposed into (1) a full set of intraseasonal characteristics (ISCs) including onset/demise (and thus duration) of the rainy season and (2) the frequency and mean duration in wet/dry spells between onset and demise dates (Moron *et al.* 2015). Wet (dry) days in the absolute or relative sense are frequently clustered into "active" ("break") spells, lasting from a few days to a few weeks across most tropical regions (Australia, India, etc.). The timing and length of active and break spells, as well as the timing of onset and demise dates of the rainy season, have important implications for a broad audience, including various end users. From a statistical point of view, these characteristics are not directly related to the intensity of rainfall, except for the threshold to define a wet day (or spell). Therefore, these characteristics may be more spatially coherent and consequently potentially predictable than the daily mean intensity of rainfall. There is a considerable interest in the predictability of the intraseasonal variability (ISV, Zhu *et al.* 2014; Vigaud *et al.* 2017), loosely defined here as between 7 days (i.e., approximately the upper limit of the weather forecasts) and 90 days (see, for example, the seasonal-to-sub-seasonal — S2S — project: http://s2sprediction.net/ Robertson *et al.* 2015; Vitart *et al.* 2017; Robertson and Vitart 2019 and the special issue of Frontiers http://journal.frontiersin.org/research topic/5124/sub-seasonal-to-seasonal-predictabil ity-and-prediction-of-monsoon-climates).

This chapter reviews recent advances and current issues related to the spatial coherence of rainfall anomalies and the predictability and prediction of several ISCs, including onset and demise dates of the rainy season across the tropics as well as regional examples of active/break spell occurrence by the MJO/MISO (India and Australia).

2. Amplitude and Sub-seasonal Modulation of Spatial Coherence of Rainfall Anomalies

The prediction of rainfall is challenging due to interactions involving a range of spatial and temporal scales. In this context, temporal filtering helps to diagnose and, if possible, predict filtered bandwidths. The relevance of this exercise for the predictability of local-scale rainfall depends on both our ability to predict accurately these timescales and on the variance of local-scale rainfall explained by each timescale.

Figure 1 (left panels) shows a decomposition of variance below 180 days (that is, removing interannual and annual/biannual variations) into three bands: 20–90 days corresponding roughly with MJO, 7–20 days to consider shorter Kelvin and Rossby waves, and below 7 days, including synoptic and mesoscales and any fast convectively coupled equatorial waves (Wheeler and Kiladis 1999). It is clear that most of the variance is explained by the shortest timescales, especially over the landmasses. The 20–90-day band shows a clear maximum over the Indian ocean and tropical Western Pacific (including the South Pacific Convergence Zone) and local maximums over eastern and southern Africa, Western India, Northeast Brazil, and parts of northern Australia. The contribution of the 7–20-day band peaks almost symmetrically around the equator across the tropical oceans, with an extension over several continents, as those described above. The contribution of the 7–20-day and 20–90-day bands is especially weak across equatorial and western Africa, the Amazon forest, and the Indonesian islands.

The right panel of Fig. 1 shows the decorrelation surfaces (DSs, i.e., the surface around each grid point where the correlations between rainfall anomalies $> e^{-1} = 0.3679$ with the target grid point) for each ISV band. DS is similar to the decorrelation distances (i.e., Ricciardulli and Sardeshmukh 2002), except that the DS considers equally isotropic and anisotropic spatially coherent surfaces around each grid point. For example, it is well known that the spatially coherent anomalies related to the ENSO phenomenon over the Central Equatorial Pacific are clearly anisotropic (i.e., respectively longer distance on the E–W than on the N–S dimensions). For an isotropic surface, a decorrelation distance of 50, 100, 500, and 1000 km corresponds roughly to a DS of $\sim 8 \times 10^{-3}$, $3 \times 104, 8 \times 10^5$ and 3×10^6 sq. km. A small DS indicates that local-scale rainfall anomalies are not spatially coherent, thus limiting their predictability at this scale, while a large DS suggests that anomalies are spatially consistent and thus may be predictable (Moron *et al.* 2019). In that context, the predictability depends on the timescale and periodic character of the synchronizing factor, which could be "fast" as temperate-tropical-troughs (TTT, Macron *et al.* 2014) or "slow" as the MJO. All panels show that DSs are systematically smaller across the continents in comparison to the oceans. Note that this is also observed in OLR data (with a larger DS on average, not shown) which are not biased by the heterogeneous validation with rain gauges as GPCP dataset. Interestingly, there is some

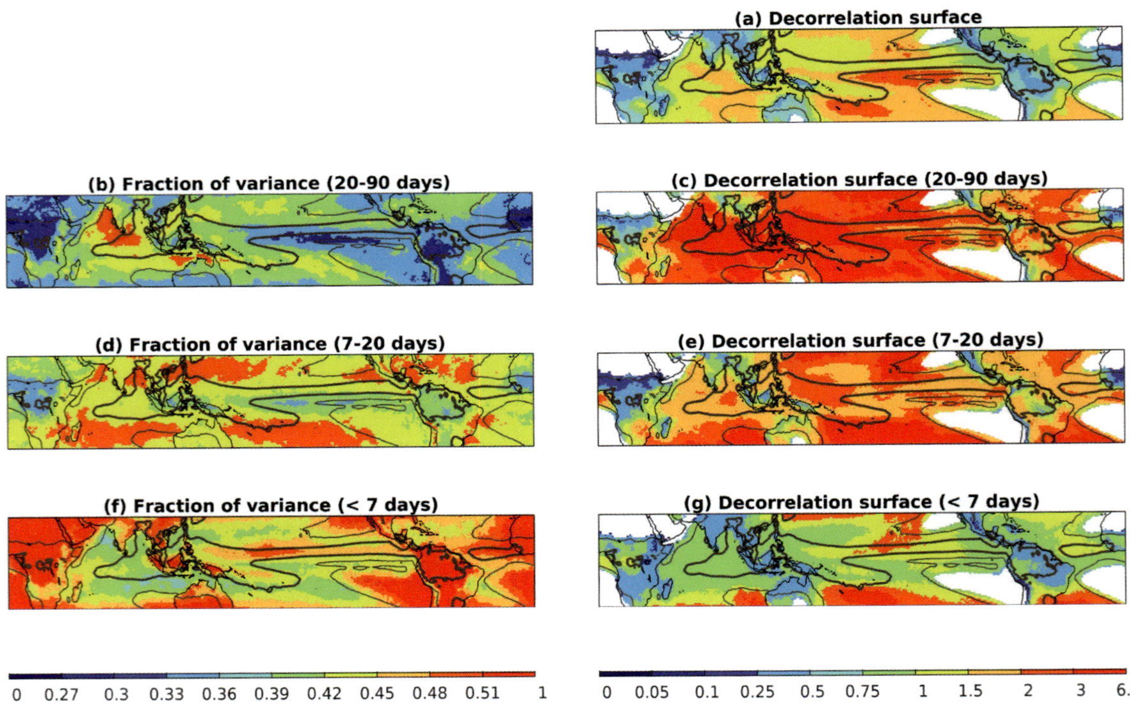

Fig. 1. Left panels: Relative partition of the 2–180-day variance into three bands from top to bottom for 20–90 days, 7–20 days, and $<$7days. The decomposition has been computed from $1^\circ \times 1^\circ$ daily GPCP rainfall on 1996–2016 time period. Right panels: area (in millions of square kilometers) of the spatial autocorrelation >0.3679 ($= e^{-1}$) around each grid point. The correlations are computed using rank of amount of rainfall for the anomalies versus the mean climatological cycle (upper panel) and the three intraseasonal bands. The correlations are computed only if at least 100 observations are above a climatological mean of 1 mm/day. The blank areas have less than 100 days receiving at least 1 mm/day. The climatological mean rainfall (1 mm/day as thin line and 5 mm/day as thick line) is superimposed on each panel.

correspondence between amplitude and DS for the 20–90-day band. In other words, relatively large amplitudes matched well with a large DS, except for the interior of the Amazon (rather large decorrelation surface despite the small amplitude). As before, the DS is very small over equatorial and western Africa. The similarity between relative amplitude and decorrelation surface is less clear for 7–20 days even though the DS peaks across tropical oceans. The DS for the shortest timescales shows opposite patterns between continents and oceans even though the surfaces are larger over the subtropical southern oceans (Fig. 1), partly related to TTTs.

Figure 2 shows the maps of the DSs for three different stages (onset, peak, and end of the rainy season) defined from the local-scale mean annual cycle. In case of double rainy seasons, as over East Africa, the computation mixes both rainy seasons (i.e., the onset stage is, in this case, the onset of both rainy seasons and the peak stage is defined as the intermediate period receiving at least 50% of the mean annual rainfall; if a season is significantly drier than another region, the peak stage may reflect only the wettest rainy season). Overall, the DS is maximum either for the onset or the end stages, especially over landmasses where the 3rd and 4th scenarios cover together less than 20% of the landmasses. Therefore, this suggests a significant seasonal modulation of spatial coherence with smaller spatial scales, inherently limiting

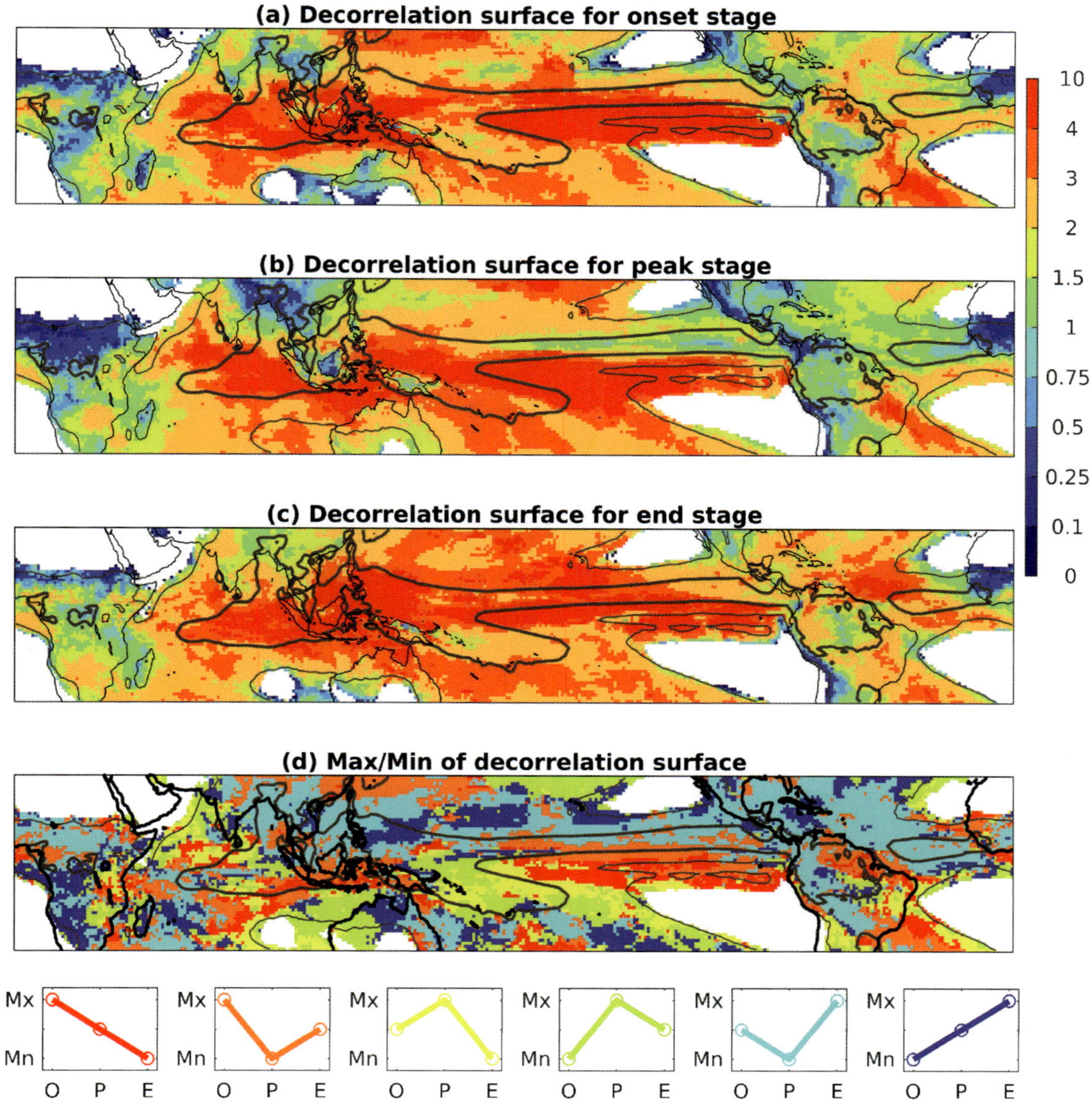

Fig. 2. (a)–(c) Decorrelation surfaces (in millions sq. km) for the onset, core, and end stages of the local monsoon. The onset and end are symmetrically defined from the mean climatological cycle and the core stage is the period when more than 50% of the mean annual rainfall occurs between onset and end stages. (d) Map of the six scenarios shown beneath panel (d). The scenarios are defined as the relative rank in decorrelation surfaces between onset (O), core (C), and end (E) stages. The climatological mean rainfall (1 mm/day as thin line and 5 mm/day as thick line) is superimposed on each panel.

the predictability around the peak of the local-scale rainy season.

This section suggests that intraseasonal predictions may not have the same degree of relevance everywhere across the tropics, since there is a larger amount of noise over the continents. The amplitude of ISV bands (7–20-day and 20–90-day bands) is also weak and/or DSs are small over the equatorial and western Africa, central America, the Maritime Continent, and the interior of the Amazon. This does not mean that specific periods of time (as onset and/or demise of the rainy season) are not predictable since ISV relative amplitude and spatial coherence (and potential predictability) may be modulated by sub-seasonal timescales.

3. Onset and Demise of the Rainy Season across the Tropics

The previous section suggests that rainfall anomalies around the early and late stages of the rainy season usually exhibit a larger DS than those around its peak. A large DS may be related to potential predictability (of either onset or demise dates), while a small DS may be related to its absence or weakness. This section analyzes the predictability of onset and demise dates across the tropical zone.

There is strong evidence that the onset dates of monsoon regions are predictable on sub-seasonal to seasonal timescales using either statistical or dynamical methods (Vellinga *et al.* 2013; Alessandri *et al.* 2015; Bombardi *et al.* 2017). Bombardi *et al.* (2017) used a hybrid approach, that combines observations and simulations, to predict the onset and demise dates of the rainy season on sub-seasonal timescales over monsoonal regions across the tropics. The onset (Fig. 3a) and withdrawal dates (not shown) are defined using precipitation data only based on the method created by Liebmann and Marengo (2001) and adapted by Bombardi and Carvalho (2009). Retrospective forecasts of four global coupled models participating in the S2S project (Robertson *et al.* 2015) are used. The observed climatology was used as the benchmark forecast for comparison against the retrospective forecasts. The authors found that this hybrid approach outperformed the climatology in predicting the onset and withdrawal dates of the rainy season by up to 20–30 days lead time. Skillful forecasts for the onset date exist as much as 1 month in advance over South America, East Asia, and Northern Australia. On the contrary, the Indian monsoon region shows low forecast skill for the onset date (Fig. 3b–h). The withdrawal dates show sub-seasonal forecast skill by up to a month in advance over parts of North and South America, East Asia, and Northern Australia (not shown). The sources of predictability of the onset and demise dates vary from region to region and need to be further investigated.

4. MISO and Break/Active Spells of Monsoonal India

India offers a nice example of a monsoonal climate where a clear link between ISV and rainfall has been established (e.g., Gadgil 2003). The main manifestation of ISV of Indian summer monsoon is the coupling between northward propagation of monsoonal intraseasonal oscillations (MISOs) and the occurrence of break and active phases of the monsoon (Chattopadhyay *et al.* 2008; Suhas *et al.* 2013). The MISO propagates northward at a phase speed of 1°/day and has a broad spectral scale between 10 and 80 days with a dominant spectral peak around 40 days. The operational prediction of MISO spells is important for several sectors of society in India, in particular for flood and disaster forecasting and agricultural planning. MISOs can cluster the high-frequency sub-seasonal modes in the synoptic scales (Goswami *et al.* 2003). The India Meteorological Department (IMD) has adopted the National Centers for Environmental Predictions (NCEP) Climate Forecast System (NCEP-CFSv2) (Saha *et al.* 2014) as its operational forecast model. Additional climate services based on extended range forecasts became operational for the 2017 monsoon season. The IMD extended range forecasts are based on a version of NCEP-CFSv2 and its atmospheric counterpart the GFS model run, adopted and developed at the Indian Institute of Tropical Meteorology (IITM). The initial

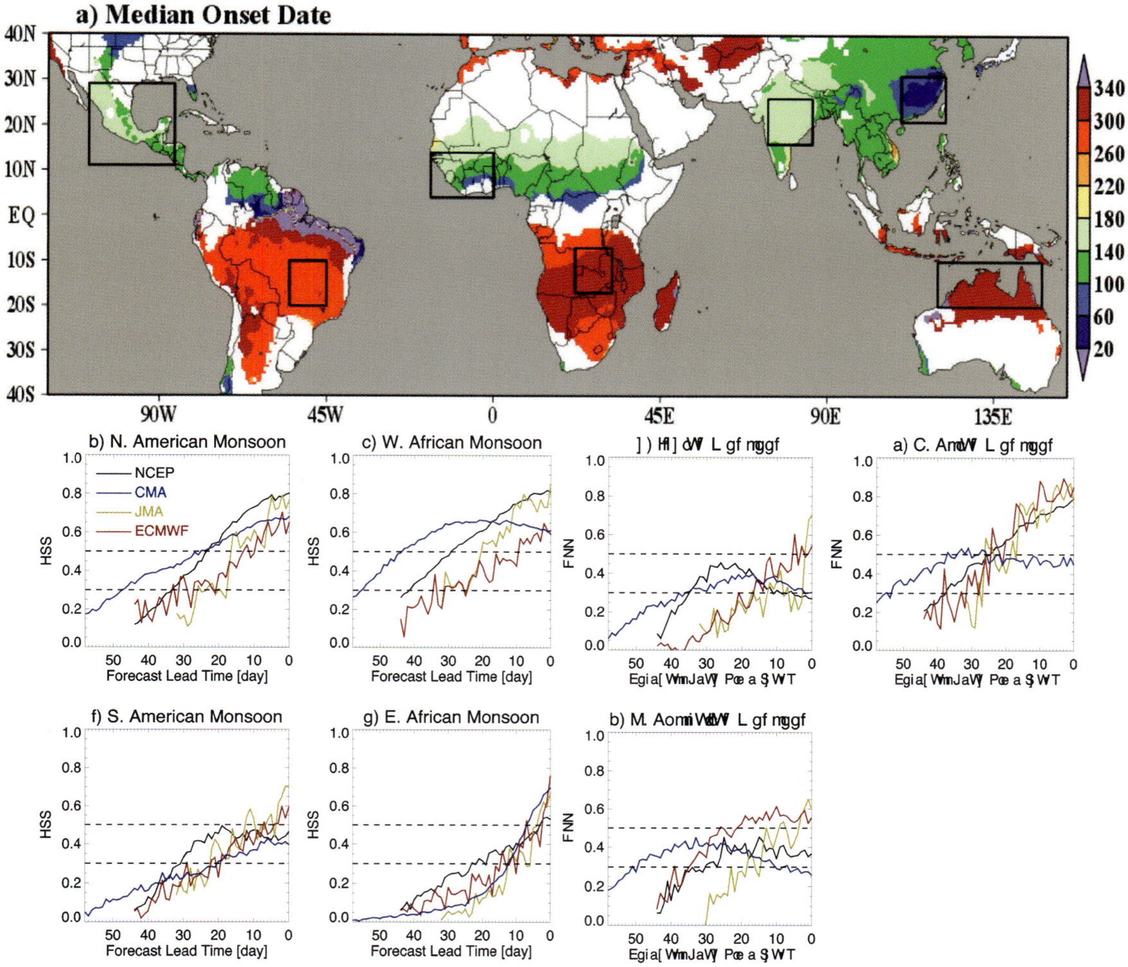

Fig. 3. Upper panel: Median onset date [DOY or Julian day] of onset dates calculated from CPC_UNI data (1979–2014). The boxes indicate monsoonal regions of interest. We mask regions where explained variance of the first harmonic of the mean annual cycle of observed precipitation is small (<30%). Lower panels: Probabilistic forecast skill (Heidke Skill Score — HSS; Wilks 2006) for the onset date forecast using four models from the S2S database. The lines show HSS for (b) North American Monsoon, (c) West African Monsoon, (d) Indian Monsoon, (e) East Asian Monsoon, (f) South American Monsoon, (g) East African Monsoon, and (h) Northern Australian Monsoon. Lead time 0 (zero) refers to the observed onset date. The multi-model ensemble (MME) was calculated using only NCEP, CMA, and ECMWF models for the period of 1999–2010. A negative HSS indicates that a forecast is correct by chance. A value of 0.5 indicates that 2 out of 3 forecasts were in the correct tercile, while a value of 0.0 indicates that 1 out of 3 forecasts was in the correct tercile. Since the climatology will always fall in the same tercile, the forecast using the climatology has an HSS value of 0.0. A value of 0.3 is commonly considered as the threshold for skillful forecasts. The dashed black lines show HSS equal to 0.3 and 0.5 (details in Bombardi *et al.* 2017).

conditions for the IITM-adopted version were generated in-house using an indigenous perturbation technique (Abhilash *et al.* 2014b) and the boundary conditions of sea surface temperature for the GFS runs were modified using a new bias-correction technique (Abhilash *et al.* 2014a) as shown in Fig. 4a.

The current hindcast status shows that forecasting of the active–break spells is considered skillful over the core monsoon zone of

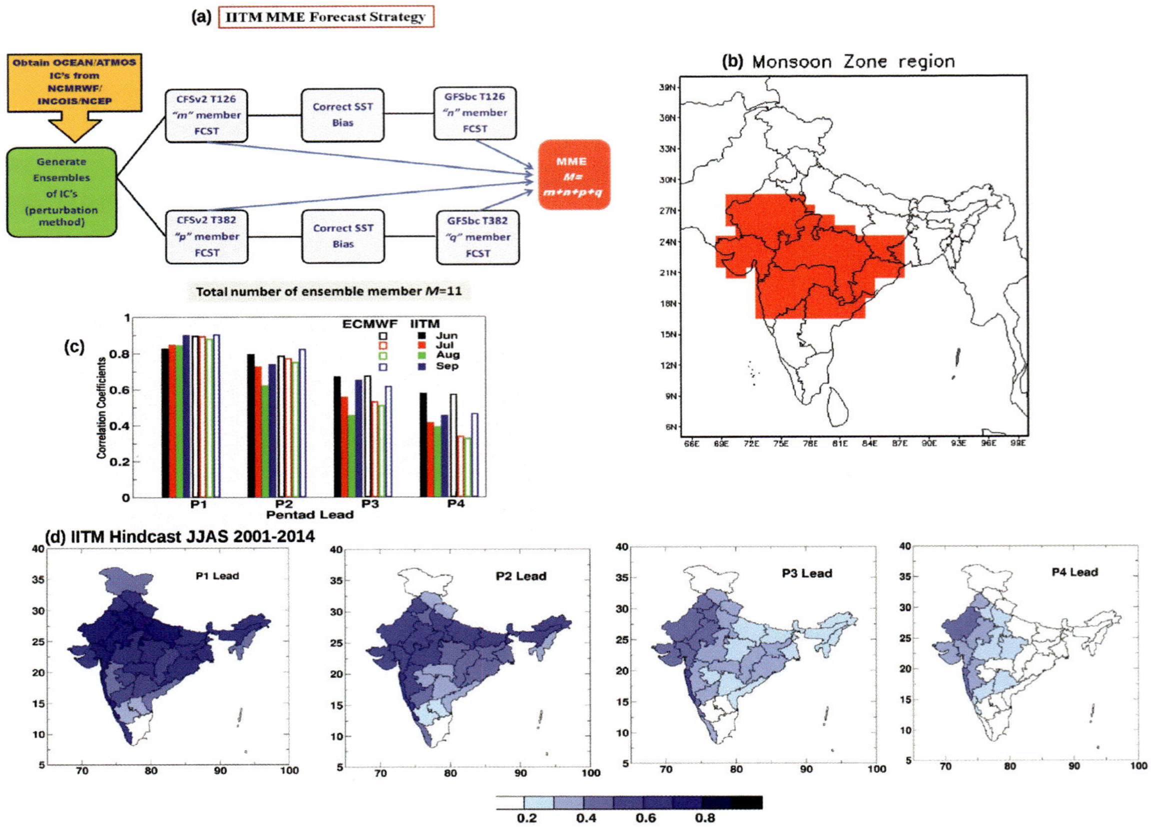

Fig. 4. (a) Schematic diagram of IMD-IITM extended range prediction system. (b) Monsoon zone of India (MZI). (c) Month-wise (June–September) correlation skill of predictions of monsoon rainfall over the MZI, given in (b) based on the IITM and the ECMWF extended range forecast system; (d) skill (anomaly correlation) of IITM extended range forecast system for the IMD defined subdivisions over India during June–September (JJAS) for the P1–P4 pentad lead time (left to right).

India (MZI, Fig. 4b) and the rainfall forecast over this region is given at extended range lead time with a pentad or 5-day averaging of daily forecasts. Forecasts of pentad 1 (P1, an average of lead day 1 to lead day 5) to pentad 4 (P4, an average of lead day 16 to lead day 20) are used for the evaluations of the current generation IITM extended range prediction system. Figure 4 shows the monthly prediction skill of the state-of-the-art IMD operational model as well as a comparison with the ECMWF operational extended range forecast based on S2S hindcasts (http://s2sprediction.net/). Figure 4d summarizes the correlation skill for the JJAS period for the meteorological subdivisions of India. For the MZI, the P3 and P4 forecasts have a minimum correlation skill of 0.4 (Fig. 4c), with higher skill during June and September in comparison to July and August. The decrease of skill during the peak season, especially over eastern and northern peninsular India (Fig. 2), may be related to the increased impact of very wet days (Stephenson *et al.* 1999; Moron *et al.* 2017) as a result of increased convective activity associated with the monsoon synoptic scale systems.

The skill of the IITM extended range forecasts for the 28 Indian subdivisions is shown in Fig. 4d. It may be seen that most of the subdivisions show a correlation skill of 0.4 and above until the third pentad. This means that the prediction of sub-seasonal variability over the Indian region is operationally skillful up to 15–20 days in advance. The comparison of IITM forecasts with ECMWF forecasts (not shown) shows that the skill is statistically comparable in both the models.

5. MISO in Tropical Australia

Tropical Australia exhibits a classical monsoonal regime with rainfall concentrated between November and March. Located close to the Maritime Continent and Equatorial Pacific, the broad-scale Australian monsoon is directly influenced by the MJO and ENSO on sub-seasonal to seasonal timescales. The role of the MJO in driving active and break episodes of the Australian monsoon is assessed in DJF from 1981 to 2010 using an index defined as the daily rainfall anomaly area-averaged over all Australian land points equatorward of 25° S. Active monsoon days are defined to occur when the index values are greater than or equal to one daily standard deviation (sd), calculated using all days in DJF. Monsoon break days are similarly defined when index values are less than or equal to −1 sd. To complement behavior captured in the rainfall index, a wind index is defined as the area-averaged 850-hPa zonal wind (u850) anomaly over the box region 110–125° E and 0–10° S, which captures the low-level zonal wind signature of active (anomalous westerly) and break (anomalous easterly) episodes of the monsoon (Hendon and Liebmann 1990). The predictive skill of active and break monsoon episodes is assessed at multi-week lead times based on a 30-year retrospective forecast dataset using Predictive Ocean Atmosphere Model for Australia (POAMA, Alves *et al.* 2003). This dataset comprises a 33-member ensemble of forecasts of nine-month duration starting on the 1st, 6th, 11th, 16th, 21st, and 26th day of each month.

Figure 5 shows composite rainfall, u850, and MSLP anomalies for all days when the monsoon rainfall index exceeds 1 standard deviation, for observations (upper left) and POAMA forecasts at 2–3 weeks of lead time (long enough away from the initial condition so that model behavior is captured; upper right). Active monsoon episodes are associated with widespread rainfall anomalies over northern Australia that extend eastward into the Coral Sea, a negative MSLP anomaly (i.e., enhanced monsoon trough) centered on the northwest Australian coast, and broadscale westerlies to the northwest, and also suppressed rainfall and easterly anomalies farther west in the Indian Ocean and farther east in the Central Pacific. The spatial distributions of rainfall, u850, and MSLP anomalies for active episodes are well depicted by POAMA, although the peak strength is somewhat exaggerated (13.4 mm $d{-}1$ for rainfall and −3.2 hPa for MSLP) compared to the observations (8.9 mm $d{-}1$ for rainfall and −2.6 hPa for MSLP).

Active episodes show a ~5-fold increase rate of occurrence in MJO phases 5–7 compared to MJO phases 8–1–2 in both observations and POAMA (Fig. 5, middle panel), consistent with a previous study of modulation of northern Australian summer rainfall demonstrated by the MJO (e.g., Wheeler *et al.* 2009). We also see a similar ~4–5-fold increased occurrence of break episodes when the MJO is in phases 8–1–2 (not shown). The forecast model fully represents this modulation by the MJO, thus providing motivation for further investigation of the multi-week predictive capability.

Correlation skill for predicting the weekly mean monsoon rainfall and u850 indices is shown in Fig. 5 (lower panels) for lead times out to 6 weeks (black bars), using the ensemble mean POAMA-2 forecast whose mid-week target date verifies during DJF 1981–2010. Correlation coefficients between predicted and observed are 0.81 for rainfall and 0.89 for u850 in week 1, falling below 0.5 after week 2 and week 5,

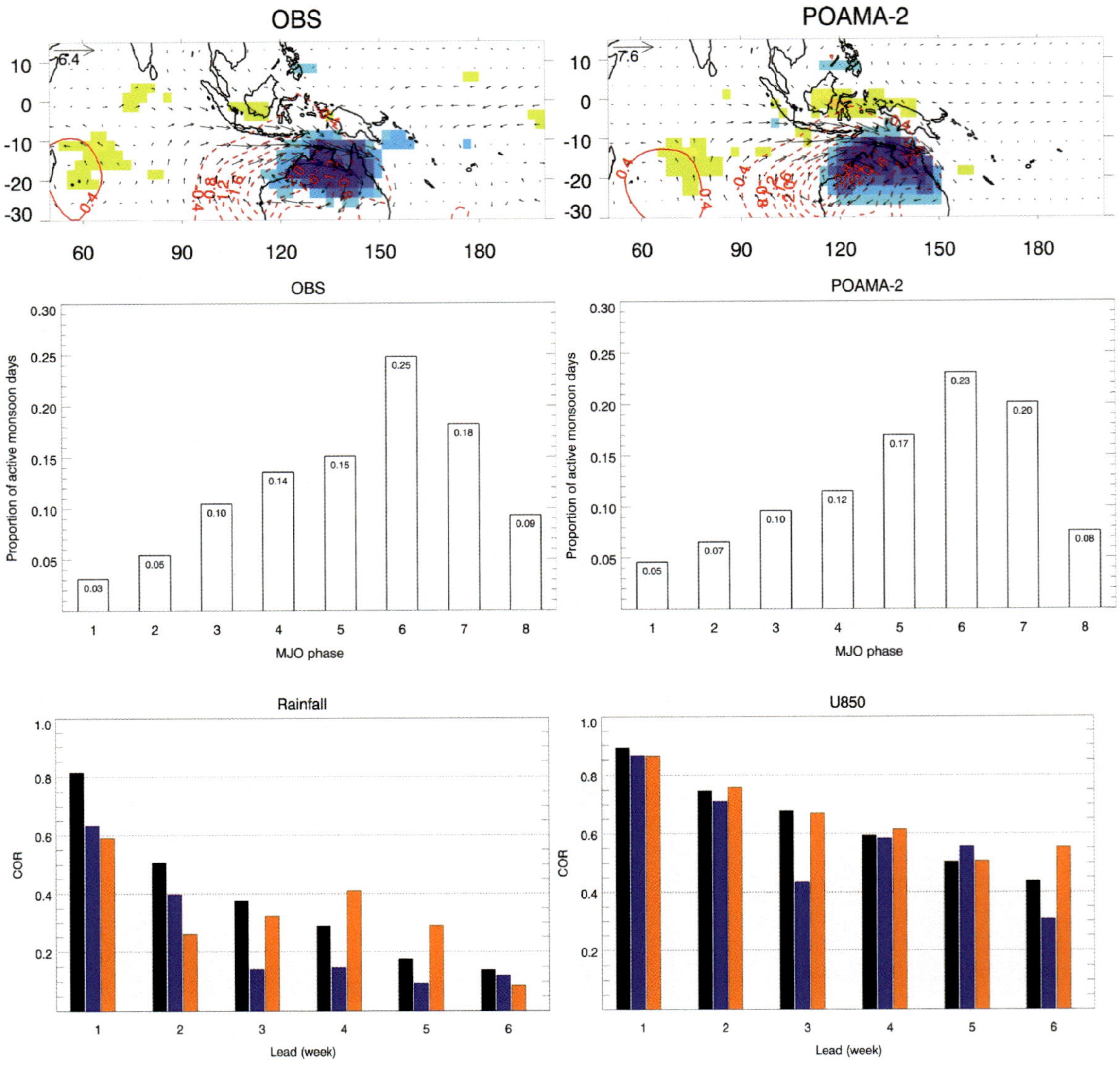

Fig. 5. Upper panels: composite rainfall (shading in mm/d), u850 (vector), and MSLP (contour in hPa) anomalies for all days when the monsoon rainfall index exceeds 1 standard deviation, for observations (upper left) and POAMA forecasts at 2–3 weeks of lead time (long enough away from the initial condition so that model behavior is captured; upper right); middle panels: proportion of active days in the MJO phase for observations (left) and POAMA forecasts (right); lower panels: skill (= correlation between observations and forecasts) of rainfall and atmospheric monsoon index for various lead times (in abscissa) and for all initializations (black bar); initialization in active (blue bar) and break (orange bar) periods.

respectively. Further analysis shows the predictability in monsoon rainfall after week 1 to be primarily provided by the MJO (Marshall and Hendon 2015). Higher skill for predicting u850 is expected since rainfall is inherently noisier and more difficult to predict than large-scale circulation.

This discrepancy in lead time suggests the potential for extending the useful prediction of monsoon rainfall to longer lead time if model errors associated with the prediction of the MJO and the local depiction of the MJO in the monsoon could be alleviated. For forecasts initialized only on monsoon active days (blue bars) and monsoon break days (orange bars), prediction skill for rainfall is reduced in the first two weeks, compared to using all forecasts. However, skill for forecasts initialized during break episodes recovers in week four, which presumably would coincide with an active phase associated with the next passage of the MJO (i.e., about half an MJO cycle later).

6. Concluding Remarks

Forecasting monsoon features, such as onset and demise dates, duration, amplitude, and active and break phases, represents an important challenge due to the societal applications associated with it, especially in agriculture and hydropower generation. This chapter summarized some current efforts to predict monsoon features on subseasonal timescales.

Spatially coherent rainfall anomalies (from daily GPCP $1° \times 1°$) are significantly smaller across the continents than across the oceans. They increase, as expected, as the timescale grows. Small spatially coherent rainfall anomalies across some of the continents (such as equatorial and northern tropical Africa) suggest that atmospheric phenomena able to efficiently synchronize anomalies do not have a strong impact on rainfall variability beyond the diurnal cycle and the synoptic timescale. Thus, potential predictability is strongly limited at intraseasonal timescales (i.e., in the 7–20 and 20–90-day timescales). It also seems that the spatial coherence usually peaks around the onset or withdrawal stages of the regional-scale monsoons across most of the landmasses (Moron *et al.* 2019) rather than near the monsoon peak stage (when the highest rainfall occurs, and thus strongly impacts the interannual variations of the seasonal amount). For example, the temporal modulation fits with the higher hindcast skill observed in June and September than in July and August across India (Fig. 4b). This modulation may be due to many different processes. First of all, the spatial coherence of northern tropical monsoons at the end of boreal summer/early autumn may be increased by the increasing amplitude of ENSO events. It can also be due to varying land–atmosphere coupling across the rainy season, the progressive moistening of surfaces and sub-surfaces (Moron *et al.* 2017; Vigaud *et al.* 2017; Halder *et al.* 2018), or the internal atmospheric dynamics due to intense convective activities associated with monsoon low-pressure systems which can contribute to the larger and faster error growth. We can hypothesize that intense rainfall due to small scales of motion, allowed by a large amount of precipitable water and favorable vertical and horizontal tropospheric structures during the peak of the season, tends to fragment the latent heat that drives the regional and large-scale monsoon (Slingo *et al.* 2003). The spatial incoherence of rainfall anomalies may then be induced by the strength and the preeminence of the smaller scales of atmospheric motions. Again, the modulation of spatial coherence observed across India (i.e., larger DS over NW India, Fig. 1, right column) fits rather well with the highest skill observed around Rajasthan and NW India in general (Fig. 4d). This relatively large spatial coherence and skill may be attributed to the large fraction of variance conveyed by intraseasonal modes of variation (Fig. 1, left column). The higher skill in this region is also noted from statistical studies (e.g., Fig. 9 of Xavier and Goswami 2007). There are less day-to-day convective activities in this region. With high ISO variance (as shown here) and less convective activity, the signal-to-noise ratio is better and the forecast models can capture the rainfall in a better way in this region. On the contrary, the rainfall observed over the Gangetic Plain and NE India may be less predictable, especially during the peak stage of the monsoon (Fig. 2b).

Bombardi *et al.* (2017) achieved forecast skill for the timing of the rainy season on sub-seasonal timescales over some monsoonal regions. However, more research is needed to further evaluate the predictability of the timing of the rainy season and its sources of forecast skill. For instance, analysis of the signal-to-noise ratio of forecasts relating to the timing of the rainy season would strengthen our confidence in these forecasts.

Prediction of active/break phases of the Australian and Indian summer monsoons is limited to less than 3 weeks of lead time with current operational S2S prediction systems. This is shorter than the limit of prediction of the MJO/MISO (Neena *et al.* 2014; Woolnough 2019), suggesting the potential for improved predictions if errors in the depiction of regional impacts of the MJO/MISO are alleviated. It is found that the improved representation of sea surface temperature is a key to improve the MISO in India in operational runs (Sharmila *et al.* 2013). Also, multiple available options (choices) of physical parameterizations and horizontal resolutions on a dynamical core improve the spread–error relation for probabilistic MISO forecast in the ensemble mode. Other sources of predictability should also be explored including improved initialization of land surface conditions and shallow upper ocean anomalies (e.g. resulting from salinity stratification), which may be especially sensitive to and potentially feed back on intraseasonal variability that drives monsoon variations.

References

Alessandri, A., A. Borrelli, A. Cherchi, *et al.*, 2015: Prediction of Indian summer monsoon onset using dynamical sub-seasonal forecasts: effects of realistic initialization of the atmosphere. *Mon. Weather Rev.*, **143**, 778–793.

Alves, O., G. Wang, A. Zhong, *et al.*, 2003: POAMA: Bureau of meteorology operational coupled model forecast system. In *Proceedings of National Drought Forum*, Brisbane, April 2003, pp. 49–56, Available from DPI Publications, Department of Primary Industries, GPO Box 46, Brisbane, Qld 4001, Australia.

Abhilash, S., A. K. Sahai, N. Borah, *et al.*, 2014a: Does bias correction in the forecasted SST improve the extended range prediction skill of active-break spells of Indian summer monsoon rainfall? *Atmos. Sci. Lett.*, **15**, 114–119.

Abhilash, S., A. K. Sahai, S. Pattnaik, *et al.*, 2014b: Extended range prediction of active-break spells of Indian summer monsoon rainfall using an ensemble prediction system in NCEP Climate Forecast System. *Int. J. Climatol.*, **34**, 98–113.

Biasutti, M. and Coauthors, 2018: Global energetics and local physics as drivers of past, present and future monsoons. *Nat. Geosci.*, **11**, 392–400.

Bombardi, R. J. and L. M. V. Carvalho, 2009: IPCC global coupled model simulations of the South American monsoon system. *Clim. Dyn.*, **33**, 893–916.

Bombardi, R. J., K. V. Pegion, J. L. Kinter, B. Cash, and J. M. Adams, 2017: Sub-seasonal predictability of the onset and demise of the rainy season over monsoonal regions. *Front. Earth Sci.*, doi:10.3389/feart.2017.00014.

Chattopadhyay, R., A. K. Sahai, and B. N. Goswami, 2008: Objective identification of nonlinear convectively coupled phases of monsoon intraseasonal oscillation: Implications for prediction. *J. Atmos. Sci.*, **65**, 1549–1569.

Drosdowsky, W. and M. C. Wheeler, 2017: Extended-range ensemble predictions of convection in the North Australian monsoon region. *Frontiers Earth Sci.*, doi:10.3389/feart.2017.00028.

Gadgil, S., 2003: The Indian monsoon and its variability. *Annu. Rev. Earth Planet. Sci.*, **31**, 429–467.

Halder, S., P. A. Dirmeyer, L. Marx, and J. L Kinter III, 2018: Impact of land-surface initialization and land-atmosphere coupling on the prediction of the Indian summer monsoon with the CFSv2. *Frontiers Env. Sci.*, https://doi.org/10.3389/fenvs.2017.00092.

Hendon, H. H. and B. Liebmann, 1990: The intraseasonal (30–50 day) oscillation of the Australian summer monsoon. *J. Atmos. Sci.*, **47**, 2909–2924.

Kiladis, G. N., M. Wheeler, P. T. Haertel, K. H. Straub, and P. E. Roundy, 2009: Convectively coupled equatorial waves. *Rev. Geophys.*, **47**, RG2003.

Liebmann, B. and J. A. Marengo, 2001: Interannual variability of the rainy season and rainfall in the Brazilian Amazon basin. *J. Clim.*, **14**, 4308–4318.

Lubis, S. W. and C. Jacob, 2015: The modulating influence of convectively coupled equatorial waves (CCEWs) on the variability of tropical precipitation. *Int. J. Climatol.*, **35**, 1465–1483.

Macron, C., B. Pohl, Y. Richard, and M. Bessafi, 2014: How do tropical temperate troughs form and develop over Southern Africa? *J. Clim.*, **27**, 1635–1647.

Madden, R. A. and P. R. Julian, 1994: Observations of the 40–50 day tropical oscillation — A review. *Mon. Wea. Rev.*, **122**, 814–837.

Marshall, A. G. and H. H. Hendon, 2015: Subseasonal prediction of Australian summer monsoon anomalies. *Geophys. Res. Lett.*, **42**, 10913–10919.

Mohan, T. S., H. Annamalai, L. Marx, *et al.*, 2018: Representation of ocean-atmosphere processes associated with extended monsoon episodes over South Asia in CFSv2. *Frontiers Earth Sci.*, 10.3389/feart.2018.00009.

Moron, V., A. W. Robertson, M. N. Ward, and P. Camberlin, 2007: Spatial coherence of tropical rainfall at the regional scale. *J. Clim.*, **20**, 5244–5263.

Moron, V., J. Boyard-Micheau, P. Camberlin, *et al.*, 2015: Ethnographic context and spatial coherence of climate indicators for farming communities — A multi-regional comparative assessment. *Clim. Risk Manag.*, **8**, 28–46.

Moron, V., A. W. Robertson, and S. Pai, 2017: On the spatial coherence of sub-seasonal to seasonal Indian rainfall anomalies. *Clim. Dynam.*, **49**, 3403–3423.

Moron, V., A. W. Robertson, and L. Wang, 2019: Weather within climate: Sub-seasonal predictability of daily rainfall characteristics of tropical daily rainfall characteristics. In *Subseasonal to Seasonal Prediction: The Gap between Weather and Climate Prediction*, Eds. A. W Robertson and F. Vitart, Amsterdam, Elsevier, pp. 47–64.

Neena, J. M., J. Y. Lee, D. Waliser, B. Wang, and X. Jiang, 2014: Predictability of the Madden–Julian oscillation in the intraseasonal variability hindcast experiment (ISVHE). *J. Clim.*, **27**, 4531–4543.

Ricciardulli, L. and P. Sardeshmukh, 2002: Local time- and space scales of organized tropical deep convection. *J. Clim.*, **15**, 2775–2790.

Robertson, A. W., A. Kumar, M. Peña, and F. Vitart, 2015: Improving and promoting subseasonal to seasonal prediction. *Bull. Amer. Meteo. Soc.*, **96**, ES49–ES53.

Robertson, A. W. and F. Vitart (Eds), 2019: *Subseasonal to Seasonal Prediction: The Gap between Weather and Climate Prediction*, Amsterdam, Elsevier, 588pp.

Saha, S., S. Moorthi, *et al.*, 2014: The NCEP climate forecast system version 2. *J. Clim.*, **27**, 2185–2208.

Sharmila, S., P. A. Pillai, S. Joseph, M. Roxy, R. P. M. Krishna, R. Chattopadhyay, S. Abilash, A. K. Sakai, and B. N. Goswami, 2013: Role of ocean–atmosphere interaction on northward propagation of Indian summer monsoon intra-seasonal oscillations (MISO). *Clim. Dynam.*, **41**, 1651–1669.

Stephenson, D. B., K. R. Kumar, F. J. Doblas-Reyes, J. Royer, F. Chauvin, and S. Pezzulli, 1999: Extreme daily rainfall events and their impact on ensemble forecasts of the Indian monsoon. *Mon. Wea. Rev.*, **127**, 1954–1966.

Suhas, E., J. M. Neena, and B. N. Goswami, 2013: An Indian monsoon intraseasonal oscillations (MISO) index for real time monitoring and forecast verification. *Clim. Dynam.*, **40**, 2605–2616.

Vellinga, M., A. Arribas, and R. Graham, 2013: Seasonal forecasts for regional onset of the West African monsoon. *Clim. Dynam.*, **40**, 3047–3070.

Vigaud N., A. W. Robertson, M. K. Tippett, and N. Acharya, 2017: Subseasonal predictability of boreal summer monsoon rainfall from ensemble forecasts. *Frontiers Env. Sci.*, **5**, doi: 10.3389/fenvs.2017.00067.

Vitart, F., C. Ardilouze, *et al.*, 2017: The subseasonal to seasonal (S2S) prediction project database. Bull. *Amer. Meteo. Soc.*, **98**, 163–173.

Wheeler, M. C. and G. N. Kiladis, 1999: Convectively coupled equatorial waves: analysis of clouds and temperature in the wavenumber-frequency domain. *J. Atmos. Sci.*, **56**, 374–399.

Wheeler, M. C., H. H. Hendon, *et al.*, 2009: Impacts of the Madden-Julian oscillation on Australian rainfall and circulation. *J. Clim.*, **22**, 1482–1498.

Wheeler, M. C. and J. L. McBride, 2011: Australasian monsoon. In *Intraseasonal Variability in the Atmosphere-Ocean Climate System*, 2nd Edn, Eds. W. K. M. Lau and D. E. Waliser, Springer, Berlin, pp. 147–198.

Wilks, D. S. 2006: *Statistical Methods in the Atmospheric Sciences*. Burlington, MA: Academic Press.

Woolnough, S. J., 2019: The Madden-Julian oscillation. *Subseasonal to Seasonal Prediction: The Gap between Weather and Climate Prediction*, Eds. A. W. Robertson and F. Vitart, Amsterdam, Elsevier, pp. 93–117.

Xavier, P. K. and B. N. Goswami, 2007: An analog method for real-time forecasting of summer monsoon subseasonal variability. *Mon. Wea. Rev.*, **135**, 4149–4160.

Zhu, H., M. C. Wheeler, A. H. Sobel, and D. Hudson, 2014: Seamless precipitation prediction skill in the tropics and extratropics using a global model. *Mon. Wea. Rev.*, **147**, 1556–1569.

Part IV

Climate Change

Chapter 27

Future Changes in Global Monsoon Precipitation and Their Uncertainty: Results from 20-km and 60-km MRI-AGCM Ensemble Simulations

Akio Kitoh*,†,‡ and Hirokazu Endo†

**Research Promotion Department,*
Japan Meteorological Business Support Center,
Tsukuba, Ibaraki, 305-0052, Japan
†Climate and Geochemistry Research Department, Meteorological Research Institute,
Tsukuba, Ibaraki, 305-0052, Japan
‡kitoh@mri-jma.go.jp

We study future changes of mean and extreme precipitations and their uncertainty, based on ensemble simulations with 20-km mesh and 60-km mesh global atmospheric general circulation models (AGCMs). In present-day climate realizations, which are forced by observed sea surface temperature (SST), the AGCM shows high performance in simulating mean and extreme precipitations, such as the annual maximum 1-day precipitation total (Rx1d), with some dependency on the choice of convection schemes. In a warmer climate, mean precipitation is projected to increase in most land monsoon regions especially in East Asia. Extreme precipitation is projected to increase with higher rates than mean precipitation except some regions around the western tropical Pacific. Uncertainty in tropical land precipitation changes originates mainly from differences in convection schemes, contrasting with the tropical oceanic precipitation uncertainty, which is mainly influenced by differences in future SST warming patterns. A significant decrease of tropical cyclone frequency in future climate over the western tropical Pacific results in little changes in Rx1d there, but their interannual variability will increase.

1. Introduction

Projections by coupled atmosphere-ocean general circulation models (AOGCMs) in the Coupled Model Intercomparison Project phase 5 (CMIP5) show that monsoon precipitation is likely to strengthen in the 21st century on the global scale, while monsoon circulation weakens (Christensen *et al.* 2013). The projected increase of global monsoon precipitation is attributed to enhanced moisture flux convergence and increased surface evaporation, which are partly offset by weakening of the monsoon circulation (Christensen *et al.* 2013; Hsu *et al.* 2013; Kitoh *et al.* 2013). However, contributions from dynamical responses differ from region to region (Endo and Kitoh 2014). For example, intensification of the subtropical North Pacific anticyclone contributes to increasing moisture transport over some parts of East Asia (Seo *et al.* 2013; Wang *et al.* 2014; Lee and Wang, 2014).

From regional perspectives, monsoon precipitation is projected to remarkably increase in the East and South Asian monsoon regions, while changes in other monsoon regions are rather modest (Christensen *et al.* 2013; Kitoh *et al.* 2013; Endo and Kitoh 2014; Lee and Wang 2014). However, large uncertainty remains in regional precipitation projections. Also, the CMIP5 models have limited capability in studies of regional precipitation changes, especially for extremes, due to their low horizontal resolution

The Multiscale Global Monsoon System
Edited by C. P. Chang et al.

(approximately 100–400 km) as well as systematic biases in their present-day climatology, largely due to their sea surface temperature (SST) biases (He and Soden 2016; Kitoh *et al.* 2016). To simulate weather and climate extremes appropriately, a high-resolution model is necessary. In this paper, we review present-day realization of monsoon precipitation and its future changes and uncertainty simulated by the Meteorological Research Institute (MRI) high-resolution atmospheric general circulation models (AGCMs) with grid sizes of 20 km and 60 km.

2. MRI-AGCM3.2

At the MRI in Japan, high-resolution AGCMs with grid sizes of 20 km and 60 km have been developed. Characteristics of simulated precipitation are strongly affected by the cumulus convection scheme. Currently, we have three options for cumulus convection schemes: the Yoshimura scheme (YS; Yoshimura *et al.* 2015), the prognostic Arakawa–Schubert scheme (AS; Randall and Pan 1993), and the Kain–Fritsch scheme (KF; Kain and Fritsch 1990). The standard version of the former model (MRI-AGCM3.1, Mizuta *et al.* 2006) was based on the AS scheme, while the current model is based on the YS scheme (MRI-AGCM3.2, Mizuta *et al.* 2012). These three cumulus convection schemes are used for multi-physics ensemble experiments.

High horizontal resolution is necessary to simulate weather and climate extremes such as tropical cyclones, precipitation extremes, terrain-induced high wind events, and so on (Kitoh *et al.* 2016). It is demonstrated that the high-resolution MRI-AGCMs perform well in reproducing climatology and extremes, and are superior to lower-resolution models, including tropical cyclones (Oouchi *et al.* 2006; Murakami *et al.* 2012a, b), precipitation extremes (Kamiguchi *et al.* 2006; Kitoh and Endo, 2016a, b; Endo *et al.* 2017), as well as the monsoon precipitation (Kusunoki *et al.* 2006; Kitoh and Kusunoki 2008; Wang *et al.* 2011; Hsu *et al.* 2012; Rajendran *et al.* 2012; Krishnan *et al.* 2013; Kusunoki 2016, 2018).

3. Present-day Simulation

3.1. *20-km mesh MRI-AGCM3.2*

First, we evaluate present-day precipitation simulated by the 20-km mesh MRI-AGCM3.2. The boundary conditions used are the observed interannually varying monthly mean SST and sea ice during 1979–2003. The model has reproduced well the spatial pattern of annual mean precipitation. Seasonal migration of monsoon precipitation zones is also well simulated (Mizuta *et al.* 2012).

Kitoh and Endo (2016a) evaluated five precipitation indices with two observed datasets, 0.25° TRMM-3B42 (Huffman *et al.* 2007) and 1° GPCP-1DD (Huffman and Bolvin 2009), both averaged for 1998–2013. The indices they used are average precipitation (Pav), simple precipitation daily intensity index (SDII), annual maximum 5-day precipitation total (Rx5d), annual maximum 1-day precipitation total (Rx1d), and annual maximum consecutive dry days (CDD). Compared to TRMM data, the simulated Pav is overestimated and SDII is underestimated. On the contrary, the magnitude of Rx5d and Rx1d averaged between 50°S and 50°N is very close to the TRMM data. Here, we show Rx1d simulated by the 20-km and 60-km mesh MRI-AGCM3.2, compared with the TRMM data in Fig. 1. Large Rx1d is observed over the western North Pacific and the Bay of Bengal. The model reproduces the observed characteristics in their spatial distribution and magnitude well, particularly in the 20-km mesh model. The 20-km model, however, underestimated the Rx1d magnitude in the North Atlantic, probably related to poor representation of tropical cyclones there (Kitoh and

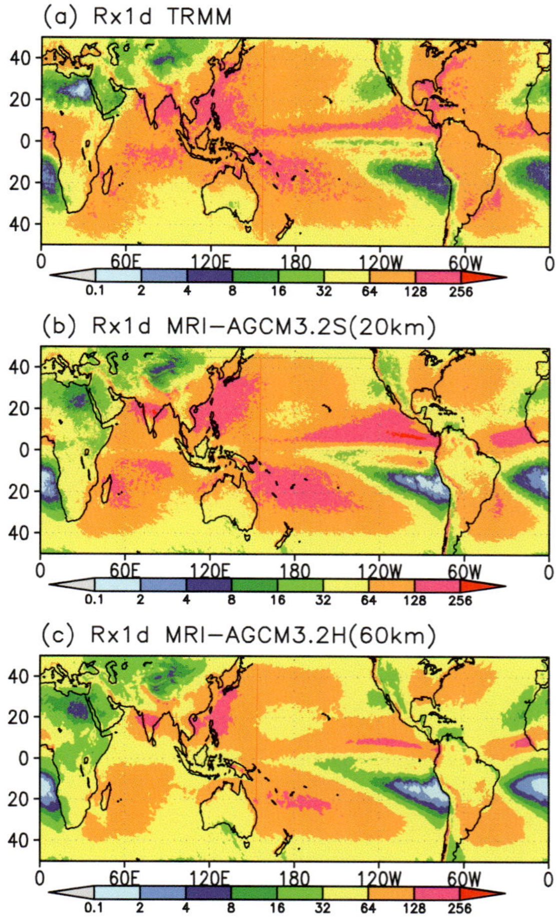

Fig. 1. Rx1d climatology (mm). (a) TRMM-3B42. (b) 20-km mesh MRI-AGCM3.2. (c) 60-km mesh MRI-AGCM3.2.

Endo 2016a). The 60-km model generally underestimates the magnitude of Rx1d about 20% compared to the 20-km version.

Kitoh and Endo (2016a) also compared regional averages of five observed and simulated precipitation indices for 22 regional domains over land. The simulated Pav is within 20% of the observations in tropical and mid-latitude wet regions. Regionally averaged Rx5d and Rx1d are within observed uncertainty, i.e., between TRMM-3B42 and GPCP-1DD. Both Rx5d and Rx1d are largest in Southeast Asia and South Asia in agreement with the observations.

3.2. *60-km mesh MRI-AGCM3.2*

Precipitation simulated by the 60-km mesh MRI-AGCM3.2 was evaluated by Endo *et al.* (2012) for Asia, and by Kusunoki and Mizuta (2013) and Kusunoki (2016) for East Asia. Three convection schemes (YS, AS, and KF) are used. The models reproduce the spatial pattern and the amount of mean and extreme precipitations very well, although the model with the AS scheme underestimated the magnitude of extreme precipitation. The skill of the 20-km and 60-km mesh MRI-AGCM3.2 is higher or comparable to that of the CMIP5 models in reproducing the seasonal average of the precipitation pattern, the seasonal march of the rainy zone, and extreme precipitation events in East Asia. Among the three convection schemes, the KF scheme was the best for this particular region. Among the CMIP5 models, a significant negative correlation is found between grid spacing and skill; the higher the resolution, the better the skill.

3.3. *Monsoon precipitation*

Figure 2 shows the monsoon precipitation intensity (MPI) defined as the annual range of precipitation (difference between May–September (MJJAS) and November–March (NDJFM) mean) divided by the annual mean (Wang *et al.* 2011) for MRI–AGCMs, CMIP5 AGCMs (AMIP experiment), and CMIP5 AOGCMs (historical experiment). The MRI–AGCMs show generally greater performance in simulating global-scale precipitation distribution, here in MPI and also in other metrics (not shown), than the CMIP5 models. Compared to the CMIP5 AOGCMs, the CMIP5 AGCMs show generally better performance. This is probably due to systematic SST biases in AOGCMs, while AGCMs are forced by an observed SST.

Regional monsoon precipitation is assessed in terms of spatial distribution. The regional domains are shown in the upper panel in Fig. 3.

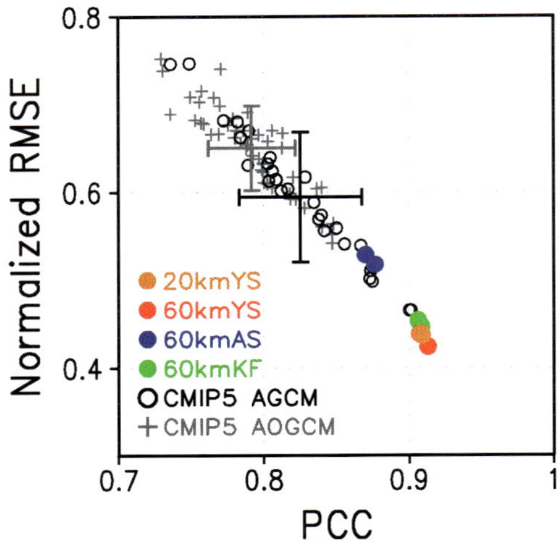

Fig. 2. Monsoon precipitation intensity defined as the annual range of precipitation divided by the annual mean for the period 1984–2003. The horizontal axis is the spatial pattern correlation coefficient (PCC) with the observation (the average) of GPCP (Huffman *et al.* 2009) and CMAP (Xie and Arkin 1997). The vertical axis is the root mean square error (RMSE) with the observation (normalized by the observed spatial standard deviation). Colored closed circles show MRI-AGCM simulations with different convection schemes. There are two experiments with different initial conditions for each model. Open black circles (gray crosses) show CMIP5 AGCM (AOGCM) simulations, for which the multi-model mean and the inter-model standard deviation are also shown. All calculations are made for common grid boxes with 2.5° horizontal resolution over 0°E–360°E, 45°S–45°N.

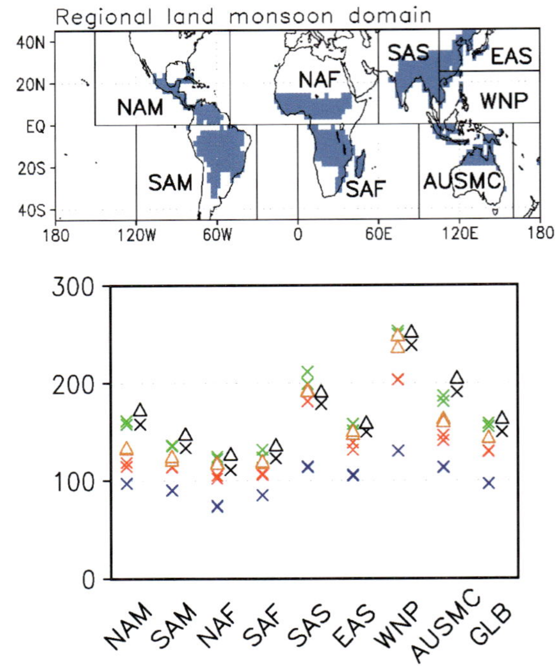

Fig. 3. Validation for Rx1d (mm) over the regional land monsoon domains. Orange triangles denote the 20-km MRI-AGCM3.2 simulations. Colored crosses denote the 60-km MRI-AGCM3.2 simulations (red, blue, and green denote the model with the YS, AS, and KF convection schemes, respectively). Black triangles (crosses) denote TRMM observations with the 0.25° (TL319 ~60 km) spatial resolution. The top panel shows the land monsoon domain where the annual range of precipitation exceeds 2.5 mm day^{-1} (Wang *et al.* 2011) and its regional divisions: NAM (North America), SAM (South America), NAF (North Africa), SAF (South Africa), SAS (South Asia), EAS (East Asia), WNP (Western North Pacific), AUSMC (Australia and maritime continent), and GLB (global). Note that GLB denotes the global land monsoon domain within 45°S–45°N.

The simulations by MRI–AGCM demonstrate generally better performance compared to those of CMIP5 models in all domains, with relatively lower skill of MRI–AGCM incorporating the AS scheme (not shown). Most models including MRI–AGCMs poorly simulate the EAS monsoon domain (not shown). This defect has been pointed out by Wang *et al.* (2011) and Kitoh *et al.* (2013). Wang *et al.* (2011) suggested that this failure could be attributed to the unique geographical background of the East Asian monsoon, such as prominent east–west land-ocean thermal contrast and the existence of the Tibetan Plateau.

Figure 3 compares regional averages of simulated and observed Rx1d for the regional and global monsoon domains over land. It is revealed that the simulations for Rx1d by the 20-km MRI–AGCM agree with the observation in the Asian monsoon region (SAS, EAS, and WNP). For the 60-km MRI–AGCM, the KF and YS models reproduce the observed magnitudes of Rx1d in the Asian monsoon region as well as those of Rx5d in most regions (not shown). On the contrary, the AS model underestimates the observations in all the monsoon regions, suggesting a strong dependence of extreme precipitation on the choice of convection schemes.

4. Future Changes in Precipitation

Using the 20-km and 60-km mesh MRI–AGCMs, we have conducted several time-slice experiments to project future climate at the end of the 21st century under several scenarios (Kitoh *et al.* 2009, 2016; Mizuta *et al.* 2017). For the future climate simulations, the boundary SST data were prepared by superposing the future changes in the multi-model ensemble or sub-ensemble of SSTs projected by CMIP3 or CMIP5 dataset to the present-day observed SST (Mizuta *et al.* 2008, 2014).

4.1. *Multi-SST experiments with 20-km mesh MRI-AGCM3.2*

Kitoh and Endo (2016a) investigated changes in precipitation extremes at the end of the 21st century. The CMIP5 multi-model SST change patterns under the RCP8.5 scenario are added to the present-day SST distribution: one with the mean SST change by 28 CMIP5 models, and the other three obtained by a cluster analysis (Mizuta *et al.* 2014). The same interannual variations in SST are assumed in the future as in the present.

Figure 4a shows projected future changes in Rx1d by the 20-km mesh MRI-AGCM3.2. The spatial pattern of Rx1d changes is similar to that of the annual mean precipitation, but the area with decreasing precipitation becomes smaller. Globally averaged change ratio, normalized by the global annual mean surface air temperature change, is found to become larger in Rx5d (5.36%/°C) and more in Rx1d (7.36%/°C) than that in Pav (2.22%/°C). Considering four ensemble members, it is found that the heavy precipitation extremes (Rx5d and Rx1d) will increase in all 22 land regional domains, even where Pav will decrease. South Asia is the domain of the largest extreme precipitation increase, with West Africa the second and Southeast Asia the third. It is also found that a different future SST change pattern changes large-scale atmospheric

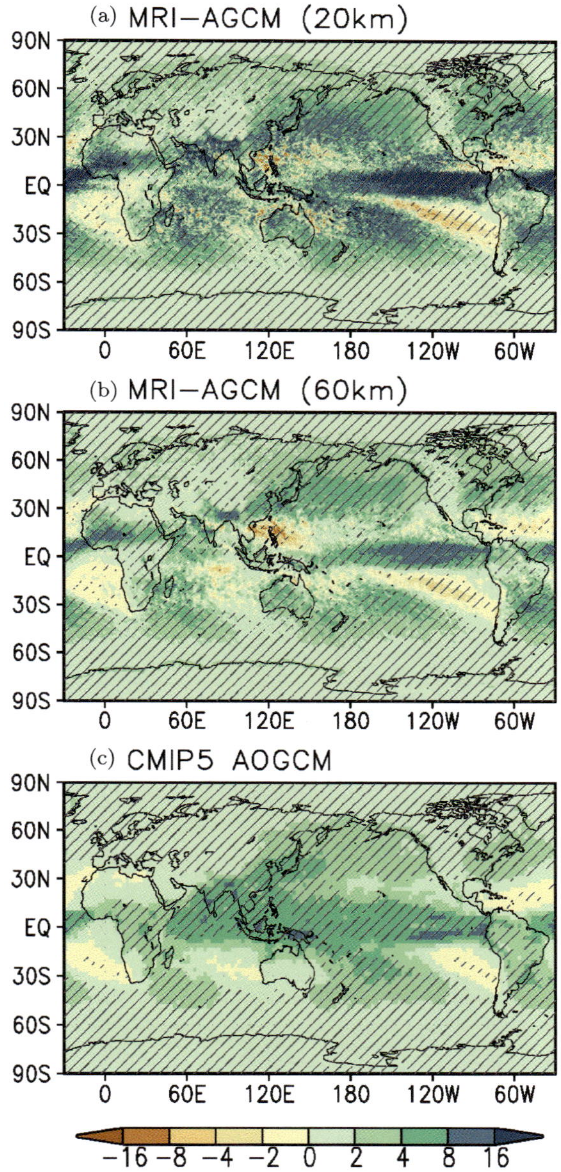

Fig. 4. Future changes in Rx1d (mm/K) by (a) 20-km MRI-AGCM3.2 (4 members), (b) 60-km MRI-AGCM3.2 (12 members), and (c) CMIP5 AOGCMs (31 models). Hatching indicates regions where more than 80% of members or models have the same sign for the change. The changes are scaled by a globally averaged surface air temperature change for each.

circulation, and then affects precipitation changes.

Kitoh and Endo (2016b) investigated changes in Rx1d associated with tropical cyclones (TCs). Precipitation within 500 km from the TC center is defined as TC-associated precipitation. Figure 5 shows the simulated TC-associated precipitation in the present and changes in the future projected by the 20-km mesh MRI-AGCM3.2. The model reproduced the observed one. It is also shown that a decrease in Rx1d in the western North Pacific is mainly due to a projected decrease in tropical cyclone frequency in the future. Further, interannual variability associated with El Niño was investigated. Mean Rx1d will increase little in the western North Pacific in the future, but its interannual variability will largely increase due to enhanced association with El Niño even without changes in El Niño intensity.

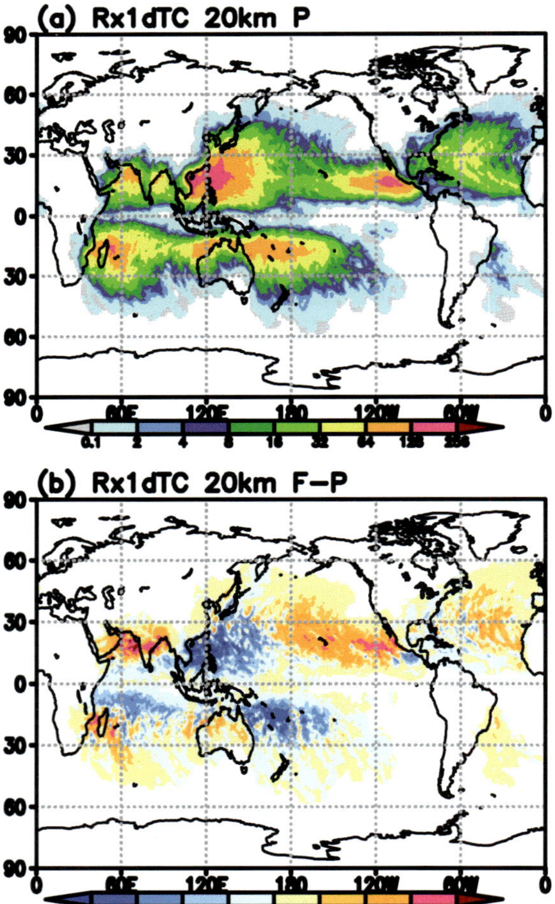

Fig. 5. (a) Present-day simulation of Rx1d (mm) associated with tropical cyclones by 20-km mesh MRI-AGCM3.2. (b) Their changes from the present (1979–2003) to the future (2075–2099).

4.2. *Large ensemble experiments with 60-km mesh MRI-AGCM3.2*

A large ensemble future climate simulation by 60-km MRI-AGCM3.2 was performed under the "Database for Policy Decision Making for Future Climate Change (d4PDF)" (Mizuta *et al.* 2017). The present climate for the period 1951–2010 is simulated with 100 members (total 6,000 years), while for the future climate a 60-year integration with 90 members (total 5,400 years) is conducted corresponding to conditions around the 2090s under the RCP8.5 scenario. This corresponds to 4.1 K warmer global surface air temperature than the pre-Industrial level or 3.6 K warmer than the present climate (1951–2010).

The Rx1d is projected to increase in the future warmer world, with a larger increase over the regions where it is climatologically large. The exception is the western North Pacific where there is only a small change or an even reduction of Rx1d is projected. This is mainly due to a projected decrease of tropical cyclone frequency in a future climate in this region. This result is consistent with the 20-km model experiment (Kitoh and Endo 2016b). Moreover, 90 or 99 percentile values of Rx1d will increase over most of the world (Kitoh and Endo, 2019), implying an increasing risk of heavier rainfall events by global warming.

Endo *et al.* (2017) examined future changes in precipitation extremes in East Asia and showed that Rx1d is projected to increase throughout East Asia with robustness and a larger rate in comparison with Rx5d and Pav.

In particular, the increase of Rx1d exceeds 30% over eastern China and northern Japan. Moisture transport from the western North Pacific and the South China Sea toward East Asia is projected to enhance due to an increase in atmospheric moisture and an intensification of the western North Pacific subtropical high (WNPSH).

4.3. *Multi-SST and multi-physics experiments with 60-km mesh MRI-AGCM3.2*

Endo *et al.* (2012) assessed future changes in mean and extreme precipitations in Asia and their uncertainty by conducting 12-member ensemble projections with the 60-km mesh MRI-AGCM3.2 under the SRES A1B scenario. Three different cumulus convection schemes (YS, AS, and KF) and four different SST change patterns obtained from the CMIP3 coupled model projections were used. This experiment setup covers the uncertainty coming from future SST change patterns and the uncertainty due to different model physics. They showed that mean and extreme precipitations generally increase in South Asia and Southeast Asia, but with large uncertainty.

It is known that the projected SST change patterns by CMIP5 models differ from those by CMIP3 models, particularly in the tropics (Mizuta *et al.* 2008, 2014). Therefore, another multi-SST and multi-physics experiment was performed where three convection schemes (YS, AS, and KF schemes) were used with four different future SST change patterns derived from CMIP5 models under the RCP8.5 scenario (Mizuta *et al.* 2014). Here, we show some results from this multi-SST and multi-physics experiment.

Figure 4 shows future changes in Rx1d for the 20-km and 60-km MRI–AGCMs as well as the CMIP5 AOGCMs. Compared to Pav changes, larger areas show positive changes. Rx1d is increased over 84.8% (83.0%) of the 45°S–45°N domain for the 20-km (60-km) MRI-AGCM, while Pav is increased over 60.7% (58.7%) of the domain for the 20-km (60-km) MRI-AGCM. This distribution combines the "wet-gets-wetter" pattern (Held and Soden 2006; Chou *et al.* 2009) and the "warmer-gets-wetter" pattern (Xie *et al.* 2010; Chadwick *et al.* 2013). In addition, we find regional details of precipitation changes that are regulated by topography, such as in the southern foot of the Himalayas and Indochinese Peninsula, as pointed out by Endo *et al.* (2012).

In comparison with the CMIP5 AOGCM ensemble (Fig. 4c), we found some discrepancies in the future change on regional scales. Specifically, the MRI–AGCM projects a decreased precipitation over the northwestern tropical Pacific, while the CMIP5 AOGCMs project an increased precipitation. The precipitation anomaly over the northwestern tropical Pacific is strongly correlated with local circulation anomaly in the lower troposphere. Thus, the MRI–AGCM projects enhanced WNPSH, while the CMIP5 AOGCMs do not reach a consensus about whether WNPSH is enhanced or weakened, as suggested by He and Zhou (2015).

There are many possible factors that could cause the different precipitation responses between the MRI–AGCM and the CMIP5 AOGCMs over the northwestern tropical Pacific. First, model formulation is different: the MRI-AGCM ensemble simulations are conducted by using a single model, although the ensemble considers various types of convective schemes and underlying SST warming patterns. Second, the AGCM does not represent air-sea interaction process. Since the northwestern tropical Pacific is a typical region subject to air-sea interaction (e.g., Wang *et al.* 2005), the precipitation response in the AGCM time-slice projection may differ from those of AOGCM projections over the region. This may have a greater effect on mean precipitation than on heavy precipitation. Third, there are differences in present-day SST climatology between the two ensembles: CMIP5-AOGCMs have systematic biases in present-day SST climatology, while an observed SST is prescribed in the MRI-AGCM. The use of different baseline SSTs may lead to different precipitation responses on a regional scale. Fourth, the horizontal resolution of MRI-AGCM is much finer than that of CMIP5-AGCMs. The former is capable of more realistically simulating tropical cyclone and convective precipitation. The resolution difference could potentially lead to discrepancies of the responses. This remains as an important subject in our future work.

Figure 6 shows future changes in Rx1d over the land monsoon regions. Rx1d is projected to increase in most of the regions, but with large uncertainty. The increase in ratio becomes large from Pav to Rx5d and further to Rx1d. In the global monsoon domain (GLB), the change rates are 6.3% (9.5%), 19.4% (17.5%), and 27.4% (19.7%) for Pav, Rx5d, and Rx1d, respectively, for the ensemble average of 20-km (60-km) MRI-AGCM. In the East Asian domain (EAS), the largest increase of Pav is projected by 20-km and 60-km MRI-AGCMs among the regional monsoons (not shown). In the western North Pacific (WNP) domain, some members project decreases of Rx1d and Rx5d. The weaker intensity of extreme precipitation in WNP could be due to weakened tropical cyclone activity in the future climate, as suggested by Kitoh and Endo (2016b).

Relative contribution of the differences in SST patterns and convection schemes to the total variation in the ensemble projection is assessed by applying the two-way analysis of variance (ANOVA) to the future changes of the ensemble simulations of 60-km MRI-AGCM (9 members combining three SST patterns obtained by a cluster analysis with three convection schemes). Figure 7 shows the result of the two-way ANOVA applied to regional averages of future changes in Rx1d. It is found that the convection scheme effect dominates over the land monsoon regions except for the South African

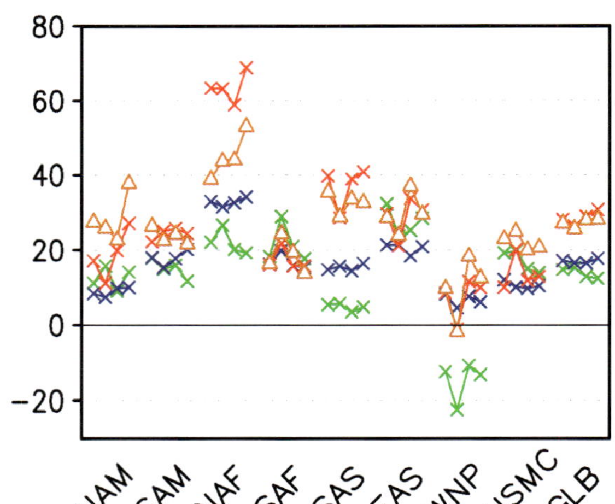

Fig. 6. Future percentage changes in Rx1d averaged over the land monsoon regions. Orange triangles denote the 20-km MRI-AGCM3.2 simulations. Colored crosses denote the 60-km MRI-AGCM3.2 simulations (red, blue, and green denote the model with the YS, AS, and KF convection schemes, respectively). Difference in the horizontal axis in each region corresponds to difference in SST patterns.

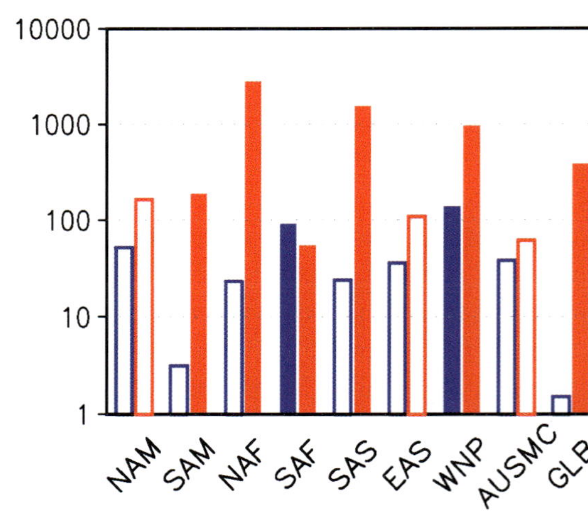

Fig. 7. The two-way ANOVA applied to the regional averages of future changes in Rx1d. Variance due to SST pattern difference ($\%^2$) is shown by blue bars, and variance due to convection scheme difference ($\%^2$) is shown by red bars. Filled bars indicate that the factor affects the variance with statistical significance at the 95% confidence level.

domain (SAF). Over the ocean, the SST pattern effect is generally larger than the convection scheme effect (not shown). Compared to Pav, the relative contribution of the convection scheme effect becomes larger in many areas for Rx1d. This result is basically consistent with Endo *et al.* (2012) whose projections are based on the CMIP3 framework.

5. Concluding Remarks

This study assessed future changes in heavy precipitation (Rx1d) around the world and discussed their uncertainty, based on ensemble projections using 20-km and 60-km mesh MRI-AGCMs. In the present-day climate simulation, we confirmed that MRI-AGCM shows high performance in simulating mean and extreme precipitations. We also found that the performance is sensitive to the choice of convective schemes.

The ensemble future projections, combining different convective parameterization schemes with different SST warming patterns, showed that mean precipitation will increase in most land monsoon regions, especially in East Asia. Applying a statistical method to the ensemble projections, we found that uncertainty in land monsoon precipitation changes is derived mainly from differences in convective schemes, while uncertainty in tropical oceanic precipitation changes originates mainly from differences in SST warming patterns.

We find that the MRI-AGCM projects larger increases in extreme precipitation (e.g., Rx5d and Rx1d) than Pav on the global scale. It is interesting to note that some members of the MRI-AGCM3.2 project smaller increases in Rx1d than Pav over the western tropical Pacific and surrounding regions. The smaller increase rates in extreme precipitation than mean precipitation may be a consequence of significant decreases of tropical cyclone frequency over the western tropical Pacific, as projected by Murakami *et al.* (2012a), Kitoh and Endo (2016b), and Endo *et al.* (2017), which are based on ensemble simulations with the 20-km mesh or 60-km mesh MRI-AGCM. Furthermore, increasing variability in Rx1d over the western tropical Pacific is also found, implying a risk of heavy precipitation there.

Acknowledgments

This work was conducted under the SOUSEI and TOUGOU (JPMXD0717935561) of the Ministry of Education, Culture, Sports, Science and Technology (MEXT) of Japan, as well as the Environment Research and Technology Development Fund (2-1503 and 2-1904) of the Environmental Restoration and Conservation Agency. The calculations were performed on the Earth Simulator of the Japan Agency for Marine-Earth Science and Technology.

References

Chadwick, R., I. Boutle, and G. Martin, 2013: Spatial patterns of precipitation change in CMIP5: Why the rich do not get richer in the tropics. *J. Clim.*, **26**, 3803–3822.

Chou, C., J. D. Neelin, C.-A. Chen, and J.-Y. Tu, 2009: Evaluating the "rich-get-richer" mechanism in tropical precipitation change under global warming. *J. Clim.*, **22**, 1982–2005.

Christensen, J. H. *et al.*, 2013: Climate phenomena and their relevance for future regional climate change. *Climate Change 2013: The Physical Science Basis. Contribution of Working Group I to the Fifth Assessment Report of the Intergovernmental Panel on Climate Change*, Stocker, T. F. *et al.* (eds.), Cambridge University Press, Cambridge, UK and New York, USA, pp. 1217–1308.

Endo, H. and A. Kitoh, 2014: Thermodynamic and dynamic effects on regional monsoon rainfall changes in a warmer climate. *Geophys. Res. Lett.*, **41**, 1704–1710.

Endo, H., A. Kitoh, T. Ose, R. Mizuta, and S. Kusunoki, 2012: Future changes and uncertainties in Asian precipitation simulated by multi-physics and multi-sea surface temperature ensemble experiments with high-resolution Meteorological Research Institute atmospheric general circulation models (MRI-AGCMs). *J. Geophys. Res.*, **117**, D16118, doi:10.1029/2012JD017874.

Endo, H., A. Kitoh, R. Mizuta, and M. Ishii, 2017: Future changes in precipitation extremes in East Asia and their uncertainty based on large ensemble simulations with a high-resolution AGCM. *SOLA*, **13**, 7–12.

He, C. and T. Zhou, 2015: Response of the western North Pacific subtropical high to global warming under RCP4.5 and RCP8.5 scenarios projected by 33 CMIP5 models: The dominance of tropical Indian Ocean-tropical western Pacific SST gradient. *J. Clim.*, **28**, 365–380.

He, J. and B. J. Soden, 2016: The impact of SST biases on projections of anthropogenic climate change: A greater role for atmosphere-only models? *Geophys. Res. Lett.*, **43**, 7745–7750.

Held, I. M. and B. J. Soden, 2006: Robust responses of the hydrological cycle to global warming. *J. Clim.*, **19**, 5686–5699.

Hsu, P.-C., T. Li, J.-J. Luo, H. Murakami, A. Kitoh, and M. Zhao, 2012: Increase of global monsoon area and precipitation under global warming: A robust signal? *Geophys. Res. Lett.*, **39**, L06701, doi:10.1029/2012GL051037.

Hsu, P.-C., T. Li, H. Murakami, and A. Kitoh, 2013: Future change of the global monsoon revealed from 19 CMIP5 models. *J. Geophys. Res. Atmos.*, **118**, 1247–1260.

Huffman, G. J. and D.T., and Bolvin, 2009: GPCP one-degree daily precipitation data set documentation. <ftp://precip.gsfc.nasa.gov/pub/1dd-v1.1/1DD_v1.1_doc.pdf>

Huffman, G. J., R. F. Adler, D. T. Bolvin, G. Gu, E. J. Nelkin, K. P. Bowman, Y. Hong, E. F. Stocker, and D. B. Wolff, 2007: The TRMM multi-satellite precipitation analysis: Quasi-global, multi-year, combined-sensor precipitation estimates at finer scale. *J. Hydrometeor.*, **8**, 38–55.

Huffman, G. J., R. F. Adler, D. T. Bolvin, and G. Gu, 2009: Improving the global precipitation record: GPCP Version 2.1. *Geophys. Res. Lett.*, **36**, L17808, doi:10.1029/2009gl040000.

Kain, J. S. and J. M. Fritsch, 1990: A one-dimensional entraining/detraining plume model and its application in convective parameterization. *J. Atmos. Sci.*, **47**, 2784–2802.

Kamiguchi, K., A. Kitoh, T. Uchiyama, R. Mizuta, and A. Noda, 2006: Changes in precipitation-based extremes indices due to global warming projected by a global 20-km-mesh atmospheric model. *SOLA*, **2**, 64–67.

Kitoh, A. and S. Kusunoki, 2008: East Asian summer monsoon simulation by a 20-km mesh AGCM. *Clim. Dynam.*, **31**, 389–401.

Kitoh, A. and H. Endo, 2016a: Changes in precipitation extremes projected by a 20-km mesh global atmospheric model. *Weather Clim. Extr.*, **11**, 41–52.

Kitoh, A. and H. Endo, 2016b: Future changes in rainfall extremes associated with El Niño projected by a global 20-km mesh atmospheric model. *SOLA*, **12A**, 1–6.

Kitoh, A. and H. Endo, 2019: Future changes in precipitation extremes associated with tropical cyclones projected by large-ensemble simulations. *J. Meteor. Soc. Jpn.*, **97**, 141–152.

Kitoh, A., T. Ose, K. Kurihara, S. Kusunoki, M. Sugi, and KAKUSHIN Team-3 Modeling Group, 2009: Projection of changes in future weather extremes using super-high-resolution global and regional atmospheric models in the KAKUSHIN Program: Results of preliminary experiments, *Hydrol. Res. Lett.*, **3**, 49–53.

Kitoh, A., H. Endo, K. Krishna Kumar, I. F. A. Cavalcanti, P. Goswami, and T. Zhou, 2013: Monsoons in a changing world: A regional perspective in a global context. *J. Geophys. Res. Atmos.*, **118**, 3053–3065.

Kitoh, A., T. Ose, and I. Takayabu, 2016: Dynamical downscaling for climate projection with high-resolution MRI AGCM-RCM. *J. Meteorol. Soc. Jpn.*, **94A**, 1–16.

Krishnan, R., T. P. Sabin, D. C. Ayantika, A. Kitoh, M. Sugi, H. Murakami, A. G. Turner, J. M. Slingo, and K. Rajendran, 2013: Will the South Asian monsoon overturning circulation stabilize any further? *Clim. Dynam.*, **40**, 187–211.

Kusunoki, S. 2016: Is the global atmospheric model MRI-AGCM3.2 better than the CMIP5 atmospheric models in simulating precipitation over East Asia? *Clim Dynam.*, doi:10.1007/s00382-016-3335-9.

Kusunoki, S. 2018: Future changes in precipitation over East Asia projected by the global atmospheric model MRI-AGCM3.2. *Clim Dynam.*, **51**, 4601–4617.

Kusunoki, S. and R. Mizuta, 2013: Changes in precipitation intensity over East Asia during the 20th and 21st centuries simulated by a global atmospheric model with a 60 km grid size. *J. Geophys. Res. Atmos.*, **118**, doi:10.1002/jgrd.50877.

Kusunoki, S., J. Yoshimura, H. Yoshimura, A. Noda, K. Oouchi, and R. Mizuta, 2006: Change of Baiu rain band in global warming projection by an atmospheric general circulation model with a 20-km grid size. *J. Meteorol. Soc. Jpn.*, **84**, 581–611.

Lee, J.-Y. and B. Wang, 2014: Future change of global monsoon in the CMIP5. *Clim. Dynam.*, **42**, 101–119.

Mizuta, R., K. Oouchi, H. Yoshimura, A. Noda, K. Katayama, S. Yukimoto, M. Hosaka, S. Kusunoki, H. Kawai, and M. Nakagawa, 2006: 20-km-mesh global climate simulations using JMA-GSM model–Mean climate states. *J. Meteorol. Soc. Jpn.*, **84**, 165–185.

Mizuta, R., Y. Adachi, S. Yukimoto, and S. Kusunoki, 2008: Estimation of future distribution of sea surface temperature and sea ice using CMIP3 multi-model ensemble mean. *Tech. Rep. Meteorological Research Institute*, 56, 28 pp., doi: 10.11483/mritechrepo.56.

Mizuta, R., H. Yoshimura, H. Murakami, M. Matsueda, H. Endo, T. Ose, K. Kamiguchi, M. Hosaka, M. Sugi, S. Yukimoto, S. Kusunoki, and A. Kitoh, 2012: Climate simulations using MRI-AGCM3.2 with 20-km grid. *J. Meteorol. Soc. Jpn.*, **90A**, 233–258.

Mizuta, R., O. Arakawa, T. Ose, S. Kusunoki, H. Endo, and A. Kitoh, 2014: Classification of CMIP5 future climate responses by the tropical sea surface temperature changes. *SOLA*, **10**, 167–171.

Mizuta, R. *et al.*, 2017: Over 5000 years of ensemble future climate simulations by 60 km global and 20 km regional atmospheric models. *Bull. Am. Meteorol. Soc.*, **98**, 1383–1398.

Murakami, H., R. Mizuta, and E. Shindo, 2012a: Future changes in tropical cyclone activity projected by multi-physics and multi-SST ensemble experiments using the 60-km-mesh MRI-AGCM. *Clim. Dyn*am., **39**, 2569–2584.

Murakami, H., Y. Wang, H. Yoshimura, R. Mizuta, M. Sugi, E. Shindo, Y. Adachi, S. Yukimoto, M. Hosaka, S. Kusunoki, T. Ose, and A. Kitoh, 2012b: Future changes in tropical cyclone activity projected by the new high-resolution MRI-AGCM. *J. Clim.*, **25**, 3237–3260.

Oouchi, K., J. Yoshimura, H. Yoshimura, R. Mizuta, S. Kusunoki, and A. Noda, 2006: Tropical cyclone climatology in a global warming climate as simulated in a 20 km-mesh global atmospheric model: Frequency and wind intensity analyses. *J. Meteorol. Soc. Jpn.*, **84**, 259–276.

Rajendran, K., A. Kitoh, J. Srinivasan, R. Mizuta, and R. Krishnan, 2012: Monsoon circulation interaction with Western Ghats orography under changing climate: Projection by an ultra-high resolution global model. *Theor. Appl. Climatol.*, **110**, 555–571.

Randall, D. A. and D.-M. Pan, 1993: Implementation of the Arakawa-Schubert cumulus parameterization with a prognostic closure, In: *The Representation of Cumulus Convection in Numerical Models.* Meteorol. Monogr., Vol. 24, No. 46, K. A. Emanuel and D. J. Raymond, (eds.), *Am. Meteor. Soc., Boston*, pp. 137–144.

Seo, K.-H., J. Ok, J.-H. Son, and D.-H. Cha, 2013: Assessing future changes in the East Asian summer monsoon using CMIP5 coupled models. *J. Clim.*, **26**, 7662–7675.

Wang, B., Q. Ding, X. Fu, I.-S. Kang, K. Jin, J. Shukla, and F. Doblas-Reyes, 2005: Fundamental challenge in simulation and prediction of summer monsoon rainfall. *Geophys. Res. Lett.*, **32**, L15711, doi:10.1029/2005GL022734.

Wang, B., H.-J. Kim, K. Kikuchi, and A. Kitoh, 2011: Diagnostic metrics for evaluation of annual and diurnal cycles, *Clim. Dynam.*, **37**, 941–955.

Wang, B., S.-Y. Yim, J.-Y. Lee, J. Liu, and K.-J. Ha, 2014: Future change of Asian-Australian monsoon under RCP4.5 anthropogenic warming scenario. *Clim. Dynam.*, **42**, 83–100.

Xie, P.-P. and P. A. Arkin, 1997: Global precipitation: A 17-year monthly analysis based on gauge observations, satellite estimates and numerical model outputs. *Bull. Amer. Meteor. Soc.*, **78**, 2539–2558.

Xie, S.-P., C. Deser, G. A. Vecchi, J. Ma, H. Teng, and A. T. Wittenberg, 2010: Global warming pattern formation: Sea surface temperature and rainfall. *J. Clim.*, **23**, 966–986.

Yoshimura, H., R. Mizuta, and H. Murakami, 2015: A spectral cumulus parameterization scheme interpolating between two convective updrafts with semi-Lagrangian calculation of transport by compensatory subsidence. *Mon. Wea. Rev.*, **143**, 597–621.

Chapter 28

Overview of the Global Monsoons Model Intercomparison Project (GMMIP): Progress and Challenges

Andrew G. Turner*,¶, Tianjun Zhou†,§,||, James Kinter III‡, Xiaolong Chen†, Lixia Zhang†, Zhun Guo†, Xin Huang†, Bo Liu†, Senfeng Liu†, Kai-Chi Wong* and Wenxia Zhang†

**National Centre for Atmospheric Science and Department of Meteorology, University of Reading, Reading, UK*

†Institute of Atmospheric Physics, Chinese Academy of Sciences, Beijing 100029, China

‡Centre for Ocean-Land-Atmosphere Studies and Department of Atmospheric, Oceanic and Earth Sciences, George Mason University, Fairfax, Virginia, USA

§University of Chinese Academy of Sciences, Beijing, China

¶a.g.turner@reading.ac.uk

||zhoutj@lasg.iap.ac.cn

The Global Monsoons Model Inter-comparison Project (GMMIP) has been endorsed by the Coupled Model Inter-comparison Project (CMIP) as one of the participating model inter-comparison projects (MIPs) in the sixth phase of CMIP (CMIP6). The focus of GMMIP is on monsoon climatology, variability, prediction and projection, which is relevant to four of the "Grand Challenges" proposed by the World Climate Research Programme. More than twenty international modelling groups are committed to joining GMMIP.

During the Sixth WMO International Workshop on Monsoons (IWM-6), 13–17 November 2017 Singapore, a discussion session was held to discuss the motivation, progress and early results of GMMIP. This paper describes the motivation for GMMIP, lists the scientific questions to be addressed and the planned GMMIP methodology, before outlining some of the early results based on the analysis of current existing numerical experiments. Finally, we summarize the discussion points for the GMMIP discussion session at IWM-6.

1. Introduction

The Global Monsoons Model Intercomparison Project (GMMIP) has been endorsed by the Coupled Model Intercomparison Project (CMIP) in its sixth phase (CMIP6) as one of the participating model intercomparison projects (MIPs). The focus of GMMIP is on monsoon climatology, variability, prediction, and projection, which are relevant to four of the "Grand Challenges" proposed by the World Climate Research Programme. More than twenty international modeling groups are committed to joining GMMIP.

During the Sixth WMO International Workshop on Monsoons (IWM-6), 13–17 November 2017, Singapore, a session was held to discuss the motivation, progress, and early results of GMMIP. This paper describes the motivation for GMMIP, lists the scientific questions to be addressed and the planned GMMIP methodology, and outlines some of the early results based on the analysis of current existing numerical experiments. Finally, we summarize the points for the GMMIP discussion session at IWM-6.

2. Background of GMMIP

Changes in monsoon precipitation are of great societal importance owing to the monsoon's significant contribution to annual rainfall and

The Multiscale Global Monsoon System
Edited by C. P. Chang et al.

thus water supply for agriculture, industry, and human consumption. Thus, monsoons are vital for food, water, and energy security, often in some of the world's most highly populated and developing regions. While each monsoon in the tropical Asia–Australia, Africa, and Americas regions (see Fig. 1) has its own characteristics, feedbacks, and drivers relating to the local topography, they are all fundamentally driven by temperature and pressure gradients and bounded together by the global divergent circulation (Trenberth *et al.* 2000). Thus, the monsoons across different parts of the world show coherent variations at variable timescales. The GMMIP project aims to provide a platform for the international monsoon community in the collaborative investigation on monsoon changes across timescales.

The global land monsoon precipitation has exhibited variations at multiple timescales during the 20th century (Zhang and Zhou 2011). Future projection of global land monsoon precipitation change also shows a significant increase in both mean state and extremes (Kitoh *et al.* 2013; Zhang *et al.* 2018). In order to fully understand the influence of anthropogenic factors on the monsoons under greenhouse gas and aerosol emissions, we must also quantify the impact of internal, multi-decadal modes of variability in the climate system on the monsoon. Numerous studies have explored the influence of modes such as the Interdecadal Pacific Oscillation (IPO), the Pacific Decadal Oscillation (PDO), and Atlantic Multi-decadal Oscillation (AMO) on the various regional monsoons, as well as on a hemisphere-wide basis (e.g. Zhou *et al.* 2008; Wang *et al.* 2013). Coordinated research is thus needed to disentangle the influence of SST variability on interannual to interdecadal variability of the monsoons from external factors such as anthropogenic forcing. In addition, we should also determine what aspects of the regional monsoons vary under the same driver.

The regional differences between monsoons arise due to the details of the local geography: the structure of the ocean basins, the distribution of mountain ranges, and the details of local land use or land cover. The influence of regional orography is most apparent for the Asian monsoon, bounded to the north of India by the Himalayas reaching 7000–8000 m and the huge expanse of the Tibetan Plateau at around 4500 m. The mountain regions may be acting as

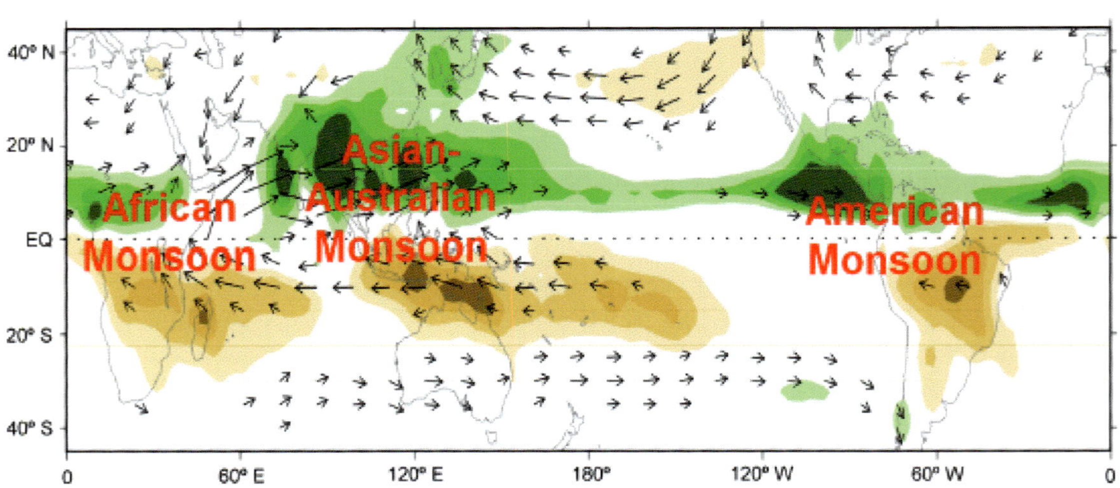

Fig. 1. Schematic figure of the global monsoon domain and its local components, indicated by the differences of 850-hPa wind and precipitation between the June–July–August and December–January–February mean, modified from Wang and Ding (2008) and taken from Zhou *et al.* (2016).

a physical barrier, isolating the moist tropical air over South and East Asia from lower-entropy air deeper into the Eurasian Continent; on the contrary, they may also be acting as an elevated heat pump to enhance the meridional overturning circulation of the monsoon. Recent works have debated the relative roles of these mechanisms (Boos and Kuang 2010, 2013; Wu *et al.* 2012).

The systematic biases present in contemporary coupled climate models (for example, the consistent and large dry biases for the South Asian monsoon shown in CMIP5 by Sperber *et al.* 2013) emphasize the need to use multi-model frameworks to determine consistent responses of the monsoons to forcing. Thus, the Global Monsoons Model Intercomparison Project (GMMIP; Zhou *et al.* 2016) was designed to examine the robustness of the impacts of modes of variability, anthropogenic forcing, and regional orography on the monsoons. In addition, the limitation of medium- to low-resolution climate models employed in the past CMIPs in presenting the narrow monsoon rainband has also hampered our understanding of the mechanism of monsoon rainfall variability through numerical modeling experiments. In CMIP6, there is the high-resolution model intercomparison project (HighResMIP), which as the output of partner MIPs will also be used in the diagnostic analysis of GMMIP.

3. Scientific Questions and Experiment Design

As a result of the motivating research mentioned above, GMMIP aims to answer four primary scientific questions:

(1) What are the relative contributions of internal processes and external forcing that are driving the historical evolution of monsoon of the late-19th through early-20th centuries?
(2) To what extent and how does atmosphere–ocean interaction contribute to the interannual variability and reproducibility of monsoons?
(3) How can high-resolution and associated improved model dynamics and physics help to reliably simulate monsoon precipitation and its variability and change?
(4) What is the effect of the orography of the Himalayan-Tibetan Plateau on the development and maintenance of the Asian monsoon? Similarly, what is the impact of orography elsewhere on other regional monsoons?

In addition to the three "tiers" of GMMIP experiments outlined below, the outputs of the CMIP6 DECK, "historical" simulation, and other endorsed MIPs will be used in the diagnostic analysis of GMMIP to give a comprehensive understanding of the roles played by different external forcings, the potential improvements in the simulation of monsoon rainfall at high resolution, and reproducibility at decadal timescales.

In order of decreasing priority, GMMIP experiments are designed to examine the following:

Tier-1: Model skill at simulating the climatology and interannual to multi-decadal variability of global monsoons forced by sea surface temperatures during the historical climate period (the so-called extended-AMIP experiment);

Tier-2: The role of the Interdecadal Pacific Oscillation (IPO) and Atlantic Multidecadal Oscillation (AMO) in driving variations of the global and regional monsoons;

Tier-3: The effects of large orographic terrain on the establishment of the monsoons (such as examining the relative influence of the Himalayas and the Tibetan Plateau on the regional components of the Asian monsoon).

The first question can be addressed by a combination of Tier-1 and Tier-2 experiments; the second can be addressed by a comparison between Tier-1 (extended AMIP) outputs with those of the historical experiments from the CMIP6 DECK. The third question will be addressed in an analysis of the HighResMIP-endorsed MIP of CMIP6. The current existing high-resolution numerical experiments from some pioneering climate modeling centers will also be used. Finally, the fourth question can be addressed by Tier-3.

4. Emerging Results

We expected the majority of GMMIP experiments to be carried out by modeling centers during 2018. However, the delay of CMIP6 in providing the needed external forcing data has affected the progress of all modeling centers in conducting both the extended AMIP-type simulation and the essential 20th century historical simulation. Nonetheless, some results are beginning to emerge based on the diagnosis of current existing numerical experiments. Below we present preliminary results that highlight the crucial role of model resolution. The dynamical/physical processes associated with ocean-atmosphere interaction in the simulation of EASM rainfall variability are also reviewed.

4.1. *Improvement of high-resolution models in the simulation of global monsoon annual cycle*

The complexity of monsoon systems presents great challenges in simulating the climatological seasonal means and annual cycles of the monsoon (Sperber *et al.* 2013), although substantial efforts have been made to improve model physics and dynamics in the past several decades. Based on a fractional accumulation method to evaluate the monsoon annual cycle, Sperber and Annamalai (2014) presented that coupled climate models in CMIP5 have delayed onsets of summer rainfall over India, the Gulf of Guinea, and South America, but early onsets for the Sahel and North America. This result is consistent with Sperber *et al.* (2013), in which Asian monsoon onset and retreat are defined based on rainfall thresholds. A recent study of Zhang *et al.* (2018) showed that the higher resolution can partly improve the simulation of the global monsoon annual cycle by comparing MetUM-GA3, MRI-AGCM3, and GFDL-HiRAM. The outputs from MetUM-GA3 at N216 (90 km) and N512 (40 km), MRI-AGCM3.2H (60 km) and MRI-AGCM3.2S (20 km), and GFDL-HiRAM-C180 (65 km) and GFDL-HiRAM-C360 (32 km) were compared to verify the sensitivity of the annual cycle of the global monsoon to resolution.

All the low-resolution AGCMs are biased toward higher annual mean precipitation and stronger solstitial and equinoctial asymmetric modes. With increasing resolution, all the AGCMs show consistent improvements in simulating the precipitation and low-level circulation of the annual mean and the first two annual-cycle modes, as measured by the pattern correlation coefficient (PCC) and equitable threat score (ETS), which can be seen from the red boxes for ANN, AC1, AC2, and Domain in Fig. 2. Improvements in simulating the summer monsoon onset and withdrawal are region dependent. No consistent response to resolution is found in simulating summer monsoon retreat (blue box for *WithD* in Fig. 2).

Regionally, increased resolution reduces the positive bias in simulated annual mean precipitation and the two annual-cycle modes over the West African monsoon and Northwestern Pacific monsoon. An overestimation of the solstitial mode and an underestimation of the equinoctial asymmetric mode of the East Asian monsoon are reduced in all high-resolution configurations. Systematic errors exist in lower-resolution models for simulating the onset and withdrawal of the summer monsoon. Higher-resolution models consistently improve the early summer monsoon

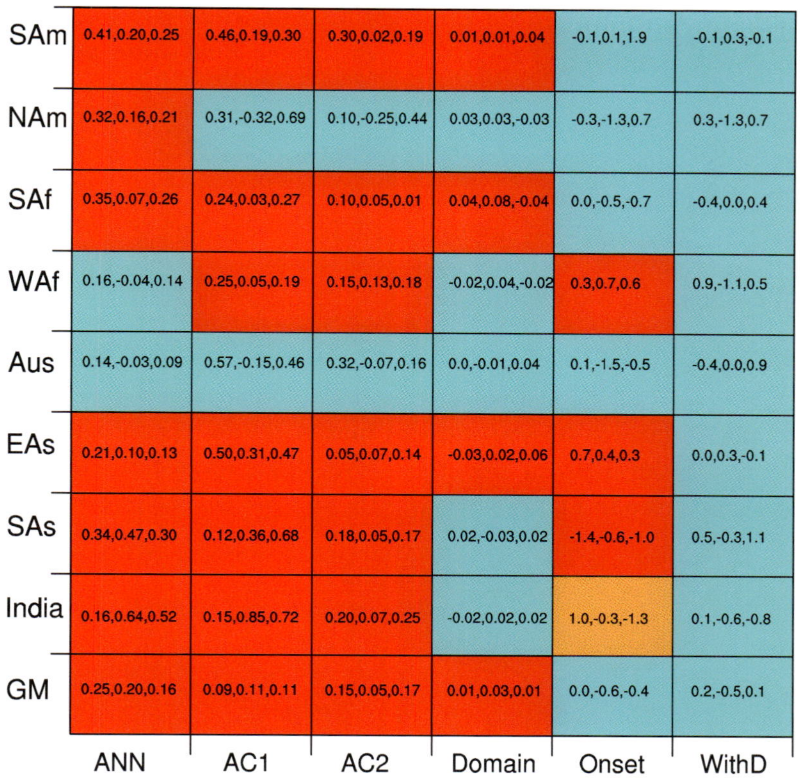

Fig. 2. Improvement in the simulation of the annual cycle of global monsoon precipitation with horizontal resolution. Red boxes indicate high-resolution consistently improves performance across all three AGCMs; blue boxes indicate improvements with resolution are inconsistent; orange boxes indicate a better simulation is shown in all three high-resolution AGCMs but the responses to resolution are inconsistent across models. The values from left to right in each box are the difference between high- and low-resolution configurations of MetUM-GA3, MRI-AGCM3, and GFDL-HiRAM, respectively. The metric for the annual modes, including annual mean (ANN), solstitial mode (AC1), and the equinoctial asymmetric mode (AC2), is the PCC with TRMM. The metrics for the monsoon domain and monsoon onset/withdrawal are the ETS and the onset/withdrawal pentad, respectively (adopted from Zhang *et al.* 2018).

onset over East Asia and West Africa, but substantial differences exist in the responses over the Indian monsoon region, where biases differ across the three low-resolution AGCMs.

4.2. *Improvement of high-resolution models in the simulation of East Asian summer monsoon*

Increasing the model resolution has been demonstrated as one of the feasible ways to improve the simulation of global monsoon, because the higher resolution leads to a better representation of atmospheric circulations (Zhang *et al.* 2018). Such an added value of high resolution is even more credible over the regions which have more complex terrains, e.g., the leeside of the Tibetan Plateau. It is partly supported by the results of Coupled Model Inter-comparison Project Phase 5 (CMIP5). If we categorize the CMIP5 atmospheric models by their horizontal resolution, it would be obvious that a model with high resolution usually reproduces a more reasonable intensity and a spatial pattern of the EASM rainfall than a model with low resolution. The diagnosis of moist static energy balance and moisture budgets can further attribute the added value to the contributions from intensified meridional convergence along the EASM rain belt in the high-resolution models (Yao *et al.* 2017).

In addition, the added value of high resolution in the monsoon modeling is also supported by the sensitivity experiments, which are configured at identical dynamic core and physics schemes but different resolutions. For example, Yao *et al.* (2017) configured CAM5 with three different resolutions (~2°, 1°, and 0.5°) and revealed that the stationary meridional eddy flow changes in high-resolution simulations are highly related to the barotropic Rossby wave downstream of the Tibetan Plateau due to increased resolution, as shown in Fig. 3.

4.3. *Enhanced monsoon rainfall over East Asia projected by higher-resolution models*

Climate models are an indispensable tool for regional climate projection in the future. However, resolutions of coupled climate models that participated in the CMIP5 were still too low to resolve some regional-scale rainfall systems, for instance, the Mei-yu rain belt along the Yangtze River valley, which impairs the reliability of their projected results. Enhanced moisture in future warmer climates is definitely in favor of an increase in rainfall, while the spread in circulation change contributes most to the

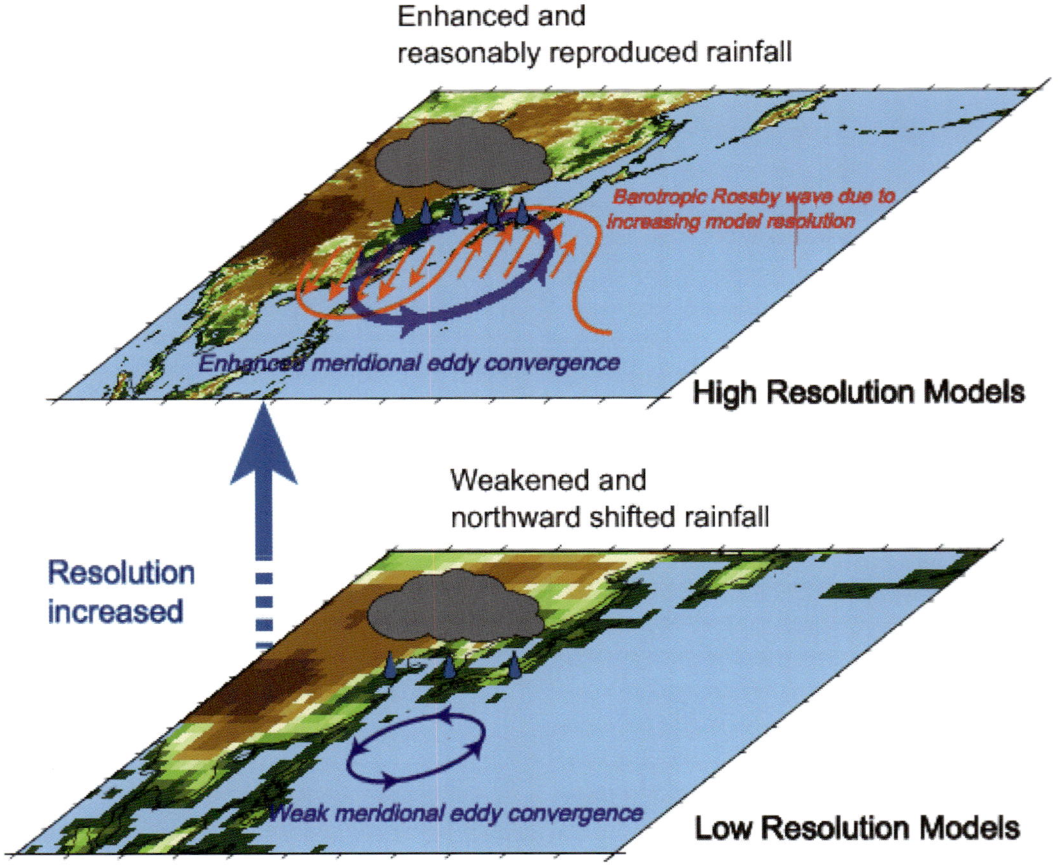

Fig. 3. Schematic of the high-resolution climate model, which improves simulation of the EASM rain belt. Due to increasing model resolution, a barotropic Rossby wave downstream of the Tibetan Plateau is generated, and further intensifies meridional convergence and moisture convergence along the EASM rain belt. Thus, the EASM rain belt is improved in the high-resolution models (after Yao *et al.* 2017).

uncertainty in regional rainfall projection, in which the role of model resolution may be not negligible.

To address possible effects of model resolution on the Mei-yu rainfall projection, the UK MetOffice Hadley Centre's latest climate model, the HadGEM3 Global Coupled Configuration 2.0 (HadGEM3-GC2), was used in a recent study by only changing the horizontal resolution of the atmosphere component (Chen *et al.* 2018). The high-resolution HadGEM3-GC2 (N216, ~60 km) projects large increases of summer rainfall under two representative concentration pathway scenarios (RCP8.5 and RCP4.5), whereas the low-resolution (N96, ~130 km) shows a decrease (Fig. 4a). A larger increase of projected Mei-yu rainfall in higher-resolution models is also observed across the CMIP5 ensemble (Fig. 4b). Based on moist static energy and moisture budget analysis,

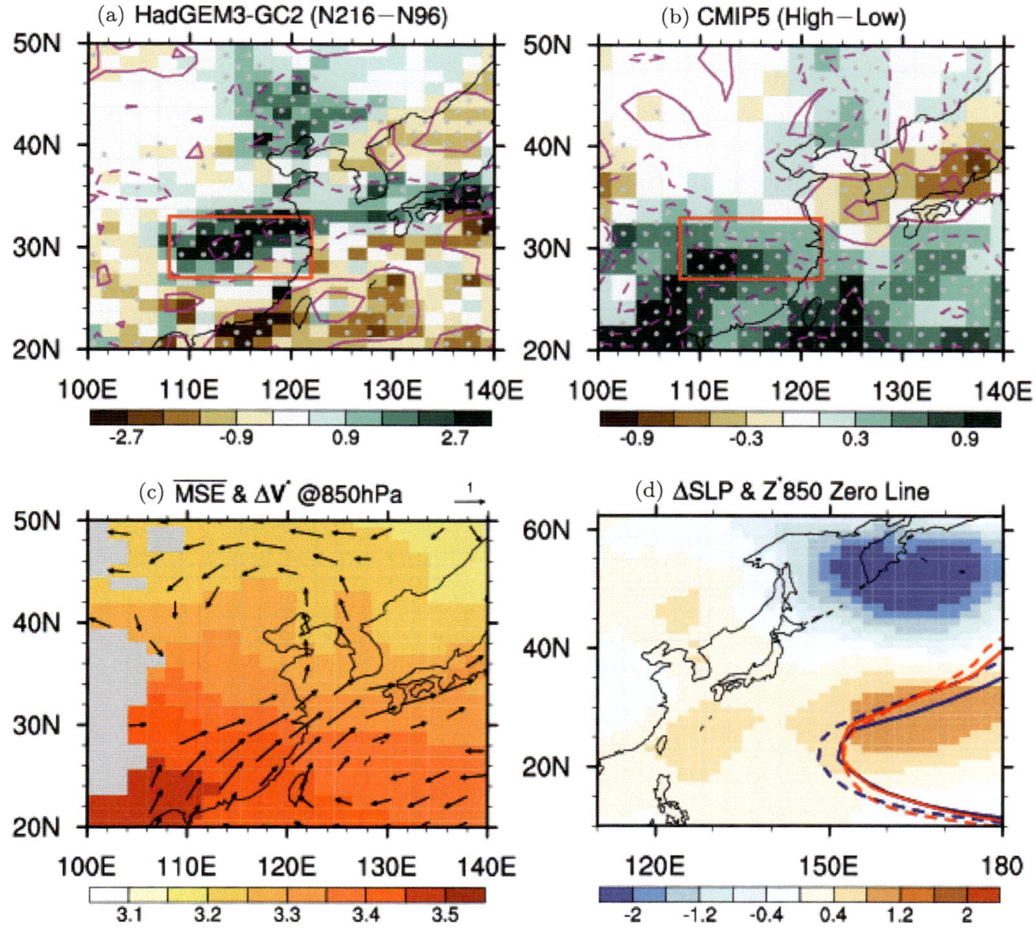

Fig. 4. (a) Differences between N216 and N96 resolutions in projected precipitation (shading; mm day^{-1}) and vertical motion (contour; 10^{-3} Pa s^{-1}) changes under RCP8.5 scenario. (b) Same as (a) but for difference between CMIP5 high (<1.5°) and low (>1.5°) resolutions. (c) Climatological moist static energy during 1976–2005 (shadings; 10^5 J kg^{-1}) and anomalous stationary eddy (vectors; m s^{-1}) at 850 hPa during 2070–2099. Winds less than 0.5 m s^{-1} are omitted. Gray shading denotes orography. (d) Same as (a) but for sea level pressure (shading; hPa) and zero isoline of geopotential height eddy at 850 hPa (m). Solid (dashed) lines denote N216 (N96) and blue (red) lines denote historical (RCP8.5) simulation. Dotted areas exceed the 5% significance level of the student's *t*-test. Red box denotes the Mei-yu region (after Chen *et al.* 2018).

deficient moist static energy advection and moisture convergence by stationary eddies (mainly in the meridional direction) in N96 (Fig. 4c) are attributed to the projected weaker upward motion with less Mei-yu rainfall.

The anomalous stationary eddies manifest the contrasting responses to the same warming scenario by the western North Pacific subtropical high (WNPSH), which is almost unchanged in N216 but evidently retreats eastward in N96 (Fig. 4d). Interactions between high-latitude wave activity and low-latitude convection are dynamical reasons for the distinct responses of WNPSH in high- and low-resolution HadGEM3-GC2. With unchanged circulation in N216, more moisture in warmer climate will naturally increase the projected Mei-yu rainfall, whereas excessive eastward retreat in N96 impedes the moist static energy and moisture being transported into the Mei-yu region. This study indicates that an increase in model resolution to resolve parameterized processes and detailed orographic features will hopefully reduce the spread in future climate projections.

4.4. *Improvement in the simulation of monsoon variability due to air-sea coupling*

Analysis of 13 Atmospheric General Circulation Models (AGCMs) from phase 3 of the Coupled Model Inter-comparison Project (CMIP3) and 19 AGCMs from CMIP5 found that the deficiencies of AGCMs in the simulation of the climatological position of the monsoon rainband are closely related to the weaker and southward shift of the western Pacific anticyclone (WPAC). For the simulation of the interannual variability of EASM rainfall, the tropical eastern Indian Ocean (IO) rainfall response to local warm SST anomalies and the associated Kelvin wave response over the Indo–western Pacific region are important to maintain the WPAC. A successful reproduction of an anomalous EASM rainfall pattern depends highly on the IO-WPAC teleconnection (Song and Zhou 2014a).

To estimate the role of air-sea coupling in simulating the mean state and interannual variability of East Asian summer monsoon, 17 CGCMs are compared to their corresponding atmospheric general circulation models (AGCMs) (Song and Zhou 2014b). The results show that the climatological low-level monsoon circulation and Mei-yu/Changma/Baiu rainfall band are improved in CGCMs from AGCMs. However, the improvement is at the cost of the local cold sea surface temperature (SST) biases in the coupled models, since the process of ocean-atmosphere coupling decreases the surface evaporation and enhances the circulation.

The 17 CGCMS are further categorized into high- and low-skill models based on their skills in reproducing the interannual EASM pattern. Composite analysis shows that the observed IO warming, tropical eastern Indian Ocean (TEIO) rainfall anomalies, and Kelvin wave response are well captured in high-skill models, but poorly reproduced in low-skill models. The skill difference of the two model groups in the simulation of interannual EASM rainfall variability is the result of the difference of the two model groups in simulating the IO warming associated with ENSO. Compared to AGCMs, in the El Niño decaying year summer, the easterly wind anomalies in the southern flank of the WPAC reduce the climatological monsoon westerlies and decrease the surface evaporation, and thus the TEIO SST anomaly is warmer in CGCMs. The warmer SST anomaly further induces a stronger precipitation anomaly and intensifies the teleconnection. The dynamical and physical processes of ocean-atmosphere coupling in the simulation of both the climatology and interannual variability of EASM are summarized in Fig. 5.

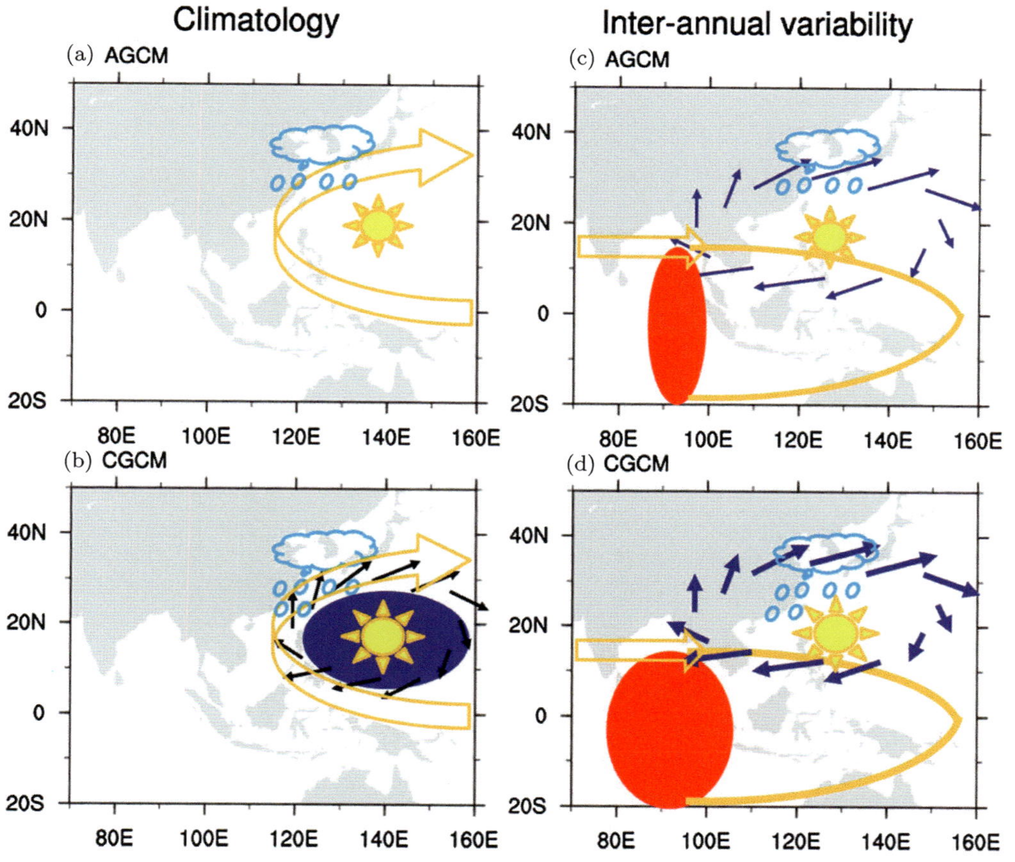

Fig. 5. The schematic plot of the air-sea coupling's role in the EASM simulation for both the (a) and (b) climatology, and (c) and (d) interannual variability. For the climatology, compared to the AGCM, the cold SST bias over the WNP in the CGCM decreases the surface evaporation and strengthens the WNPSH, inducing more rainfall over the Mei-yu/Changma/Baiu and less rainfall over the NWP. For the interannual variability, the WPAC extends to the eastern Indian Ocean, and the associated easterly anomalies on the southern flank weaken the climatological monsoon westerlies over the TEIO. Hence, the TEIO warming anomaly is stronger in the CGCMs than in the AGCMs. The stronger TEIO warming will induce stronger WPAC through IO–WPAC teleconnection, forming a positive feedback (after Song and Zhou 2014b).

4.5. *Himalayan/Tibetan Plateau orography experiments (GMMIP Tier-3)*

A collaboration between IAP and the University of Reading has led to identical experiments removing the Tibetan Plateau and Himalayas (reducing the orographic height to 500 m in those regions; named *NoTP*) in the IAP FGOALS model and UKMO HadGEM3 model (Wong *et al.* 2017). Each *NoTP* experiment is run in the AMIP mode for 20 years at N96 resolution and compared to a *control* over the same period.

Figure 6 shows the difference between JJA monsoon climatologies, i.e., *NoTP* minus *control.* While flattening the regional orography has broadly the same impact on the monsoon circulation in both models (the Somali Jet of the South Asian monsoon circulation is weakened; the Western North Pacific High is also weakened), the impact on precipitation over India is rather different. There is little change in the precipitation over India in FGOALS,

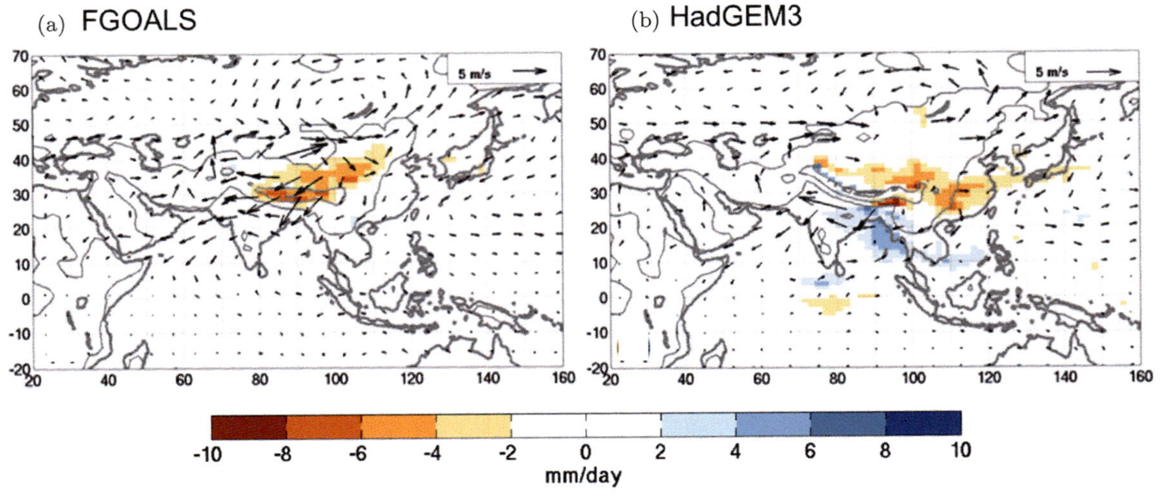

Fig. 6. Change in 20-year mean experiments forced with AMIP SST of NoTibetan Plateau experiment minus Control for June–August precipitation and 850-hPa winds in (a) IAP FGOALS and (b) UKMO HadGEM3 models. All orography in the Tibetan Plateau and Himalayan regions are flattened to 500 m (adapted from Wong *et al.* 2017).

whereas precipitation increases in HadGEM3 due to enhanced moisture convergence over the subcontinent. These models were chosen for this comparison since they have rather different mean state biases in their monsoon simulations: FGOALS is too wet over the South Asia region, while HadGEM3 is too dry. This result emphasizes the importance of performing a multi-model inter-comparison of orographic forcing experiments as in GMMIP Tier-3.

5. Summary of Panel Discussion

The following points were deliberated in the panel discussion session of IWM-6:

- Introduction of the motivation behind GMMIP including background work on the subject;
- Description of the scientific questions that GMMIP intends to answer;
- Issuing an update of the current state of play of participating modeling groups in GMMIP;
- Revealing emerging results from trial GMMIP experiments.

In the discussion section, we highlighted that this was the first time in the 20-year history of CMIP that we had a specific project that focused on the simulation of global monsoons. We hope that the GMMIP will provide a useful platform to the international monsoon community in the collaboration of monsoon simulation, prediction, and projection. To achieve these goals, we need to foster the establishment of international monsoon communities to perform diagnostic analyses of the global monsoon, various regional monsoons, and planetary-scale features of the global energy and water cycles in the GMMIP experiments. Depending on the progress of the diagnosis, coordinated experiments additional to standard GMMIP experiments may need to be developed.

Acknowledgments

This work is supported by the International Partnership Program of Chinese Academy of Sciences under Grant No. 134111KYSB20160031.

References

Boos, W. R. and Z. M. Kuang, 2010: Dominant control of the South Asian monsoon by orographic insulation vs. plateau heating. *Nature*, **463**, 218–223.

Boos, W. R. and Z. M. Kuang, 2013: Sensitivity of the South Asian monsoon to elevated and non-elevated heating. *Sci. Rep.*, **3**, 1192.

Chen Xiaolong, Peili Wu, Malcolm J. Roberts, and Tianjun Zhou, 2018: Potential underestimation of future Mei-yu rainfall with coarse resolution climate models. *J. Clim.*, https://doi.org/10.1175/JCLI-D-17-0741.1.

Kitoh, A., H. Endo, K. Krishna Kumar, I. F. A. Cavalcanti, P. Goswami, and T. Zhou, 2013: Monsoons in a changing world: a regional perspective in a global context. *J. Geophys. Res. Atmos.*, **118**, doi:10.1002/jgrd.50258.

Kosaka, K. and S.-P. Xie, 2013: Recent global-warming hiatus tied to equatorial Pacific surface cooling. *Nature*, **501**, 403–407.

Song, F., T. Zhou, 2014a: Interannual variability of East Asian summer monsoon simulated by CMIP3 and CMIP5 AGCMs: Skill dependence on Indian Ocean-Western pacific anticyclone teleconnection. *J. Clim.*, **27**, 1679–1697.

Song, F., T. Zhou, 2014b: The climatology and interannual variability of East Asian summer monsoon in CMIP5 coupled models: Does air-sea coupling improve the simulations? *J. Clim.*, **27**, 8761–8777.

Sperber, K. R. and H. Annamalai, 2014: The use of fractional accumulated precipitation for the evaluation of the annual cycle of monsoons. *Clim. Dynam.*, **43**, 3219–3244, https://doi.org/10.1007/s00382-014-2099-3.

Sperber, K. R., H. Annamalai, I.-S. Kang, A. Kitoh, A. Moise, A. Turner, B. Wang, and T. Zhou, 2013: The Asian summer monsoon: An intercomparison of CMIP5 vs. CMIP3 simulations of the late 20th century. *Clim. Dynam.*, **41**, 2711–2744, doi:10.1007/s00382-012-1607-6.

Trenberth, K. E., D. Stepaniak, and J. Caron, 2000: The global monsoon as seen through the divergent atmospheric circulation. *J. Clim.*, **13**, 3969–3993.

Watanabe, M., H. Shiogama, H. Tatebe, M. Hayashi, M. Ishii, and M. Kimoto, 2014: Contribution of natural decadal variability to global warming acceleration and hiatus. *Nat. Clim. Change*, **4**(10), 893.

Wang, B. and Q. Ding, 2006: Changes in global monsoon precipitation over the past 56 years. *Geophys. Res. Lett.*, **33**, L06711.

Wang, B., J. Liu, H. J. Kim, P. J. Webster, S. Y. Yim, and B. Xiang, 2013: Northern Hemisphere summer monsoon intensified by mega-El Niño/southern oscillation and Atlantic multidecadal oscillation. *P. Natl. Acad. Sci.*, **110**, 5347–5352.

Wong, K.-C., S. Liu, A. G. Turner, R. K. Schiemann, and G. Wu, 2017: Contrasting impacts of Tibetan Plateau/Himalaya-region orography on the Asian monsoon in the FGOALS and HadGEM3 models. *Adv. Atmos. Sci.*, in preparation.

Wu, G., Y. Liu, B. He, Q. Bao, A. Duan, and F.-F. Jin, 2012: Thermal controls on the Asian summer monsoon. *Sci. Rep.*, **2**, 404.

Yao, J., T. Zhou, Z. Guo, X. Chen, L. Zou, and Y. Sun, 2017: Improved performance of high-resolution atmospheric models in simulating the East-Asian Summer Monsoon Rainbelt. *J. Clim.*, **30**(21), 8825–8840, https://doi.org/10.1175/JCLI-D-16-0372.1.

Zhang L. and T. Zhou, 2011: An assessment of monsoon precipitation changes during 1901–2001. *Clim. Dynam.*, **37**, 279–296, doi:10.1007/s00382-011-0993-5.

Zhang L., T. Zhou, N. P. Klingaman, *et al.* 2018: Effect of horizontal resolution on the representation of the global monsoon annual cycle in AGCMs. *Adv. Atmos. Sci.*, 10.1007/s00376-018-7273-9, https://link.springer.com/article/10.1007/s00376-018-7273-9.

Zhang, W., T. Zhou, L. Zou, L. Zhang, and X. Chen, 2018: Reduced exposure to extreme precipitation from 0.5C less warming in global land monsoon regions. *Nature Commun.* **9**, Article number: 3153, doi:10.1038/s41467-018-05633-3.

Zhou T., R. Yu, H. Li, and B. Wang, 2008: Ocean forcing to changes in global monsoon precipitation over the recent half-century. *J. Clim.*, **21**(15), 3833–3852.

Zhou, T., A. G. Turner, J. Kinter, B. Wang, Y. Qian, X. Chen, B. Wu, B. Wang, B. Liu, L. Zou, and B. He, 2016: GMMIP (v1.0) contribution to CMIP6: Global Monsoons Model Inter-comparison Project. *Geosci. Model. Dev.*, **9**, 3589–3604.

Chapter 29

Grand Challenges in Asian Summer Monsoon Modeling — Representation of Processes and Sources of Model Error*

H. Annamalai[†], William R. Boos[‡], Gill Martin[§], Brian Mapes[¶], Yi Ming[‖],
P. Mukhopadhyay[**], Tieh-Yong Koh[††] and Suryachandra Rao[**]

[†] *IPRC/Department of Oceanography, University of Hawaii at Manoa, 1680 East-West Road, Honolulu, HI 96822, USA*
[‡] *University of California, Berkeley, USA*
[§] *Met Office Hadley Centre, UK*
[¶] *University of Miami, USA*
[‖] *GFDL/NOAA, Princeton, USA*
[**] *Indian Institute of Tropical Meteorology, Pune, India*
[††] *Singapore University of Social Sciences, Singapore*
[†] *hanna@hawaii.edu*

Simulating the monsoon precipitation climatology in climate models remains a grand challenge. Multi-model-mean (MMM) errors for the Asian–Australian Monsoon (AAM) precipitation climatology have shown little improvement in 7 years (from CMIP3 to CMIP5), relative to observations. Bias in future projections of seasonal-mean AAM rainfall may therefore, logically, also not have been reduced. Might there be fundamental limits to realistically simulating the AAM? Or can a concerted observational and modeling effort enhance the fidelity of models in simulating the monsoon? The persistence of model errors suggests that they are not easy to fix, perhaps arising from multiple processes and their interactions. However, the constancy of their spatial pattern may suggest a robust source of error, perhaps in model numerics (e.g., treatments of topography at coarse resolution) or in a fundamental assumption behind physical parameterizations (such as convection) that, perhaps, persists because of a lack of detailed observations to constrain model physics. A systematic effort on this problem would require, but might also serve to foster, a better understanding of the interacting dynamic, thermodynamic, orographic, and oceanographic processes that shape the monsoon precipitation climatology, and the representation of these processes in climate models.

Toward these goals, members of the CLIVAR-GEWEX Working Group on the Asian–Australian monsoons have diagnosed processes in CMIP3/5 models, performed idealized numerical experiments with coupled and uncoupled models, and identified a few possible sources of model errors. In this paper, we present key results from these recent studies and discuss how errors in one variable cascade into others. For improved process representation and monsoon simulation, we stress the need for coordinated efforts including focused process-based diagnostics and high-quality observations to constrain model physics.

*Contribution from CLIVAR-GEWEX Working Group on Asian–Australian Monsoons.

The Multiscale Global Monsoon System
Edited by C. P. Chang et al.

1. Introduction

In South and Southeast Asia, farming and agriculture-related markets employ about two-thirds of the work force, so the seasonal-mean monsoon precipitation during boreal summer and winter exerts significant impacts on socio-economic conditions in the region. As populations grow and as weather forecasts become more widely used, demand is increasing for accurate prediction of precipitation on seasonal timescales and for accurate projection of long-term precipitation trends. Despite recognition of the societal need for such precipitation forecasts, skill in monsoon precipitation prediction by dynamical climate models remains low (DelSole and Shukla 2002; Johnson *et al.* 2017), perhaps due to model errors or biases in simulating the annual cycle of the monsoon (Sperber *et al.* 2013). Can we identify the source(s) of model errors and suggest pathways for model improvements? In an effort to understand the causes of error in dynamical climate models, members of the CLIVAR-GEWEX Working Group on the Asian–Australian monsoons (AAM) have diagnosed processes in CMIP3/5 models and performed idealized numerical experiments with coupled and uncoupled models to identify the possible sources of model errors in the simulation of seasonal-mean monsoon precipitation.

Over the past few decades, sustained research using observations, models, and theory has demonstrated how monsoons arise due to complex interactions among ocean, atmosphere, and land components of the climate system (Charney and Shukla 1981; Hoskins and Rodwell 1995). In the annual cycle, in conjunction with land–surface warming, the AAM can be regarded as the seasonal displacement of the Inter-Tropical Convergence Zone (ITCZ) into the warm (summer) hemisphere. At a finer level of detail, however, simulating the monsoon precipitation climatology with global climate models remains a grand challenge. Despite years of development at modeling centers, the multi-model-mean (MMM) errors for the AAM precipitation climatology (ΔP) in CMIP5 (Taylor *et al.* 2012) have shown little improvement over CMIP3 (Fig. 1; Sperber *et al.* 2013).

During boreal summer (Fig. 1a and b), for example, positive rainfall errors (~3–4 mm/day) cover a large area of the tropical Indian Ocean, while a dry bias exists along the monsoon trough (low-pressure area extending from central India to the plains of Indo-China). A closer look at Fig. 1a and b suggests that the structure of the wet bias is prominent all along the low-level climatological cross-equatorial flow path (Annamalai *et al.* 2017) with a local maximum centered near 60°E. In addition to not getting smaller in *magnitude*, this climatology error is remarkably constant in its *pattern*, suggesting that it must be anchored to particular geographical boundary conditions. The reasons may be nonlocal in time, as errors generated in one season can persist throughout the annual cycle (Levine *et al.* 2013; Annamalai *et al.* 2017), showing how the misrepresentation of a process in one model component can lead to errors in other components. This same logic complicates interpretation of the boreal winter wet bias over the Maritime Continent region (Fig. 1c and d). Compared to observations, modeled wet bias is persistent over the near-equatorial western-southern Indian Ocean, a region that is climatically "dry", and the dry bias is prominent in climatically "wet" spots (Annamalai 2018). In short, present-day climate simulations do not fully capture the basic annual cycle of the AAM (Sperber and Annamalai 2014). Since that mean bias has not improved between CMIP3 and CMIP5, it follows logically that uncertainties in subtler aspects of climate, such as future projections of seasonal changes to AAM rainfall, have likely also not have been reduced between CMIP3 and CMIP5.

This lack of progress in model simulations leads us to wonder if there are fundamental limits to realistically simulating the AAM with

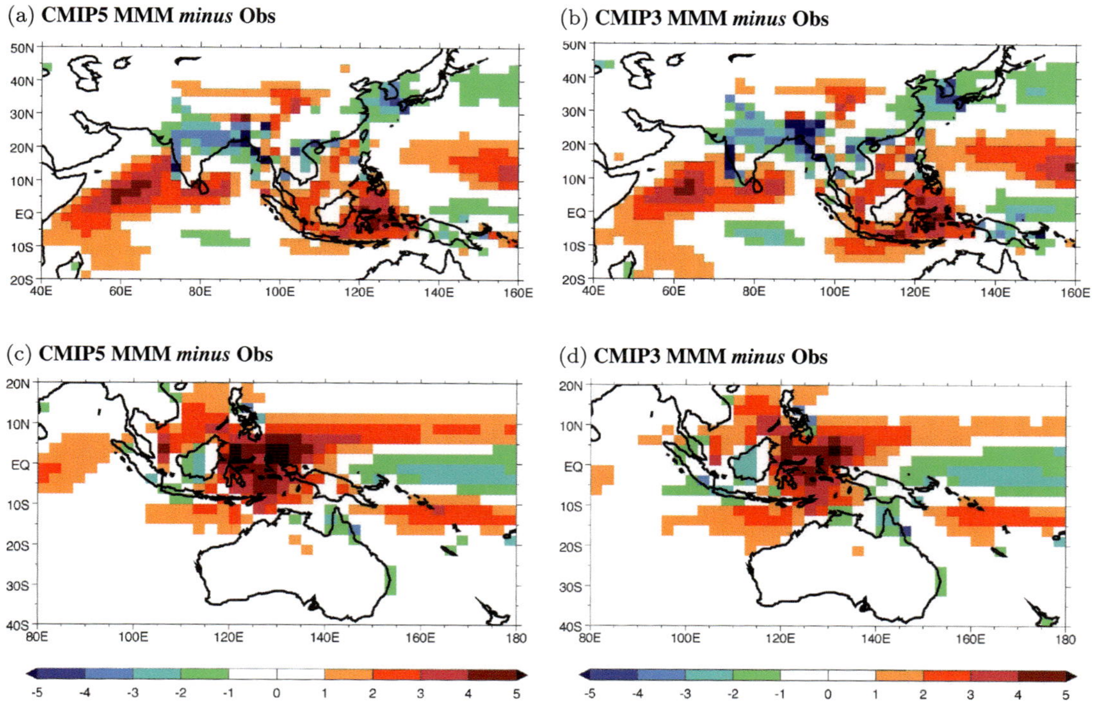

Fig. 1. Seasonal-mean precipitation climatology difference (mm/day) between CMIP5/3 multi-model-mean (MMM) and GPCP observations: (a) and (b): boreal summer; (c) and (d): boreal winter (adapted from Sperber *et al.* 2013).

CMIP-class models (such as unresolved topographic or oceanographic phenomena and their atmospheric effects), or if a concerted observational and modeling effort might be able to enhance the fidelity of these simulations. The persistence of model error patterns suggests that they arise due to multiple processes and their interactions, and/or in fundamental assumptions underlying physical parameterizations (particularly for convection) which may need to be revisited. An effort to understand these biases would require, but might also serve to foster, an improved understanding of the interacting dynamic, thermodynamic, orographic, and oceanographic processes that shape precipitation climatology everywhere: this grand challenge is indeed a worthy goal. Toward that end, this paper summarizes recent works on understanding model errors, emphasizing prospects for future coordinated activities among observationalists, modelers, and diagnosticians to assess the representation of model processes. Success may require sustained observations, both locally in challenge areas and regionally, to constrain model representations of processes.

2. Sources of Model Errors

Here, we review efforts to identify the source(s) of errors in atmospheric processes, the representation of orography, and coupled processes in the Equatorial Indian Ocean (EIO).

2.1. *Atmospheric processes*

To study the development of model errors in the HadGEM1 configuration (Martin *et al.*, 2006) of the Unified Model (MetUM), Martin *et al.* (2010) utilized an initialized forecast approach, as is used in numerical weather prediction (NWP; Klinger and Sardeshmukh 1992), the transpose Atmosphere Model Intercomparison Project (Willamson and Olsen 2007), and the Cloud-Associated Parameterizations Testbed (Phillips *et al.* 2004). The Met Office employs a seamless approach to development of the MetUM, allowing regular comparison of model biases across timescales. Figure 2 illustrates such a comparison: The MetUM-simulated precipitation climatology from the atmosphere-only configuration HadGEM2-A (HadGEM2 Development Team, 2011) during June–July–August (JJA; Fig. 2a) compared to observational estimates shows the prominent "wet bias" over the western EIO and "dry bias" along the monsoon trough (Fig. 2b). The precipitation biases are similar to those from a set of

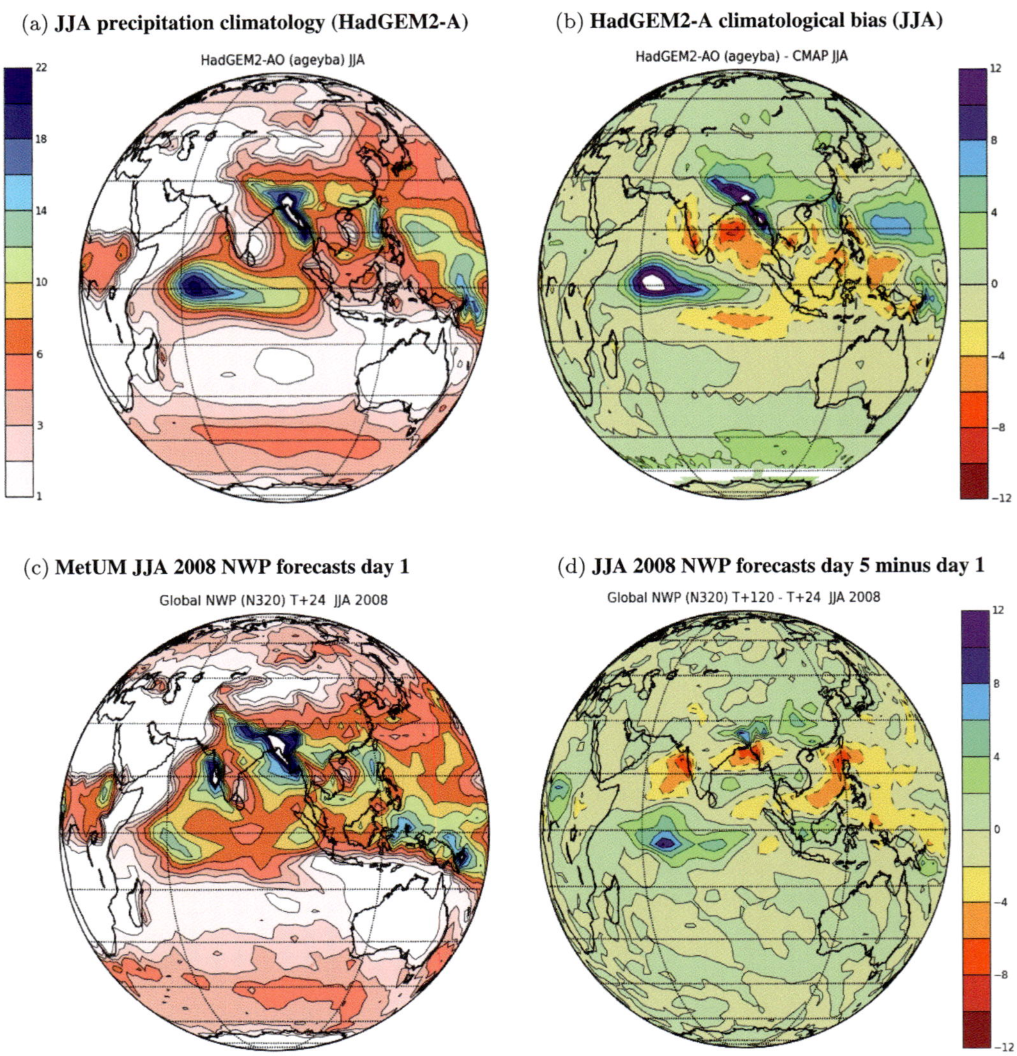

Fig. 2. (a) Boreal summer (JJA) MetUM model precipitation climatology from HadGEM2-A atmosphere-only configuration; (b) model biases (against CMAP observations; Xie *et al.* 2007) in precipitation climatology during JJA; (c) model precipitation averaged over day 1 of a set of initialized NWP forecasts at N320 resolution (∼60 km at the equator) from JJA 2008; (d) model precipitation difference: day 5 minus day 1 of the JJA 2008 NWP forecasts. Units are in mm/day. Figure supplied by Sean Milton.

NWP case studies for 2008 initialized at various times through the JJA season, which show a rapid growth of precipitation errors (ΔP) over the western EIO in the first 5 days of the simulation (Fig. 2c and d). This suggests that the biases over the western EIO are a direct impact of parameterizations (errors due to "fast atmospheric processes") and not dependent on coupled feedback processes operating on longer timescales.

This tactic of prescribing observed sea surface temperature (SST) and initializing the atmosphere with observational estimates generated by data assimilation systems minimizes initial biases in large-scale circulation, facilitating isolation of the growth of model errors due to parameterization shortcomings (Rodwell and Palmer 2007). Generalizing the conclusion of Martin *et al.* (2010) that these tactics of the "seamless" approach lead to parameterization-relevant understanding of rainfall errors on a range of temporal and spatial scales, Ma *et al.* (2014) show comparable success through simple interpolation of analyzed atmospheric states onto the grids of climate models without assimilation systems.

2.1.1. *Entrainment in cumulus schemes*

Mixing coefficients in cumulus convection schemes are always at the top of the list of sensitive uncertain parameters, and thus are fruitful for experimentation. In a set of experiments, Bush *et al.* (2015) multiplied entrainment (ε) and detrainment (δ) rates by a factor of 1.5 in the Global Atmosphere 3.0 (GA3.0) configuration of the MetUM (Walters *et al.* 2011). Precipitation fields and their bias (Fig. 3) show that this configuration exhibits similar systematic errors to those in previous UM configurations (compare Figs. 2b and 3a). While the enhanced entrainment strongly reduced the wet bias over the western EIO (Fig. 3b), modeled precipitation showed a greatly enhanced wet bias over the tropical western Pacific, and the dry bias over the Maritime Continent also intensified. The study went on to show how this pattern of change developed with lead time: more entrainment reduces convective precipitation almost instantaneously, while the compensatory rainfall enhancements elsewhere develop somewhat more gradually. The Asian monsoon consists of multiple regional heat sources, so at low latitudes, perturbations to one are quickly (but not instantly) communicated to others by equatorial waves (Annamalai and Sperber 2005) and by the transport of water vapor decrements or increments that did or did not get condensed out upstream.

Unfortunately, the pattern of cumulus entrainment impacts in the UM in Fig. 3b is very model dependent. The total precipitation difference in Fig. 3c indicates the impact, averaged over lead times of 11–15 days in a set of daily initialized JJAS 2008 hindcasts, of doubling entrainment rates in the SAS cumulus scheme of the U.S. NOAA Global Forecast System (GFS) model, version circa 2014. Because the basic entrainment rate is so small in the GFS model, the pattern of impacts of multiplying entrainment by 2 and by 0 was found to be almost equal and opposite in the tropics, and independent of a second factor varied in the set of experiments, within sampling uncertainty. For this reason, the grand composite from all experiments in panel 3c) may be regarded as twice the effect of doubling entrainment. The main point of Fig. 3c is that *the pattern of GFS entrainment impacts in the Indian sector is totally unlike the pattern of MetUM entrainment impacts in panel 3b*, with increased rainfall over the EIO, and decreased rainfall all around India aside from regions with steep topography.

A third comparable experiment is reported in Zhao *et al.* (2018), using Geophysical Fluid Dynamics Laboratory (GFDL) model AM4.0, whose systematic biases in precipitation over the monsoon region have persisted in three generations (since AM2.0). In these experiments, the pattern again appears monotonic in entrainment

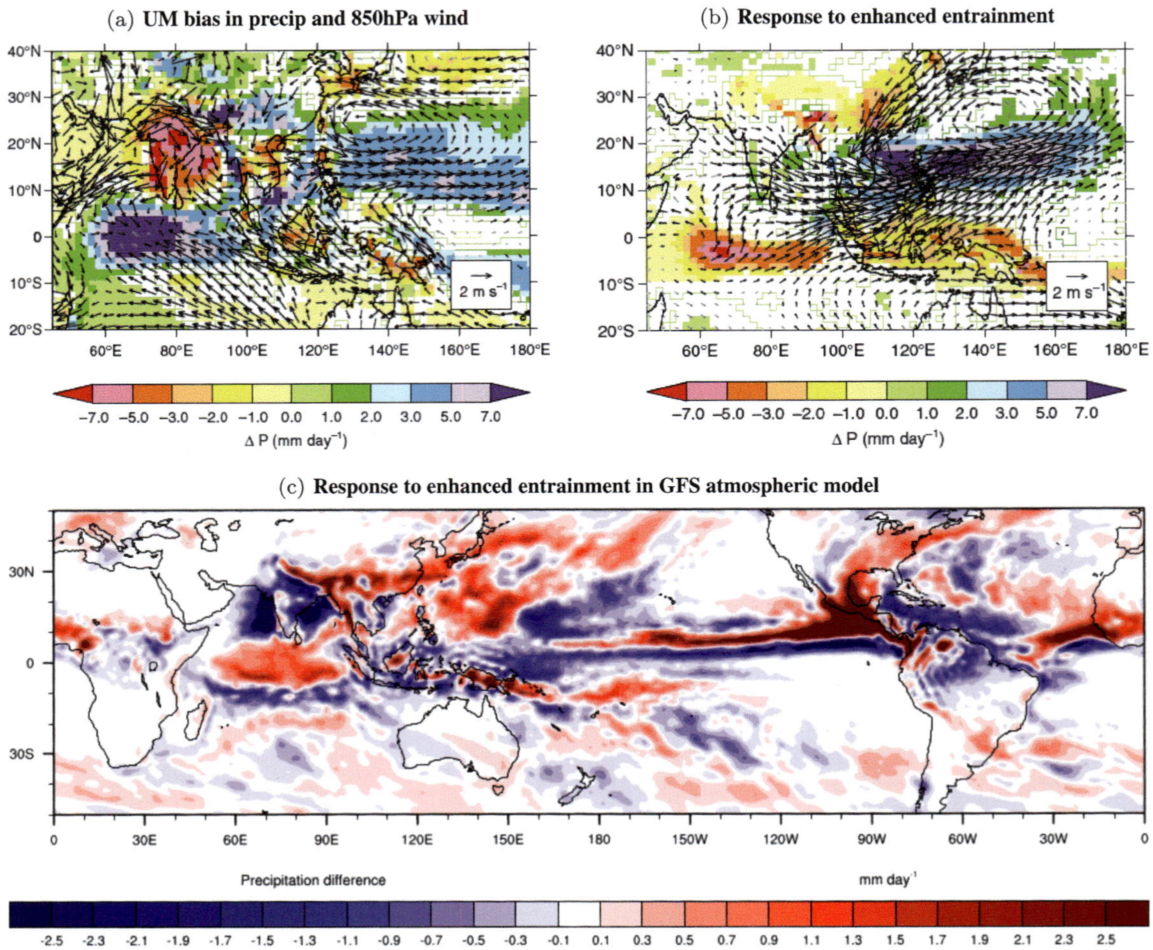

Fig. 3. (a) Unified Model precipitation and 850hPa wind bias during JJAS, (b) precipitation and 850 hPa wind response to 1.5× enhanced cumulus entrainment rates (enhanced entrainment minus control runs), (c) precipitation response to 2× enhanced entrainment rates minus the response to 0× (zero) entrainment rates. Panels a,b adapted from Bush *et al.* (2015); panel c is from unpublished results by Mapes *et al.* pers. comm.

rate. Simulated precipitation increases over the tropical western Pacific with increasing entrainment rates (Fig. 4, bottom panels), consistent with the results in both MetUM and GFS (Fig. 3b and c; noting reversed cool–warm color scales in Fig. 3b). In the Indian Ocean sector, the response is complex, but broadly opposite to the Pacific response, but with an intra-Indian-Ocean pattern that matches neither Fig. 3b nor 3c in any simple sense.

In part these disparate entrainment sensitivities seen in Figs. 3 and 4 may reflect different background biases in each model — as Bush *et al.* (2015) noted, "several aspects of a region's mean state, including moisture content, sea surface temperature and circulation, play a role in local feedbacks which determine the response to increased entrainment and detrainment." But in addition, the stark differences among Figs. 3b, 3c, and 4d may reflect in a cross-model manner what Mapes and Neale (2011) postulated in a single model: that univariate treatments of mixing in convection are subject to a "dilemma" (simultaneously too great and

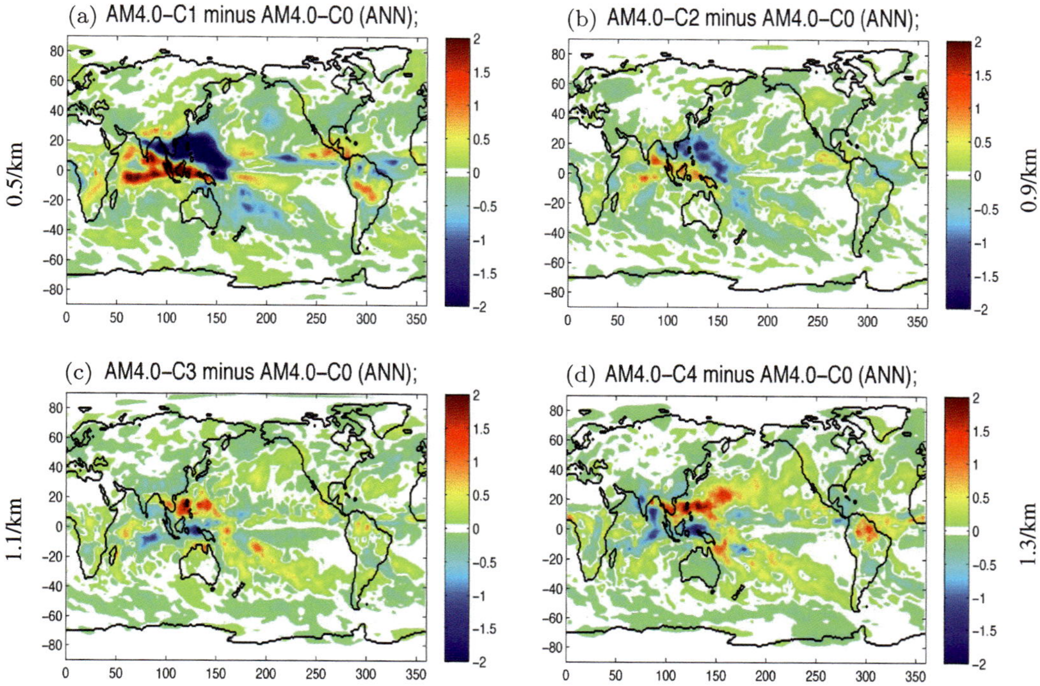

Fig. 4. Geographical distribution of the differences in long-term annual mean surface precipitation (mm/day) for (a) C1 minus C0; (b) C2 minus C0; (c) C3 minus C0, and (d) C4 minus C0. Entrainment values used for the different experiments are C0 (0.9/km); C1 (0.5/km); C2 (0.7/km); C3 (1.1/km), and C4 (1.3/km) (adapted from Zhao *et al.* 2018).

too small), arguably indicative of the univariate framing of the problem being too simplistic (reiterated in Chen and Mapes 2018).

2.1.2. *Why is the WEIO dry in nature but rainy in models?*

Perhaps a focus on one local error could be more clarifying: In climate models, why do large values of ΔP persist over the western EIO? Consider the following scenario. Climatologically, the western EIO receives much less rainfall (dotted line in Fig. 5a) compared to the eastern EIO throughout the annual cycle (thick line in Fig. 5a), with a consistent pattern in vertical velocity (not shown). Therefore, the western EIO is climatically a "dry" region where an increase in near-surface moist static energy (MSE) would be required to initiate convection and maintain precipitation.

Climatologically during May, observed SST peaks around 30°C (Fig. 5b) and promotes local evaporation (not shown), and a weak cross-equatorial flow develops (Fig. 5c) that transports moisture into the Northern Hemisphere. These processes lead to the accumulation of column water vapor (CWV) in the western EIO. As a result, a local maximum in precipitation (~4.0 mm/day) is observed (Fig. 5a). Immediately thereafter in June, SST drops to ~26°C and rainfall decreases subsequently. It is during this "time window" of the annual cycle that the model errors begin to emerge, with the wet bias being associated with warm SST bias over the western EIO (Fig. 5d), and easterly wind stress ($\Delta\tau$) in the EIO (Annamalai *et al.* 2017). Then, due to nonlinear atmospheric feedbacks, ΔP amplifies and the coupled nature of the monsoon–Indian Ocean Climate systems anchors its persistence.

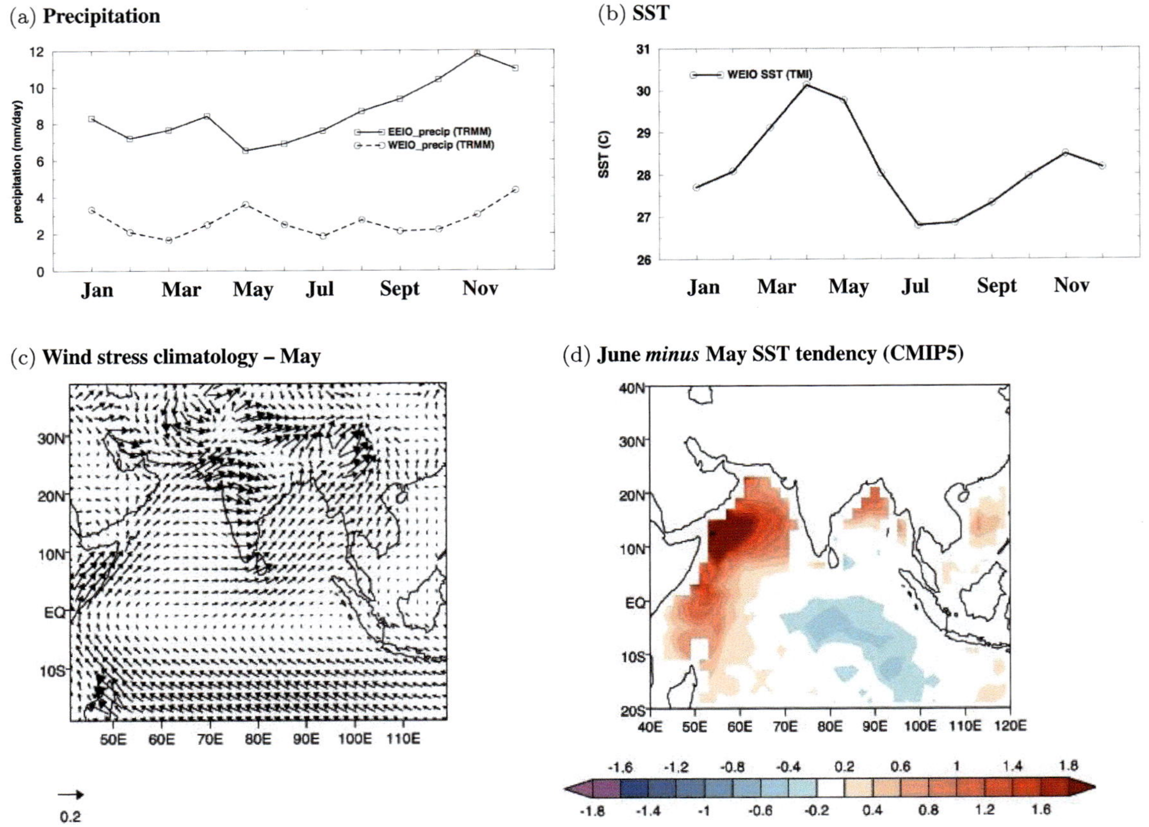

Fig. 5. (a) Observed monthly precipitation (mm/day) climatology averaged over eastern EIO (5°S–5°N; 90–105°E) and western EIO (5°S–5°N; 50–60°E); (b) observed monthly SST (°C) climatology averaged over western EIO; (c) observed wind stress (N/m^2) climatology during May; and (d) June *minus* May SST (°C) tendency bias in CMIP5 MMM (adapted from Annamalai *et al.* 2017).

Hanf and Annamalai (2020) diagnosed a series of NCAR CAM4/5 model solutions and reanalysis products. Their MSE budget analyses from ERA-Interim revealed that after the "burst" of precipitation during late May–early June over the western EIO, accumulation of dry and cold air in the lower troposphere by horizontal advection is the dominant term in anchoring the demise of convection in mid-June, processes not adequately represented in models. The source of this "dry–cold air couplet" is the upwelling region off the Somali–Oman coasts. Noting insufficient representation of convection scheme interactions with environmental air, Hanf and Annamalai (2020) performed sensitivity experiments by increasing entrainment rates in the convection scheme employed in NCAR-CAM4. Their solutions showed a reduction of convective precipitation over the western EIO, and an increase of large-scale precipitation along the monsoon trough. Over the broader Asian monsoon region, a reduction in wet bias over the Maritime Continent is noted but wet bias over the tropical west Pacific is further enhanced.

In cumulus convection schemes, entrainment and detrainment coefficients have notable imprints on cloud properties and precipitation partitioning (Neale *et al.* 2008; Martin *et al.* 2010; Zhao *et al.* 2018), and models whose convection schemes are most responsive to free tropospheric moisture perform better (Lin *et al.* 2012; Neale *et al.* 2008). While inter-model differences are prominent to changes in mixing coefficients, Hanf and Annamalai (2020) also note an increased wet bias over the tropical western Pacific region, consistent with Bush *et al.* (2015) and Zhao *et al.* (2018).

To capture realistic time–mean tropical precipitation, adjustment schemes are one major class of convective parameterizations that simulate the net effect of entrainment and detrainment in releasing the potential instability of the environment rather than the details of these processes themselves. In a recent effort by Fonseca *et al.* (2015), the reference vertical profiles of potential temperature and specific humidity in the Betts–Miller scheme were perturbed to demonstrate how such a scheme can be simply and effectively adapted to improve latent heat release in the global tropics at mesoscale-pentad space–time resolution. Because traditional adjustment schemes do not simulate convective clouds, Koh and Fonseca (2016) developed a rainfall-based Precipitating Convective Cloud scheme to permit cloud–radiation interaction when adjustment schemes are used. The modification of radiative heating is a factor important for linking monsoonal convection to circulation other than the release of latent heat from rainfall itself. The two heating processes, radiative and latent, associated with monsoonal convection are crucial for the longer-term climate simulated over 27 years in Fonseca *et al.* (2018).

2.2. *Orography and resolution*

Boos and Kuang (2010) used observations and numerical model simulations to show that the high humidity and MSE of the South Asian monsoon, and thus its high precipitation, are maintained because a series of mountain ranges insulate that region from dry extratropical air. Refining that view, Boos and Hurley (2013) noted that the largest horizontal gradients in MSE are observed along the Hindu Kush range west of Tibet, between the moist regions of northwest India and the dry desert air of Afghanistan and southwestern Asia. They hypothesized that if the fine-scale orography of the Hindu Kush (near 73°E; Fig. 6a) is not resolved in climate models, then those models will simulate too much transport of low-MSE air across the overly smoothed orography, producing a dry bias over India. Consistent with this hypothesis, they showed that the CMIP3 and CMIP5 multi-model-mean negative bias in South Asian monsoon precipitation is accompanied by a negative bias in near-surface MSE over northwest India, adjacent to the Hindu Kush, and by a southeastward bias in near-surface wind across the Hindu Kush. In CMIP3/5 models, the peak height of orography along 73°E is furthermore around 3 km compared to ∼5 km in observations (Fig. 6b). The idea is that low-MSE air (meaning dry air) can more easily cross over the lower, smoother model topography than it can in observations, ventilating the monsoon and decreasing its strength. To validate their hypothesis, Boos and Hurley (2013) truncated the height of the Hindu Kush in the NCAR coupled model (NCAR-CESM) to ∼1.0 km and found that this caused a negative bias in precipitation over India and surrounding regions of similar magnitude to the CMIP3/CMIP5 bias (Fig. 6b, compare to Fig. 1a and b). Their orographic perturbation also produced negative biases in near-surface MSE and upper-tropospheric temperature that had many similarities to the CMIP3/CMIP5 biases in those thermodynamic fields. While encouraging, the Boos and Hurley (2013) hypothesis and simulations do not capture the CMIP3/CMIP5 wet biases over the adjoining oceans; this raises the question of whether the entire spatial pattern of precipitation bias in the CMIP models is caused by a single model attribute or arises from multiple attributes that perhaps interact.

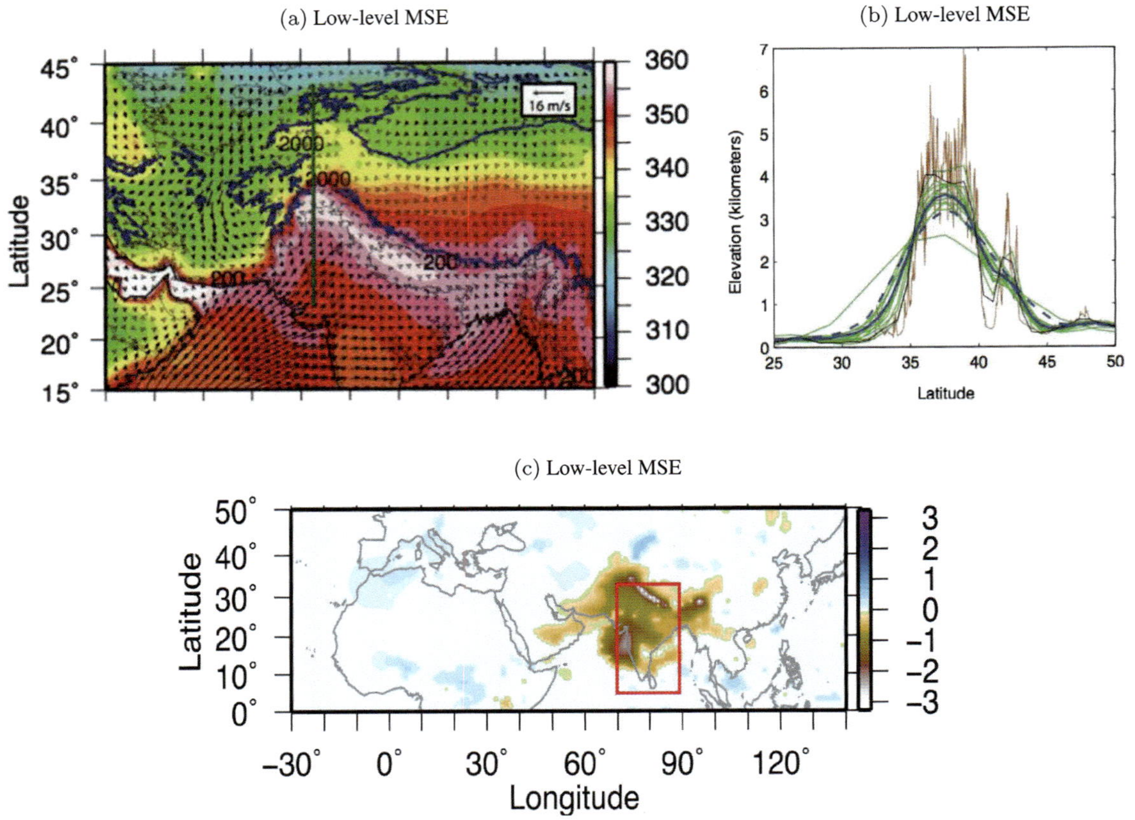

Fig. 6. (a) Observed low-level moist static energy (shaded in color), surface wind and topography outline (blue); (b) Observed topography height (brown line), and individual CMIP model topography (light green lines), CMIP3/5 multi-model-mean topography (thick dashed and boldbluelines); (c) NCARCESM coupled model precipitation response for Hindu Kush orography height truncated to 1 km (adapted from Boos and Hurley 2013).

Increasing horizontal resolution (i.e., reducing the grid spacing) is one of the most straightforward ways of improving the representation of orography in climate models, and there is generally a hopeful expectation that simulations of tropical climate will dramatically improve as models move toward the O(1 km) grid spacings needed to eventually eliminate convective parameterizations. Such increases in resolution come with a high computational cost, but even refining horizontal grid spacing to the comparatively modest value of 50 km seems to greatly improve the representation of synoptic-scale disturbances in South Asia (Praveen *et al.* 2015; Sandeep *et al.* 2018; Hunt and Turner 2017) and in other monsoons (e.g., Pascale *et al.* 2016). Substantial improvement in the simulation of time–mean South Asian rainfall was demonstrated as resolution was refined from roughly 200 km to 50 km in the GFDL model (CM2.1 and CM2.5), with only modest changes in subgrid-scale physics (Delworth *et al.* 2012).

Yet resolution refinement alone is not a panacea, as some of the highest-resolution CMIP5 models had relatively poor simulations of the time–mean boreal summer monsoon, while some of the best-performing CMIP5 models had horizontal grid spacings coarser than 100 km (Sperber *et al.* 2013). Johnson *et al.* (2016) reviewed prior studies of resolution sensitivity, stating that orographic precipitation windward of the Western Ghats and the Indo-China peninsula generally improved in models as grid spacing decreased, but mean rainfall bias over continental India did not change consistently with resolution. They also showed that reducing grid spacing from 200 km to 40 km in the UK MetUM failed to eliminate most of the monsoon biases, but had a number of small, beneficial impacts such as a slight reduction in the negative rainfall bias over peninsular India. By imposing low-resolution topography in a high-resolution version of the Community Atmospheric Model (CAM) version 5.1, Mishra *et al.* (2018) found that many of the improvements in simulated South Asian climate that accompanied increased resolution were caused by the use of a higher-resolution atmosphere rather than higher-resolution surface topography. Simulations of East Asian monsoon rainfall systematically improve at finer resolution, seemingly because of improvements in the stationary eddy transports that are fundamentally forced by orography and land-sea contrast (Yao *et al.* 2017). However, expectations of model improvement with resolution should be tempered by the understanding that tropical rainfall bias can originate in other model components, such as the prescription of land surface albedo (Levine and Boos 2017). Goswami *et al.* (2015), Ramu *et al.* (2016), and Abhik *et al.* (2017) illustrated this well by performing two integrations of the CFSv2 coupled model at ~100-km and ~38-km horizontal grid spacing: They found that changing resolution produced little change in precipitation bias in the tropics in general, with a strong negative bias over India persisting at high resolution (Fig. 7).

2.3. *Coupled processes*

Annamalai *et al.* (2017) examined how misrepresentation of fast oceanic processes could lead to model errors. In CMIP5 models, their analyses indicate that an easterly wind stress bias ($\Delta\tau$) along the EIO begins during April-May and peaks during November (Fig. 8a); the severity of the $\Delta\tau$ is that the Wyrtki Jets (WJs; Wyrtki 1973), eastward-flowing equatorial currents during the intermonsoon seasons (April–May and October–November), are almost eliminated (not shown). As a consequence, an erroneous, east-to-west SST gradient (warm west and cold east) develops in June. An examination of SST tendency errors (June *minus* May) in CMIP5 MMM clearly shows a warming (cooling) tendency over western (eastern) EIO and the emergence of an erroneous SST gradient along the EIO (Fig. 5d). The structure of the model errors indicates that they arise from Bjerknes feedback in the EIO. In CMIP5 models, the reversed gradient in precipitation along the EIO and its persistence result from errors in the coupled processes.

To test their hypothesis that the ΔP over South Asia is due to weakened WJs leading to a too strong Bjerknes feedback, Annamalai *et al.* (2017) forced the coupled model, namely,

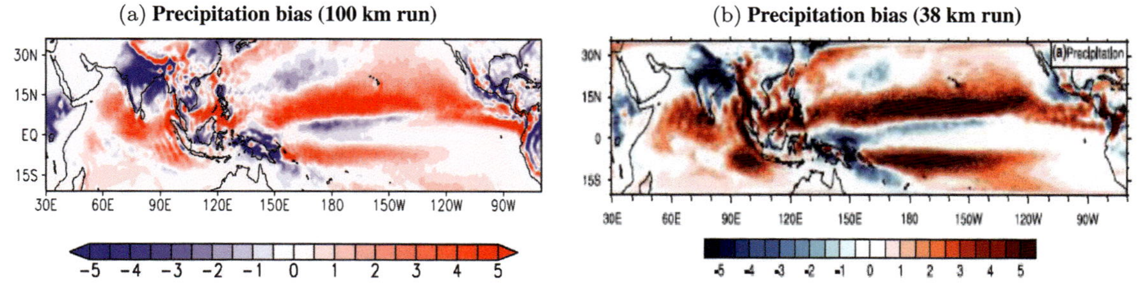

Fig. 7. Boreal summer precipitation bias in CFSv2 coupled model simulations for horizontal resolutions of (a) 100 km and (b) 38 km (adapted from Goswami *et al.* 2015).

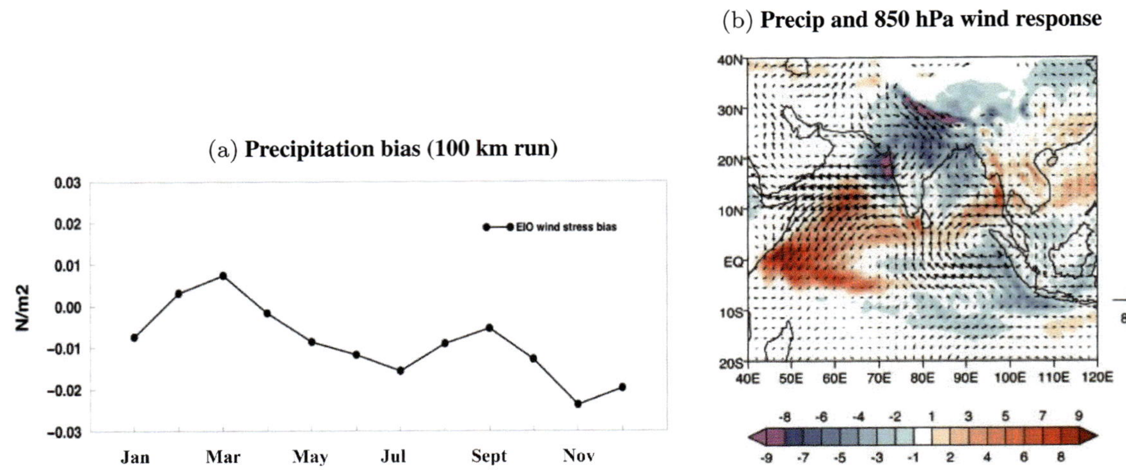

Fig. 8. (a) CMIP5 MMM wind stress along the equatorial Indian Ocean (3°S–3°N; 40°–100°E) and (b) precipitation and 850-hPa wind response in CFES coupled model to suppression of coupled processes along the equatorial Indian Ocean (adapted from Annamalai *et al.* 2017).

Coupled model for Earth Simulator (CFES) developed at JAMSTEC, Japan, with the easterly $\Delta\tau$ noted along the EIO. In one of the sensitivity solutions where the easterly $\Delta\tau$ is introduced only during intermonsoon seasons (April–May; October–November), the sensitivity run develops westward equatorial currents throughout the year, suppressing the WJs in comparison to the control run.

The anomalous forcing also excites oceanic Rossby and Kelvin waves that tend to deepen (shallow) the thermocline depth in the western EIO (eastern EIO). In response, warm (cool) SST and wet (dry) conditions develop in each region, respectively. Additionally, the simulated cold SST bias along the equatorial central Indian Ocean closely resembles errors in the CMIP5 models (Levine *et al.* 2013). Strikingly, the spatial patterns in ΔP and low-level $\Delta\tau$ (Fig. 8b) look identical to the CMIP5 MMM biases (Figs. 1a, b and 3a). Since errors in the sensitivity experiment are driven only from the equator, it follows that the South Asian precipitation errors are as well. More details on the dynamical linkages are discussed in Annamalai *et al.* (2017).

3. Summary and Way Forward

Here, we summarize the recent findings (Section 3.1), and suggest pathways for improving representation of model processes through coordinated modeling and observational efforts (Section 3.2).

3.1. *Summary*

Sources of model errors include fast atmospheric processes (Martin *et al.* 2010) and prominently deep convection, with its large dependence on entrainment and detrainment rates assumed in model convective parameterizations. However, these effects can be subtle: for instance, Bush *et al.* (2015) showed that increased entrainment rate can actually increase precipitation in parts of the monsoon (like the Philippine sea), while decreasing it elsewhere (such as the western Indian ocean). Their results showed that several aspects of a region's mean state, including moisture content, SST, and circulation, play a role in local feedbacks that determine the response to increased entrainment and detrainment. The partitioning between convection scheme and cloud scheme (so-called stratiform or large-scale) precipitation may also be involved. The GFDL AM4.0 and NCAR CAM4 models also have enhanced precipitation over the tropical west Pacific monsoon region for enhanced entrainment rates (Zhao *et al.* 2018; Hanf and Annamalai 2020), an issue to be investigated more systematically in a perturbed physics ensemble. For a given atmospheric treatment of moist processes, the resolution-dependent representation of steep orography also has a strong impact on low-level MSE and rainfall (Boos and Hurley 2013), in a pattern with a distinct spatial fingerprint. Oceanic as well as atmospheric and land processes may also be implicated, with Annamalai *et al.* (2017) arguing that the cause of coupled model errors is the models' near-equatorial, coupled processes (Bjerknes feedback) being too strong. Diagnosing reanalysis products, Hanf and Annamalai (2020) identified horizontal advection of a "dry–cold air couplet" arising near the Somali–Oman coasts as a possible cause in the demise of convection over the western EIO in mid-June, with this near-surface process not adequately represented in NCAR models. Given that the Indian Ocean and monsoon climate systems are inherently coupled, in CMIP5 models, precipitation errors over the Bay of Bengal are attributed to simulated errors in mixed-layer depth over the Arabian Sea (Nagura *et al.* 2018).

3.2. *Way forward*

Despite dedicated efforts by the modeling community to improve process representation, the persistent nature of model precipitation errors over the AAM region leads us to wonder if there are fundamental limits to realistically simulating the monsoon, or if a concerted observational and modeling effort can enhance the fidelity of models in simulating the monsoon. To improve the simulation of the monsoon climatological annual cycle, particularly the heterogeneous regional distribution of precipitation structure and amplitude, the important questions we need to address include the following: At regional spatial scales, do we have sufficient observations for a comprehensive understanding of how dynamic and thermodynamic processes, flow-orography interactions, and ocean-atmosphere and land–atmosphere interactions determine precipitation structure and strength? Moreover, what are the specific observations (land, ocean, and atmosphere) needed to constrain model physics? In modeling efforts, what are the priority areas that need improvements to reduce the systematic errors? For example, do we need to dig down to the convective timescale and examine the scale interactions that make up the climatology?

In the CMIP database, irrespective of horizontal and vertical resolutions and a spectrum of varied physical parameterizations employed, the structure of the climatological precipitation errors is robust across all models. Such a behavior across models suggests that the errors may arise due to multiple processes and their interactions, and/or that certain processes are missing in all models. Since "fast atmospheric" processes have been identified as a leading source of model errors, improvements to cloud-cumulus schemes and cloud–radiation feedbacks stand out as top priorities. However, no detailed process-based diagnostics at a level close enough to parameterization schemes has been performed yet to identify the limitations in the representation of individual processes and their interactions. For example, do errors emerge due to (a) convective triggering processes (possibly associated with inadequate representations of boundary layer processes), (b) lack of convection schemes' interactions with environmental humidity (poorly representing moisture–convection feedbacks), (c) lack of communication between convection-cloud schemes (large-scale precipitation) leading to misrepresentation of cloud–radiation feedbacks, or (d) communication deficiencies between shallow and deep convection schemes? A thorough examination of modeled vertical processes and their interactions is needed to identify the initial source in the chain of errors and to provide pathways for model improvements.

Could the precipitation errors in the CMIP5 models arise from errors due to coupled rather than fast atmospheric processes (a view supported by the results of Annamalai *et al.* 2017)? To check this idea, Fig. 9 compares precipitation and SST climatologies in the AFES (Fig. 9a) and CFES (Fig. 9b) control simulations. In both runs the atmospheric model is the same, the two solutions differing in that AFES is forced with daily-varying, observed SST for the period 1979–2010. The simulated precipitation in AFES is biased with higher precipitation >4–8 mm/day ~west of 65°E, a bias similar to that in the CMIP5 models and in other atmosphere-only models (Ma *et al.* 2014). In contrast, this erroneous feature is clearly rectified by the coupling in the CFES solution: While the structure of precipitation along the EIO is present in the AFES and CFES solutions, its amplitude distribution is more realistic in CFES, implying the need for the correct representation of EIO coupled processes. Similar results are noticeable in MIROC5 coupled (Sperber *et al.* 2013) *versus* uncoupled solutions (Ma *et al.* 2014).

Motivated by this idea, Hanf and Annamalai (2020) diagnosed coupled and uncoupled simulations of NCAR family models and concluded that realistic representations of regional air-sea interactions and boundary layer processes along the western EIO are also important for the improved simulation of precipitation and reduction of systematic errors over the western EIO. In this scenario, errors in either atmospheric or oceanic variables could lead to errors in coupled processes due to feedbacks. In other words, even if the fast atmospheric processes are realistically represented initially, errors in the representation of near-surface wind could cascade into surface fluxes, ocean currents, and thermocline properties with subsequent effects on SST and

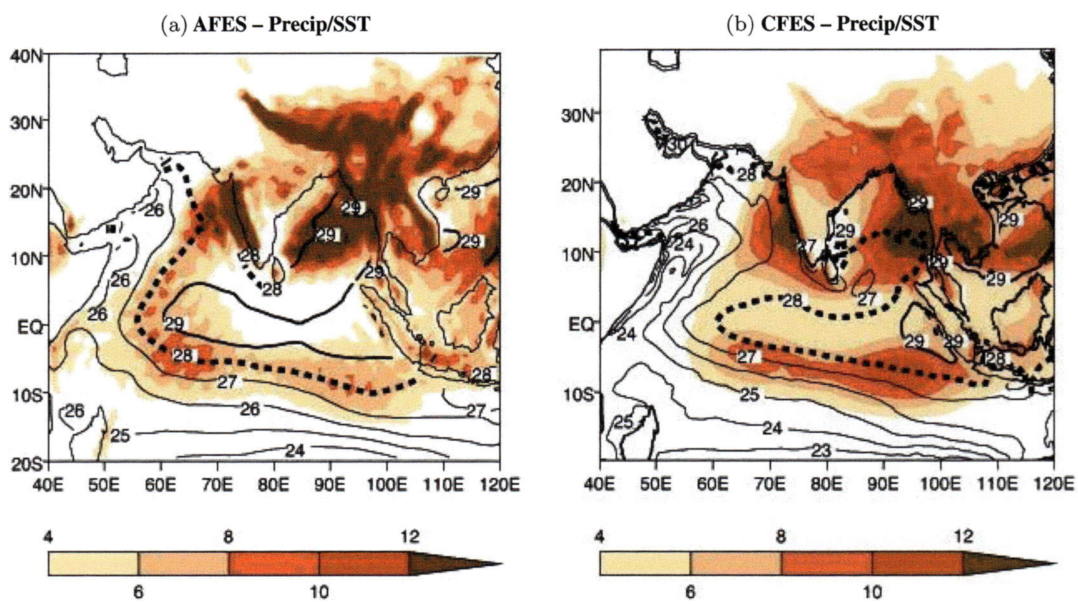

Fig. 9. The seasonal (June–September) mean precipitation (shaded, mm/day) and SST (contour, °C) climatology from control solutions for (c) AFES and (d) CFES. Note that AFES is forced with climatologically varying daily-observed SST (from Annamalai *et al.* 2017).

atmospheric convection. Given that in CMIP5-era coupled models, cold SST bias persists over the northern Indian Ocean (Levine *et al.* 2013), it is important to perform systematic process-based diagnostics to assess representation of upper-oceanic processes (e.g., mixed-layer and salinity stratification). Specifically over regions of high mean precipitation such as the northern Bay of Bengal, modeling the thin mixed-layer and upper-ocean stratifications in temperature and salinity, and their impact on SST evolution, remains a challenge (Annamalai *et al.* 2019). In these regions, the ocean model component optimal vertical resolution needed to resolve the upper-ocean physics is not clear (Kelvin Richards 2018, personal communication).

Land surface processes were not discussed in detail here, although their role in setting the moist energy content of the lower troposphere and thus in influencing moist convection was discussed in the thermodynamic analyses of Boos and Hurley (2013) and Levine and Boos (2017). While land–atmosphere coupling has been identified as an important factor for summer precipitation over South Asia (Koster *et al.* 2004), large inter-model uncertainty exists at local and regional scales (Dirmeyer *et al.* 2006; Douville *et al.* 2001). Due to lack of high-quality observations of surface fluxes of latent and sensible heat as well as soil moisture over the Asian monsoon region, it is challenging to model the land–atmosphere interactions accurately.

Both for comprehensive understanding of the heterogeneous regional distribution of precipitation and to constrain model physics new observations of the atmosphere, land surface, and ocean are needed. Recent field studies such as ASIRI (An Ocean-Atmosphere Initiative for Bay of Bengal; 2013–2017) have collected high-resolution observations from multiple instruments and have identified features in the northern Bay of Bengal where shallow, salinity-controlled mixed layers result from high river runoff and heavy rainfall (Wijesekera *et al.* 2016); mixed layers there are only about $\sim$20 m deep, and lateral mixing due to oceanic features contributes to upper-ocean salinity stratification (Mahadevan *et al.* 2016). The ASIRI program in the Bay of Bengal was complemented by an Arabian Sea campaign that deployed autonomous sensors during 2015–2018: the Northern Arabian Sea Circulation autonomous research (NASCar) initiative (Centurioni *et al.* 2017). In conjunction with the above field campaigns, sustained observations over the historically data-sparse tropical Indian Ocean are being collected, with more planned as part of IndOOS 2020–2050 (Indian Ocean Observation System). As part of RAMA (Research Array of African–Asian–Australian Monsoon Analysis and Prediction), moored buoys cover the EIO, southern Indian Ocean, and parts of the Bay of Bengal (around 90°E), and coordinated through IndOOS nearly 1000 Argo floats and drifters have been deployed over the Indian Ocean. Yet, RAMA buoys do not cover regions in the northern bay (poleward of 14°N) where upper-ocean stratification is particularly important. As part of the planned IndOOS 2020–2050, Annamalai *et al.* (2019) strongly recommended the deployment of a few buoys in the northern bay and creation of an intensive observational array (covering 12°–22°N; 80°–100°E) that contains Argo floats equipped with auxiliary surface temperature salinity (STS) sensors ($\sim$10-cm vertical resolution in the top 2–4 m) similar to those implemented in a few regions of the Bay of Bengal (Anderson and Riser 2014). Sustained and adequate observations are needed to resolve and model the freshwater-induced upper-ocean stratification that is deemed important for maintaining the warm pool over the Bay of Bengal (e.g., Shenoi *et al.* 2002; Mahadevan *et al.* 2016).

Very recent field campaigns such as INCOMPASS (Interaction of Convective Organization with Monsoon Precipitation, Atmosphere, Surface and Sea) and BoBBLE (Bay of Bengal Boundary Layer Experiment) collected airborne and ground measurements of the 2016 Indian monsoon. These campaigns also performed detailed modeling at a range of resolutions, generated multiple forecast products from a variety of model configurations, diagnosed the dynamic and thermodynamic structures of the monsoon over southern India, examined the land–atmosphere coupling during monsoon onset, and studied the air-sea interface over the Bay of Bengal (a series of papers has been submitted to a QJRMS special collection). While encouraging, high-frequency (hourly to daily) sustained observations related to precipitation (e.g., satellite-derived hourly precipitation estimates from the Global Precipitation Mission), representation of vertical processes in the free troposphere (e.g., vertical profiles of vertical velocity, cloud amount, Q1, Q2, and MSE), and boundary layer structure are required to assess model performance during different convective life cycles. Satellite-derived estimates of convective and large-scale precipitation and Q1 profiles need to be verified against *in situ* observations. To constrain the convection tendencies Q1, Q2 in models, DOE-ARM (Department of Energy-Atmosphere Radiation Measurement) observations in the West Pacific are being currently used (Neale 2019, personal communication). A further network of ARM-based observations in other regions of the AAM (e.g., along the monsoon trough over the plains of India and Indo-China) would immensely benefit the modeling community. *Our conclusion will continue to be that without such an observational effort, improvement of the representation of physical processes in numerical models will be severely limited.*

New efforts to better observe land surface processes and reduce land model bias also provide great potential for improvement of monsoon simulations. The recently launched Land Surface, Snow and Soil Moisture Model Intercomparison Project (LS3MIP; van den Hurk *et al.* 2016) aims to assess the sources of bias in the land surface components of current Earth system models; it will assess land surface model biases in an offline mode (where the land model is integrated with prescribed atmospheric forcings) and in coupled frameworks. Since the LS3MIP community has a global focus, engagement of the monsoon research community will be key to leveraging its methods for the improvement of monsoon precipitation simulation. Numerous other land surface modeling and observational efforts are being coordinated, such as land surface fluxes of sensible and latent heat estimated by the Land Flux project (McCabe *et al.* 2016) and the Global Soil Wetness Project (GSWP3; Dirmeyer *et al.* 2006).

Regional coupled modeling at very high spatial resolution is beginning to emerge for the Asian monsoon region. Therefore, high-quality sustained observations, process-based diagnostics at a level close enough to parameterizations, and continued coordinated activities across disciplines, and among diagnostics, modeling, and observational personnel, are very much required to make progress in monsoon modeling. Note that the Asian monsoon comprises heterogeneous regional precipitation centers; given their close proximity to the Equator, perturbations to one of the precipitation centers are quickly communicated to others by the equatorial wave response (Annamalai and Sperber 2005). In other words, if one of the model-simulated precipitation centers is erroneous, so is the entire Asian monsoon. So, future efforts need to focus on improving the entirety of the Asian monsoon, including high mean precipitation regions of the equatorial Indian Ocean and the tropical West Pacific.

Acknowledgments

H. Annamalai acknowledges the funding from the National Science Foundation (NSF) under grant 1460742. W.R. Boos was supported by U.S. Department of Energy, Office of Science, Biological and Environmental Research under Award Number DE-SC0019367. Gill Martin was supported by the Met Office Hadley Centre Climate Programme funded by BEIS and Defra. The Indian Institute of Tropical Meteorology (IITM), Pune, is fully funded by the Ministry of Earth Sciences (MoES), Government of India. Model runs reported by PM are carried out on the MoES Aditya High Performance Computing (HPC) system at IITM. The authors sincerely thank the two anonymous reviewers for their comments/suggestions that led to the improvement of Section 3.2.

References

Abhik S., R. P. M. Krishna, M. Mahakur, M. Ganai, P. Mukhopadhyay, and J. Dudhia, 2017: Revised cloud processes to improve the mean and intraseasonal variability of Indian summer monsoon in climate forecast system: Part 1. *J. Adv. Model Earth Syst.* **9**, 1–28, doi:10.1002/2016MS000819

Anderson, J. E. and S. C. Riser, 2014: Near-surface variability of temperature and salinity in the near-tropical ocean: Observations from profiling floats. *J. Geophys. Res. Oceans*, **119**, 7433–7448, doi: 10.1002/2014JC010112.

Annamalai, H., B. Taguchi, J. P. McCreary, M. Nagura, and T. Miyama, 2017: Systematic errors in South Asian monsoon simulation: Importance of equatorial Indian Ocean processes. *J. Clim.*, **30**, 8159–8178, doi: 10.1175/JCLI-D-16-0573.1.

Annamalai, H., M. Nagura, and K. Richards, 2019: Indian Ocean and South-Asian monsoon: upper-ocean processes relevant to monsoon annual cycle. A White Paper chapter in IndOOS(2020-50) Review.

Annamalai, H. and K. R. Sperber, 2005: Regional heat sources and the active and break phases of boreal summer intraseasonal (30–50 day) variability. *J. Atmos. Sci.*, **62**, 2726–2748.

Annamalai, H. 2018: Model systematic errors in the annual cycle of monsoon: Inferences from process-based diagnostics. In '*Current trends in the Representation of Physical Processes in Weather and Climate Models*,' eds. Randall *et al.*

Boos, W. R. and J. V. Hurley, 2013: Thermodynamic bias in the multi-model mean boreal summer monsoon. *J. Clim.*, **26**, 2279–2287.

Boos, W. R. and Z. Kuang, 2010: Dominant control of the South Asian monsoon by orographic insulation versus plateau heating. *Nature*, **463**(7278), 218–222, http://doi.org/10.1038/nature08707.

Bretherton, C. S., M. E. Peters, and L. E. Back, 2004: Relationships between water vapor path and precipitation over the tropical oceans. *J. Clim.*, **17**, 1517–1528.

Bush, S. J., A. G. Turner, S. J. Woolnough, G. M. Martin, and N. P. Klingaman, 2015: The effect of increased convective entrainment on Asian monsoon biases in the MetUM general circulation model. *Q.J.R. Meteorol. Soc.*. **141**, 311–326, doi: 10.1002/qj.2371.

Centurioni, L., A. Horanyi, C. Cardinali, E. Charpentier, and R. Lumpkin, 2017: A global ocean observing system for measuring sea level atmospheric pressure: Effects and impacts on numerical weather prediction. *Bull. Amer. Met. Soc.*, https://doi.org/10.1175/BAMS-D-15-00080.1.

Charney, J. G. and J. Shukla, 1981: Monsoon predictability. In *Monsoon Dynamics*, Eds. Lighthill and Pearce, R.P, Cambridge Press, pp. 99–109.

Chen, B. and B. E. Mapes, 2018: Effects of a simple convective organization scheme in a two-plume GCM. *J. Adv. Model. Earth Syst.*, **10**, 867–880, https://doi.org/10.1002/2017MS001106.

Del Sole, T. and J. Shukla, 2002: Linear prediction of Indian monsoon rainfall. *J. Clim.*, **15**, 3645–3658.

Delworth, T. L., A. Rosati, W. Anderson, A. J. Adcroft, V. Balaji, R. Benson, and R. Zhang, 2012: Simulated climate and climate change in the GFDL CM2.5 high-resolution coupled climate model. *J. Clim.*, **25**(8), 2755–2781, http://doi.org/10.1175/JCLI-D-11-00316.1.

Dirmeyer, P. A., Z. Guo, and X. Gao, 2004: Comparison, validation, and transferability of eight multi-year global soil wetness products. *J. Hydrometeor.*, **5**, 1011–1033.

Douville, H., F. Chauvin, and H. Broqua, 2001: Influence of soil moisture on the Asian and African monsoons. Part I: Mean monsoon and daily precipitation. *J. Clim.*, **14**, 2381–2403.

Fonseca, R. M., T. Zhang, and K.-T. Yong, 2015: Improved simulation of precipitation in the tropics using a modified BMJ scheme in the WRF model. *Geosci. Model Dev.*, **8**, 2915–2928.

Fonseca, R., T.-Y. Koh, and C.-K. Teo, 2018: Multi-scale interactions in a high-resolution tropical-belt experiment and observations. *Clim. Dyn.*, in press.

Goswami, B. B., R. P. M. Krishna, P. Mukhopadhyay, M. Khairoutdinov, and B. N. Goswami,

2015: Simulation of the Indian summer monsoon in the superparameterized Climate Forecast System version 2: Preliminary results. *J. Clim.*, **28**(22), 8988–9012, doi:10.1175/837 JCLI-D-14-00607.1.

Hanf, F. and H. Annamalai, 2020: Systematic errors in South Asian monsoon precipitation: Process-based diagnostics and sensitivity to entrainment in NCAR models. *J. Clim.*, 33, 2817–2840.

Hoskins, B. J. and M. J. Rodwell, 1995: A model of the Asian summer monsoon. Part I: The global scale. *J. Atmos. Sci.*, **52**, 1329–1340.

Hunt, K. M. R. and A. G. Turner, 2017: The effect of horizontal resolution on Indian monsoon depressions in the Met Office NWP model. *Quart. J. Roy. Meteorol. Soc.*, **143**, 1756–1771, http://doi.org/10.1002/qj.3030.

Johnson, S. J., R. C. Levine, A. G. Turner, G. M. Martin, S. J. Woolnough, R. Schiemann, *et al.*, 2016: The resolution sensitivity of the South Asian monsoon and Indo-Pacific in a global 0.35° AGCM. *Clim. Dynam.*, http://doi.org/10.1007/s00382-015-2614-1.

Johnson, S. J., A. G. Turner, S. J. Woolnough, G. M. Martin, and C. MacLachlan, 2017: An assessment of Indian monsoon seasonal forecasts and mechanisms underlying monsoon interannual variability in the Met Office GloSea5-GC2 system. *Clim. Dynam.*, **48**(5), pp. 1447–1465, ISSN 14320894, doi: https://doi.org/10.1007/ s0038201631512.

Klinker, E. and P. D. Sardesmukh, 1992: The diagnosis of mechanical dissipation in the atmosphere from large-scale balance requirements. *J. Atmos. Sci.*, **49**, 608–627.

Koh, T.-Y. and R. M. Fonseca, 2016: Subgrid-scale cloud-radiation feedback for the Betts–Miller-Janjić convection scheme. *Q.J.R. Meteorol. Soc.*, **142**, 989–1006.

Koster, R. D., P. A. Dirmeyer, Z. Guo, G. Bonan, E. Chan, P. Cox, H. Davies, T. Gordon, S. Kanae, E. Kowalczyk, D. Lawrence, P. Liu, S. Lu, S. Malyshev, B. McAvaney, K. Mitchell, T. Oki, K. Oleson, A. Pitman, Y. Sud, C. Taylor, D. Verseghy, R. Vasic, Y. Xue, and T. Yamada, 2004: Regions of strong coupling between soil moisture and precipitation. *Science*, **305**, 1138–1140, doi: 10.1126/science.1100217.

Levine, R. C., A. G. Turner, D. Marathayil, and G. M. Martin, 2013: The role of northern Arabian Sea surface temperature biases in CMIP5 model simulations and future projections of Indian summer monsoon rainfall. *Clim. Dynam.*, **41**, 155–172.

Levine, X. J. and W. R. Boos, 2017: Land surface albedo bias in climate models and its association with tropical rainfall. *Geophys. Res. Lett.*, **44**(12), 1–10, http://doi.org/10.1002/2017GL072510.

Lin, *et al.*, 2012: TWP-ICE global atmospheric model intercomparison: Convection responsiveness and resolution impact. *J. Geo. Phys. Res.*, **117**, D09111, doi:10.1029/2011JD017018.

Ma, H.-Y. *et al.*, 2104: On the correspondence between mean forecast errors and climate errors in CMIP5 models. *J. Clim.*, **27**, 1781–1798.

Mahadevan, A., G. Spiro Jaeger, M. Freilich, M. Omand, E. L. Shroyer, and D. Sengupta, 2016: Freshwater in the Bay of Bengal: Its fate and role in air-sea heat exchange. *Oceanography* **29**(2), 72–81, http://dx.doi.org/10.5670/oceanog.2016.40.

Mapes, B. E. and R. B. Neale, 2011: Parameterizing convective organization to escape the entrainment dilemma. *J. Adv. Model. Earth Syst.*, **3**, M06004, https://doi.org/10.1029/2011MS000042.

Martin, G. M., M. A. Ringer, V. D. Pope, A. Jones, C. Dearden, and T. J. Hinton, 2006: The physical properties of the atmosphere in the new Hadley Centre Global Environmental Model (HadGEM1). Part I: Model description and global climatology. *J. Clim.*, **19**, 1274–1301.

Martin, G. M., S. F. Milton, C. A. Senior, M. E. Brooks, S. Ineson, T. Reichler, and J. Kim, 2010: Analysis and reduction of systematic errors through a seamless approach to modeling weather and climate. *J. Clim.*, **23**, 5933–5957.

McCabe, M. F., A. Ershadi, C. Jimenez, D. G. Miralles, D. Michel, and E. F. Wood, 2016: The GEWEX Land Flux project: evaluation of model evaporation using tower-based and globally gridded forcing data. *Geosci. Model Dev.*, **9**, 283–305

Mishra, S. K., A. Anand, J. Fasullo, S. Bhagat, S. K. Mishra, A. Anand, *et al.*, 2018: Importance of the Resolution of Surface Topography in Indian Monsoon Simulation. *J. Clim.*, http://doi.org/10.1175/JCLI-D-17-0324.1.

McCreary, J. P., P. K. Kundu, and R. Molinari, 1993: A numerical investigation of dynamics, thermodynamics and mixed-layer processes in the Indian Ocean. *Prog. Oceanogr.*, **31**, 181–244.

Nagura, M., J. P. McCreary, and H. Annamalai, 2018: Origins of coupled-model biases in the Arabian-Sea climatological state. *J. Climate*, **31**(5), 2005–2029, doi:10.1175/JCLI-D-17-0417.1.

Neale, R. B., J. H. Richter, and M. Jochum, 2008: The Impact of Convection on ENSO: From a Delayed Oscillator to a Series of Events. *J. Clim.*, **21**, 5904–5924.

Pascale, S., S. Bordoni, S. B. Kapnick, G. A. Vecchi, L. Jia, T. L. Delworth, *et al.*, 2016: The impact of horizontal resolution on North American monsoon Gulf of California moisture surges in a suite of coupled global climate models. *J. Clim.*, **29**(21), 7911–7936, http://doi.org/10.1175/JCLI-D-16-0199.1.

Phillips, T. J. and Coauthors, 2004: Evaluating parameterizationsin General Circulation Models: Climate simulation meets weather prediction. *Bull. Amer. Meteor. Soc.*, **85**, 1903–1915.

Praveen, V., S. Sandeep, and R. S. Ajayamohan, 2015: On the relationship between mean monsoon precipitation and low pressure systems in climate model simulations. *J. Clim.*, **28**(13), 5305–5324, http://doi.org/10.1175/JCLI-D-14-00415.1.

Ramu D. A., C. T. Sabeerali, R. Chattopadhyay, D. N. Rao, G. George, A. R. Dhakate, K. Salunke, A. Srivastava, and S. A. Rao, 2016: Indian summer monsoon rainfall simulation and prediction skill in the CFSv2 coupled model: Impact of atmospheric horizontal resolution. *J. Geophy. Res.*, **121**, 1–17, doi:10.1002/2015JD024629.

Rodwell, M. J. and T. N. Palmer, 2007: Using numerical weather prediction to assess climate models. *Quart. J. Roy. Meteor. Soc.*, **133**, 129–146.

Sandeep, S., R. S. Ajayamohan, W. R. Boos, T. P. Sabin, and V. Praveen, 2018: Decline and poleward shift in Indian summer monsoon synoptic activity in a warming climate. *Proc. Nat. Acad. Sci. USA*, **115**(11), http://doi.org/10.1073/pnas.1709031115.

Shenoi, S. S. C., D. Shankar and S. R. Shetye, 2002: Differences in heat budgets of the near surface Arabian Sea and Bay of Bengal: Implications for the summer monsoon. *J. Geo. Phy.*, **107**(C6), 3052, doi: 10.1029/2000JC000679.

Sperber, K. R. *et al.*, 2013: The Asian monsoon: an intercomparison of CMIP3vs CMIP5 simulations of the late 20th century. *Clim. Dynam.*, **41**(9–10), 2711–2744, doi:10.1007/s00382-012-1607-6.

Sperber, K. R. and H. Annamalai, 2014: The use of fractional accumulated precipitation for the evaluation of the annual cycle of monsoons. *Clim. Dynam.*, **43**(12), 3219–3244, doi:10.1007/s00382-014-2099-3.

Taylor, K. E., R. J. Stouffer, and G. A. Meehl, 2012: An Overview of CMIP5 and the Experiment Design. *Bull. Amer. Meteor. Soc.*, **93**, 485–498.

van den Hurk, B. H. and Coauthors, 2016: LS3MIP (v1.0) contribution to CMIP6: The land surface, snow and soil moisture model intercomparison project — Aims, setup and expected outcome. *Geosci. Model. Dev.*, **6**, 2809–2832.

Walters, *et al.*, 2011: The Met Office Unified Model Global Atmosphere 3.0/3.1 and JULES Global Land 3.0/3.1 configurations. *Geosci. Model Dev.*, **4**, 919–941, https://doi.org/10.5194/gmd-4-919-2011.

Wijesekera, *et al.*, 2016: ASIRI: An ocean-atmosphere initiative for Bay of Bengal. *Bull. Amer. Met. Soc.*, 1859–1884

William, D. L. and J. G. Olsen, 2007: A comparison of forecast errors in CAM2 and CAM3 at the ARM Southern Great Plains site. *J. Clim.*, **20**, 4572–4585.

Xie P., P. A. Arkin, and J. E. Janowiak, 2007: *CMAP: The CPC Merged Analysis of Precipitation.* In: *Measuring Precipitation From Space. Advances in Global Change Research*, Eds. Levizzani V., Bauer P., Turk F.J. Vol 28. Springer, Dordrecht.

Yao, J., T. Zhou, Z. Guo, X. Chen, L. Zou, and Y. Sun, 2017: Improved performance of high-resolution atmospheric models in simulating the East Asian summer monsoon rain belt. *J. Clim.*, http://doi.org/10.1175/JCLI-D-16-0372.1.

Zhao, M., J. C. Golaz, I. M. Held, H. Guo, V. Balaji, R. R. Benson, and Coauthors, 2018: The GFDL global atmospheric and land model AM4.0/LM4.0: Model description, sensitivity studies, and tuning strategies. *J. Adv. Model. Earth Syst.*, **10**, 735–769, doi:10.1002/2017MS001209.

Part V
Field Experiments

Chapter 30

The Southern China Monsoon Rainfall Experiment (SCMREX)

Yali Luo*,¶, Xinghua Bao*, Hui Wang*, Rudi Xia*, Liping Liu*, Weitao Lu*, Mengwen Wu*, Kalli Furtado†, Yanluan Lin‡ and Xiantong Liu§

**State Key Laboratory of Severe Weather, Chinese Academy of Meteorological Sciences, Beijing, China*
†Met Office, Exeter, UK
‡Center for Earth System Science, Tsinghua University, Beijing, China
§Institute of Tropical and Marine Meteorology/Guangdong Provincial Key Laboratory of Regional Numerical Weather Prediction, China Meteorological Administration, Guangzhou, China
¶ylluo@cma.gov.cn

During the pre-summer rainy season (April–June), Southern China often experiences frequent occurrences of extreme rainfall, leading to severe flooding and inundations (Ding 1994). Although scientific understanding of the processes responsible for heavy rainfall production continues to advance (Schumacher 2017), quantitative precipitation forecasting is still a great challenge over the world including Southern China (Zhang *et al.* 2011). To improve the QPF skill of the pre-summer rainy season rainfall, the China Meteorological Administration (CMA) initiated a nationally coordinated research project, the Southern China Monsoon Rainfall Experiment (SCMREX; Luo *et al.* 2017). With the participation and strong support of international experts, the SCMREX was approved by the World Meteorological Organization (WMO) as a World Weather Research Programme (WWRP) Research and Development Project (RDP).

The scientific objectives of the SCMREX are 1) to better understand the physical mechanisms governing the initiation, evolution, morphology, organization, and duration of the heavy-rain-producing MCSs that determine the precise timing and location and accumulative amount of rainfall; 2) to better understand the microphysical and kinematic structures of the heavy-rain-producing MCSs that determine the instantaneous rainfall amount and are of relevance when validating parameterization schemes representing cloud-precipitation microphysical processes in numerical weather prediction (NWP) models; and 3) to improve fine-scale quantitative precipitation forecasting (QPF) skills by better understanding the multi-scale precipitation processes, assessing the impact of assimilating high-resolution observations into convection-permitting (horizontal grid spacing of 1–4 km) numerical models, and evaluating and improving cloud microphysical parameterization schemes in convection-permitting models. This article reviews the progress of SCMREX in 2016-17 and briefly discusses future research opportunities.

1. Observations

The field campaign component of SCMREX aims to obtain integrated observations to detect the atmospheric environmental conditions and internal structures of rainstorms during the pre-summer rainy season of southern China. The intensive observation periods (IOPs) of 2013 covered 15 days (8–17 and 24–28 May 2013), while the IOPs of 2014–2017 ran continuously from 1 May to 15 June. The observation regions covered Guangdong, southeastern Guangxi, Hainan, Hong Kong, and the offshore South China Sea (SCS) (Fig. 1a). During each IOP, several sounding stations in Guangdong, Guangxi, Hainan, and Hong Kong

The Multiscale Global Monsoon System
Edited by C. P. Chang et al.

launched two extra soundings per day (i.e., at 0600 and 1800 UTC) in addition to the routine observations at 0000 and 1200 UTC.

Many instruments were deployed in Guangdong during the 2016–2017 IOPs and three super-observation sites were built at Longmen, Nanhai, and Conghua (Fig. 1b), to obtain the cloud-precipitation microphysical properties and vertical kinematic structure of the rainstorms, the 3D kinematic structure of the fast-evolving convective storm, and the lightning processes, respectively. The major instruments deployed in the 2017-IOP, such as the C-band frequency-modulated continuous wave radar (C-FMCW) at Longmen, the X-band phased array radar (X-PAR) at Nanhai, and the lightning low-frequency E-field detection array (LFEDA) at Conghua, are listed in Table 1. Most of the instruments were operated simultaneously and continuously during the IOPs, while the phased array radar and the NJU C-POL radar started/ended operation when the operational radar network observed appearance/disappearance of precipitation echoes in Guangdong. The observational datasets are archived at the Chinese Academy of Meteorological Sciences (CAMS) and available through application online at http://exps.camscma.cn/scmrex.

2. Process Understanding

2.1. *Initiation and maintenance of coastal extreme-rain-producing MCSs*

The initiation and maintenance of the extreme-rain-producing MCSs in coastal areas of Guangdong during SCMREX appear to be attributable to the characteristic environmental conditions around the monsoon onset in the SCS. MCSs develop mostly in southwesterly/southerly/southeasterly near-surface oceanic flows with little convective inhibition (about 10 J kg^{-1} or less), low level of free convection, large convective available potential energy (about 1,600–2,300 J kg^{-1}), and high moisture content (precipitable water of about 60 mm). The small, yet necessary, lift for convective initiation is provided by the combined effects of land-sea frictional contrast and mountains near the coastline (Wang *et al.* 2014), and sometimes a shallow cold pool left behind by previous convection (Wu and Luo 2016). MCSs are maintained through the interaction between the convective feedback (i.e., near-surface cold outflow) and the ambient air. Specifically, a mesoscale outflow boundary (MOB) between a storm-generated cold pool and the high-θ_e surface airflow from the ocean provides persistent lift, leading to convective backbuilding near the MOB. Meso-β-scale band-shaped areas of large reflectivity (>35 dBZ), referred to as rainbands in Wang *et al.* (2014) and Wu and Luo (2016), are subsequently formed under the combined influence of southwesterly environmental flow in the low and middle troposphere and continuous convective initiation along the MOB. As the MOB is quasi-stationary, mainly due to the weak speed of cold outflow at the surface resulting from high moisture content, the MCSs consisting of the multiple rainbands tend to move very slowly. The MOBs in coastal Guangdong during SCMREX possess both similarities (shallowness and long duration) and differences (the presence or lack of a contrast in wind directions across the MOB) compared with those in southwest Taiwan offshore (Davis and Lee, 2012; Xu *et al.*, 2012).

2.2. *A record-breaking rain-producing storm influenced by urban effects*

Up to 524-mm accumulated rainfall in 12 h with the maximum hourly rainfall rate of 180 mm h^{-1} was produced over Guangzhou Megacity during the early hours of 7 May 2017. Both the integrated observations and the numerical experiments with a sub-kilometer grid spacing show

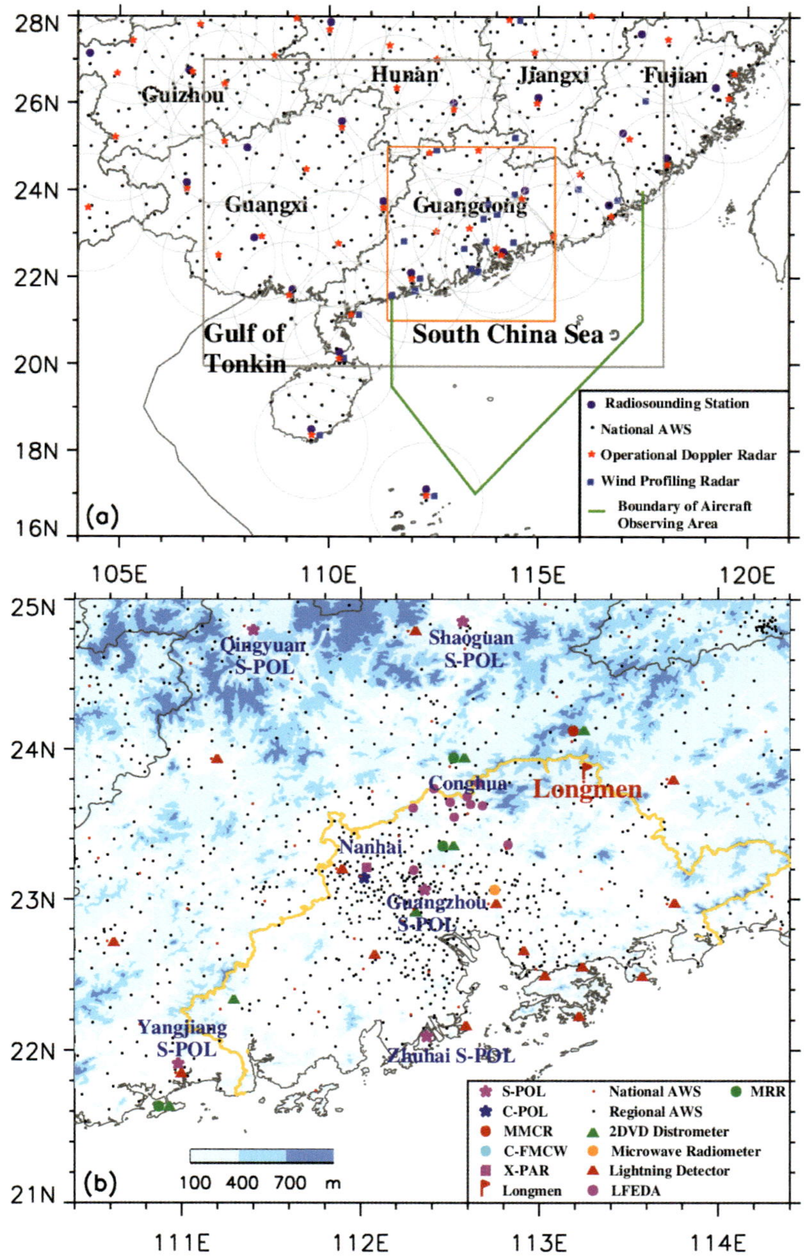

Fig. 1. Distribution of major observing facilities that participated during the SCMREX pilot field campaigns over (a) southern China and its vicinity and (b) central Guangdong Province where the portable instruments were placed during the 2016–2017 IOPs. In (a), orange and green lines highlight the enhanced observing area (21°–25° N, 111.4°–115.4° E) and the aircraft observing area, respectively, and the circles with a radius of 150 km are centered at the operational S-band radars approximately representing their PPI scanning range. In (b), blue shading represents topography and the orange line denotes the boundary of the Pearl River Delta area. All the facilities are consistent with markers in the legend.

Table 1. Major instruments deployed during the SCMREX-2017 field campaign.

Instrument or station	Number	Coverage or location
Automated weather station	1952	110°–117°E, 21–25°N
Radiosonde station with additional launches	8	See Fig. 1a (Guangdong 3, Hong Kong 1, southeast Guangxi 2, Hainan 2)
Operational Doppler radar	14	110°–117° E, 21°–25°N
Polarimetric radar	7	Zhuhai S-POL (113.37°E, 22.03°N) Guangzhou S-POL (113.36°E, 23.00°N) Qingyuan S-POL (112.40°E, 24.73°N) Shaoguan S-POL (113.57°E, 24.79°N) Yangjiang S-POL (111.98°E, 21.85°N), Nanhai (NJU) C-POL (113.02°E, 23.08°N) Foshan X-POL (113.01°E, 23.15°N)
Wind-profiling radar	17	See Fig. 1a
X-band phased array radar	1	Nanhai (113.02°E, 23.08°N)
2DVD Disdrometer	6	Xinfeng (114.19°E, 24.06°N) Fogang (113.52°E, 23.88°N) Maofengshan (113.46°E, 23.29°N) Shunde (113.25°E, 22.85°N) Enping (112.23°E, 22.27°N) Hailingdao (111.87°E, 21.57°N)
8-mm-wavelength cloud radar	1	Xinfeng (114.19°E, 24.06°N)
Micro rain radar	3	Fogan (113.52°E,23.88°N) Maofengshan (113.46°E,23.29°N) Hailingdao (111.87°E,21.57°N)
Lightning detector	17	15 are shown in Fig. 1b
Longmen (114.27°E, 23.78°N)		8-mm-wavelength cloud radar C-band frequency-modulated continuous wave radar Microwave radiometer 2DVD Disdrometer Micro rain radar Ceilometer Precipitation Particle Imager (PPI) Cloud Condensation Nuclei Counter (CCNC) Aerodynamic Particle Sizer (APS) Lidar for water vapor, aerosol, and cloud observation

the initiation and subsequent organization of convective cells in a meso-β-scale convergence region on the northern edge of the city cluster in the Pearl River Delta region, where an onshore low-level stream with ample moisture decelerated while moving across the urban areas, and then converged when hitting a mountainous region to the north. Numerical sensitivity experiments (by removing the small mountains or removing the urban effect) indicate that the enhanced urban friction accounted for the formation of such a convergence region, while the urban heat island effect helped maintain conditionally unstable columns during this nighttime period with a relatively deeper boundary layer on the downstream side of the city cluster. Because of the presence of deep moist columns with precipitable water values of about 60 mm, very weak downdrafts and cold outflows were observed.

Examination of the VDRAS (Sun and Crook 1997) output reveals the presence of a moderately deep, meso-γ-scale, circular convective core (C^3) within a meso-β-scale vortex. Vorticity diagnosis indicates that the tilting of environmental horizontal vorticity plays a key role in the early development of the vortex, while the stretching effect associated with low-level convergence is responsible for the vorticity maintenance. The thermal buoyancy resulting from latent heat release is the primary forcing and the dynamic vertical perturbation pressure gradient force is secondary in supporting the upward motion in the C^3. A large number of raindrops of moderate to large sizes are produced mostly through active warm rain processes (as suggested by vertical distributions of the dual-polarimetric radar measurements and lack of lightning flashes from the lightning detectors), which might be attributed to the presence of large sea salt aerosols in the southerly inflow acting as cloud condensation nuclei.

2.3. *Cloud and precipitation properties*

Based on the raindrop size distribution (DSD) measurements made by the PARSIVEL disdrometers in Yangjiang and Zhuhai during April–June of 2014–16 and by the 2DVideo disdrometer (2DVD) in Longmen during May-September of 2016, the mass-weighted mean diameter D_m (mm) and the generalized intercept parameter N_w($\mathrm{mm^{-1}\ m^{-3}}$) are estimated following Brandes *et al.* (2002). The results are separated into convective and stratiform rain types (Fig. 2). The averaged values are compared to statistics in the literature for the Meiyu precipitation in East China (Chen *et al.* 2013; Jin *et al.* 2015; Wen *et al.* 2016), Taiwan (Chen *et al.* 2009), and Japan (Bringi *et al.* 2006), as well as the convective precipitation in the maritime and continental regimes reported by Bring *et al.* (2003). Relatively higher values of D_m but slightly lower values of N_w are observed during SCMREX compared with observations in East China (Fig. 2a and b), while the SCMREX results are quite close to those found in Taiwan and Japan (Fig. 2b). It is hypothesized that the similar rain DSD properties between SCMREX, Taiwan, and Japan, in contrast to inland East China, are related to lower aerosol pollution (Zhou *et al.* 2016) and the presence of sea salt aerosols (Zhang *et al.* 2012) in these coastal and island regions.

At the Longmen observation site, vertical structures of nonprecipitating and precipitating clouds have been observed during SCMREX using a Ka-band cloud radar (CR) with three working modes (CR1, CR2, CR3), a C-band frequency-modulated continuous wave (FMCW) vertically pointing radar (CVPR), a microwave radiometer, and a laser ceilometer (CEIL). The CR plays a key role in observing high-level clouds, while the CVPR is important for detecting low- and mid-level clouds and heavy precipitation. The CEIL helps diminish the effects of "clear sky" in the planetary boundary layer. An algorithm has been developed to construct merged cloud-precipitation radar datasets that provide more complete information about the vertical distribution and internal dynamics of cloud and precipitation (Liu *et al.* 2017).

Moreover, a hydrometeor classification algorithm (HCA) for the US Weather Surveillance Radar 88 Doppler (WSR-88D) dual-polarization radars (Elmore 2011) has been adjusted using a statistics-based optimization method for the polarimetric radars in Southern China (Wu *et al.* 2018). The optimization includes calibration checking, datasets selection, membership functions modification, computation thresholds modification, and effect verification. The adjusted HCA is applied to the Zhuhai S-POL measurements in May 2014, and the results show a reasonable hydrometeor distribution in a squall line passing over the S-POL on 11 May 2014, e.g., the presence of graupel/hail and heavy-rainfall particles in the deep convective

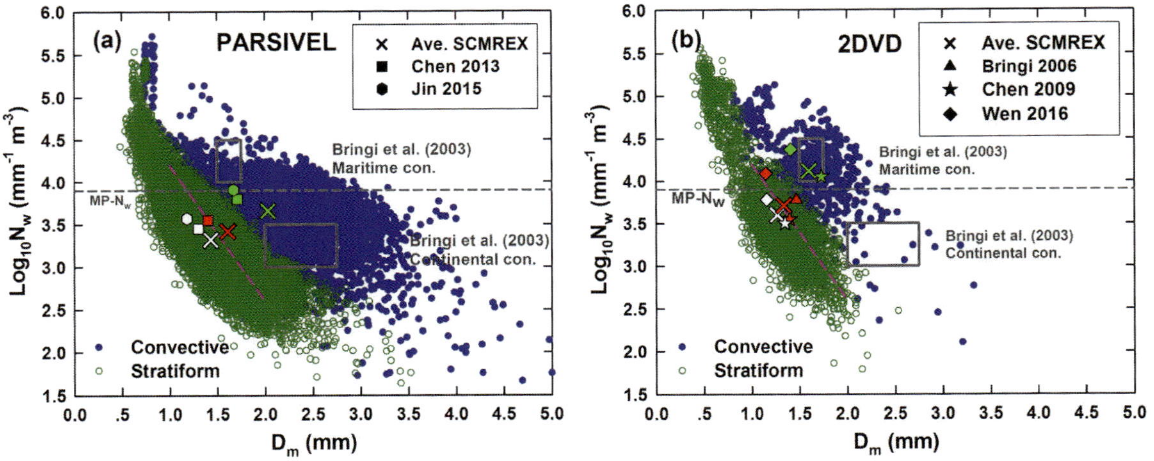

Fig. 2. Scatter plots of $\log_{10}N_w$ versus D_m for convective (blue filled circles) and stratiform (green hollow circles) precipitation types based on measurements made by (a) the PARSIVEL disdrometers in Yangjiang and Zhuhai and (b) the 2DVD disdrometer in Longmen during the SCMREX field campaigns of 2014–2016 and 2016, respectively. The crosses in green, white, and red denote the averaged values for convective, stratiform, and all rain types, respectively. The two gray boxes correspond to the results for the maritime and continental convective clusters reported by Bringi *et al.* (2003). The filled square boxes and diamonds in panel (a) represent the averaged values from Chen *et al.* (2013) and Jin *et al.* (2015), respectively. In panel (b), the filled triangle, stars, and diamonds represent the results in Bringi *et al.* (2006), Chen *et al.* (2009) and Wen *et al.* (2016), respectively, while the magenta dashed line denotes the least squares fit of the averaged value for stratiform precipitation in Bringi *et al.* (2003).

cores but of melting particles and light rainfall drops in the stratiform precipitation regions.

2.4. *Lightning activities associated with the rainy storms*

The spatial and temporal variation of cloud-to-ground (CG) lightning and rainfall over Southern China during the pre-summer rainy season of a 6-year period from 2008 to 2013 was examined (Xia *et al.* 2015). The highest flash densities are found in the northern Pearl River Delta region. The flash density maximal does not coincide with the heavy-rainfall centers in Guangdong, reflective of the importance of warm rain processes in the heavy-rainfall centers. Both the flash density and rainfall amount rise from April to June, which coincides with the onset and prevalence of the summer monsoon and the increase in surface heating. An analysis of diurnal cycles shows that there are double peaks in hourly rainfall amount, one in the afternoon and the other during the early morning, with a minimum at midnight. In contrast, a single-peak diurnal cycle of flash density dominates Southern China, except for a weak secondary early morning peak during the month of April and May. This is consistent with the higher percentage of stratiform rainfall amount during the early morning hours.

3. Numerical Weather Prediction

3.1. *Data assimilation impact on convection-permitting simulation of storms*

Zhang *et al.* (2016) assessed the impact of assimilating horizontal wind data from the wind profiling radars (WPRs) in Guangdong using the operational partial-cycle data assimilation (DA) system at the Institute of Tropical and Marine Meteorology. The analyses from the DA system are used as initial conditions for the convection-permitting Global/Regional Assimilation and Prediction System (GRAPES) model (Chen *et al.* 2008). The impact of assimilating WPR data on the QPF was evaluated during the SCMREX period in May 2014, by comparing the results of a control experiment with WPR data assimilated and a denial experiment without WPR data. The positive impact of WPRDA was significant on the forecast of atmospheric variables in the vertical and diagnostic fields at the surface, especially those of surface wind fields in the 0–6-h range. The inclusion of WPR data also improved the QPF skill of light and heavy rainfall throughout the 12-h forecast period by reducing the predicted spurious precipitation (thereby alleviating overestimations and false alarms), with the largest improvement in 6-h heavy-rainfall forecasts. WPRDA considerably alleviated the spin-up problem, remarkably improving the QPF of heavy rainfall (especially extreme rainfall). The improved representation of wind and moisture at lower levels in the analyses due to WPR DA was the physical cause of the QPF improvement, as is illustrated using a case study.

Bao *et al.* (2017) investigated the impact of assimilating Doppler radar radial velocity observations using a WRF (Skamarock *et al.* 2008)-based EnKF system (WRF-EnKF; Zhang *et al.* 2009) for convection-permitting prediction of convection evolution in the high-impact heavy-rainfall event over coastal areas of Southern China on 8 May 2013. An ensemble of 40 deterministic forecast experiments (40 DADF) with data assimilation (DA) was conducted, in which the DA started at the same time but lasted for different timespans (up to 2 h) and with different time intervals of 6, 12, 24, and 30 min. The reference experiment was conducted without DA (NODA). To show more clearly the impact of radar DA on mesoscale convective system (MCS) forecasts, two sets of 60-member ensemble experiments (NODA EF and exp37 EF) were performed using the same 60-member perturbed-ensemble initial fields but with the radar DA being conducted every 6 min in the exp37 EF experiments from 0200 to 0400 BST. It was found that the DA experiments generally improved the convection prediction. The 40 DADF experiments can forecast a heavy-rain-producing MCS over land and an MCS over the ocean with high probability, despite slight displacement errors. The exp37 EF improves the probability forecast of inland and offshore MCSs more than the NODA EF. The experiment with the longest DA time span and shortest time interval shows the best performance. However, a shorter DA time interval (e.g., 12 min) or a longer DA timespan does not always help.

3.2. *Assessment and improvement of cloud microphysics schemes*

To understand the effects of cloud microphysics complexity on simulations of organized convection during the East Asian Summer Monsoon, we evaluated a set of convection-permitting model configurations of the Met Office Unified Model against radar observations of the 20 May 2016 heavy-rainfall case from the SCMREX, and against remotely sensed estimates of clouds, top-of-atmosphere radiation, and surface precipitation (Furtado *et al.* 2018). In particular, the relative importance of the number of prognostic moments was assessed, in comparison to cloud fraction diagnosis and the microphysical properties of hydrometeor particles. The sensitivity tests showed differences in rainfall and the radiative properties of clouds, which are related to the distribution of condensed water between hydrometeor species. The models showed biases in outgoing shortwave radiation that are dominated by low-level liquid clouds, outgoing longwave biases that are dominated by high clouds, and radar reflectivity biases that are dominated by rain and snow. By "transplanting" individual parameterizations between the microphysics configurations, it was shown that only a small subset of the process parameterizations is responsible for almost all the variations in model performance with physics configuration. In particular, biases related to low clouds are dominated by the cloud fraction scheme, and biases in high clouds are controlled by ice particle sedimentation rates. It was shown that differences in radar reflectivity factor between the configurations were due to the concentration of rain droplets in single- and double-moment representations of the rain droplet size spectrum. By contrast, the role of the number of prognostic variables was much less important for understanding model biases. This information is used to improve the performance of a single-moment scheme and to make judgments about the cost-to-benefit of higher-order microphysics schemes, for predicting organized convection.

Meanwhile, various bulk microphysical parameterizations (BMPs) were used in simulations of the squall line on 22 May 2014 over Southern China (Qian *et al.* 2018). Although most BMPs are able to capture the basic features of the squall line, movement morphology and length of the simulated squall lines differ significantly among different BMPs. Such differences are closely related to the rain evaporation parameterization and rain size distribution assumed in these schemes. Stronger rain evaporation generally contributes to stronger cold pools and thus faster movement and length of the simulated squall lines. Failure of the SBU-YLIN scheme (Lin and Colle 2011) in capturing the squall line was related to the turnoff of rain evaporation once environmental relative humidity is greater than 90%. After some modifications, including adding a density factor to the precipitating ice category, modifying rain evaporation calculations, and the saturation adjustment method, the scheme is able to simulate this squall line reasonably well. This study suggests that rain evaporation is critical for the simulation of both the movement and length of the squall line. A better simulation of the trailing stratiform precipitation requires both realistic ice microphysical processes and rain evaporation parameterization.

4. Concluding Remarks

The SCMREX has obtained integrated observations that have been used to improve not only process-level understanding but also the prediction of heavy-rain-producing storms in Southern China during the pre-summer rainy season. In the future, efforts will be made to enhance the observational capability, for example, by using instrumented aircraft to measure the pre-storm environmental thermodynamics/aerosol properties and storm internal microphysics, conducting simultaneous observations of the vertical fine structure of rainy storms using C-POL RHI and 3D lightning processes using LFEDA, and adding observations of surface sensible and latent heat fluxes at inland and offshore locations in Guangdong and over the South China Sea. Efforts will also be made to use the measurements, e.g., dual-polarimetric radars, to further assess and improve NWP model physics schemes. Data impact studies will be extended to include the newly launched FY-4 satellite observations. Aerosol–convection interactions and urban–mountain effects on a storm's development will be further investigated.

Acknowledgments

We would like to thank Dr. Weixin Xu (Colorado State University) for providing constructive comments that helped improve the quality of the manuscript. This research has been supported by the National Natural Science Foundation of China (41775050), the Meteorology Public Welfare Scientific Research Projects (GYHY201406013) and the Basic Research Funds of the Chinese Academy of Meteorological Sciences (2016Z005, 2017Z006, and 2017Z014).

References

Bao, X., Y., Luo, J. Sun, and coauthors, 2017: Assimilating Doppler radar observations with an ensemble Kalman filter for convection-permitting prediction of convective development in a heavy rainfall event during the pre-summer rainy season of South China. *Sci. China Earth Sci.*, **60**, 1866–1885.

Brandes, E. A., G. Zhang, and J. Vivekanandan, 2002: Experiments in rainfall estimation with a polarimetric radar in a subtropical environment. *J. Appl. Meteorol.*, **41**(6), 674–685.

Bringi, V. N., V. Chandrasekar, J. Hubbert, E. Gorgucci, W. L. Randeu, M. Schoenhuber, 2003: Raindrop size distribution in different climatic regimes from disdrometer and dual-polarized radar analysis. *J. Atmos. Sci.*, **60**(2), 354–365.

Bringi, V. N., M. Thurai, K. Nakagawa, G. J. Huang, T. Kobayashi, A. Adachi, H. Hanado, S. Sekizawa, 2006: Rainfall estimation from C-band polarimetric radar in Okinawa, Japan: Comparisons with 2D-Video disdrometer and 400 MHz wind profiler. *J. Meteor. Soc. Japan. Ser. II*, **84**(4), 705–724.

Chen, D. H. and Coauthors, 2008: New generation of multiscale NWP system (GRAPES): General scientific design. *Chin. Sci. Bull.*, **53**, 3433–3445, doi:10.1007/s11434-008-0494-z.

Chen, Y. and P. Lin, 2009: The characteristic of drop size distribution during SoWMEX [in Chinese], MS thesis, 68 pp., Inst. of Atmos. Phys., Natl. Cent. Univ., Chungli, Taiwan.

Chen, B., J. Yang, and J. Pu, 2013: Statistical characteristics of raindrop size distribution in the Meiyu season observed in Eastern China. *J. Meteor. Soc. Japan*, **91**, 215–227.

Davis, C. and W. Lee, 2012: Mesoscale analysis of heavy rainfall episodes from SoWMEX/ TiMREX. *J. Atmos. Sci.*, **69**, 521–537.

Ding, Y. H., 1994: *Monsoons over China.* Kluwer Academic Publishers, 419 pp.

Elmore, K. L. 2011: The NSSL hydrometeor classification algorithm in winter surface precipitation: Evaluation and future development. *Wea. Forecasting*, **26**, 756–765, doi:10.1175/ WAF-D-10-05011.1.

Furtado, K., P. R. Field, Y. Luo, X. Liu, Z. Guo, and T. Zhou, 2018: Cloud-microphysical factors affecting simulations of deep convection during the presummer rainy-season in southern China. *J. Geophys. Res.*, Under revision.

Lin, Y. and B. A. Colle, 2011: A new bulk microphysical scheme that includes riming intensity and temperature-dependent ice characteristics. *Mon. Wea. Rev.*, **139**(3), 1013–1035.

Liu, L., Z. Ruan, J. Zheng, and W. Gao, 2017: Constructing a cloud–precipitation radar dataset by merging reflectivity and velocity from a Ka-band cloud radar, C-band frequency modulated continuous wave radar and a ceilometer for analysis of summertime cloud characteristics in South China. *J. Atmos. Oceanic Tech.*, Under revision.

Luo, Y., 2017: Advances in understanding the early-summer heavy rainfall over south China. In *The Global Monsoon System: Research and Forecast, III.* C. P. Chang *et al.*, Eds., World Scientific Series on Asia-Pacific Weather and Climate, Vol. 9, World Scientific, 215–226.

Qian, Q., Y. Lin, Y. Luo, and coauthors, 2018: Sensitivity of a simulated squall line during SCMREX to parameterization of microphysics. *J. Geophys. Res.*, online.

Schumacher, R. S., 2017: Heavy rainfall and flash flooding. In *Oxford Research Encyclopedia of Natural Hazard Science*, Oxford University Press, USA, doi:10.1093/acrefore/ 9780199389407.013.132.

Skamarock, W. C., *et al.* 2008: A description of the advanced research WRF version 3, NCAR Tech. Note NCAR/TN/u2013475.

Sun, J. and N. A. Crook, 1997: Dynamical and microphysical retrieval from Doppler radar observations using a cloud model and its adjoint. Part I: Model development and simulated data experiments. *J. Atmos. Sci.*, **54**, 1642–1661.

Wang, H., Y. Luo, and B. Jou, 2014: Initiation, maintenance, and properties of convection in an extreme rainfall event during SCMREX: Observational analysis. *J. Geophys. Res. Atmos.*, **119**, 13-206–13-232, doi:10.1002/2014JD022339.

Wen, L., K. Zhao, G. Zhang, and Co-authors, 2016: Statistical characteristics of raindrop size distributions observed in East China during the Asian summer monsoon season using 2D video disdrometer and Micro Rain Radar data. *J. Geophys. Res. Atmos.*, **121**, 2265–2282.

Wu, C., L. Liu, and coauthors, 2018: Statistics-based optimization of the polarimetric radar hydrometeor classification algorithm and its application for a squall line in South China. *Adv. Atmos. Sci.*, **35**, 296–316.

Wu, M., and Y. Luo, 2016: Mesoscale observational analysis of lifting mechanism of a warm-sector convective system producing the maximal daily precipitation in China mainland during pre-summer rainy season of 2015. *J. Meteor. Res.*, **30**, 719–736, doi:10.1007/s13351-016-6089-8.

Xia, R., D.-L. Zhang, and B. Wang, 2015: A 6-yr cloud-to-ground lightning climatology and its relationship to rainfall over central and eastern china. *J. Appl. Meteor. Climatol.*, **54**, 2443–2460, doi:10.1175/JAMC-D-15-0029.1.

Xu, W., E. Zipser, Y. Chen, and coauthors, 2012: An orography–associated extreme rainfall event during TiMREX: Initiation, storm evolution, and maintenance. *Mon. Wea. Rev.*, **140**, 2555–2574.

Zhang, R., Y. Ni, L. Liu, Y. Luo, and Y. Wang, 2011: South China Heavy Rainfall Experiments (SCHeREX). *J. Meteor. Soc. Japan*, **89A**, 153–166, doi:10.2151/jmsj.2011-A10.

Zhang, X., Y. Luo, Q. Wan, W. Ding, and J. Sun, 2016: Impact of assimilating wind profiling radar observations on convection-permitting quantitative precipitation forecasts during SCMREX. *Wea. Forecast.*, **31**, 1271–1292, doi:10.1175/WAF-D-15-0156.1.

Zhang, X., Y. Wang, T. Niu, X. Zhang, S. Gong, Y. Zhang, and J. Sun, 2012: Atmospheric aerosol compositions in China: Spatial/temporal variability, chemical signature, regional haze distribution and comparisons with global aerosols. *Atmos. Chem. Phys.*, **12**, 779–799, doi:10.5194/acp-12-779-2012.

Zhang, F., Y. Weng, J. Sippel, Z. Meng, and C. Bishop, 2009: Cloud resolving hurricane initialization and prediction through assimilation of Doppler radar observations with an ensemble Kalman filter. *Mon Wea. Rev.*, **137**, 2105–2125.

Zhou, C., X. Zhang, S. Gong, Y. Wang, and M. Xue, 2016: Improving aerosol interaction with clouds and precipitation in a regional chemical weather modeling system. *Atmos. Chem. Phys.*, **16**, 145–160, doi:10.5194/acp-16-145-2016.

Chapter 31

Years of the Maritime Continent (YMC): Multi-year Field Campaigns in the Largest Archipelago on Earth

Kunio Yoneyama*,‡ and Chidong Zhang†,§

**Japan Agency for Marine-Earth Science and Technology, Yokosuka, Japan*
†National Oceanic and Atmospheric Administration/Pacific Marine Environmental Laboratory, Seattle, USA
‡yoneyamak@jamstec.go.jp
§chidong.zhang@noaa.gov

The Years of the Maritime Continent (YMC) is a multi-year international project starting from July 2017 through 2021, which is designed to improve our understanding and prediction skill of the weather–climate systems over the Maritime Continent (MC) and their global impact through observations and numerical modeling. During YMC, several Intensive Observation Periods (IOPs) of field campaigns have been designed to collect observations focusing on specific research topics. In this Chapter, a summary of the planned and proposed IOPs and a basic strategy of data management are provided by referring to one IOP as an example.

1. Introduction

The Maritime Continent (MC), a unique mixture of land and ocean straddling the equator from the eastern Indian Ocean to the western Pacific, hosts various weather and climate systems and plays a role as a heat engine for the global atmospheric circulation owing to its vigorous convective activity (Chang and Lau 1982; Lau *et al.* 1983; Nitta 1987; Yamanaka *et al.* 2018). While previous studies have revealed its convective features from various temporal and spatial-scale viewpoints, our knowledge of them is still very limited due to its complicated multiscale interactions under the conditions of complex land-sea distribution and topography (Slingo *et al.* 2003). For instance, state-of-the-art climate models and weather prediction models are still suffering from systematic errors of precipitation amount, with common overestimates over the land and underestimation over the ocean (Love *et al.* 2011). This may affect weather forecasting skill even in higher latitudes (Neale and Slingo 2003). Recent studies suggest that accurate knowledge of the diurnal cycle in rain is the key to correct such inconsistent precipitation in simulations (Love *et al.* 2011; Peatman *et al.* 2014). In addition, interaction between the diurnal cycle and large-scale phenomena such as the Madden–Julian oscillation (MJO) is also a clue (Peatman *et al.* 2014; Birch *et al.* 2016; Hagos *et al.* 2016; Vincent and Lane 2016; Zhang and Ling 2017; Hung and Sui 2018). The behavior of the MJO over the MC, however, is still a big mystery. Recent studies on impacts of the stratospheric quasi-biennial oscillation (QBO) on the MJO have also added further evidence of the difficulty of fully understanding the behavior of the MJO over the MC region (Yoo and Son 2016; Son *et al.* 2017; Zhang and Zhang 2018). These studies indicate that mitigating the bias of the diurnal cycle in rain is not straightforward. To overcome this difficulty, the Years of the Maritime Continent (YMC) has been planned as a multi-year international project starting from July 2017 through early

The Multiscale Global Monsoon System
Edited by C. P. Chang et al.

2021 that fully covers entire monsoon cycles. It is worth noting that the campaign period was initially set to 2 years, but then has been extended to 2021 by considering funding status. In addition, YMC itself will continue beyond 2021 as a feedback phase to follow the current activity. Its overarching goal is to expedite the progress of improving understanding and prediction of local multiscale variability of the MC weather and climate systems, and the global impact. It is expected that researchers from over 70 institutes will join this project.

YMC includes five science themes: (1) atmospheric convection, (2) upper-ocean processes and air-sea interaction, (3) stratosphere–troposphere interaction, (4) aerosols, and (5) prediction improvement. YMC is designed to bring new knowledge into these science themes through five major activities: (a) data sharing, (b) field campaigns, (c) modeling, (d) prediction applications, and (e) outreach/capacity building. More details on these science themes and activities can be found in the YMC science plan, which is available from http://www.jamstec.go.jp/ymc/.

In this Chapter, we introduce the YMC activity focusing on field campaigns by demonstrating an example of its first intensive observation period (IOP). It should be noted that YMC is ongoing and part of the description herein might be dated when this book is published. This description is based on information as of June 2019.

2. Observational Approach

YMC observations consist of two parts: (1) Operational atmospheric and oceanic measurements made by the local meteorological agencies and (2) IOPs of field campaigns focusing on specific research topic at specific locations and over specific time periods.

Currently, the following MC agencies are involved in the YMC project: Indonesian Agency for Meteorology Climatology and Geophysics (BMKG), Meteorological Service Singapore (MSS), Philippine Atmospheric, Geophysical and Astronomical Services Administration (PAGASA), the Meteorological Department of Malaysia (MetMalaysia), and the Australian Bureau of Meteorology (BoM). They are responsible for operational observations in their respective countries. BMKG offers the secretariat office as well as the YMC official website (http://www.bmkg.go.id/ymc/). In addition, BMKG is one of the YMC data archive centers as described in a later section. These agencies operate various basic meteorological observations including radiosonde soundings, weather radars, and surface meteorology and climatology stations. Figure 1 shows a summary of radiosonde stations. The sites basically obtain sounding data twice daily and send their data to the numerical weather prediction centers through global telecommunication system (GTS), so that the data can be used for weather forecasting and (re)analysis products. When they enhance the sounding daily frequency during IOPs, such additional data will also be sent via GTS. Currently, JAMSTEC, which is another data archive center, has started to collect operational "original high-resolution" data courtesy of the MC agencies. While original high-resolution (every 1–2 s during ascent) data are archived for future use, data at several conventional pressure levels are available to use near real time, as each site sends their data via GTS immediately after soundings.

Under the framework of YMC, several IOPs focusing on specific topics of the five major science themes are planned. These specific topics include the diurnal cycle of rain, the MJO, the boreal summer intraseasonal oscillations (BSISO), equatorial waves in the ocean, troposphere and stratosphere, extreme events associated with cold surges, boreal and austral summer monsoons, lower stratosphere–upper troposphere interaction, the Indonesian through flow, and others. Basically, they

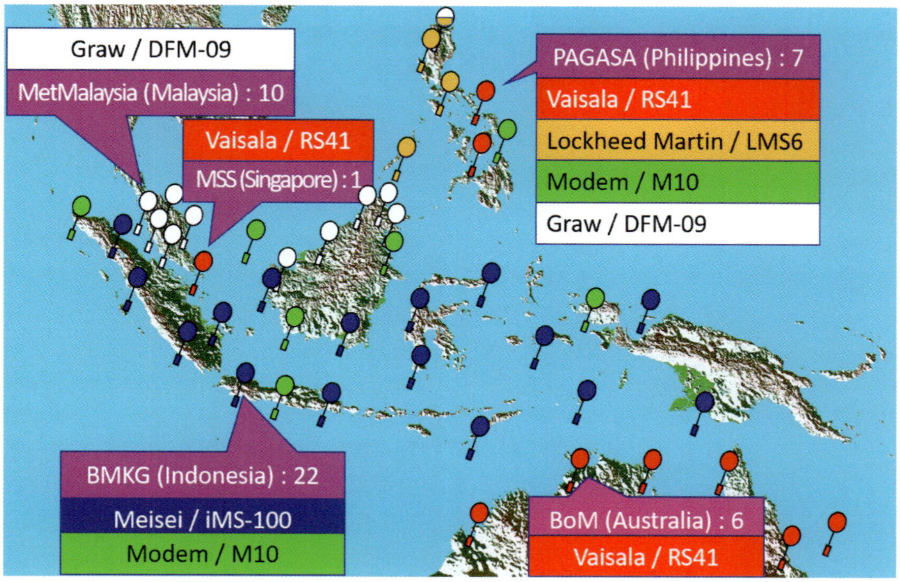

Fig. 1. Radiosonde sounding stations operated by the MC meteorological agencies. Different colors indicate different radiosonde types (manufacturers). Each box indicates total number of sounding sites managed by each agency, their sounding manufacture name and transmitter type as of June 2019.

cover the temporal variations from the diurnal cycle to the seasonal scale. Figure 2 summarizes the funded IOPs, which are expected to be carried out as multinational cooperation. Thus, even if a box indicates one country or region, the project is coordinated with other countries or projects. For example, the Taiwanese project SCSTIMX (South China Sea Two-Island Monsoon Experiment) tightly coordinates with US projects PISTON (Propagation of Intra-Seasonal Tropical Oscillations) and CAMP^{2}Ex (Clouds, Aerosol, Monsoon Processes-Philippines Experiment). PISTON also works with the SALICA (Sea–Air–Land Interactions in the Context of Archipelago) project.

It should be noted that the completion of a funded project critically depends on research permits from the respective MC country, which is possible only with strong support from the MC agencies. Such arrangement and coordination among participants are the keys to the success of YMC. In other words, the idea of YMC was born based on the good relationship among the MC and international community through the many field projects in the past (Yamanaka 2016). As shown in Fig. 2, the first multinational IOP started in November 2017, which will be described in detail in the next section. In 2018–2019, many IOPs were planned including the study of the monsoons, BSISO, and the diurnal cycle (DC). In addition, several IOPs will take place from late 2019 to early 2021. UK's TerraMeris works with Australian research vessel (R/V) Investigator cruise and US's Equatorial Line Observations (ELO) and Diurnal Cycle Interactions with MJO Propagation (DIMOP) projects. These projects aim to understand the diurnal cycle of rain, the MJO, CCKW (convectively coupled Kelvin wave), and their interactions by deploying oceanic (ship, autonomous underwater vehicle) and atmospheric (aircraft, land-based) observation systems. Several Indonesian agencies are closely involved in these projects. The above-mentioned plans are summarized in and/or linked from the YMC ancillary website at JAMSTEC (http://www.jamstec.go.jp/ymc/).

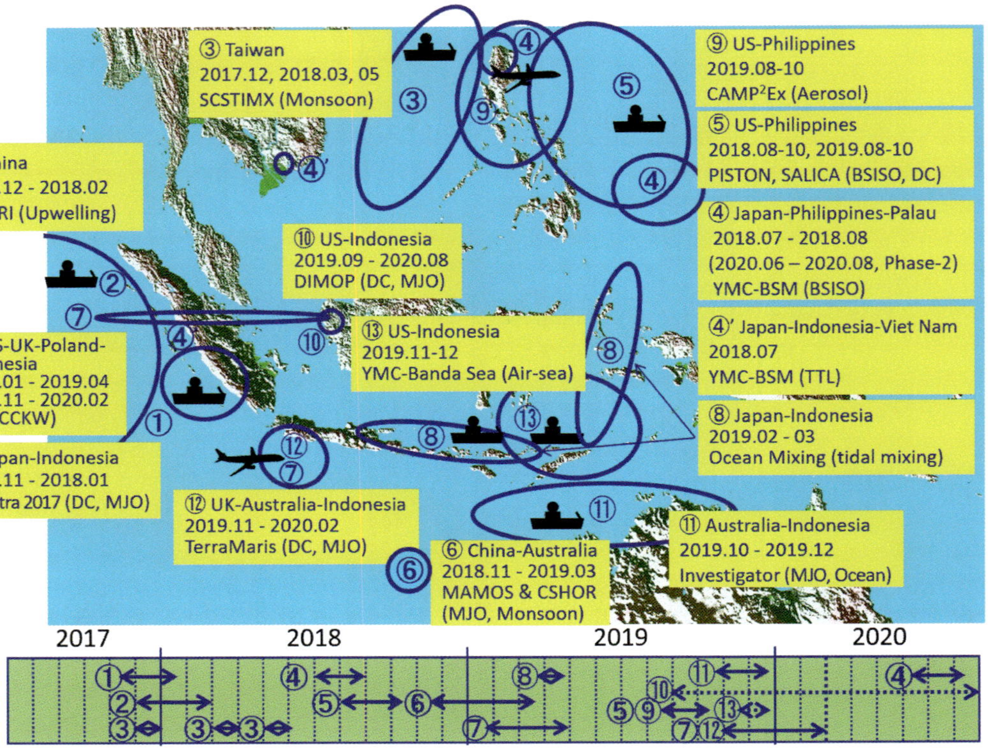

Fig. 2. Funded IOPs as of June, 2019. Each box includes information about participating countries/regions, period, project name, and main science targets. Acronyms are spelled out in the text.

3. An Example of Multinational Cooperation

The first YMC IOP, "YMC–Sumatra 2017", was conducted which aimed to understand the diurnal cycle of convective rain and its interaction with the MJO. This campaign is coordinated among many institutes and universities from Indonesia, Japan, and US. Recent studies have shown the dominance of precipitation in the coast line and this fact plays a key role in the water cycle of the entire globe (Ogino *et al.* 2016, 2017). Thus, the first IOP took place on and off the west coast of Sumatra Island, and an observation network was formed by the R/V MIRAI and a land-based site at Bengkulu to capture the diurnal cycle in rain along the coast and its offshore propagation (Fig. 3). Prior to that, an Indonesia–Japan joint team had

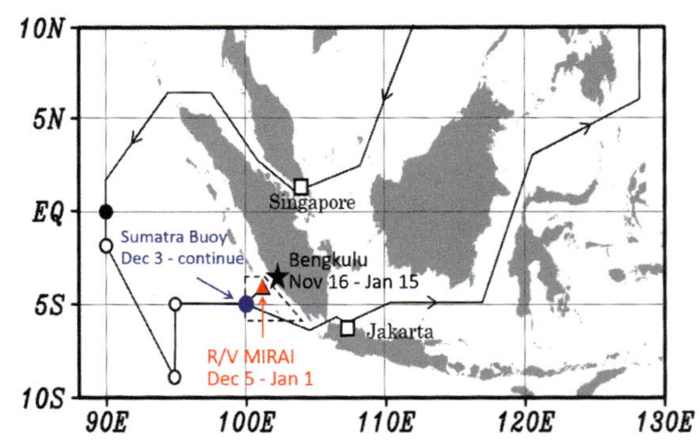

Fig. 3. Configuration of observation sites during YMC-Sumatra 2017 with the R/V MIRAI cruise tracks. Solid and open circles indicate sub-surface ADCP mooring and surface m-TRITON buoy recovery and redeployment sites.

already conducted a pilot study in the same region in November–December 2015, and captured the above-mentioned convective activity (Wu *et al.* 2017; Yokoi *et al.* 2017). Based on this pilot study, they proposed a hypothesis that the gravity wave is the key to offshore propagation. Other (not yet published) work also suggests that the ocean surface condition might affect this propagation. Based on this experience, the observation network in 2017 was slightly modified compared to the pilot study. For example, the ship kept her position one month in December off the coast by 100 km (50 km for the pilot study), so that the two scanning weather radars deployed at Bengkulu by BMKG and onboard the R/V MIRAI by JAMSTEC could cover the area from the coast to the edge of the open ocean. In addition, a buoy and an autonomous unmanned surface vehicle (ASV) were deployed to study air-sea interaction in detail. A buoy, called the Sumatra Buoy, was deployed at 5°S, 100°E under the framework of the collaborative project EIOURI (Eastern Indian Ocean Upwelling Research Initiative), which is one of the core projects of the international program IIOE-2 (2nd International Indian Ocean Expedition; http://www.iioe-2.incois.gov.in/). It measures surface meteorology as well as underwater conditions of temperature and salinity. This buoy was deployed on 3 December 2017. With this buoy, an observation line from the coast to the open ocean was formed. The ASV was deployed near the Sumatra Buoy for intercomparison of surface meteorology after about one week, then she cruised to the stationary site of the R/V MIRAI at 4.24°S, 101.52°E. The ASV collected observations near the MIRAI from 4.27°S, 101.80°E–4.52°S, 101.55°E during 14–30 December.

Onboard the R/V MIRAI, oceanic and atmospheric observations were collected from November 22 through January 1 (5–31 December on station off Bengkulu) using CTD, ADCP, a turbulent measurement system, a polarimetric C-band Doppler radar, radiosondes, a ceilometer, a lidar, a micro-rain radar, GNSS (Global Navigation Satellite System)-based water vapor measurement, surface meteorology stations, etc. Meanwhile, at the Bengkulu land-based site, observations of radiosondes, C/X-band scanning radars, ozone and highly accurate water vapor sondes, GNSS-based water vapor measurements, and surface meteorology were collected from November 16 to January 15. At both MIRAI and Bengkulu, radiosonde soundings were enhanced every 3 hours during the IOP to capture the diurnal cycle, and data at the mandatory levels and significant levels were sent to GTS.

During the IOP, numerical forecasting was made using global cloud system resolving model NICAM (non-hydrostatic icosahedral atmospheric model) as for the past field campaigns (Nasuno *et al.* 2017). Two types of forecasts were performed: 14-day forecasts on 7-km mesh updated daily and 30-day forecasts on 14-km mesh updated weekly. Their products were shared among participants in near real time, and also provided through the website http://nicamfcst.jamstec.go.jp/.

4. Data Policy and Management

One major activity of YMC is data sharing. It covers collecting, archiving, and sharing data from the YMC observation networks including the operational stations and IOPs. In addition, numerical products as well as other monitoring data such as satellites, ocean floats, and moorings will also be archived. Only such a comprehensive database will enable us to study multiscale variability of the MC weather and climate systems. Currently, BMKG and JAMSTEC take the role of the YMC data archive centers, so they collect and archive quality-controlled datasets. Such data collection and sharing would be impossible without full mutual understanding from the MC agencies and international participants. Since YMC adopts timely release and an open/free data-sharing policy, participating researchers are required to make their quality-controlled data open within one year after the completion of YMC. Here, one example of this activity is introduced.

As shown in Fig. 1, various types of radiosondes are used at different sites. Atmospheric profiles are the foundation for studying weather and climate systems in the MC region. Unfortunately, their original data quality often shows bias among different types of radiosondes, especially for humidity data. Producing the same quality datasets is crucial for the field campaign (Ciesielski *et al.* 2012). Figure 4 shows one example of intercomparison of relative humidity between two different types of radiosondes. It clearly shows bias above the 200-hPa level. One possible reason for this difference is the different correction schemes for solar radiation-induced dry biases by the two manufacturers, as they show such differences in only the upper troposphere during daytime sounding (Vömel *et al.* 2007b). To mitigate the possible errors or biases, intercomparison should be performed. During the IOP of YMC Sumatra-2017 campaign, Vaisala RS41 was used onboard the R/V MIRAI, while Meisei iMS100 was used at the BMKG Bengkulu meteorological station. Thus, intercomparison was conducted 18 times at the Bengkulu site apart from 3-h sounding operation (Fig. 5). In particular, to obtain a reference for radiosonde data, different measurement systems were adopted. The first one is a cryogenic frost point hygrometer (Vömel *et al.* 2007a), which measures dew point temperature directly without any assumption that standard radiosonde adopts. In addition, we deployed a receiver of GNSS to derive water vapor (Rocken *et al.* 1993). Currently, a correction scheme has been studied based on this intercomparison. But it is offered only for the case of the Vaisala and Meisei radiosondes.

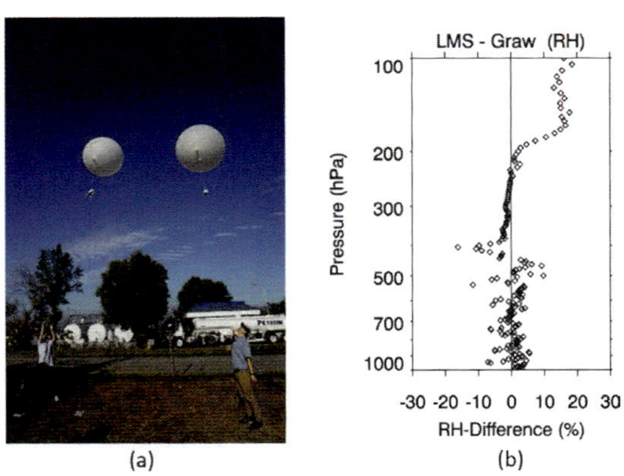

Fig. 4. (a) An example of radiosonde intercomparison between Lockheed Martin LMS6 and Graw DFM-09 done at Laoag station on 22 February 2017 (courtesy of PAGASA). (b) Differences of relative humidity between two the sensors as a function of pressure.

Fig. 5. Intercomparison among Vaisala RS41, two Meisei iMS100, and cryogenic frost-point hygrometer during YMC-Sumatra 2017 campaign.

To develop such bias correction schemes for all types of radiosondes used in the MC countries, JAMSTEC conducted another intercomparison among the Vaisala, Lochheed Martin, and Graw radiosondes at Laoag in the Philippines in 2018 with the help of PAGASA. A GNSS receiver was also deployed to measure water vapor. Colleagues from the University of the Philippines joined this experiment. Through these experiments, we intended not only to develop a correction scheme for different radiosonde types but also to transfer such knowledge and methods to local staff and young scientists, so that quality-controlled data will be released as standard products from sites beyond YMC.

5. Concluding Remarks

In this Chapter, we showed only the observation side of YMC except a small note on forecasting during the YMC–Sumatra 2017 campaign case. As a matter of fact, more than half of the expected participants will conduct numerical modeling. While they provide forecast and simulation results, observations help to improve numerical modeling. Thus, when we developed an Implementation Plan for YMC, input from numerical model was also a key part. As noted in the Introduction, applications and outreach are also major activities of YMC. Through YMC, interaction between observations and modeling will be incorporated into different stages, and new knowledge should feed back to local agencies. It is impossible to fully understand weather and climate systems over the MC within the short period of YMC. However, through YMC we strive to establish a pathway to move forward with more collaboration with the MC countries. In this sense, YMC is a catalyst for a long-term international collaboration with the MC countries in future.

The YMC field campaign phase has been in effect since July 2017, and IOPs have been/will be conducted. More details on each IOP including associated numerical modeling activity can be found in the YMC ancillary site (http://www.jamstec.go.jp/ymc/). This site will also provide field reports with a brief glimpse of observations.

Acknowledgments

We would like to thank the MC agencies (BMKG/Indonesia, MSS/Singapore, PAGASA/Philippines, MetMalaysia/Malaysia, and BoM/Australia) for their support and leadership role in YMC. KY, as the principal investigator of the YMC–Sumatra 2017 campaign, would like to thank the Ministry of Research, Technology, and Higher Education (RISTEKDIKTI), Bengkulu State Government, BMKG, Agency for the Assessment and Application of Technology (BPPT), and the Embassy of Japan in Indonesia for their understanding and support, in particular for getting research permits and conducting observations. This is Pacific Marine Environmental Laboratory Contribution Number 4787.

References

Birch, C. E., S. Webster, S. C. Peatman, D. J. Parker, A. J. Matthews, Y. Li, and M. E. Hassim, 2016: Scale interactions between the MJO and the western Maritime Continent. *J. Clim.*, **29**, 2471–2492.

Chang, C.-P. and K.-M. Lau, 1982: Short-term planetary-scale interactions over the tropics and midlatitudes during northern winter. Part I: Constrasts between active and inactive periods. *Mon. Wea. Rev.*, **110**, 933–946.

Ciesielski, P. E., P. T. Haertel, R. H. Johnson, J. Wang, and S. Loehrer, 2012: Developing high-quality field program sounding data. *Bull. Amer. Meteor. Soc.*, **93**, 325–336.

Hagos, S. M., C. Zhang, Z. Feng, C. D. Burleyson, C. De Mott, B. Kerns, J. J. Benedict, and M. N. Martini, 2016: The impact of the diurnal cycle on the propagation of Madden-Julian Oscillation convection across the Maritime Continent. *J. Adv. Model. Earth Syst.*, **8**, 1552–1564.

Hung, C.-S. and C.-H. Sui, 2018: A diagnostic study of the evolution of the MJO from Indian Ocean to Maritime Continent: Wave dynamics versus advective moistening processes. *J. Clim.*, **31**, 4095–4115.

Lau, K.-M., C.-P. Chang, and P. H. Chan, 1983: Short-term planetary-scale interactions over the tropics and midlatitudes during northern winter. Part II: Winter-MONEX period. *Mon. Wea. Rev.*, **111**, 1372–1388.

Love, B. S., A. J. Matthews, and G. M. S. Lister, 2011: The diurnal cycle of precipitation over the Maritime Continent in a high-resolution atmospheric model. *Quart. J Roy. Meteor. Soc.*, **137**, 934–947.

Nasuno, T., K. Kikuchi, M. Nakano, Y. Yamada, M. Ikeda, and H. Taniguchi, 2017: Evaluation of the near real-time forecasts using a global nonhydrostatic model during the CINDY2011/DYNAMO. *J. Meteor. Soc. Jpn.*, **95**, 345–368.

Neale, R. and J. Slingo, 2003: The Maritime Continent and its role in the global climate: A GCM study. *J. Clim.*, **16**, 834–848.

Nitta, T., 1987: Convective activities in the tropical western Pacific and their impact on the Northern Hemisphere summer circulation. *J. Meteor. Soc. Jpn.*, **65**, 373–390.

Ogino, S.-Y., M. D. Yamanaka, S. Mori, and J. Matsumoto, 2016: How much is the precipitation amount over the tropical coastal region? *J. Clim.*, **29**, 1231–1236.

Ogino, S.-Y., M. D. Yamanaka, S. Mori, and J. Matsumoto, 2017: Tropical coastal dehydrator in global atmospheric water circulation. *Geophys. Res. Lett.*, **44**, 11636–11643.

Peatman, S. C., A. J. Matthews, and D. P. Stevens, 2014: Propagation of the Madden–Julian Oscillation through the Maritime Continent and scale interaction with the diurnal cycle of precipitation. *Quart. J. Roy. Meteor. Soc.*, **140**, 814–825.

Rocken, C., R. Ware, T. van Hove, F. Solheim, C. Alber, J. Johnson, M. Bevis, and S. Businger, 1993: Sensing atmospheric water vapor with the global positioning system. *Geophys. Res. Lett.*, **20**, 2631–2634.

Slingo, J. M., P. Innes, R. Neale, S. Woolnough. and G.-Y. Yang, 2003: Scale interactions on diurnal to seasonal timescales and their relevance to model systematic errors. *Annals Geophys.*, **46**, 139–155.

Son, S.-W., Y. Lim, C. Yoo, H. H. Hendon, and J. Kim, 2017: Stratospheric control of the Madden–Julian oscillation. *J. Clim.*, **30**, 1909–1922.

Vincent, C. L. and T. P. Lane, 2016: Evolution of the diurnal precipitation cycle with the passage of a Madden–Julian Oscillation event through the Maritime Continent. *Mon. Wea. Rev.*, **144**, 1983–2005.

Vömel, H., D. David, and K. Smith, 2007a: Accuracy of tropospheric and stratospheric water vapor measurements by the cryogenic frost point hygrometer: Instrumental details and observations. *J. Geophys. Res.*, **112**, D08305, doi:10.1029/2006JD007224.

Vömel, H., H. Selkirk, L. Miloshevich, J. Valverde-Canossa, J. Valdés, E. Kyrö, R. Kivi, W. Stolz, G. Peng, and J. A. Diaz, 2007b: Radiation dry bias of the Vaisala RS92 humidity sensor. *J. Atmos. Oceanic Technol.*, **24**, 953–963.

Wu, P., D. Ardiansyah, S. Yokoi, S. Mori, F. Syamsudin, and K. Yoneyama, 2017: Why torrential rain occurs on the western coast of Sumatra Island at the leading edge of the MJO westerly wind bursts. *SOLA*, **13**, 36–40.

Yamanaka, M. D., 2016: Physical climatology of Indonesian maritime continent: An outline to comprehend observational studies. *Atmos. Res.*, **178–179**, 231–259.

Yamanaka, M. D., S.-Y. Ogino, P. Wu, J.-I. Hamada, S. Mori, J. Matsumoto, and F. Syamsudin, 2018: Maritime continent coastlines controlling Earth's climate. *Prog. Earth Planet. Sci.*, **5**, 21, doi:10.1186/s40645-018-0174-9.

Yokoi, S., S. Mori, B. Geng, K. Yasunaga, M. Katsumata, F. Syamsudin, Nurhayati, and K. Yoneyama, 2017: Diurnal cycle of precipitation observed in the western coastal area of Sumatra Island: Offshore preconditioning by gravity waves. *Mon. Wea. Rev.*, **145**, 3745–3761.

Yoo, C. and S.-W. Son, 2016: Modulation of the boreal wintertime Madden–Julian oscillation by the stratospheric quasi-biennial oscillation. *Geophys. Res. Lett.*, **43**, 1392–1398.

Zhang, C. and J. Ling, 2017: Barrier effect of the Indo-Pacific Maritime Continent on the MJO: Perspectives from tracking MJO precipitation. *J. Climate*, **30**, 3439–3459.

Zhang, C. and B. Zhang, 2018: QBO-MJO connection. *J. Geophys. Res. Atmos.*, **123**, 2957–2967.